THIRD EDITION

Beginning Algebra

W9-CCD-102

R. David Gustafson

Peter D. Frisk

Rock Valley College

 Brooks/Cole Publishing Company
Pacific Grove, California

Brooks/Cole Publishing Company
A Division of Wadsworth, Inc.

Printed in the United States of America

10   9   8   7   6   5   4   3   2   1

Library of Congress Cataloging-in-Publication Data

Gustafson, R. David (Roy David), [date]
  Beginning algebra / R. David Gustafson, Peter D. Frisk. — 3rd ed.
     p.   cm.
  Includes index.
  ISBN 0-534-16392-0
    1. Algebra.   I. Frisk, Peter D., [date].   II. Title.
  QA152.2.G85   1991
  512.9—dc20   91-3767
                                                            CIP

Sponsoring Editor: Paula-Christy Heighton
Editorial Assistant: Carol Ann Benedict
Production Editor: Ellen Brownstein
Manuscript Editor: David Hoyt
Permissions Editor: Carline Haga
Interior and Cover Design: E. Kelly Shoemaker
Cover Photo: Lee Hocker
Art Coordinator: Lisa Torri
Interior Illustration: Lori Heckelman
Photo Editor: Ruth Minerva
Typesetting: Syntax International
Cover Printing: Lehigh Press Lithographers
Printing and Binding: R. R. Donnelley & Sons, Crawfordsville

Photo Credits: **P. 1,** Comstock; **p. 68,** Rick Rusing/Tony Stone
Worldwide; **p. 123,** Sylvain Grandadam/Tony Stone Worldwide;
p. 170, Michael Thompson/Comstock; **p. 216,** Michael Stuckey/
Comstock; **p. 271,** Andy Sacks/Tony Stone Worldwide; **p. 328,** David Joel/
Tony Stone Worldwide; **p. 372,** John Telford/Index Stock
Photography; **p. 415,** Rene Sheret/Tony Stone Worldwide.

To
Caitlin Mallory Barth
Nicholas Connor Barth
Prescott Alexander Heighton
Laurel Marie Heighton
Daniel Mark Voeltner
and the new generation of mathematics students

This textbook is intended for a course in beginning or introductory algebra. We assume that the majority of students taking this course will enroll in a subsequent mathematics or statistics course, although for some students beginning algebra may be their only exposure to mathematics in college.

Our goal is to write a book that prepares students to succeed in subsequent mathematics courses while holding the attrition rate in the current course to a minimum. We present comprehensive, in-depth, and precise coverage of the topics of beginning algebra, incorporated into a framework of tested teaching strategy combined with carefully selected pedagogical features. We believe that *Beginning Algebra,* Third Edition, accomplishes our goal because of the successful blending of these elements.

Beginning Algebra, Third Edition, developed from dual concerns we have as educators: appropriate level and the ability of students to master the material. As teachers, we wrestle with these concerns. Are the course and its text at the right level, or too high, or too low? Are students prepared to work at the level needed for this course? Will we prepare them sufficiently well for the next course? By what criteria do we determine the proper level?

It is a challenge to agree on level because individuals define level in different ways. We use the following criteria to determine level: topical coverage, which can range from partial to comprehensive; depth of coverage, which can range from an overview to in-depth treatment; and ease of use, which can include numerous components, such as structure of the content's presentation, reading level, use of examples, and a host of pedagogical elements also known as features.

During the last two decades, mathematics instructors have been offered two different types of texts. The books considered to be lower level either limit the number of topics they include or limit the depth at which they cover them. They also use many pedagogical elements, such as summary boxes, to help students remember the material. These books concentrate on manipulation and drill. They may be successful at helping a higher percentage of students get through the course but students later find they are inadequately prepared to succeed in the next course. The higher-level books, which take a second approach, are both comprehensive and in-depth. These books provide a solid foundation in concepts and manipulative skills for those few students whose mathematical backgrounds are strong enough to enable them to survive a tough book.

PHILOSOPHY AND APPROACH

Because we found that neither of these approaches met our needs, we chose yet another. It has three components: complete and in-depth coverage, a teaching strategy based on our successful classroom experiences, and carefully selected pedagogical features.

We believe the integrity of mathematical standards must be maintained in order to prepare students for subsequent courses.

The amount of mathematics required for degrees in an ever-growing number of majors has increased. Therefore, it has become more likely that our students will take a math course beyond beginning algebra. Skills must be learned thoroughly in beginning algebra if students are to employ these techniques successfully in applied courses such as statistics or finite mathematics. They need a solid foundation in concepts as well if they are to continue on to calculus.

Consequently, we have written a book that contains comprehensive coverage and complete, precise explanations of the concepts and processes used in algebra. We present new material in the context of concrete examples, then generalize the ideas and write them as rules or theorems, after which we present additional examples. We motivate mathematical concepts by need and explain them through discussion and example. We also want to motivate students by showing them the utility of mathematics in many applications and by including examples and exercises through which they can discover some of the intrinsic beauty of mathematics.

We believe every student has the right to experience success, both in this course and the next one.

Some students come to us after long absences from the classroom and others previously had unhappy experiences with mathematics. They give us an opportunity to get them off to a good start and to revise their poor impressions of mathematics by developing their confidence and their mastery of beginning algebra.

Because our book offers a substantial treatment of algebra, it is our responsibility to offer more than the usual amount of support to the student. We believe that previous editions of our book have been successful at a wide range of schools because we see such a wide range of abilities among our own students. We organized our book based on what has worked for our students over the years. The writing is clear and direct, examples are heavily annotated with comments similar to what we say in class, and many pedagogical elements are used to help students learn and retain the material.

Review is an integral part of the text. Of special note are the twelve cumulative Review Exercises that follow the exercise set for each section.

Full color is used to make the book easier to use and to learn from by guiding the students' studying. Red is used consistently to point to terms and expressions that we point to in class discussion and to annotate the examples. All definition boxes are beige, theorem boxes are lavender, and strategy boxes are green. Colors are used in graphs to highlight important elements. Bright blue appears only in the Annotated Instructor's Edition (AIE) and is reserved exclusively for comments intended for the instructor.

Changes for the Third Edition

Beginning Algebra, Third Edition, is the culmination of more than 65 years of combined teaching experience, dozens of peer reviews, many conversations with mathematics instructors around the country, and countless student comments and suggestions.

The overall effect of the changes we have made has been

- to increase our emphasis on visualizing mathematics through the use of geometry and graphing,
- to make more use of mathematics in the world around us through realistic applications, and
- to fine-tune the presentation of certain topics.

Some of the specific changes we have made are listed below:

1. We emphasize topics in geometry throughout the text, particularly the concepts of perimeter, area, and volume.

2. Many additional applications throughout the text illustrate how the mathematical topics are useful. As in our earlier editions, beginning in the first chapter and continuing throughout the text, students must apply newly acquired skills to solve word problems. Chapter 2 now includes two sections that allow more time to introduce and practice setting up a wide variety of word problems.

3. The treatment of exponents is rewritten and divided into two sections. Section 3.1 now covers the properties of natural-number exponents. Section 3.2 covers integer exponents.

4. The section on factoring by grouping appears earlier in the text in Section 4.2. You now can use factoring by grouping as an aid in factoring trinomials.

5. More emphasis had been placed on factoring −1 out in the numerators and denominators of fractions. A section on applications of fractional equations is now included. Applications involving similar triangles have been added to the section on ratio and proportion.

6. Systems of three equations in three variables are now included for the many instructors who requested it.

7. Cube roots now receive more emphasis. The section on fractional exponents now appears before the section discussing the distance formula.

8. The section on imaginary numbers has been expanded to include the arithmetic of complex numbers.

9. Improved exercise sets and additional worked examples help students apply skills to the more difficult exercises.

ILLUSTRATING OUR APPROACH

Our textbooks are noted for a straight-forward, no-nonsense approach to teaching mathematics. New material is presented in the context of concrete examples. The ideas are then generalized and written in rule or theorem form. Following each theorem are several examples that illustrate the use of the theorem. Each section ends with a set of carefully graded exercises that give students ample practice to develop confidence in their skills, as well as some real challenge.

Many topics in *Beginning Algebra* go beyond what is found in other elementary algebra texts. For example, we cover exponential expressions with variable exponents early in the course. Additional topics we include are: the distance formula, rational exponents, a thorough treatment of cube and higher roots, and complex numbers.

Each chapter ends with a chapter summary that includes key terms and key ideas, a set of chapter review exercises that are keyed to each section, and a sample chapter test.

We believe in continuous review. This is apparent in the chapter review exercises, the chapter tests, the twelve review exercises at the end of each exercise set, the practice final examination, and the important formulas listed for reference inside the covers of the book.

Review also is built into the text itself. For example, Chapter 4 covers the topic of factoring. These factoring skills are then used throughout the rest of the text. They are used in Section 4.8 to solve quadratic equations, throughout Chapter 5 to work with fractions, in Chapter 8 to solve radical equations, and in Chapter 9 to solve quadratic equations.

Thorough Treatment of Graphing

Although we live in a world of graphing calculators, we believe that students first need to learn the basics of drawing graphs by hand. Our Chapter 6 on graphing begins with the discussion of graphing lines. We thoroughly discuss graphing both by plotting points and by the intercept method. We cover vertical and horizontal lines in detail.

A complete presentation of the concept of slope and the method of graphing a line using its slope and y-intercept are included.

Section 6.3 includes a complete discussion of writing equations of lines with known properties and Section 6.5 provides an introduction to functions.

Thorough Treatment of Word Problems

The treatment of word problems is often a distinguishing feature among textbooks. We therefore take great care in presenting word problems. Students begin solving word problems as early as page 8. Translating from English to mathematics, and from mathematics to English, begins in earnest in Section 1.3, and a problem-solving theme continues throughout the text.

We believe the key to solving word problems is the ability to set up word equations. Thus, we include a written analysis before the solution of almost every word problem. A word equation is always formulated.

All the traditional word problems are included but many more-current applications, such as break-even analysis, depreciation, and appreciation also are included.

STUDENT LEARNING AIDS

A recurring comment, made by faculty and students alike, in reviews of our books is: "The book is well written and easy to read." We believe this comment reflects our many years of teaching experience and our genuine concern for helping students. Because our goal always has been to help as many students succeed as possible, we have pioneered several student-oriented features, such as the functional use of second color, now found in virtually all mathematics textbooks.

SUPPLEMENTS FOR THE INSTRUCTOR

Annotated Instructor's Edition

The Annotated Instructor's Edition contains answers next to all problems in the exercise sets; margin notes with commentary relating material in one section to that in another; tips on how to avoid common student difficulties; teaching points; historical information; and creative problems and discussion questions designed to help students integrate material and to enhance students' discovery experiences.

Instructor's Resource Manual
Darrell Ropp

The Instructor's Resource Manual contains lecture outlines and page-referenced guides correlating the software, videos, Study Guide, and other supplements to the text.

Transparency Masters
Darrell Ropp

The set of Transparency Masters covers all topics covered in the course.

Answer Book
Diane Koenig

The Answer Book contains answers to all even-numbered exercises.

Test Manual
William Hinrichs

The Test Manual contains three ready-to-use forms of every chapter test.

Classroom Manager
(IBM) Software

This record-keeping program handles records for up to 1000 students, performs various calculations, imports Scantron data, and exports to most word processors and spreadsheets.

Computer Testing
Software

Available with our text are two extensive electronic question banks, one short-answer and one multiple-choice. Each contains approximately 1700 test items. Both are available for IBM and compatible machines and for Macintosh machines. The testing programs give you all the features of state-of-the-art word processors and more, including the ability to see all technical symbols, fonts, and formatting on the screen just as they will appear when printed. The test banks can be edited.

ExamBuilder® runs on Macintosh machines.
EXPTEST® runs on IBM and compatible computers.

SUPPLEMENTS FOR THE STUDENTS

Conquering Math Anxiety
Cynthia Arem

Conquering Math Anxiety speaks to the student about math anxiety and how to recognize it. It includes strategies and practice exercises to cope with it, and offers tips on conquering test anxiety.

Study Guide
George Grisham and
Robert Eicken

For each chapter in the text, the Study Guide contains a message to the student, completely worked examples for each section, additional exercises, cautions and hints, and completely worked selected odd-numbered problems from the text.

Spanish Study Guide
Myriam Steinback

This is a complete Study Guide in the Spanish language with step-by-step worked examples, study tips, and additional practice exercises.

Geometry
Karl J. Smith

Geometry provides a complete review of measurement and plane geometry.

Student Solutions Manual
Diane Koenig

The Student Solutions Manual contains complete solutions to alternate even-numbered exercises from the text.

Success! **Software**
(IBM)
Lawrence Sowers

Success! Software covers all topics in the course. It is an interactive program keyed, in the software itself, to our book. It reviews key definitions, formulas, and examples, provides practice problems, checks work, and gives appropriate help. Both the site-license and single-user versions are available.

Algebra Homework Tutor
(Macintosh) Software
Ken Tilton

This interactive tutorial allows students to learn in three ways: by the trial-and-error method, by example, and by specific hints. *Algebra Homework Tutor* covers signed number review and all topics and concepts found in beginning algebra. Both site license and single copies are available.

Algebra Mentor
Version 2.0
(IBM and Apple II series)
Software
John Miller

Algebra Mentor is an "intelligent" interactive tutorial that covers all topics and concepts found in beginning algebra, along with an introduction to quadratic equations, graphing equations, and the quadratic formula. *Algebra Mentor* contains five levels of difficulty. A site license is available.

Videos

These videos review the key concepts in each chapter.

ACCURACY

Gustafson/Frisk textbooks, published by Brooks/Cole, have the reputation of being error-free. This reputation has been earned by the hard work of many people, including very competent production and design departments, and by the diligence of many reviewers and proofreaders. Each exercise has been worked independently by both authors and several other problem checkers.

ACKNOWLEDGMENTS

We appreciate the excellent input and comments provided by the contributors to the Annotated Instructor's Edition:

Kent Aeschliman, Oakland Community College–Orchard Ridge
Jim Fryxell, College of Lake County
Michael Contino, California State University, Hayward
Ann Smallen, Mohawk Community College

We are grateful to the following people who have reviewed the text at various stages of its development:

Elaine D. Bouldin
Middle Tennessee State University

Baruch Cahlon
Oakland University

Patricia Cooper
St. Louis Community College at
    Park Forest

Russel M. Day
Illinois Central College

Elias Deeba
University of Houston

Edward Doran
Front Range Community College

Arthur Dull
Diablo Valley College

Robert Eicken
Illinois Central College

George Grisham
Bradley University

Robert G. Hammond
Utah State University

Robert Keicher
Delta College

Katherine McLain
Cosumnes River College

Myrna Mitchell
Pima Community College

John Monroe
University of Akron

Carol M. Nessmith
Georgia Southern University

Michael Rosenthal
Florida International University

Jack W. Rotman
Lansing Community College

Erik A. Schreiner
Western Michigan University

Kenneth Seydel
Skyline College

David Sicks
Olympia College

Willie Taylor
Texas Southern University

Lynn E. Tooley
Bellevue Community College

Gerry C. Vidrine
Louisiana State University

Clifton T. Whyburn
University of Houston

Hette Williams
Broward Community College

We also thank the authors who prepared supplementary materials to accompany our book:

George Grisham and Robert Eicken for the *Study Guide*
Diane Koenig for the *Student Solutions Manual* and the *Answer Book*
William Hinrichs for the electronic question banks for EXPTEST®

Darrell Ropp for the *Instructor's Resource Manual* and for the *Transparency Masters*
Cynthia Arem for *Conquering Math Anxiety*
Myriam Steinback for the *Spanish Study Guide*
Karl J. Smith for *Geometry*, a title in his One Unit Series
Lawrence Sowers for Beginning Algebra *Success!* Software
Ken Tilton for *Algebra Homework Tutor* software
John Miller for *Algebra Mentor* software

We give special thanks to Diane Koenig, who read the entire manuscript and worked every problem, to Gary Schultz, Bill Hinrichs, James Yarwood, Jerry Frang, David Hinde, and Darrell Ropp for their helpful suggestions, to Betty Fernandez for her assistance in the preparation of the manuscript, and, at Brooks/Cole, Paula-Christy Heighton, Ellen Brownstein, Kelly Shoemaker, Lisa Torri, Faith Stoddard, and David Hoyt for their fine work.

R. David Gustafson
Peter D. Frisk

TO THE STUDENT

This state-of-the-art textbook has been written especially for you. We, the authors of the text, have more than 65 years of combined teaching experience and, in that time, have taught more than 32,000 students. More than a million others have learned their mathematics from Gustafson/Frisk textbooks. Based on our teaching experience, and countless teacher and student comments and suggestions, we have prepared an educational package that is designed solely for your success.

You've taken an important step by enrolling in beginning algebra and we want to do all we can to help you succeed. For you to get the most out of your algebra course, you must read and study the textbook properly. The text includes carefully written narrative and a large number of worked examples. We have written in clear, direct language that should make sense to you if you study it. Read the examples slowly and with a pencil in your hand. We recommend that you work the examples on paper and be sure you understand them before attempting to work the exercises.

Here are some of the features in the book that should make it easy for you to use.

STUDENT SUPPORT

Refer to sample pages 131, 396, 237, and the sample end papers for illustrations of some of the teaching techniques that we have included to help you use the text.

3.2 ZERO AND NEGATIVE INTEGRAL EXPONENTS   131

$$\frac{6^2}{6^5} = 6^{2-5} = 6^{-3}$$

However, we know that

$$\frac{6^2}{6^5} = \frac{\overset{1}{\cancel{6}} \cdot \overset{1}{\cancel{6}}}{\underset{1}{\cancel{6}} \cdot \underset{1}{\cancel{6}} \cdot 6 \cdot 6 \cdot 6} = \frac{1}{6^3}$$

To make the results of 6^{-3} and $\frac{1}{6^3}$ consistent, we shall define 6^{-3} to be equal to $\frac{1}{6^3}$. In general,

> **DEFINITION.** If x is any nonzero number and n is a natural number, then
>
> $$x^{-n} = \frac{1}{x^n}$$

◄ Definition boxes are beige.

EXAMPLE 2 Write each quantity without using negative exponents or parentheses. Assume that no denominators are zero.

a. $3^{-5} = \frac{1}{3^5}$ 　　　**b.** $x^{-4} = \frac{1}{x^4}$ 　　　**c.** $(2x)^{-2} = \frac{1}{(2x)^2}$

$$= \frac{1}{4x^2}$$

d. $2x^{-2} = 2\left(\frac{1}{x^2}\right)$ 　　**e.** $(-3a)^{-4} = \frac{1}{(-3a)^4}$ 　　**f.** $(x^3x^2)^{-3} = (x^5)^{-3}$

$$= \frac{2}{x^2}$$ 　　　　　　　　　$$= \frac{1}{81a^4}$$ 　　　　　　　$$= \frac{1}{(x^5)^3}$$

$$= \frac{1}{x^{15}} ■$$

Because of the definition of negative and zero exponents, the product, power, and quotient rules are also true for all integral exponents. We restate the properties of exponents for integral exponents.

> **Properties of Exponents.** If m and n are natural numbers and there are no divisions by 0, then
>
> $$x^m x^n = x^{m+n} \qquad (x^m)^n = x^{mn} \qquad (xy)^n = x^n y^n \qquad \left(\frac{x}{y}\right)^n = \frac{x^n}{y^n}$$
>
> $$x^0 = 1 \quad (x \neq 0) \qquad x^{-n} = \frac{1}{x^n} \qquad \frac{x^m}{x^n} = x^{m-n}$$

Theorem and rule boxes are lavender. ▲

Strategy boxes, which are used to ► summarize many processes, are green.

XPRESSIONS

$$\sqrt{x + 2} = 3$$
$$\sqrt{7 + 2} \overset{?}{=} 3 \qquad \text{Replace } x \text{ with 7.}$$
$$\sqrt{9} \overset{?}{=} 3$$
$$3 = 3$$

solution checks. Because no solutions are lost in this process, 7 is the only tion of the original equation. ■

e the equation $\sqrt{x + 1} + 5 = 3$.

rrange the terms to isolate the radical on one side of the equation. Then pro-
as follows:

$$\sqrt{x + 1} + 5 = 3$$
$$\sqrt{x + 1} = -2 \qquad \text{Add } -5 \text{ to both sides.}$$
$$(\sqrt{x + 1})^2 = (-2)^2 \qquad \text{Square both sides.}$$
$$x + 1 = 4 \qquad \text{Simplify.}$$
$$x = 3 \qquad \text{Add } -1 \text{ to both sides.}$$

Check this solution by substituting 3 for x in the original equation.

$$\sqrt{x + 1} + 5 = 3$$
$$\sqrt{3 + 1} + 5 \overset{?}{=} 3 \qquad \text{Replace } x \text{ with 3.}$$
$$\sqrt{4} + 5 \overset{?}{=} 3$$
$$2 + 5 \overset{?}{=} 3$$
$$7 = 3$$

Because $7 = 3$ is a *false* result, the number 3 is *not* a solution of the given equation. Thus, the original equation has *no* solution. ■

Example 2 illustrates that squaring both sides of an equation can lead to **extraneous solutions.** These solutions do not satisfy the original equation and must be discarded.

Follow these steps to solve an equation containing radical expressions.

> **1.** Whenever possible, rearrange the terms to isolate a single radical on one side of the equation.
> **2.** Square both sides of the equation and solve the resulting equation.
> **3.** Check the solution in the original equation. This step is required.

EXAMPLE 3 Solve the equation $\sqrt{x + 12} = 3\sqrt{x + 4}$.

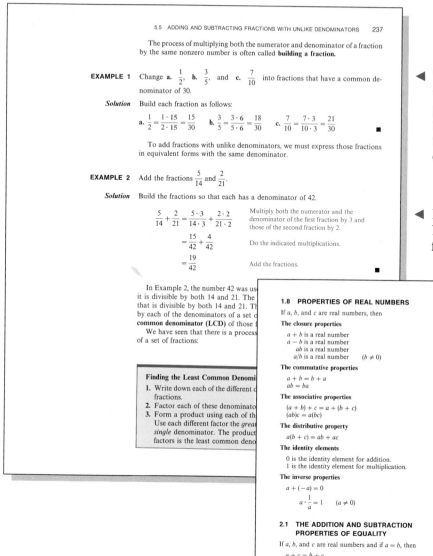

The process of multiplying both the numerator and denominator of a fraction by the same nonzero number is often called **building a fraction.**

EXAMPLE 1   Change **a.** $\frac{1}{2}$, **b.** $\frac{3}{5}$, and **c.** $\frac{7}{10}$ into fractions that have a common denominator of 30.

Solution   Build each fraction as follows:

a. $\frac{1}{2} = \frac{1 \cdot 15}{2 \cdot 15} = \frac{15}{30}$ **b.** $\frac{3}{5} = \frac{3 \cdot 6}{5 \cdot 6} = \frac{18}{30}$ **c.** $\frac{7}{10} = \frac{7 \cdot 3}{10 \cdot 3} = \frac{21}{30}$ ∎

To add fractions with unlike denominators, we must express those fractions in equivalent forms with the same denominator.

EXAMPLE 2   Add the fractions $\frac{5}{14}$ and $\frac{2}{21}$.

Solution   Build the fractions so that each has a denominator of 42.

$$\frac{5}{14} + \frac{2}{21} = \frac{5 \cdot 3}{14 \cdot 3} + \frac{2 \cdot 2}{21 \cdot 2}$$ Multiply both the numerator and the denominator of the first fraction by 3 and those of the second fraction by 2.

$$= \frac{15}{42} + \frac{4}{42}$$ Do the indicated multiplications.

$$= \frac{19}{42}$$ Add the fractions.   ∎

In Example 2, the number 42 was us[...] it is divisible by both 14 and 21. The [...] that is divisible by both 14 and 21. Th[...] by each of the denominators of a set [...] **common denominator (LCD)** of those [...] We have seen that there is a process [...] of a set of fractions:

Finding the Least Common Denomi[...]
1. Write down each of the different [...] fractions.
2. Factor each of these denominato[...]
3. Form a product using each of th[...] Use each different factor the *grea[...] single* denominator. The product [...] factors is the least common deno[...]

◀ The text contains more than 370 worked examples, many with several parts. The color red, when used in text, indicates terms and expressions that we would point to in a classroom discussion. (See page 237.)

◀ Extensive use of explanatory Author's Notes makes the examples easy to follow.

Key formula and ideas are listed ▶ inside the front and back covers for easy reference.

1.8 PROPERTIES OF REAL NUMBERS

If a, b, and c are real numbers, then

The closure properties

$a + b$ is a real number
$a - b$ is a real number
ab is a real number
a/b is a real number $(b \neq 0)$

The commutative properties

$a + b = b + a$
$ab = ba$

The associative properties

$(a + b) + c = a + (b + c)$
$(ab)c = a(bc)$

The distributive property

$a(b + c) = ab + ac$

The identity elements

0 is the identity element for addition.
1 is the identity element for multiplication.

The inverse properties

$a + (-a) = 0$
$a \cdot \dfrac{1}{a} = 1$ $(a \neq 0)$

2.1 THE ADDITION AND SUBTRACTION PROPERTIES OF EQUALITY

If a, b, and c are real numbers and if $a = b$, then

$a + c = b + c$
$a - c = b - c$

2.2 THE DIVISION AND MULTIPLICATION PROPERTIES OF EQUALITY

If a, b, and c are real numbers and if $a = b$, then

$\dfrac{a}{c} = \dfrac{b}{c}$ $(c \neq 0)$
$ca = cb$

2.3 SIMPLIFYING EXPRESSIONS TO SOLVE EQUATIONS

To combine like terms, add their numerical coefficients and keep the same variables and exponents.

3.1 NATURAL-NUMBER EXPONENTS

If n is a natural number, then

$$x^n = \overbrace{x \cdot x \cdot x \cdot \cdots \cdot x}^{n \text{ factors of } x}$$

If m and n are integers and there are no divisions by 0, then

$x^m x^n = x^{m+n}$ $(x^m)^n = x^{mn}$ $(xy)^n = x^n y^n$
$\left(\dfrac{x}{y}\right)^n = \dfrac{x^n}{y^n}$ $\dfrac{x^m}{x^n} = x^{m-n}$

3.2 ZERO AND NEGATIVE INTEGRAL EXPONENTS

$x^0 = 1$ $x^{-n} = \dfrac{1}{x^n}$ $(x \neq 0)$

3.6 MULTIPLYING POLYNOMIALS

$(x + y)^2 = x^2 + 2xy + y^2$
$(x - y)^2 = x^2 - 2xy + y^2$
$(x + y)(x - y) = x^2 - y^2$

3.7 DIVIDING POLYNOMIALS BY MONOMIALS

$\dfrac{a}{b} = \dfrac{1}{b} \cdot a$ $(b \neq 0)$

4.1–4.6 FACTORING POLYNOMIALS

$ax + bx = x(a + b)$
$(a + b)x + (a + b)y = (a + b)(x + y)$
$ax + ay + cx + cy = a(x + y) + c(x + y)$
$ = (x + y)(a + c)$
$x^2 - y^2 = (x + y)(x - y)$
$x^2 + 2xy + y^2 = (x + y)(x + y)$
$x^2 - 2xy + y^2 = (x - y)(x - y)$
$x^3 + y^3 = (x + y)(x^2 - xy + y^2)$
$x^3 - y^3 = (x - y)(x^2 + xy + y^2)$

4.8 SOLVING EQUATIONS BY FACTORING

Zero factor theorem: If a and b represent two real numbers and $ab = 0$, then

$a = 0$ or $b = 0$

Review information is integrated throughout the text. Each chapter ends with a chapter summary that includes key terms and key ideas (see page 166), a set of chapter review exercises that are keyed to each section, and a sample chapter test.

In Review Exercises 9–10, simplify each expression.

9. $3(2x^2 - 4x + 5) + 2(x^2 + 3x - 7)$

10. $-2(y^3 + 2y^2 - y) - 3(3y^3 + y)$

11. The approximate area of the ring between two circles of radius r and R (see Illustration 1) is given by

$$A = \frac{22}{7}(R + r)(R - r)$$

If $r = 3$ inches and $R = 17$ inches, find A.

ILLUSTRATION 1

12. Laura bought a color TV set for $502.90, which included a 7% sales tax. Find the selling price of the TV before the tax was added.

CHAPTER SUMMARY

Key Words

algebraic terms (3.4)	*exponent* (3.1)	*quotient* (3.8)
base (3.1)	*FOIL* (3.6)	*scientific notation* (3.3)
binomial (3.4)	*like terms* (3.5)	*special products* (3.6)
degree of a monomial (3.4)	*minuend* (3.5)	*standard notation* (3.3)
degree of a polynomial (3.4)	*monomial* (3.4)	*subtrahend* (3.5)
dividend (3.8)	*polynomial* (3.4)	*trinomial* (3.4)
divisor (3.8)	*power* (3.1)	

Key Ideas

(3.1)–(3.2) **Properties of exponents.** If n is a natural number, then

$$x^n = \underbrace{x \cdot x \cdot x \cdot x \cdots \cdot x}_{n \text{ factors of } x}$$

If m and n are integers, then

$$x^m x^n = x^{m+n} \qquad (x^m)^n = x^{mn} \qquad (xy)^n = x^n y^n$$

$$\left(\frac{x}{y}\right)^n = \frac{x^n}{y^n} \qquad \text{provided } y \neq 0$$

$$\frac{x^m}{x^n} = x^{m-n} \qquad \text{provided } x \neq 0$$

$$x^0 = 1 \qquad \text{provided } x \neq 0$$

$$x^{-n} = \frac{1}{x^n} \qquad \text{provided } x \neq 0$$

(3.3) A number is written in scientific notation if it is written as the product of a number between 1 (including 1) and 10 and an integer power of 10.

(3.4) If $P(x)$ is a polynomial in x, then $P(r)$ is the value of the polynomial when $x = r$.

(3.5) When adding or subtracting polynomials, combine like terms by adding or subtracting the numerical coefficients and using the same variables and the same exponents.

(3.6) To multiply two monomials, first multiply the numerical factors and then multiply the variable factors.

To multiply a polynomial with more than one term by a monomial, multiply each term of the polynomial by the monomial, and simplify.

To multiply one polynomial by another, multiply each term of one polynomial by each term of the other polynomial, and simplify.

To multiply two binomials, use the **FOIL** method.

We believe that a process must be repeated several times before it is learned. Therefore, we believe in continuous review. This review is apparent in the material at the end of each chapter and in the important formulas listed for reference inside the covers of the book. However, it also is built into the text itself. For example, Chapter 4 covers the topic of factoring. These factoring skills are then used throughout the rest of the text. They are used in Section 4.8 to solve quadratic equations, throughout Chapter 5 to work with fractions, in Chapter 8 to solve radical equations, and in Chapter 9 to solve quadratic equations.

François Vieta (Viête) (1540–1603)
By using letters in place of
unknown numbers, Vieta simplified
algebra and brought its notation
closer to the notation that we use
today. The one symbol he didn't
use was the equal sign.

Because mathematics is important and improves our daily lives, it is fitting that we honor pioneers in the field. For this reason, we have included portraits of several mathematicians and short captions about them. See page 69 to learn something about François Vieta.

As we have said, this text has been written for you to read and we think you will find the explanations helpful. Get off to a good start. Read the following article by the noted psychologist, Wayne Weiten. Then read the text carefully, do the exercises, and check your progress with the review exercises and the chapter tests.

When you finish this course, we suggest that you consider keeping your book. It is the single reference source that will keep the information that you have learned at your fingertips. You may need this reference material in future mathematics, science, or business courses.

We wish you well.

R. David Gustafson
Peter D. Frisk

Answer the following true or false.

1. It's a good idea to study in as many different locations (your bedroom or kitchen, the library, lounges around school, and so forth) as possible.
2. If you have a professor who delivers chaotic, hard-to-follow lectures, there is little point in attending class.
3. Cramming the night before an exam is an efficient way to study.
4. In taking lecture notes, you should try to be a "human tape recorder" (that is, take down everything exactly as said by your professor).
5. Outlining reading assignments is a waste of time.

As you will soon learn, all of the statements above are false. If you answered them all correctly, you may already have acquired the kinds of skills and habits that lead to academic success. If so, however, you are not typical. Today, a huge number of students enter college with remarkably poor study skills and habits—and it's not entirely their fault. Our educational system generally does not provide much in the way of formal instruction on good study techniques. We will try to remedy this oversight to some extent by sharing with you some insights on how to improve your academic performance.

DEVELOPING SOUND STUDY HABITS

Effective study is crucial to success in college. You may run into a few classmates who boast about getting good grades without studying. But you can be sure that if they perform well on exams—they study. Students who claim otherwise simply want to be viewed as extremely bright rather than studious.

Learning can be immensely gratifying, but studying usually involves hard work. The first step toward effective study habits is to face this reality. You don't have to feel guilty if you don't look forward to studying. Most students don't.

SOURCE: Weiten, W., Lloyd, M. A., and Lashley, R. L. (1991). *Psychology applied to modern life: Adjustment in the 90s,* 3rd ed. Pacific Grove, CA: Brooks/Cole. Adapted by permission.

Once you accept the premise that studying doesn't come naturally, it should be clear that you need to set up an organized program to promote adequate study. Such a program should include the following steps.

1. *Set up a schedule for studying.* If you wait until the urge to study hits you, you may still be waiting when the exam rolls around. Thus, it is important to allocate definite times to studying. Review your time obligations (work, housekeeping, and so on) and figure out in advance when you can study. In allotting certain times to studying, keep in mind that you need to be wide awake and alert. It won't do you much good to plan on studying when you're likely to be very tired. Be realistic, too, about how long you can study at one time before you wear down from fatigue. Allow time for study breaks; they can revive sagging concentration.

It's important to write down your study schedule. Writing it down serves as a reminder and increases your commitment to the schedule. Begin by setting up a general schedule for the quarter or semester. Then, each week plan the specific assignments that you intend to work on during each study session. This should help you to avoid cramming for exams at the last minute.

In planning your weekly schedule, try to avoid putting off working on major tasks such as term papers and reports. Many of us tend to tackle simple, routine tasks first, while saving larger tasks for later, when we supposedly will have more time. This tendency leads many of us to delay working on major assignments until it's too late to do a good job. You can avoid this trap by breaking major assignments into smaller component tasks that you schedule individually.

2. *Find a place to study where you can concentrate.* Where you study is also important. The key is to find a place where distractions are likely to be minimal. Most people cannot study effectively while watching TV, listening to the stereo, or overhearing conversations. Don't depend on willpower to carry you through. It's much easier to plan ahead and avoid the distractions altogether.

3. *Reward your studying.* One of the reasons it is so difficult to motivate yourself to study regularly is that the payoffs for studying often lie in the distant future. The ultimate reward, a degree, may be years away. Even more short-term rewards, such as an A in the course, may be weeks or months away. To combat this problem, it helps to give yourself immediate rewards for studying. It is easier to motivate yourself to study if you reward yourself with a tangible payoff when you finish. Thus, you should set realistic study goals for yourself and then reward yourself when you meet them.

IMPROVING YOUR READING

Much of your study time is spent reading and absorbing information. *These efforts must be active.* If you engage in passive reading, the information will pass right through you. Many students deceive themselves into thinking that they are studying by running a marker through a few sentences here and there in their books. If this isn't done selectively, the student is simply turning a textbook into a coloring book. Underlining in your text can be useful, but you have to distinguish between important ideas and mere supportive material.

There are a number of ways of actively attacking your reading assignments. One of the more worthwhile strategies is the SQ3R method. **SQ3R is a study system designed to promote effective reading that includes five steps: survey, question, read, recite, and review.** Its name is an abbreviation for the five steps in the procedure:

- *Step 1: Survey.* Before you plunge into the actual reading, glance over the topic headings in the chapter and try to get an overview of the material. Try to understand how the various chapter segments are related. Consult the chapter summary to get a feel for the chapter. If you know where the chapter is going, you can better appreciate and organize the information you are about to read.
- *Step 2: Question.* Once you have an overview of your reading assignment, proceed through it one section at a time. Take a look at the heading of the first section and convert it into a question. This is usually quite simple. If the heading is "Factoring by Grouping," your question should be "What is the procedure to factor by grouping?" Asking these questions gets you actively involved in your reading and helps you to identify the main ideas.
- *Step 3: Read.* Only now are you ready to sink your teeth into the reading. Read only the specific section that you have decided to tackle. Read it with an eye toward answering the question that you just formulated. If necessary, reread the section until you can answer that question. Work the examples in the section as you study them. Decide whether the segment addresses any other important questions and answer these as well.
- *Step 4: Recite.* Now that you can answer the key question for the section, recite it out loud to yourself in your own words. Use your own words because that requires understanding instead of simple memorization. Now work the exercises at the end of the section. Don't move on to the next section until you understand the main idea(s) of the present section. You may want to write down these ideas for review later. When you have fully digested the first section, go on to the next. Repeat steps 2 through 4 with the next section. Once you have mastered the crucial points there, you can go on again. Keep repeating steps 2 through 4 until you finish the chapter.
- *Step 5: Review.* When you have read the chapter, test and refresh your memory by going back over the key points. Repeat your questions and try to answer them without consulting your book or notes. This review should fortify your retention of the main ideas and should alert you to any key ideas that you haven't mastered. It should also help you to see the relationships between the main ideas.

The SQ3R method does not have to be applied rigidly. For example, it is often wise to break your reading assignment down into smaller segments than those separated by section headings. In fact, you should probably apply SQ3R to mathematics texts on a paragraph by paragraph basis. Obviously, this will require you to formulate some questions without the benefit of topic headings. However, the headings are not absolutely necessary to use this technique. If you don't have enough headings, you can simply reverse the order of steps 2 and 3.

Read the paragraph first and then formulate a question that addresses the basic idea of the paragraph. The point is that you can be flexible in your use of the SQ3R technique. *What makes SQ3R effective is that it breaks a reading assignment down into manageable segments and requires understanding before you move on.* Any method that accomplishes these goals should enhance your reading.

It is easier to use the SQ3R method when your textbook has plenty of topic headings. This brings up another worthwhile point about improving your reading. It pays to take advantage of the various learning aids incorporated into many textbooks. If a book provides a chapter outline or chapter summary, don't ignore them. They can help you to recognize the important points in the chapter and to understand how the various parts of the chapter are interrelated.

GETTING MORE OUT OF LECTURES

Although lectures are sometimes boring and tedious, it is a simple fact that poor class attendance is associated with poor grades. For example, one study found that absences from class were much more common among "unsuccessful" students (grade average: C− or below) than among "successful" students (grade average: B or above). Even when you have an instructor who delivers hard-to-follow lectures from which you learn virtually nothing, it is still important to go to class. If nothing else, you'll get a feel for how the instructor thinks. This can help you to anticipate the content of exams and to respond in the manner your professor expects.

Fortunately, most lectures are reasonably coherent. Research indicates that accurate note taking is related to better test performance. Good note taking requires you to actively process lecture information in ways that should enhance both memory and understanding. Books on study skills offer a number of suggestions on how to take good lecture notes. Some of these are summarized here.

- Extracting information from lectures requires *active listening procedures.* Focus full attention on the speaker. Try to anticipate what's coming and search for deeper meanings. Pay attention to nonverbal signals that may serve to further clarify the lecturer's intent or meaning.
- When course material is especially complex and difficult, it is a good idea to prepare for the lecture by reading ahead on the scheduled subject in your text. Then you have less information to digest that is brand-new.
- You should not try to be a human tape recorder. Instead, try to write down the lecturer's thoughts in your own words. This forces you to organize the ideas in a way that makes sense to you. In taking notes, pay attention to clues about what is most important. Many instructors give subtle and not-so-subtle clues about what is important. These clues may range from simply repeating main points to saying things like "You'll run into this again."
- Asking questions during lectures can be very helpful. This keeps you actively involved in the lecture. It also allows you to clarify points you may have misunderstood. Many students are bashful about asking questions. They don't realize that most professors welcome questions.

APPLYING THE PRINCIPLES OF MEMORY

Scientific investigation of memory processes dates back to 1885, when Hermann Ebbinghaus published a series of insightful studies. Thus, memory has been an important topic in psychology for over a century. As a result, a number of principles have been formulated that are relevant to effective study.

Engage in Adequate Practice

Practice makes perfect, or so you've heard. In reality, practice is not likely to guarantee perfection, but repeatedly reviewing information usually leads to improved retention. Studies show that retention improves with increased rehearsal. Continued rehearsal may also pay off by improving your *understanding* of assigned material. As you go over information again and again, your increased familiarity with the material may permit you to focus selectively on the most important points, thus enhancing your understanding.

There is evidence that it even pays to overlearn material. *Overlearning* **refers to continued rehearsal of material after you first appear to master it.** The implication of this finding is simple: you should not quit rehearsing material as soon as you appear to have mastered it.

Use Distributed Practice

Let's assume that you are going to study 9 hours for an exam. Is it better to "cram" all of your study into one 9-hour period (massed practice) or distribute it among, say, three 3-hour periods on successive days (distributed practice)? The evidence indicates that retention tends to be greater after distributed practice than massed practice, especially if the intervals between practice periods are fairly long, such as 24 hours.

The inefficiency of massed practice means that cramming is an ill-advised study strategy for most students. Cramming will strain your memorization capabilities and tax your energy level. It may also stoke the fires of test anxiety.

Minimize Interference

Interference **occurs when people forget information because of competition from other learned material.** Research suggests that interference is a major cause of forgetting, so you'll probably want to think about how you can minimize interference. This is especially important for students because memorizing information for one course can interfere with retaining information in another course. It may help to allocate study for specific courses to specific days. Thus, the day before an exam in a course, it is probably best to study for that course only. If demands in other courses make that impossible, study the test material last.

Of course, studying for other classes is not the only source of interference in a student's life. Other normal waking activities also produce interference. Therefore, it is a good idea to conduct one last, thorough review of material as close to exam time as possible. This last-minute review helps you to avoid memory loss due to interference from intervening activities.

Organize Information

Retention tends to be greater when information is well organized. Studies have shown that hierarchical organization is particularly helpful. Hence, one of the most potent weapons in your arsenal of study techniques is to *outline* reading assignments. Outlining is probably too time-consuming to do for every class. However, it is worth the effort in particularly important or particularly difficult classes, as it can greatly improve retention.

Use Verbal Mnemonics

People retain information better when they make the information more meaningful. A very useful strategy is to make material *personally* meaningful. When you read your textbooks, try to relate information to your own life and experience. For example, if you're reading in your psychology text about the personality trait of assertiveness, you can think of someone you know who is very assertive.

Of course, it's not always easy to make something personally meaningful. When you study mathematics, you may have a hard time relating to polynomials at a personal level. This problem has led to the development of many **mnemonic devices, or strategies for enhancing memory,** that are designed to make abstract material more meaningful.

Acrostics and Acronyms

Acrostics are phrases (or poems) in which the first letter of each word (or line) functions as a cue to help you recall the abstract words that begin with the same letter. For instance, you may remember the order of operations—Parentheses, Exponents, Multiplication, Division, Addition, and Subtraction—with the acrostic "Please Excuse My Dear Aunt Sally." A variation on acrostics is the *acronym*—a word formed out of the first letters of a series of words. You may use the acronym FOIL to remember to multiply the First, Outer, Inner, and Last terms of two binomials.

Narrative Methods

Another useful way to remember a list of words is to create a story that includes each of the words in the right order. The narrative increases the meaningfulness of the words and links them in a specific order.

Why—and how—would you use this method? Let's assume that you always manage to forget to put one item in your gym bag on your way to the pool. Short of pasting a list on the inside of the bag, how can you remember everything you need? You could make up a story that includes the items you need.

> The wind and rain in COMBINATION LOCKed out the rescue efforts—nearly. CAP, the flying ace, TOWELed the SOAP from his eyes, pulled his GOGGLES from his SUIT pocket, and COMBed the BRUSH for survivors.

Rhymes

Another verbal mnemonic that we often rely on is rhyming. You've probably repeated, "I before E except after C" thousands of times. Perhaps you also remember the number of days in each month with the old standby, "Thirty days hath September . . . ." Rhyming is an old and very useful trick.

Use Visual Imagery

Memory can be improved through the use of visual imagery. One influential theory proposes that visual images create a second memory code and that two codes are better than one. Many popular mnemonic devices depend on visual imagery, including the following examples.

Link Method

The *link method* involves forming a mental image of items to be remembered in a way that links them together. For instance, suppose that you are going to stop at the drugstore on the way home and you need to remember to pick up a news magazine, shaving cream, film, and pens. To remember these items, you might visualize a public figure likely to be in the magazine shaving with a pen while being photographed. There is evidence that the more bizarre you make your image, the more helpful it will be.

Loci Method

The *method of loci* involves taking an imaginary walk along a familiar path where you have associated images of items you want to remember with certain locations. The first step is to commit to memory a series of loci, or places along a path. Usually these loci are specific locations in your home or neighborhood. Then envision each thing you want to remember in one of these locations. Try to form distinctive, vivid images. When you need to remember the items, imagine yourself walking along the path. The various loci on your path should serve as retrieval cues for the images that you formed. The method of loci assures that items are remembered in their correct order because the order is determined by the sequence of locations along the pathway.

IMPROVING TEST-TAKING STRATEGIES

Let's face it: Some students are better than others at taking tests. ***Testwiseness*** **is the ability to use the characteristics and formats of an exam to maximize one's score.** Students clearly vary in testwiseness, and these variations influence performance on exams. Testwiseness is not a substitute for knowledge of the subject matter. However, skill in taking tests can help you to show what you know when it is critical to do so.

General Tips

- If efficient time use appears crucial, set up a mental schedule for progress through the test. Make a mental note to check whether you're one-third finished when one third of your time is gone. You may want to check again at the two-thirds time mark.
- On troublesome, difficult-to-answer items, do not waste time by pondering them excessively. If you have no idea at all, just guess and go on. If you think you need to devote a good deal of time to the item, skip it and mark it so you can return to it later if time permits.

- If you complete all of the questions and still have some time remaining, review the test. Make sure that you have recorded your answers correctly. If you were unsure of some answers, go back and reconsider them.
- Adopt the appropriate level of sophistication for the test. Don't read things into questions. Sometimes, students make things more complex than they were intended to be. Often, simple-looking questions are just what they appear to be—simple.
- Unless it is explicitly forbidden, don't hesitate to ask the examiner to clarify a question when necessary. Many examiners will graciously provide a great deal of useful information.

Tips for Multiple-Choice Exams

Sound test-taking strategies are especially important on multiple-choice (and true-false) exams. These types of questions often include clues that may help you to converge on the correct answer. You may be able to improve your performance on such tests by considering the following points.

- As you read the stem of each multiple-choice question, *anticipate* the answer if you can, before looking at the options. If the answer you anticipated is found among the options, there is a high probability that it is correct.
- Even if you find your anticipated answer among the options, you should always continue through and read all the options. There may be another option farther down the list that encompasses the one you anticipated. You should always read each question completely.
- Learn to quickly eliminate options that are highly implausible. Many questions have only two plausible options, accompanied by "throwaway" options for filler. You should work at spotting these implausible options so you can quickly discard them and narrow your search.
- Be alert to the fact that examiners sometimes "give away" information relevant to one question in another test item.
- On items that have "all of the above" as an option, if you know that just two of the options are correct, you should choose "all of the above." If you are confident that any one of the options is incorrect, you should eliminate both the incorrect option and "all of the above," and choose from the remaining options.
- Although there will always be exceptions, options that are more detailed than the others tend to be correct. Hence, it's a good idea to pay special attention to options that are extra-long or highly specific.

In summary, sound study skills and habits are crucial to academic success. Intelligence alone won't do the job (although it certainly helps). Good academic skills do not develop overnight. They are acquired gradually, so be patient. Fortunately, tasks such as reading textbooks and taking tests get easier with practice. Ultimately, you'll find that the rewards—knowledge, a sense of accomplishment, and progress toward a degree—are worth the effort.

CONTENTS

1

Real Numbers and Their Basic Properties

Algebra is an extension of arithmetic. In algebra the operations of addition, subtraction, multiplication, and division are performed on both numbers and letters, with the understanding that the letters can be replaced by numbers. This idea is simple but powerful. It leads to solutions of problems that would be difficult or impossible for us to solve by using arithmetic alone.

Algebra is not new. It has been used for thousands of years in China, India, Persia, and Arabia. The origins of algebra are found in a papyrus written before 1000 B.C. by an Egyptian priest named Ahmes. This papyrus contains 80 algebra problems and their solutions. Because the Egyptians did not have a suitable system of notation, however, they were unable to develop algebra completely.

Further development of algebra had to wait until the ninth century and the rise of the Moslem civilization. In A.D. 830 one of the greatest mathematicians of Arabian history, al-Khowarazmi, wrote a book called *Ihm al-jabr wa'l muqabalah.* This imposing title was soon shortened to *al-Jabr.* The spelling has changed over the centuries, and we now know the subject as *algebra.* The French mathematician François Vieta (1540–1603) later simplified the study of algebra by developing the symbolic notation that we use today.

We begin the study of algebra by discussing the properties of numbers.

1.1   SETS OF NUMBERS AND THEIR GRAPHS

The most basic **set** of numbers is the set of **natural numbers** that we use for counting. These are the numbers 1, 2, 3, 4, 5, 6, 7, 8, and so on.

DEFINITION.   The **natural numbers** are the numbers

   1, 2, 3, 4, 5, 6, 7, 8, 9, 10, . . .

The three dots, called the **ellipsis,** in the previous definition indicate that the list continues forever. There is no largest natural number.

The natural numbers that can be divided exactly by 2—that is, the division has a remainder of 0—are called the **even natural numbers.** They are the numbers

   2, 4, 6, 8, 10, 12, . . .

The natural numbers that cannot be divided exactly by 2—that is, the division has a remainder of 1—are called the **odd natural numbers.** They are the numbers

   1, 3, 5, 7, 9, 11, 13, . . .

Some natural numbers can be divided exactly only by 1 and the number itself. For example, the only two divisors of 3 are 1 and 3, and the only two divisors of 17 are 1 and 17. Natural numbers with exactly two divisors are called **prime numbers.**

DEFINITION.   A **prime number** is a natural number that is larger than 1 and is exactly divisible only by 1 and by itself.

The prime numbers are the numbers

2, 3, 5, 7, 11, 13, 17, 19, 23, 29, 31, . . .

The number 2 is the only even prime number. All of the other prime numbers are odd natural numbers.

The natural number 12 can be divided exactly by several numbers—1, 2, 3, 4, 6, and 12. Because 12 has more than two divisors, it is not a prime number. Rather, it is an example of a number called a **composite number.**

Leonardo Fibonacci (late 12th and early 13th cent.) Fibonacci, an Italian mathematician, is also known as Leonardo da Pisa. In his work *Liber abaci,* he advocated the adoption of Arabic numerals, the numerals that we use today. He is best known for a sequence of numbers that bears his name. Can you find the pattern in this sequence?

1, 1, 2, 3, 5, 8, 13, . . .

> **DEFINITION.**   A **composite number** is a natural number greater than 1 that is not a prime number.

The composite numbers are the numbers

4, 6, 8, 9, 10, 12, 14, 15, 16, 18, 20, 21, . . .

The natural number 1 is neither a prime number nor a composite number. The natural numbers together with 0 form the set of **whole numbers.** Thus, the whole numbers are the numbers

0, 1, 2, 3, 4, 5, 6, 7, 8, 9, 10, . . .

Braces are often used to enclose a list of the members of a set. Each pair of braces is read as "the set of."

The set of natural numbers is {1, 2, 3, 4, 5, 6, . . .}.
The set of prime numbers is {2, 3, 5, 7, 11, 13, . . .}.
The set of composite numbers is {4, 6, 8, 9, 10, 12, . . .}.
The set of whole numbers is {0, 1, 2, 3, 4, 5, . . .}.

Each member of a set is called an **element** of the set. The number 3, for example, is an element of the set of natural numbers, an element of the set of prime numbers, and an element of the set of whole numbers. However, 3 is *not* an element of the set of composite numbers.

EXAMPLE 1   Classify the numbers   **a.**   7,   **b.**   6,   and   **c.**   0 into the categories of natural number, even natural number, odd natural number, prime number, composite number, and whole number.

Solution   **a.** The number 7 is a natural number, an odd natural number, a prime number, and a whole number.
b. The number 6 is a natural number, an even natural number, a composite number, and a whole number.
c. The number 0 is a whole number.   ■

Graphing Sets of Numbers

Various sets of numbers can be pictured on the **number line.** To construct the number line shown in Figure 1-1, we pick some point on the line, label it 0, and call it the **origin.** We then pick some unit length, mark off points to the right of the origin, and label these points with the natural numbers, as shown in the figure.

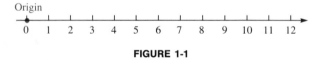

FIGURE 1-1

The point labeled 3 on the number line, for example, is 3 units to the right of the origin. The point labeled 6 is 6 units to the right of the origin.

Sets of numbers can be represented, or graphed, on the number line. For example, Figure 1-2 shows the graph of the even natural numbers from 2 to 8. The point corresponding to the number 4, for example, is called the **graph of 4.** The number 4 is called the **coordinate** of its corresponding point.

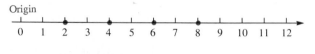

FIGURE 1-2

EXAMPLE 2   Graph the set of prime numbers between 1 and 20 on the number line.

Solution   The prime numbers between 1 and 20 are 2, 3, 5, 7, 11, 13, 17, and 19. The graph of this set of numbers is shown in Figure 1-3.

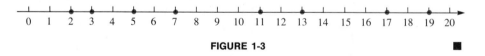

FIGURE 1-3   ■

One of the most common symbols in mathematics is the **equals sign,** written as =. It is used to indicate that two expressions represent the same number. Because 4 + 5 and 9 represent the same number, we can write

$4 + 5 = 9$ Read as "the sum of 4 and 5 is 9," or "4 plus 5 equals 9."

Likewise, we can write

$5 - 3 = 2$ Read as "the difference between 5 and 3 is 2," or "5 minus 3 equals 2."

$4 \cdot 5 = 20$ Read as "the product of 4 and 5 is 20," or "4 times 5 equals 20."

and

$30 \div 6 = 5$ Read as "the quotient obtained when 30 is divided by 6 is 5," or "30 divided by 6 equals 5."

Other symbols, called **inequality symbols,** are used to indicate that expressions are *not* equal.

Symbol	Read as
≠	"is not equal to"
<	"is less than"
>	"is greater than"
≤	"is less than or equal to"
≥	"is greater than or equal to"

EXAMPLE 3   **a.** $6 \neq 9$ is read as "6 is not equal to 9."

b. $8 < 10$ is read as "8 is less than 10."

c. $12 > 1$ is read as "12 is greater than 1."

d. $5 \leq 5$ is read as "5 is less than or equal to 5."
(Because 5 is equal to 5, this is a true statement.)

e. $9 \geq 7$ is read as "9 is greater than or equal to 7."
(Because 9 is greater than 7, this is a true statement). ■

Statements of inequality can be written so that the inequality symbol points in the opposite direction. For example, the inequality statement

$$5 < 7$$

is read as "5 is less than 7," and the statement

$$7 > 5$$

is read as "7 is greater than 5." Both indicate that 5 is a smaller number than 7. Likewise,

$$12 \geq 3 \qquad \text{Read as "12 is greater than or equal to 3."}$$

and

$$3 \leq 12 \qquad \text{Read as "3 is less than or equal to 12."}$$

are equivalent statements.

If one point is to the *right* of a second point on the number line, its coordinate is the *greater*. For example, the point with coordinate 5 in Figure 1-3 lies to the right of the point with coordinate 2. Thus, 5 > 2. If one point is to the *left* of another, its coordinate is the *smaller*. The point with coordinate 11, for example, is to the left of the point with coordinate 19. Thus, 11 < 19.

EXERCISE 1.1

In Exercises 1–8, list the numbers in the set {0, 1, 2, 4, 13, 15} *that satisfy the given condition.*

1. natural number

2. even natural number

3. odd natural number

4. prime number

5. composite number

6. whole number

7. even prime number

8. odd composite number

In Exercises 9–16, list the numbers in the set {0, 1, 2, 6, 11, 12} that satisfy the given condition.

9. whole number

10. prime number

11. even natural number

12. composite number

13. odd prime number

14. even composite number

15. neither prime nor composite

16. whole number but not a natural number

In Exercises 17–26, graph each set of numbers on the number line.

17. The natural numbers between 2 and 8.

18. The prime numbers from 10 to 20.

19. The composite numbers from 20 to 30.

20. The whole numbers less than 6.

21. The even natural numbers greater than 10 but less than 20.

22. The even natural numbers that are also prime numbers.

23. The numbers that are whole numbers but not natural numbers.

24. The prime numbers between 20 and 30.

25. The natural numbers between 11 and 25 that are exactly divisible by 6.

26. The odd natural numbers between 14 and 28 that are exactly divisible by 3.

In Exercises 27–40, place one of the symbols =, <, or > in the box to make a true statement.

27. $5 \,\boxed{}\, 3 + 2$

28. $9 \,\boxed{}\, 7$

29. $25 \,\boxed{}\, 32$

30. $2 + 3 \,\boxed{}\, 17$

31. $5 + 7 \,\boxed{}\, 10$

32. $3 + 3 \,\boxed{}\, 9 - 3$

33. $3 + 2 + 5 \,\boxed{}\, 5 + 2 + 3$

34. $8 - 5 \,\boxed{}\, 5 - 2$

35. $3 + 9 \,\boxed{}\, 20 - 8$

36. $19 - 3 \,\boxed{}\, 8 + 6$

37. $4 \cdot 2 \,\boxed{}\, 2 \cdot 4$

38. $7 \cdot 9 \,\boxed{}\, 9 \cdot 6$

39. $8 \div 2 \,\boxed{}\, 4 + 2$

40. $0 \div 7 \,\boxed{}\, 1$

In Exercises 41–46, write each statement as a mathematical expression.

41. Seven is greater than three.

42. Five is less than thirty-two.

43. Seventeen is less than or equal to seventeen.

44. Twenty-five is not equal to twenty-three.

45. The result of adding three and four is equal to seven.

46. Thirty-seven is greater than or equal to the result of multiplying three and four.

In Exercises 47–58, rewrite each inequality statement as an equivalent inequality in which the inequality symbol points in the opposite direction.

47. $3 \le 7$ **48.** $5 > 2$ **49.** $6 > 0$ **50.** $34 \le 40$

51. $3 + 8 > 8$ **52.** $8 - 3 < 8$ **53.** $6 - 2 < 10 - 4$ **54.** $8 \cdot 2 \ge 8 \cdot 1$

55. $2 \cdot 3 < 3 \cdot 4$ **56.** $8 \div 2 \ge 9 \div 3$ **57.** $\dfrac{12}{4} < \dfrac{24}{6}$ **58.** $\dfrac{2}{3} \le \dfrac{3}{4}$

In Exercises 59–66, graph each pair of numbers on a number line. Indicate which number in the pair is the greater and which number lies to the right of the other number on the number line.

59. 3, 6   **60.** 4, 7

61. 11, 6   **62.** 12, 10

63. 0, 2   **64.** 4, 10

65. 8, 0   **66.** 20, 30

67. Explain why there is no greatest natural number.

68. Explain why 2 is the only even prime number.

69. Explain why no natural numbers are both even and odd.

70. Find the only natural number that is neither a prime number nor a composite number.

REVIEW EXERCISES

In Review Exercises 1–8, perform the indicated operations.

1. Add: 132
      45
      73

2. Add: 261
      79
      31

3. Subtract: 321
        173

4. Subtract: 532
        437

5. Multiply: 437
        38

6. Multiply: 529
        42

7. Divide: $37 \overline{)\, 3885}$

8. Divide: $53 \overline{)\, 11607}$

In Review Exercises 9–10, perform the indicated operations.

9. $235 + 517 - 26$ **10.** $135 + 22 - 156$

11. Sally bought four paintings for $350, $900, $820, and $1000. One year later, she sold them for $750, $850, $1250, and $2700, respectively. Find her total profit.

12. An airplane is cruising at 29,000 feet. The pilot receives permission to descend 7000 feet and later to climb 11,000 feet. At what altitude is the plane then cruising?

1.2 FRACTIONS

We continue the review of the number system by discussing **fractions.** In fractions such as

$$\frac{1}{2}, \quad \frac{3}{5}, \quad \frac{2}{17}, \quad \text{and} \quad \frac{37}{7}$$

the number above the fraction bar is called the **numerator,** and the number below is called the **denominator.**

A fraction can be used to indicate parts of a whole. In this case, the denominator indicates the total number of equal parts into which the whole is divided, and the numerator indicates the number of these equal parts that are being considered. For example, if a test contains 25 problems and a student completes 21 of them, the student has completed $\frac{21}{25}$ (twenty-one twenty-fifths) of the test. In Figure 1-4**a**, a rectangle has been divided into 5 equal parts and 3 of the parts are shaded. Thus, $\frac{3}{5}$ of the figure is shaded. In Figure 1-4**b**, $\frac{5}{7}$ of the rectangle is shaded.

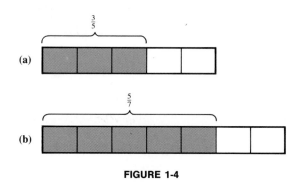

FIGURE 1-4

A fraction can also be interpreted as an indicated division. For example, the fraction $\frac{8}{2}$ indicates that 8 is to be divided by 2:

$$\frac{8}{2} = 8 \div 2 = 4$$

We note that $\frac{8}{2} = 4$ because $4 \cdot 2 = 8$, and that $\frac{0}{7} = 0$ because $0 \cdot 7 = 0$. However, the fraction $\frac{6}{0}$ is undefined because no number multiplied by 0 gives 6. The fraction $\frac{0}{0}$ is also undefined, because any number multiplied by 0 gives 0. Thus, it is understood that division by 0 is impossible. We must always remember that

> **The denominator of a fraction can never be 0.**

A fraction is said to be written in **lowest terms** if no natural number greater than 1 will divide both the numerator and the denominator of the fraction exactly. The fraction $\frac{6}{11}$, for example, is written in lowest terms because no number other than 1 divides both 6 and 11 exactly. The fraction $\frac{6}{8}$ is not in lowest terms because the number 2 divides both 6 and 8 exactly.

Simplifying Fractions

To **simplify,** or **reduce,** a fraction means to write the fraction in lowest terms. If a fraction is not in lowest terms, we can simplify it by dividing both the numerator and denominator of the fraction by the same number. For example,

$$\frac{6}{8} = \frac{6 \div 2}{8 \div 2} = \frac{3}{4} \quad \text{and} \quad \frac{15}{18} = \frac{15 \div 3}{18 \div 3} = \frac{5}{6}$$

From Figure 1-5 we see that the fractions $\frac{6}{8}$ and $\frac{3}{4}$ are equal because each fraction represents the same part of the rectangle.

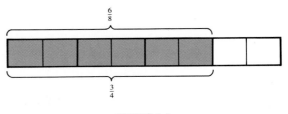

FIGURE 1-5

To develop another way to simplify fractions, we use the fact that composite numbers can be written as the product of other natural numbers. For example, the composite number 12 can be written as the product of 4 and 3:

$$12 = 4 \cdot 3$$

When a composite number has been written as the product of other natural numbers, we say that it has been **factored.** The numbers 3 and 4 are called **factors** of 12. When a composite number is written as the product of prime numbers, we say that the number is written in **prime-factored form.**

EXAMPLE 1   Write 210 in prime-factored form.

Solution   Begin by writing the number 210 as a product such as 21 times 10, even though 21 and 10 are not prime numbers, and proceed as follows:

$$210 = \mathbf{21 \cdot 10}$$
$$210 = \mathbf{3 \cdot 7} \cdot \mathbf{2 \cdot 5} \qquad \text{Factor 21 as } 3 \cdot 7 \text{ and factor 10 as } 2 \cdot 5.$$

Thus,

$$210 = 2 \cdot 3 \cdot 5 \cdot 7$$

Because 210 is now written as the product of prime numbers, it is written in prime-factored form.   ■

To **simplify** a fraction in algebra, we factor both its numerator and denominator, and then divide out all factors that appear in both the numerator and denominator. To simplify the fractions $\frac{6}{8}$ and $\frac{15}{18}$, for example, we proceed as follows:

$$\frac{6}{8} = \frac{3 \cdot 2}{4 \cdot 2} = \frac{3 \cdot \overset{1}{\cancel{2}}}{4 \cdot \underset{1}{\cancel{2}}} = \frac{3}{4} \qquad \text{and} \qquad \frac{15}{18} = \frac{5 \cdot 3}{6 \cdot 3} = \frac{5 \cdot \overset{1}{\cancel{3}}}{6 \cdot \underset{1}{\cancel{3}}} = \frac{5}{6}$$

A fraction is written in lowest terms only when its numerator and denominator share no common factors.

EXAMPLE 2   **a.** To simplify the fraction $\frac{6}{30}$, we factor both the numerator and denominator and divide out the common factor of 6 to obtain

$$\frac{6}{30} = \frac{6 \cdot 1}{6 \cdot 5} = \frac{\overset{1}{\cancel{6}} \cdot 1}{\underset{1}{\cancel{6}} \cdot 5} = \frac{1}{5}$$

b. To show that the fraction $\frac{33}{40}$ is written in lowest terms, we must show that the numerator and the denominator share no common factors. To do so, we write both the numerator and denominator in prime-factored form.

$$\frac{33}{40} = \frac{3 \cdot 11}{2 \cdot 2 \cdot 2 \cdot 5}$$

Because there are no factors common to numerator and denominator, the fraction $\frac{33}{40}$ is in lowest terms.   ■

The previous examples illustrate the **fundamental property of fractions.** To state this property, we will use letters, called **variables,** to represent numbers.

The Fundamental Property of Fractions.   If a represents a number, and b and x represent nonzero numbers, then

$$\frac{a \cdot x}{b \cdot x} = \frac{a}{b}$$

Multiplying Fractions

To *multiply* two fractions, we use the following rule:

> **Multiplying Fractions.** To multiply two fractions, we multiply the numerators and multiply the denominators. In symbols, if a, b, c, and d are numbers, then
>
> $$\frac{a}{b} \cdot \frac{c}{d} = \frac{a \cdot c}{b \cdot d} \qquad (b \neq 0 \text{ and } d \neq 0)$$

For example,

$$\frac{2}{3} \cdot \frac{4}{7} = \frac{2 \cdot 4}{3 \cdot 7} = \frac{8}{21} \qquad \text{and} \qquad \frac{4}{5} \cdot \frac{13}{9} = \frac{4 \cdot 13}{5 \cdot 9} = \frac{52}{45}$$

To justify the rule for multiplying fractions, we consider the square in Figure 1-6 with each side 1 unit in length. Because the area of a square is the product of the lengths of two sides, the square in Figure 1-6 has an area of 1 square unit. If this square is divided into 3 equal parts vertically and 7 equal parts horizontally, it is divided into 21 equal parts, and each of these parts represents $\frac{1}{21}$ of the total area. The area of the shaded rectangle in the square is $\frac{8}{21}$ because it consists of 8 of the 21 parts. Since the shaded rectangle has a length, l, of $\frac{2}{3}$ and a width, w, of $\frac{4}{7}$, its area A is the product of l and w:

$$A = l \cdot w$$

$$\frac{8}{21} = \frac{2}{3} \cdot \frac{4}{7}$$

Thus, we can find the product of

$$\frac{2}{3} \qquad \text{and} \qquad \frac{4}{7}$$

by multiplying their numerators and multiplying their denominators.

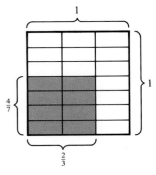

FIGURE 1-6

Fractions such as $\frac{8}{21}$, in which the numerator is smaller than the denominator, are called **proper fractions.** Fractions such as $\frac{52}{45}$, in which the numerator is larger than the denominator, are called **improper fractions.**

EXAMPLE 3   **a.** $\dfrac{3}{7} \cdot \dfrac{13}{5} = \dfrac{3 \cdot 13}{7 \cdot 5}$ Multiply the numerators and multiply the denominators

$$= \frac{39}{35}$$

b. $\dfrac{3}{7} \cdot \dfrac{14}{5} = \dfrac{3 \cdot 14}{7 \cdot 5}$ Multiply the numerators and multiply the denominators.

$\qquad\qquad = \dfrac{3 \cdot 2 \cdot 7}{7 \cdot 5}$ To attempt to simplify the fraction, factor the numerator.

$\qquad\qquad = \dfrac{3 \cdot 2 \cdot \overset{1}{\cancel{7}}}{\underset{1}{\cancel{7}} \cdot 5}$ Divide out the common factor of 7.

$\qquad\qquad = \dfrac{6}{5}$

c. $5 \cdot \dfrac{3}{15} = \dfrac{5}{1} \cdot \dfrac{3}{15}$ Write 5 as the improper fraction $\frac{5}{1}$.

$\qquad\qquad = \dfrac{5 \cdot 3}{1 \cdot 15}$ Multiply the fractions.

$\qquad\qquad = \dfrac{5 \cdot 3}{1 \cdot 5 \cdot 3}$ To attempt to simplify the fraction, factor the denominator.

$\qquad\qquad = \dfrac{\overset{1}{\cancel{5}} \cdot \overset{1}{\cancel{3}}}{1 \cdot \underset{1}{\cancel{5}} \cdot \underset{1}{\cancel{3}}}$ Divide out the common factors of 3 and 5.

$\qquad\qquad = \dfrac{1}{1}$

$\qquad\qquad = 1$ ∎

Dividing Fractions

One number is called the **reciprocal** of another if their product is 1. For example, the reciprocal of 2 is $\frac{1}{2}$, because $2 \cdot \frac{1}{2} = 1$. Likewise, the reciprocal of $\frac{3}{5}$ is $\frac{5}{3}$, because

$$\frac{3}{5} \cdot \frac{5}{3} = \frac{15}{15} = 1$$

To *divide* one fraction by a second fraction, we use the following rule:

Dividing Fractions.   To divide two fractions, we multiply the first fraction by the reciprocal of the second fraction. In symbols, if a, b, c, and d are numbers, then

$$\frac{a}{b} \div \frac{c}{d} = \frac{a}{b} \cdot \frac{d}{c} = \frac{a \cdot d}{b \cdot c} \qquad (b \neq 0,\ c \neq 0,\ \text{and}\ d \neq 0)$$

For example,

$$\frac{7}{8} \div \frac{5}{11} = \frac{7}{8} \cdot \frac{11}{5}$$ Multiply $\frac{7}{8}$ by the reciprocal of $\frac{5}{11}$.

$$= \frac{7 \cdot 11}{8 \cdot 5}$$ Multiply the fractions.

$$= \frac{77}{40}$$ Because the numerator and denominator share no common factors, the fraction does not simplify.

EXAMPLE 4   **a.** $\dfrac{3}{5} \div \dfrac{6}{5} = \dfrac{3}{5} \cdot \dfrac{5}{6}$ Multiply $\frac{3}{5}$ by the reciprocal of $\frac{6}{5}$.

$$= \frac{3 \cdot 5}{5 \cdot 6}$$ Multiply the fractions.

$$= \frac{3 \cdot 5}{5 \cdot 2 \cdot 3}$$ Factor the denominator.

$$= \frac{\overset{1}{\cancel{3}} \cdot \overset{1}{\cancel{5}}}{\underset{1}{\cancel{5}} \cdot 2 \cdot \underset{1}{\cancel{3}}}$$ Divide out the common factors of 3 and 5.

$$= \frac{1}{2}$$

b. $\dfrac{7}{9} \div 8 = \dfrac{7}{9} \div \dfrac{8}{1}$ Write 8 as the fraction $\frac{8}{1}$.

$$= \frac{7}{9} \cdot \frac{1}{8}$$ Multiply $\frac{7}{9}$ by the reciprocal of $\frac{8}{1}$.

$$= \frac{7}{72}$$

c. $10 \div \dfrac{15}{7} = \dfrac{10}{1} \div \dfrac{15}{7}$ Write 10 as the fraction $\frac{10}{1}$.

$$= \frac{10}{1} \cdot \frac{7}{15}$$ Multiply $\frac{10}{1}$ by the reciprocal of $\frac{15}{7}$.

$$= \frac{2 \cdot 5 \cdot 7}{3 \cdot 5}$$ Multiply the fractions and factor.

$$= \frac{2 \cdot \overset{1}{\cancel{5}} \cdot 7}{3 \cdot \underset{1}{\cancel{5}}}$$ Divide out the common factor of 5.

$$= \frac{14}{3}$$ ■

Adding Fractions

To *add* two fractions, we use the following rule:

> **Adding Fractions with the Same Denominator.**   To add two fractions with the same denominator, we add the numerators and keep the common denominator. In symbols, if a, b, and d are numbers, then
>
> $$\frac{a}{d} + \frac{b}{d} = \frac{a+b}{d} \qquad (d \neq 0)$$

For example,

$$\frac{3}{7} + \frac{2}{7} = \frac{3+2}{7} = \frac{5}{7}$$

Figure 1-7 shows pictorially why $\frac{3}{7} + \frac{2}{7} = \frac{5}{7}$.

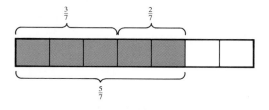

FIGURE 1-7

To add two fractions with unlike denominators, we must use this important fact:

> **If a fraction is multiplied by 1, its value is not changed.**

For example, if the fraction $\frac{1}{3}$ is multiplied by 1 written in the form $\frac{5}{5}$, we obtain an equal fraction with a different denominator:

$$\frac{1}{3} \cdot 1 = \frac{1}{3} \cdot \frac{5}{5} = \frac{1 \cdot 5}{3 \cdot 5} = \frac{5}{15}$$

This example suggests that anytime we multiply both the numerator and denominator of a fraction by the same nonzero number, the value of the fraction is unchanged.

To add the fractions $\frac{1}{3}$ and $\frac{1}{5}$, we write each fraction as an equivalent fraction having a denominator of 15 and add the results:

$$\frac{1}{3} + \frac{1}{5} = \frac{1 \cdot 5}{3 \cdot 5} + \frac{1 \cdot 3}{5 \cdot 3} = \frac{5}{15} + \frac{3}{15} = \frac{5+3}{15} = \frac{8}{15}$$

Because 15 is the smallest number that can serve as a common denominator of the previous fractions, it is called the **least** or **lowest common denominator.**

EXAMPLE 5   Add the fractions $\dfrac{3}{10}$ and $\dfrac{5}{28}$.

Solution   To add these fractions, write them as fractions having the same denominator. To find the least common denominator (LCD), find the prime factorization of both denominators and use each prime factor the greatest number of times it appears in either factorization:

$$10 = 2 \cdot 5$$
$$28 = 2 \cdot 2 \cdot 7$$

$$LCD = 2 \cdot 2 \cdot 5 \cdot 7 = 140$$

where's the 2 3

Because 140 is the smallest number that 10 and 28 divide exactly, write both fractions as fractions with the least common denominator of 140. Proceed as follows:

$$\frac{3}{10} + \frac{5}{28} = \frac{3 \cdot 14}{10 \cdot 14} + \frac{5 \cdot 5}{28 \cdot 5} \qquad \text{Write each fraction as a fraction with a denominator of 140.}$$

$$= \frac{42}{140} + \frac{25}{140}$$

$$= \frac{42 + 25}{140} \qquad \text{Add numerators and keep the denominator.}$$

$$= \frac{67}{140}$$

Because 67 and 140 share no common factors, this fraction is in lowest terms and cannot be simplifed.   ■

Subtracting Fractions

To *subtract* two fractions with the same denominator, we use the following rule:

> **Subtracting Fractions with the Same Denominator.**   To subtract two fractions with the same denominator, we subtract their numerators and keep their common denominator. In symbols, if a, b, and d are numbers, then
>
> $$\frac{a}{d} - \frac{b}{d} = \frac{a - b}{d} \qquad (d \neq 0)$$

For example,

$$\frac{7}{9} - \frac{2}{9} = \frac{7 - 2}{9} = \frac{5}{9}$$

To subtract two fractions with unlike denominators, we write them as equivalent fractions with a common denominator. For example, to subtract $\frac{2}{5}$ from $\frac{3}{4}$, both fractions must be written as fractions with the same denominator. Because 20 is the smallest number that is divisible by both 5 and 4 exactly, the least common denominator is 20. To subtract $\frac{2}{5}$ from $\frac{3}{4}$, we write $\frac{3}{4} - \frac{2}{5}$ and proceed as follows:

$$\frac{3}{4} - \frac{2}{5} = \frac{3 \cdot 5}{4 \cdot 5} - \frac{2 \cdot 4}{5 \cdot 4} = \frac{15}{20} - \frac{8}{20} = \frac{15 - 8}{20} = \frac{7}{20}$$

EXAMPLE 6   Subtract 5 from $\dfrac{23}{3}$.

Solution   The subtraction indicated by $\frac{23}{3} - 5$ represents the difference between two fractions, because 5 can be written as the fraction $\frac{5}{1}$. Thus,

$$\begin{aligned}
\frac{23}{3} - 5 &= \frac{23}{3} - \frac{5}{1} \\
&= \frac{23}{3} - \frac{5 \cdot 3}{1 \cdot 3} \qquad \text{Write } \tfrac{5}{1} \text{ as a fraction with a denominator of 3.} \\
&= \frac{23}{3} - \frac{15}{3} \\
&= \frac{23 - 15}{3} \\
&= \frac{8}{3}
\end{aligned}$$

■

Mixed Numbers

A number such as $3\frac{1}{2}$ is called a **mixed number.** It represents the sum of the natural number 3 and the proper fraction $\frac{1}{2}$. We can write $3\frac{1}{2}$ as an improper fraction as follows:

$$\begin{aligned}
3\frac{1}{2} &= 3 + \frac{1}{2} \\
&= \frac{6}{2} + \frac{1}{2} \qquad 3 = \tfrac{6}{2}. \\
&= \frac{6 + 1}{2} \qquad \text{Add the numerators and keep the denominator.} \\
&= \frac{7}{2}
\end{aligned}$$

A fraction such as $\frac{19}{5}$ can be written as a mixed number. We begin by dividing 19 by 5. The answer is 3, with a remainder of 4. Thus,

$$\frac{19}{5} = 3 + \frac{4}{5} = 3\frac{4}{5}$$

EXAMPLE 7   Add $2\frac{1}{4}$ and $1\frac{1}{3}$.

Solution   Begin by changing each mixed number into an improper fraction.

$$2\frac{1}{4} = 2 + \frac{1}{4} = \frac{8}{4} + \frac{1}{4} = \frac{9}{4}$$

$$1\frac{1}{3} = 1 + \frac{1}{3} = \frac{3}{3} + \frac{1}{3} = \frac{4}{3}$$

Add the mixed numbers by adding these improper fractions.

$$2\frac{1}{4} + 1\frac{1}{3} = \frac{9}{4} + \frac{4}{3} = \frac{9 \cdot 3}{4 \cdot 3} + \frac{4 \cdot 4}{3 \cdot 4} = \frac{27}{12} + \frac{16}{12} = \frac{43}{12}$$

Finally, change the resulting fraction back to a mixed number.

$$\frac{43}{12} = 3 + \frac{7}{12} = 3\frac{7}{12}$$ ∎

EXAMPLE 8   The three sides of a triangular piece of land measure $33\frac{1}{4}$, $57\frac{3}{4}$, and $72\frac{1}{2}$ meters. How much fencing will be needed to enclose the area?

Solution   Find the sum of the lengths of the three sides by adding the whole number parts and the fractional parts of the dimensions separately:

$$33\frac{1}{4} + 57\frac{3}{4} + 72\frac{1}{2} = 33 + 57 + 72 + \frac{1}{4} + \frac{3}{4} + \frac{1}{2}$$

$$= 162 + \frac{1}{4} + \frac{3}{4} + \frac{2}{4} \qquad \text{Change } \tfrac{1}{2} \text{ to } \tfrac{2}{4} \text{ to obtain a common denominator.}$$

$$= 162 + \frac{6}{4} \qquad \text{In the fractions, add the numerators and keep the denominator.}$$

$$= 162 + \frac{3}{2} \qquad \frac{6}{4} = \frac{2 \cdot 3}{2 \cdot 2} = \frac{\cancel{2} \cdot 3}{\cancel{2} \cdot 2} = \frac{3}{2}.$$

$$= 162 + 1\frac{1}{2} \qquad \text{Change } \tfrac{3}{2} \text{ to a mixed number.}$$

$$= 163\frac{1}{2}$$

It will require $163\frac{1}{2}$ meters of fencing to enclose the triangular area.   ∎

EXERCISE 1.2

In Exercises 1–12, write each fraction in lowest terms. If the fraction is already in lowest terms, so indicate.

1. $\dfrac{6}{12}$ **2.** $\dfrac{3}{9}$ **3.** $\dfrac{4}{16}$ **4.** $\dfrac{7}{21}$

5. $\dfrac{15}{20}$ **6.** $\dfrac{22}{77}$ **7.** $\dfrac{24}{18}$ **8.** $\dfrac{35}{14}$

9. $\dfrac{72}{64}$ **10.** $\dfrac{26}{21}$ **11.** $\dfrac{45}{49}$ **12.** $\dfrac{45}{30}$

In Exercises 13–28, perform each multiplication. Simplify each result when possible.

13. $\dfrac{1}{2} \cdot \dfrac{3}{5}$ **14.** $\dfrac{3}{4} \cdot \dfrac{5}{7}$ **15.** $\dfrac{5}{9} \cdot \dfrac{4}{11}$ **16.** $\dfrac{7}{10} \cdot \dfrac{3}{10}$

17. $\dfrac{4}{3} \cdot \dfrac{6}{5}$ **18.** $\dfrac{7}{8} \cdot \dfrac{6}{15}$ **19.** $\dfrac{5}{12} \cdot \dfrac{18}{5}$ **20.** $\dfrac{5}{4} \cdot \dfrac{12}{10}$

21. $\dfrac{17}{34} \cdot \dfrac{3}{6}$ **22.** $\dfrac{21}{14} \cdot \dfrac{3}{6}$ **23.** $\dfrac{21}{30} \cdot \dfrac{10}{7}$ **24.** $\dfrac{15}{16} \cdot \dfrac{8}{9}$

25. $12 \cdot \dfrac{5}{6}$ **26.** $9 \cdot \dfrac{7}{12}$ **27.** $\dfrac{10}{21} \cdot 14$ **28.** $\dfrac{5}{24} \cdot 16$

In Exercises 29–44, perform each division. Simplify each result when possible.

29. $\dfrac{3}{5} \div \dfrac{2}{3}$ **30.** $\dfrac{4}{5} \div \dfrac{3}{7}$ **31.** $\dfrac{4}{7} \div \dfrac{3}{11}$ **32.** $\dfrac{2}{11} \div \dfrac{7}{10}$

33. $\dfrac{3}{4} \div \dfrac{6}{5}$ **34.** $\dfrac{3}{8} \div \dfrac{15}{28}$ **35.** $\dfrac{2}{13} \div \dfrac{8}{13}$ **36.** $\dfrac{4}{7} \div \dfrac{20}{21}$

37. $\dfrac{21}{35} \div \dfrac{3}{14}$ **38.** $\dfrac{23}{25} \div \dfrac{46}{5}$ **39.** $\dfrac{42}{30} \div \dfrac{21}{15}$ **40.** $\dfrac{34}{8} \div \dfrac{17}{4}$

41. $6 \div \dfrac{3}{14}$ **42.** $23 \div \dfrac{46}{5}$ **43.** $\dfrac{42}{30} \div 7$ **44.** $\dfrac{34}{8} \div 17$

In Exercises 45–80, perform each addition or subtraction. Simplify each result when possible.

45. $\dfrac{3}{5} + \dfrac{3}{5}$ **46.** $\dfrac{4}{7} - \dfrac{2}{7}$ **47.** $\dfrac{4}{13} - \dfrac{3}{13}$ **48.** $\dfrac{2}{11} + \dfrac{9}{11}$

49. $\dfrac{13}{17} - \dfrac{11}{17}$ **50.** $\dfrac{23}{15} + \dfrac{7}{15}$ **51.** $\dfrac{1}{6} + \dfrac{1}{24}$ **52.** $\dfrac{17}{25} - \dfrac{2}{5}$

53. $\dfrac{3}{5} + \dfrac{2}{3}$ **54.** $\dfrac{4}{3} + \dfrac{7}{2}$ **55.** $\dfrac{9}{4} - \dfrac{5}{6}$ **56.** $\dfrac{2}{15} + \dfrac{7}{9}$

57. $\dfrac{7}{10} - \dfrac{1}{14}$ **58.** $\dfrac{7}{25} + \dfrac{3}{10}$ **59.** $\dfrac{5}{14} - \dfrac{4}{21}$ **60.** $\dfrac{2}{33} + \dfrac{3}{22}$

61. $3 - \dfrac{3}{4}$ **62.** $5 + \dfrac{21}{5}$ **63.** $\dfrac{17}{3} + 4$ **64.** $\dfrac{13}{9} - 1$

65. $\dfrac{3}{15} + \dfrac{6}{10}$ **66.** $\dfrac{7}{5} - \dfrac{2}{15}$ **67.** $\dfrac{5}{12} - \dfrac{4}{15}$ **68.** $\dfrac{3}{16} + \dfrac{2}{27}$

69. $\dfrac{7}{24} + \dfrac{8}{30}$ **70.** $\dfrac{9}{40} + \dfrac{13}{54}$ **71.** $\dfrac{15}{36} - \dfrac{7}{60}$ **72.** $\dfrac{11}{48} - \dfrac{5}{36}$

73. $4\frac{3}{5} + \frac{3}{5}$ **74.** $2\frac{1}{8} + \frac{3}{8}$ **75.** $3\frac{1}{3} - 1\frac{2}{3}$ **76.** $5\frac{1}{7} - 3\frac{2}{7}$

77. $3\frac{3}{4} - 2\frac{1}{2}$ **78.** $15\frac{5}{6} + 11\frac{5}{8}$ **79.** $8\frac{2}{9} - 7\frac{2}{3}$ **80.** $3\frac{4}{5} - 3\frac{1}{10}$

81. Each side of a triangle measures $2\frac{3}{7}$ centimeters. Find the sum of the lengths of its three sides.

82. Each side of a square field measures $30\frac{2}{5}$ meters. How much fence is needed to enclose the field?

83. The four sides of a garden measure $7\frac{2}{3}$ feet, $15\frac{1}{4}$ feet, $19\frac{1}{2}$ feet, and $10\frac{3}{4}$ feet. Find the length of the fence needed to enclose the garden.

84. A clothing designer requires $3\frac{1}{4}$ yards of material for each dress he makes. How much material will be used to make 14 dresses?

REVIEW EXERCISES

1. Is every natural number equal to a fraction?

2. Is every fraction equal to a natural number?

3. Simplify the fraction $\frac{26}{2}$ and tell whether it represents a prime number.

4. Graph the even natural numbers less than 20.

5. Graph the odd prime numbers less than 7.

6. Graph the even prime numbers.

In Review Exercises 7–10, indicate whether the statement is true.

7. $5 < 3$ **8.** $7 \geq 9 - 3$ **9.** $3 - 2 \neq 8 \div 8$ **10.** $2 \leq 2$

11. Find the sum of the even numbers between 1 and 10.

12. Find the product of the prime numbers that are less than 10.

1.3 ALGEBRAIC TERMS

Variables and numbers can be combined with the operations of arithmetic to produce **algebraic expressions.** For example, if x and y are variables, the algebraic expression $x + y$ represents the **sum** of x and y, and the algebraic expression $x - y$ represents their **difference.**

There are many ways to read the sum $x + y$. Some of them are

- the sum of x and y,
- x increased by y,
- x plus y,
- y more than x, and
- y added to x.

There are many ways to read the difference $x - y$. Some of them are

- the result obtained when y is subtracted from x,
- the result of subtracting y from x,
- x less y,
- y less than x,
- x decreased by y, and
- x minus y.

EXAMPLE 1   Let x represent a certain number. Write an expression that represents **a.** the number that is 5 more than x and   **b.** the number 12 decreased by x.

Solution   **a.** The number "5 more than x" is the number found by adding 5 to x. It is represented by $x + 5$.
b. The number "12 decreased by x" is the number found by subtracting x from 12. It is represented by $12 - x$.   ∎

Because the times sign, $\times$, looks like the letter x, it is seldom used in algebra. Instead, a dot, parentheses, or no symbol at all is used to denote multiplication. Each of the following expressions indicates the **product** obtained when x and y are multiplied.

$$x \cdot y \qquad (x)(y) \qquad x(y) \qquad (x)y \qquad xy$$

There are several ways to indicate the product xy in words. Some of them are

- x multiplied by y,
- the product of x and y, and
- x times y.

EXAMPLE 2   Let x represent a certain number. Denote a number that is **a.** twice as large as x,   **b.** 5 more than 3 times x,   and   **c.** 4 less than $\frac{1}{2}$ of x.

Solution   **a.** The number "twice as large as x" is found by multiplying x by 2. It is represented by $2x$.
b. The number "5 more than 3 times x" is found by adding 5 to the product of 3 and x. It is represented by $3x + 5$.
c. The number "4 less than $\frac{1}{2}$ of x" is found by subtracting 4 from the product of $\frac{1}{2}$ and x. It is represented by $\frac{1}{2}x - 4$.   ∎

EXAMPLE 3   Jim has x dimes, y nickels, and 3 quarters. **a.** How many coins does Jim have?
b. What is the value of the coins?

Solution   **a.** Because there are x dimes, y nickels, and 3 quarters, the total number of coins is $x + y + 3$.
b. The value of x dimes is $10x$ cents, the value of y nickels is $5y$ cents, and the value of 3 quarters is $3 \cdot 25$, or 75 cents. The total value of the coins is $(10x + 5y + 75)$ cents.   ∎

If x and y represent two numbers and $y \neq 0$, the **quotient** obtained when x is divided by y is denoted by each of the following expressions:

$$x \div y, \qquad x/y, \qquad \text{and} \qquad \frac{x}{y}$$

EXAMPLE 4   Let x and y represent two numbers. Write an algebraic expression that represents the sum obtained when 3 times the first number is added to the quotient obtained when the second number is divided by 6.

Solution   Three times the first number x is denoted as $3x$. The quotient obtained when the second number y is divided by 6 is the fraction $\frac{y}{6}$. Their sum is expressed as $3x + \frac{y}{6}$.   ■

EXAMPLE 5   A 5-foot section is cut from the end of a rope that is l feet long. The remaining rope is then divided into three equal pieces. Find the length of each of the equal pieces.

Solution   After a 5-foot section is cut from one end of l feet of rope, the rope that remains is $(l - 5)$ feet long. When that remaining rope is cut into 3 equal pieces, each piece will be $\frac{l-5}{3}$ feet long.   ■

Because variables represent numbers, algebraic expressions represent numbers also. We can evaluate an algebraic expression if we know the numbers that its variables represent. For example, if x represents the number 12 and y represents the number 5, we can evaluate the expression $x + y + 2$ as follows:

$$\begin{aligned} x + y + 2 &= 12 + 5 + 2 && \text{Substitute 12 for } x \text{ and 5 for } y. \\ &= 17 + 2 && \text{Do the addition from left to right.} \\ &= 19 \end{aligned}$$

EXAMPLE 6   If x represents the number 8 and y represents the number 10, evaluate the expressions **a.** $x + y$, **b.** $y - x$, **c.** $3xy$ and **d.** $\frac{5x}{y-5}$.

Solution   Substitute 8 for x and 10 for y in each expression and simplify.

a. $\begin{aligned} x + y &= 8 + 10 \\ &= 18 \end{aligned}$ 　　　　　**b.** $\begin{aligned} y - x &= 10 - 8 \\ &= 2 \end{aligned}$

c. $\begin{aligned} 3xy &= (3)(8)(10) \\ &= (24)(10) && \text{Do the multiplications from left to right.} \\ &= 240 \end{aligned}$

d. $\begin{aligned} \frac{5x}{y-5} &= \frac{5 \cdot 8}{10 - 5} \\ &= \frac{40}{5} && \text{Simplify the numerator and denominator separately.} \\ &= 8 && \text{Simplify the fraction.} \end{aligned}$

After numbers are substituted for the variables in a product, a dot or parentheses are often needed to indicate the multiplication. This is to ensure that (3)(8)(10), for example, is not mistaken for 3810, or that $5 \cdot 8$ is not mistaken for 58.   ∎

Numbers that do not contain variables, such as 7, 21, and 23, are called **constants.** Expressions such as 37, xyz, or $32t$, which are constants, variables, or products of constants and variables, are called **algebraic terms.** Because the expression $3x + 5y$ denotes the sum of the two algebraic terms $3x$ and $5y$, the expression $3x + 5y$ contains two terms. Likewise, the expression $xy - 7$ contains two terms. The expression $3 + x + 2y$ contains three terms. Its first term is 3, its second term is x, and its third term is $2y$.

Numbers that are part of an indicated product are called **factors** of that product. For example, the product $7x$ has two factors, 7 and x. Either factor in this product is called the **coefficient** of the other factor. When we speak of the coefficient in a term such as $7x$, however, we generally mean the **numerical coefficient,** which in this case is 7. For example, the numerical coefficient of the term $12xyz$ is 12. The coefficient of such terms as x, ab, or rst is understood to be 1. Thus,

$$x = 1x, \qquad ab = 1ab, \qquad \text{and} \qquad rst = 1rst$$

EXAMPLE 7 **a.** The expression $5x + y$ has two terms. The numerical coefficient of its first term is 5. The numerical coefficient of its second term is 1.

b. The expression $17wxyz$ has one term, which contains the five factors 17, w, x, y, and z. Its numerical coefficient is 17.

c. The expression 37 has one term, the constant 37. Its numerical coefficient is 37.   ∎

████████████████████████   **EXERCISE 1.3**   ████████████████

In Exercises 1–18, let x, y, and z represent three numbers. Write an algebraic expression to denote each given quantity.

1. The sum of x and y.

2. The product of x and y.

3. The product of x and twice y.

4. The sum of twice x and twice z.

5. The difference obtained when x is subtracted from y.

6. The difference obtained when twice x is subtracted from y.

7. The quotient obtained when y is divided by x.

8. The quotient obtained when the sum of x and y is divided by z.

9. The sum obtained when the quotient of x divided by y is added to z.

10. y decreased by x.

11. z less the product of x and y.

12. z less than the product of x and y.

13. The product of 3, x, and y.

14. The quotient obtained when the product of 3 and z is divided by the product of 4 and x.

15. The quotient obtained when the sum of x and y is divided by the sum of y and z.

16. The quotient obtained when the product of x and y is divided by the sum of x and z.

17. The sum of the product xy and the quotient obtained when y is divided by z.

18. The number obtained when x decreased by 4 is divided by the product of 3 and the variable y.

19. George is enrolled in college for c hours of credit, and Kia's course load is 4 hours greater than George's. Write an expression that represents the number of hours Kia is taking.

20. Sam's antique Ford has 25,000 more miles on its odometer than does his father's new car. If his father's car has traveled m miles, find an expression that represents the mileage on Sam's Ford.

21. Write an expression that represents the value of t pencils, each worth 22 cents.

22. Write an expression that represents the value of a pounds of candy worth 95 cents per pound.

23. A rope x feet long is cut into 5 equal pieces. Find the length of each piece.

24. A rope 18 feet long is cut into x equal pieces. How long is each piece?

25. Kris has d dollars, and her brother Steven has 5 dollars more than three times that amount. Write an expression that represents the amount Steven has.

26. Wendy is now x years old. Her sister Bonnie is 2 years less than twice Wendy's age. Write an expression that represents Bonnie's age.

In Exercises 27–38, write each algebraic expression as an appropriate English phrase.

27. $x + 3$

28. $y - 2$

29. $\dfrac{x}{y}$

30. xz

31. $2xy$

32. $\dfrac{x + y}{2}$

33. $\dfrac{5}{x + y}$

34. $\dfrac{3x}{y + z}$

35. $\dfrac{3 + x}{y}$

36. $3 + \dfrac{x}{y}$

37. $\dfrac{xy}{x + y}$

38. $\dfrac{x + y + z}{xyz}$

In Exercises 39–50, let x represent the number 8, y represent the number 4, and z represent the number 2. Evaluate each algebraic expression. In a fraction, simplify the numerator and denominator separately. Then perform a division whenever possible.

39. $x + z$

40. xyz

41. $y - z$

42. $\dfrac{y}{z}$

43. $3yz$

44. $7xy$

45. $\dfrac{3xy}{z}$

46. $\dfrac{10z}{4}$

47. $\dfrac{x + y + z}{7z}$

48. $\dfrac{y + x + x}{x + z}$

49. $\dfrac{8y}{y - z}$

50. $\dfrac{3z}{x - y}$

In Exercises 51–60, state the number of terms in each algebraic expression and also state the numerical coefficient of the first term.

51. $6d$

52. $4c + 3d$

53. $xy - 4t + 35$

54. xy

55. $3ab + bc - cd - ef$

56. $2xyz + cde - 14$

57. $4xyz + 7xy - z$

58. $5uvw - 4uv + 8uw$

59. $3x + 4y + 2z + 2$

60. $7abc - 9ab + 2bc + a - 1$

In Exercises 61–64, consider the algebraic expression $29xyz + 23xy + 19x$.

61. What are the factors of the third term?

62. What are the factors of the second term?

63. What are the factors of the first term?

64. What factor is common to all three terms?

In Exercises 65–68, consider the algebraic expression $3xyz + 2 \cdot 3xy + 2 \cdot 3 \cdot 3xz$.

65. What are the factors of the first term?

66. What are the factors of the second term?

67. What are the factors of the third term?

68. What factors are common to all three terms?

In Exercises 69–72, consider the algebraic expression $5xy + xt + 8xyt$.

69. Determine the numerical coefficients of each term.

70. What factor is common to all three terms?

71. What factors are common to the first and third terms?

72. What factors are common to the second and third terms?

In Exercises 73–76, consider the algebraic expression $3xy + y + 25xyz$.

73. Determine the numerical coefficient of each term and find the product of the coefficients.

74. Determine the numerical coefficient of each term and find the sum of the coefficients.

75. What factors are common to the first and the third terms?

76. What factor is common to all three terms?

In Review Exercises 1–4, perform the indicated operation and classify the result into the categories of even natural number, odd natural number, and prime number.

1. $7 - 6$

2. $32 + 0$

3. $\dfrac{31}{7} + \dfrac{18}{7}$

4. $\dfrac{14}{5} \cdot \dfrac{25}{7}$

5. Simplify the fraction $\dfrac{15}{75}$.

6. Simplify the fraction $\dfrac{14}{105}$.

7. Write 150 in prime-factored form.

8. Write 165 in prime-factored form.

In Review Exercises 9–12, perform the indicated operation and simplify if possible.

9. $\dfrac{52}{7} \cdot \dfrac{14}{13}$

10. $\dfrac{25}{12} \div 15$

11. $2\dfrac{1}{6} - 1\dfrac{1}{5}$

12. $\dfrac{25}{15} + 3$

1.4 EXPONENTS AND ORDER OF OPERATIONS

The number 5 in the product 5(4) indicates that 4 is to be used five times in a sum:

$$5(4) = 4 + 4 + 4 + 4 + 4 = 20$$

Likewise, in the algebraic term $5x$, the coefficient 5 indicates that the variable x is to be used five times in a sum:

$$5x = x + x + x + x + x$$

To show how many times a number is to be used as a *factor* in a product, we use a symbol called an **exponent.** For example, the number 5 in the expression x^5 indicates that x is to be used as a factor five times:

$$x^5 = \overbrace{x \cdot x \cdot x \cdot x \cdot x}^{5 \text{ factors of } x}$$

In the expression x^5, 5 is called the **exponent,** x is called the **base,** and the entire term itself is called an **exponential expression** or a **power** of x. In terms such as x or y, the exponent is understood to be 1:

$$x = x^1 \qquad \text{and} \qquad y = y^1$$

In general, we have the following definition:

DEFINITION. If n is a natural number, then

$$x^n = \overbrace{x \cdot x \cdot x \cdot \cdots \cdot x}^{n \text{ factors of } x}$$

EXAMPLE 1   **a.** $4^2 = 4 \cdot 4 = 16$ Read 4^2 as "4 squared" or as "4 to the second power."

b. $5^3 = 5 \cdot 5 \cdot 5 = 125$ Read 5^3 as "5 cubed" or as "5 to the third power."

c. $6^4 = 6 \cdot 6 \cdot 6 \cdot 6 = 1296$ Read 6^4 as "6 to the fourth power."

d. $\left(\dfrac{2}{3}\right)^5 = \dfrac{2}{3} \cdot \dfrac{2}{3} \cdot \dfrac{2}{3} \cdot \dfrac{2}{3} \cdot \dfrac{2}{3} = \dfrac{32}{243}$ Read $(\frac{2}{3})^5$ as "$\frac{2}{3}$ to the fifth power." ∎

In the next example, the base of each exponential expression is a variable.

EXAMPLE 2   **a.** $y^6 = y \cdot y \cdot y \cdot y \cdot y \cdot y$ Read y^6 as "y to the sixth power."

b. $x^3 = x \cdot x \cdot x$ Read x^3 as "x cubed" or as "x to the third power."

c. $z^2 = z \cdot z$ Read z^2 as "z squared" or as "z to the second power."

d. $a^1 = a$ Read a^1 as "a to the first power."

As part **d** suggests, any number raised to the first power is just the number itself. ∎

Order of Operations

The order in which we do arithmetic is important. For example, if we multiply first to evaluate $2 + 3 \cdot 4$, we obtain

$$2 + 3 \cdot 4 = 2 + 12 = 14$$

However, if we add first, we get a different result:

$$2 + 3 \cdot 4 = 5 \cdot 4 = 20$$

To eliminate the possibility of different answers, we will agree to do multiplications before additions. Thus, the correct calculation of the expression $2 + 3 \cdot 4$ is

$$2 + 3 \cdot 4 = 2 + 12 = 14$$

To indicate that additions should be done before multiplications, we must use **grouping symbols** such as parentheses ( ), brackets [ ], or braces { }. In the expression $(2 + 3)4$, the parentheses indicate that the addition should be done first:

$$(2 + 3)4 = 5 \cdot 4 = 20$$

Unless grouping symbols indicate otherwise, exponential expressions should be evaluated before multiplications. Thus, the expression $5 + 4 \cdot 3^2$ should be evaluated as follows:

$$5 + 4 \cdot 3^2 = 5 + 4 \cdot 9 \qquad \text{Evaluate the exponential expression first.}$$
$$= 5 + 36 \qquad \text{Perform the multiplication.}$$
$$= 41 \qquad \text{Perform the addition.}$$

To guarantee that calculations like those in the previous example will have a single correct result, we must use the following set of **priority rules:**

> **Order of Mathematical Operations.**   If an expression does not contain grouping symbols,
> 1. Find the values of any exponential expressions.
> 2. Do all multiplications and divisions as they are encountered while working from left to right.
> 3. Do all additions and subtractions as they are encountered while working from left to right.
>
> If an expression contains grouping symbols, use the rules above to perform the calculations within each pair of grouping symbols, working from the innermost pair to the outermost pair.
> The bar of a fraction is a grouping symbol. Thus, in a fraction, simplify the numerator and the denominator separately. Then simplify the fraction, whenever possible.

Because of these priority rules, the expression $4(2)^3$, for example, is not equal to the expression $(4 \cdot 2)^3$:

$$4(2)^3 = 4 \cdot 2 \cdot 2 \cdot 2 = 4(8) = 32 \qquad \text{but} \qquad (4 \cdot 2)^3 = 8^3 = 8 \cdot 8 \cdot 8 = 512$$

Likewise, $4x^3 \neq (4x)^3$ because

$$4x^3 = 4xxx \qquad \text{but} \qquad (4x)^3 = (4x)(4x)(4x) = 64x^3$$

In the expression $4x^3$, the number 4 is the coefficient of x^3.

EXAMPLE 3   Evaluate $5^3 + 2(8 - 3 \cdot 2)$.

Solution   Do the work within the parentheses first and then simplify.

$$5^3 + 2(8 - 3 \cdot 2) = 5^3 + 2(8 - 6) \qquad \text{Do the multiplication within the parentheses.}$$
$$= 5^3 + 2(2) \qquad \text{Do the subtraction within the parentheses.}$$
$$= 125 + 2(2) \qquad \text{Find the value of the exponential expression.}$$
$$= 125 + 4 \qquad \text{Do the multiplication.}$$
$$= 129 \qquad \text{Do the addition.} \qquad \blacksquare$$

EXAMPLE 4   Evaluate $\dfrac{3(3 + 2) + 5}{17 - 3(4)}$.

Solution   Simplify the numerator and the denominator of the fraction separately. Then simplify the fraction.

$$\frac{3(3+2)+5}{17-3(4)} = \frac{3(5)+5}{17-3(4)}$$ Do the addition within the parentheses.

$$= \frac{15+5}{17-12}$$ Do the multiplications.

$$= \frac{20}{5}$$ Do the addition and the subtraction.

$$= 4$$ Simplify the fraction.   ■

EXAMPLE 5   If $x = 3$ and $y = 4$, evaluate **a.** $3y + x^2$, **b.** $3(y + x^2)$, and **c.** $3(y + x)^2$.

Solution   **a.** $3y + x^2 = 3(4) + 3^2$ Substitute 3 for x and 4 for y.

$= 3(4) + 9$ Evaluate the exponential expression.

$= 12 + 9$ Perform the multiplication.

$= 21$ Perform the addition.

b. $3(y + x^2) = 3(4 + 3^2)$ Substitute 3 for x and 4 for y.

$= 3(4 + 9)$ Evaluate the exponential expression.

$= 3(13)$ Parentheses indicate that the addition is done next.

$= 39$ Perform the multiplication.

c. $3(y + x)^2 = 3(4 + 3)^2$ Substitute 3 for x and 4 for y.

$= 3(7)^2$ Parentheses indicate that the addition is done next.

$= 3(49)$ Evaluate the exponential expression.

$= 147$ Perform the multiplication.   ■

EXAMPLE 6   Evaluate $\dfrac{3x^2 - 2y}{2(x + y)}$ if $x = 4$ and $y = 3$.

Solution   Substitute 4 for x and 3 for y and simplify.

$$\frac{3x^2 - 2y}{2(x + y)} = \frac{3(4^2) - 2(3)}{2(4 + 3)}$$

$$= \frac{3(16) - 2(3)}{2(7)}$$ Find the value of 4^2 in the numerator and do the addition in the denominator.

$$= \frac{48 - 6}{14}$$ Do the multiplications.

$$= \frac{42}{14}$$ Do the subtraction.

$$= 3$$ Simplify the fraction.   ■

Geometry

Substituting numbers for variables is often required to find *perimeters* and *areas* of geometric figures. The **perimeter** of a figure is the distance around it, and the **area** of a figure is the amount of surface that it encloses. The perimeter of a circle is called its **circumference.**

The following chart shows the formulas for the perimeter and area of several geometric figures:

Figure	Name	Perimeter/ Circumference	Area
Square	Square	$P = 4s$	$A = s^2$
Rectangle	Rectangle	$P = 2l + 2w$	$A = lw$
Triangle	Triangle	$P = a + b + c$	$A = \dfrac{1}{2}bh$
Trapezoid	Trapezoid	$P = a + b + c + d$	$A = \dfrac{1}{2}h(b + d)$
Circle	Circle	$C = \pi D$ (π is approximately $\frac{22}{7}$.)	$A = \pi r^2$

EXAMPLE 7 Find **a.** the circumference   and   **b.** the area of a circle with a diameter of 14 centimeters.

Solution **a.** Find the circumference of the circle shown in Figure 1-8 by using the formula

$$C = \pi D$$

FIGURE 1-8

where C is the circumference, π is a number that is close to $\frac{22}{7}$, and D is the **diameter**—the distance through the center of the circle. We can find a good approximation of the circumference by substituting $\frac{22}{7}$ for π and 14 for D in the formula for circumference and simplifying.

$$C = \pi D$$

$$C \approx \frac{22}{7} \cdot 14 \qquad \text{Read } \approx \text{ as "is approximately equal to."}$$

$$C \approx \frac{22 \cdot \overset{2}{\cancel{14}}}{\underset{1}{\cancel{7}} \cdot 1} \qquad \text{Multiply the fractions and simplify.}$$

$$C \approx 44$$

The circumference is approximately 44 centimeters.
   To use a calculator, we would press these keys:

$$\boxed{\pi} \quad \boxed{\times} \quad 14 \quad \boxed{=}$$

The display will read 43.98229 . . . . This answer is not 44 because a calculator uses a better approximation of π than $\frac{22}{7}$.

b. The area, A, of the circle is given by the formula

$$A = \pi r^2$$

where r, called the **radius** of the circle, is one-half of the diameter. We can find a good approximation of the area by substituting $\frac{22}{7}$ for π and $\frac{1}{2}(14)$, or 7 for r in the formula for area and simplifying.

$$A = \pi r^2$$

$$A \approx \frac{22}{7} \cdot 7^2$$

$$A \approx \frac{22}{7} \cdot \frac{49}{1} \qquad \text{Evaluate the exponential expression.}$$

$$A \approx \frac{22 \cdot \overset{7}{\cancel{49}}}{\underset{1}{\cancel{7}} \cdot 1} \qquad \text{Multiply the fractions and simplify.}$$

$$A \approx 154$$

The area is approximately 154 square centimeters.
   To use a calculator, we would press these keys:

$$\boxed{\pi} \quad \boxed{\times} \quad 7 \quad \boxed{x^2} \quad \boxed{=} \qquad \blacksquare$$

The **volume** of a three-dimensional geometric solid is the amount of space it encloses. The following chart shows the formulas for the volume of several solids.

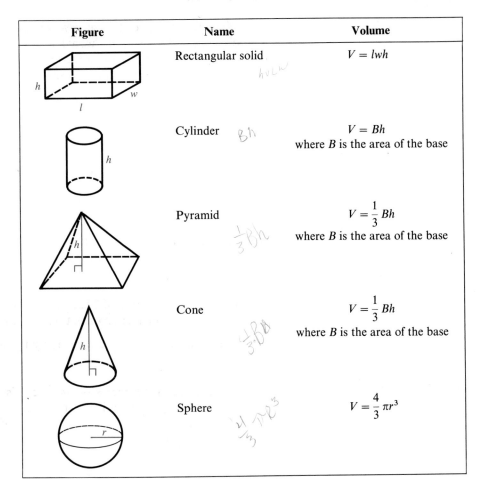

Figure	Name	Volume
	Rectangular solid	$V = lwh$
	Cylinder	$V = Bh$ where B is the area of the base
	Pyramid	$V = \dfrac{1}{3} Bh$ where B is the area of the base
	Cone	$V = \dfrac{1}{3} Bh$ where B is the area of the base
	Sphere	$V = \dfrac{4}{3} \pi r^3$

EXAMPLE 8   Find the number of cubic feet of salt in a conical pile that is 18 feet high and covers a circular area 28 feet in diameter.

Solution   First find the area, B, of the circular base of the cone shown in Figure 1-9 by substituting $\frac{22}{7}$ for π and $\frac{1}{2}(28)$, or 14, for the radius.

$$B = \pi r^2 \approx \frac{22}{7}(14)^2 = 616$$

Then substitute 616 for B and 18 for h in the formula for the volume of a cone.

$$V = \frac{1}{3} Bh \approx \frac{1}{3}(616)(18) = 3696$$

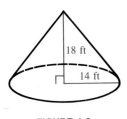

FIGURE 1-9

Thus, there are approximately 3696 cubic feet of salt in the pile.   ■

EXERCISE 1.4

In Exercises 1–6, find the value of each expression.

1. 4^3
2. 5^2
3. 6^2
4. 7^3

5. $\left(\dfrac{1}{10}\right)^4$
6. $\left(\dfrac{1}{2}\right)^6$

In Exercises 7–14, write each expression as the product of several factors.

7. x^2
8. y^3
9. $3z^4$
10. $5t^2$
11. $(5t)^2$
12. $(3z)^4$
13. $5(2x)^3$
14. $7(3t)^2$

In Exercises 15–22, find the value of each expression if $x = 3$ and $y = 2$.

15. $4x^2$
16. $4y^3$
17. $(5y)^3$
18. $(2y)^4$
19. $2x^y$
20. $3y^x$
21. $(3y)^x$
22. $(2x)^y$

In Exercises 23–66, simplify each expression by performing the indicated operations.

23. $3 \cdot 5 - 4$
24. $4 \cdot 6 + 5$
25. $3(5 - 4)$
26. $4(6 + 5)$

27. $3 + 5^2$
28. $4^3 - 2^2$
29. $(3 + 5)^2$
30. $(5 - 2)^3$

31. $2 + 3 \cdot 5 - 4$
32. $12 + 2 \cdot 3 + 2$
33. $64 \div (3 + 1)$
34. $16 \div (5 + 3)$

35. $3(2 + 4)5$
36. $(7 + 9) \div 2 \cdot 4 - 15$

37. $(14 + 7) \div 3 \cdot 4$
38. $(12 + 8) \div (2 \cdot 5) + 2 \cdot 5$

39. $24 \div 4 \cdot 3 + 3$
40. $36 \div 9 \cdot 4 - 2$

41. $49 \div 7 \cdot 7 + 7$
42. $100 \div 10 \cdot 10 + 10$

43. $100 \div 10 \cdot 10 \div 100$
44. $100 \div (10 \cdot 10) \div 100$

45. $(100 \div 10) \cdot (10 \div 100)$
46. $100 \div [10 \cdot (10 \div 100)]$

47. $[(14 + 7) \div 3]4$
48. $4[3 + 2(5 - 2)] - 1$

49. $6[2 + 3(8 - 3 \cdot 2) + 2]$
50. $5[10 - (3 \cdot 2 - 3)] - (5 + 2)$

51. $3^2 + 2(1 + 4) - 2$
52. $4 \cdot 3 + 2(5 - 2) - 2^3$

53. $5^2 - (7 - 3)^2$
54. $3^3 + (3 - 1)^3$
55. $(2 \cdot 3 - 4)^3$
56. $(3 \cdot 5 - 2 \cdot 6)^2$

57. $\dfrac{3}{5} \cdot \dfrac{10}{3} + \dfrac{1}{2} \cdot 12$
58. $\dfrac{15}{4}\left(1 + \dfrac{3}{5}\right)$
59. $\left[\dfrac{1}{3} - \left(\dfrac{1}{2}\right)^2\right]^2$
60. $\left[\left(\dfrac{2}{3}\right)^2 - \dfrac{1}{6}\right]^2$

61. $\dfrac{(3 + 5)^2 + 2}{2(8 - 5)}$
62. $\dfrac{25 - (2 \cdot 3 - 1)}{2 \cdot 9 - 8}$
63. $\dfrac{(5 - 3)^2 + 2}{4^2 - (8 + 2)}$
64. $\dfrac{(4^2 - 2) + 7}{5(2 + 4) - 3^2}$

65. $\dfrac{2[4 + 2(3 - 1)]}{3[3(2 \cdot 3 - 4)]}$
66. $\dfrac{6[3 \cdot 7 - 5(3 \cdot 4 - 11)]}{2[4(3 + 2) - 3^2 + 5]}$

In Exercises 67–92, evaluate each expression, given that $x = 3$, $y = 2$, and $z = 4$.

67. $2x - y$
68. $2z + y$
69. $10 - 2x$
70. $15 - 3z$

71. $5z \div 2 + y$
72. $5x \div 3 + y$
73. $4x - 2z$
74. $5y - 3x$

75. $x + yz$

76. $3z + x - 2y$

77. $3(2x + y)$

78. $4(x + 3y)$

79. $(3 + x)y$

80. $(4 + z)y$

81. $(z + 1)(x + y)$

82. $3(z + 1) \div x$

83. $(x + y) \div (z + 1)$

84. $(2x + 2y) \div (3z - 2)$

85. $xyz + z^2 - 4x$

86. $zx + y^2 - 2z$

87. $3x^2 + 2y^2$

88. $3x^2 + (2y)^2$

89. $\dfrac{2x + y^2}{y + 2z}$

90. $\dfrac{2z^2 - y}{2x - y^2}$

91. $\dfrac{2x^3 - (xy - 2)}{2(3y + 5z) - 27}$

92. $\dfrac{x^2[14 - y(x + 2)] - 6}{5[xy - z(5y - 9)]}$

In Exercises 93–96, insert parentheses in the expression $3 \cdot 8 + 5 \cdot 3$ to make its value equal to the given number.

93. 39

94. 117

95. 87

96. 69

In Exercises 97–100, insert parentheses in the expression $4 + 3 \cdot 5 - 3$ to make its value equal to the given number.

97. 14

98. 10

99. 32

100. 16

In Exercises 101–104, find the perimeter of each figure.

101.
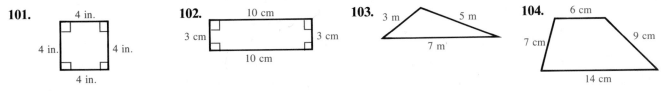

102.

103.

104.

In Exercises 105–108, find the area of each figure.

105.
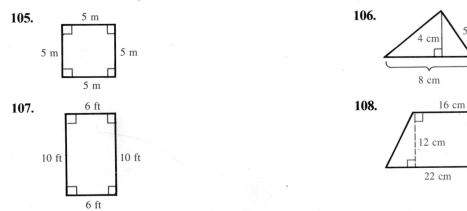

106.

107.

108.

⊞ *In Exercises 109–110, find the circumference of each circle. Approximate π as $\frac{22}{7}$. Then check your answer with a calculator.*

109.
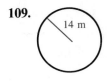

110.
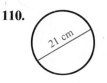

In Exercises 111–112, find the area of each circle. Approximate π as $\frac{22}{7}$. Then check your answer with a calculator.

111.

112.

49 m

In Exercises 113–118, find the volume of each solid. Then check your answer with a calculator.

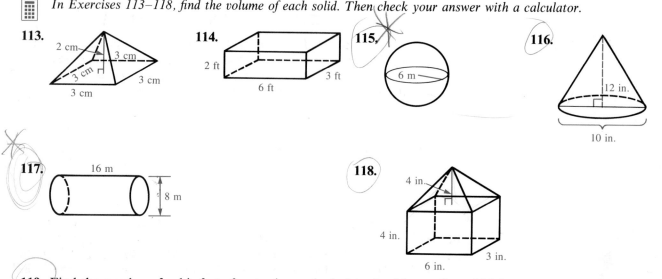

113.   2 cm   3 cm   3 cm   3 cm   3 cm

114.   2 ft   6 ft   3 ft

115.   6 m

116.   12 in.   10 in.

117.   16 m   8 m

118.   4 in.   4 in.   6 in.   3 in.

119. Find the number of cubic feet of water in a spherical tank with a radius of 21 feet.

120. Forty students are in a classroom with dimensions of 40 feet by 40 feet by 9 feet. Find how many cubic feet of air there is for each student.

REVIEW EXERCISES

In Review Exercises 1–2, place one of the symbols $\leq$ or $>$ in the box to make the statement true.

1. $3 \;\boxed{}\; \frac{6}{9}$

2. $11 - 3 \;\boxed{}\; 1 + 7$

In Review Exercises 3–4, place one of the symbols $\geq$ or $<$ in the box to make the statement true.

3. $\frac{7}{3} \;\boxed{}\; \frac{21}{9}$

4. $\frac{20 - 5}{3} \;\boxed{}\; 1 + 3 + 5$

In Review Exercises 5–8, let x represent 6, y represent 8, and z represent 0. Evaluate each expression.

5. $x + y - z$

6. xyz

7. $\frac{1}{2}y + \frac{2}{3}z$

8. $\frac{3 + x}{y + 1}$

9. Write the coefficient of the second term of the expression $15x^3y + 134xy + 12$ in prime-factored form.

10. A rectangle is 3 feet longer than it is wide. If its width is w feet, find an expression that represents its length.

11. The focal length, f, of a double-convex thin lens is given by the formula

$$f = \frac{rs}{(r+s)(n-1)}$$

If $r = 8$, $s = 12$, and $n = 1.6$, find f.

12. The total resistance, R, of two resistors in parallel is given by the formula

$$R = \frac{rs}{r+s}$$

If $r = 170$, and $s = 255$, find R.

1.5 REAL NUMBERS AND THEIR ABSOLUTE VALUES

When we graphed the set of whole numbers in Section 1.1, we chose an origin, marked off points at equal distances to the right of the origin, and labeled them with whole-number coordinates. We could also mark off points at equal distances to the left of the origin as shown in Figure 1-10.

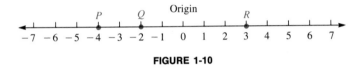

FIGURE 1-10

The coordinates of points to the left of the origin are called **negative numbers.** Because point P is 4 units to the *left* of the origin, the coordinate of P is -4 (read as "negative 4"). Because point Q is 2 units to the *left* of the origin, the coordinate of Q is -2. Because point R is 3 units to the *right* of the origin, the coordinate of R is 3. The coordinates of points to the right of the origin are called **positive numbers.** The number 3, for example, could be called *positive* 3 and could be written as $+3$. Likewise,

$$4 = +4 \qquad \text{and} \qquad 7 = +7$$

The coordinates of points shown in Figure 1-10 come from a set of numbers called the **integers.**

> **DEFINITION.** The set of **integers** is the set
> $$\{\ldots, -5, -4, -3, -2, -1, 0, 1, 2, 3, 4, 5, \ldots\}$$

The integers

$$\{\ldots, -6, -4, -2, 0, 2, 4, 6, \ldots\}$$

are even integers and the integers

$$\{\ldots, -7, -5, -3, -1, 1, 3, 5, 7, \ldots\}$$

are odd integers.

Because the set of natural numbers is included within the set of integers, we say that the set of natural numbers is a **subset** of the set of integers. The set of whole numbers is also a subset of the set of integers.

Because the coordinates of points on the number line become greater as we move from left to right, the negative numbers are all less than 0 and the positive numbers are all greater than 0. The number 0 is neither a positive number nor a negative number. Numbers that are either positive or negative are called **signed numbers.** Each of the following is a true statement about signed numbers:

$$-5 < 3 \qquad \text{Read as "negative 5 is less than 3."}$$
$$-4 < -2 \qquad \text{Read as "negative 4 is less than negative 2."}$$
$$1 \geq -1 \qquad \text{Read as "1 is greater than or equal to negative 1."}$$

Not all numbers are integers. Fractions such as $\frac{3}{2}$, $\frac{17}{12}$, and $\frac{-43}{8}$ are examples of numbers that are called **rational numbers.**

DEFINITION.   A **rational number** is any number that can be written in the form $\frac{a}{b}$ where a and b are integers and $b \neq 0$.

A decimal such as 0.5 is a rational number because it can be written as the quotient of two integers:

$$0.5 = \frac{5}{10} = \frac{1}{2}$$

Because every integer can be written as a fraction with a denominator of 1, the set of integers is a subset of the rational numbers. For example, the integers 6, -4, and 0 are rational numbers because each can be written as a fraction with a denominator of 1:

$$6 = \frac{6}{1}, \qquad 0 = \frac{0}{1}, \qquad \text{and} \qquad -4 = \frac{-4}{1}$$

Because

$$-4 = -\frac{4}{1} \qquad \text{and} \qquad -4 = \frac{-4}{1}$$

it follows that

$$-\frac{4}{1} = \frac{-4}{1}$$

This result suggests that a negative fraction can have its $-$ sign in either in front of the fraction or in the numerator. Later, we will see that

$$-\frac{a}{b} = \frac{-a}{b}$$

for all values of a and nonzero values of b.

There are numbers that are not rational numbers—numbers such as $\sqrt{2}$ and π. These numbers are called **irrational numbers.** We will study this set of numbers later in this book.

> **DEFINITION.** A **real number** is any number that is either a rational number or an irrational number.

EXAMPLE 1 Classify the numbers **a.** 5, **b.** -12, and **c.** 0.25 into the categories of positive number, negative number, integer, rational number, and real number.

Solution **a.** The number 5 is a positive number, an integer, a rational number, and a real number.
b. The number -12 is a negative number and an integer. Because it can be written as $\frac{-12}{1}$, it is a rational number. It is also a real number.
c. The number 0.25 can be written as the fraction $\frac{1}{4}$. Thus, 0.25 is a positive number, a rational number, and a real number. ■

Graphing Real Numbers

Many points on the number line do not have integer coordinates. The point midway between 0 and 1, for example, has the coordinate $\frac{1}{2}$. The point with coordinate $-\frac{3}{2}$ lies midway between -2 and -1 (see Figure 1-11).

FIGURE 1-11

Graphs of many sets of real numbers are intervals on the number line instead of isolated points. For example, the graph of the set of real numbers between -2 and 4 is shown in Figure 1-12.

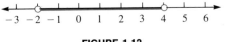

FIGURE 1-12

The open circles at -2 and 4 indicate that these points are not included in the graph. However, all the numbers between -2, and 4, such as -1, $-\frac{1}{2}$, 0, $\frac{2}{3}$, and 3.99, are included in the graph.

EXAMPLE 2 Graph all real numbers less than -3 or greater than 1.

Solution The graph of all real numbers less than -3 includes all points on the number line that are to the left of -3. The graph of all real numbers greater than 1 includes all points on the number line that are to the right of 1. The graph of these points is shown in Figure 1-13.

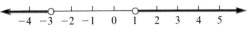

FIGURE 1-13 ■

EXAMPLE 3   Graph the set of all real numbers from -5 to -1.

Solution   The set of all real numbers from -5 to -1 includes both -5 and -1 and all the numbers in between. Use solid circles at -5 and at -1 to indicate that these points are included in the graph. The graph appears in Figure 1-14.

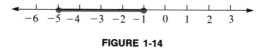

FIGURE 1-14   ∎

If two real numbers are the same distance from 0 on a number line but on opposite sides of 0, they are called **negatives** or **opposites** of each other. For example,

the negative of 5, written as $-(5)$, is -5

and

the negative of -5, written as $-(-5)$, is 5

Likewise,

the negative of 9, written as $-(9)$, is -9

and

the negative of -9, written as $-(-9)$, is 9

The results of $-(-5) = 5$ and $-(-9) = 9$ suggest that the negative of the negative of a number is the number itself.

The Double Negative Rule.   If x represents a number, then
$$-(-x) = x$$

EXAMPLE 4   If $r = 7$ and $t = -3$, evaluate **a.** $-r$, **b.** $-t$, **c.** $-(-r)$, and **d.** $-(-t)$.

Solution   **a.** $-r = -(7)$ **b.** $-t = -(-3)$
                    $= -7$ $= 3$

c. $-(-r) = -(-7)$ **d.** $-(-t) = -[-(-3)]$
                $= 7$ $= -(3)$
                                               $= -3$ ∎

Absolute Value of a Number

The distance on the number line between a given number and the origin is called the **absolute value** of that number. For example, the distance on the number line between the point with coordinate 5 and the origin is 5 units (see Figure 1-15). Thus, the absolute value of 5, denoted as $|5|$, is 5:

$|5| = 5$ Read as "the absolute value of 5 is 5."

Also, the distance on the number line between the point with coordinate -5 and the origin is 5. Again, see Figure 1-15. Thus,

$|-5| = 5$ Read as "the absolute value of -5 is 5."

Because $|-5|$ and $|5|$ are both equal to 5, $|-5| = |5|$. It is always true that the absolute values of a number and its negative are equal:

$|x| = |-x|$

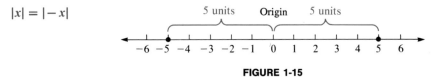

FIGURE 1-15

The absolute value of a number can be defined another way:

> **DEFINITION.**   If $x \geq 0$, then $|x| = x$.
> If $x < 0$, then $|x| = -x$.

This definition indicates that if x is positive or 0, then x is its own absolute value. However, if x is negative, then $-x$ (which is positive) is the absolute value of x. Thus $|x|$ always represents a number that is either positive or 0:

$|x| \geq 0$

EXAMPLE 5   Evaluate **a.** $|6|$,   **b.** $|-6|$,   **c.** $-|7|$,   **d.** $-|-7|$,   **e.** $|0|$,   and   **f.** $|8 - 3|$.

Solution   **a.** $|6| = 6$ **b.** $|-6| = -(-6) = 6$

   c. $-|7| = -(7) = -7$ **d.** $-|-7| = -(7) = -7$

   e. $|0| = 0$ **f.** $|8 - 3| = |5| = 5$ ∎

EXAMPLE 6   Evaluate $\left|\dfrac{5}{2} + \dfrac{1}{3}\right|$.

Solution   Begin by adding the fractions within the absolute value symbols.

$$\left|\frac{5}{2} + \frac{1}{3}\right| = \left|\frac{5 \cdot 3}{2 \cdot 3} + \frac{1 \cdot 2}{3 \cdot 2}\right| \qquad \text{Write the fractions as equivalent fractions with the common denominator 6.}$$

$$= \left|\frac{15}{6} + \frac{2}{6}\right|$$

$$= \left|\frac{15 + 2}{6}\right| \qquad \text{Add the fractions.}$$

$$= \left|\frac{17}{6}\right|$$

$$= \frac{17}{6} \qquad \text{Find the absolute value of } \tfrac{17}{6}.$$

∎

EXAMPLE 7   If $x = 9$ and $y = 12$, evaluate $3|-x| + 5|y - x|$.

Solution $3|-x| + 5|y - x| = 3|-9| + 5|12 - 9|$ Substitute values for the variables.

$\qquad\qquad\qquad\quad = 3|-9| + 5|3|$ Evaluate the expression within the second pair of absolute value symbols.

$\qquad\qquad\qquad\quad = 3 \cdot 9 + 5 \cdot 3$ Determine the absolute values.

$\qquad\qquad\qquad\quad = 27 + 15$ Do the multiplications.

$\qquad\qquad\qquad\quad = 42$ Do the addition.   ∎

EXERCISE 1.5

In Exercises 1–12, list the numbers in the set $\{0, 3, -4, \frac{1}{2}\}$ that satisfy the given condition.

1. positive number

2. negative number

3. integer

4. rational number

5. odd integer

6. even integer

7. prime number

8. real number

9. whole number

10. neither positive nor negative

11. less than 3

12. greater than 0

In Exercises 13–30, graph each set of numbers on a number line.

13. The numbers $\frac{1}{3}, \frac{5}{2}$, and $-\frac{4}{3}$.

14. The numbers $\frac{2}{3}, \frac{5}{3}$, and $-\frac{7}{3}$.

15. The numbers $-\frac{9}{2}, -\frac{5}{2}$, and $\frac{8}{2}$.

16. The numbers $\frac{1}{4}, -\frac{1}{4}$, and $\frac{0}{4}$.

17. All real numbers greater than 2 and less than 5.

18. All real numbers less than -2 or greater than -1.

19. All real numbers greater than 2.

20. All real numbers less than 2.

21. All real numbers less than -5.

22. All real numbers greater than -5.

23. All real numbers from -7 to -2.

24. All real numbers from -5 to 2.

25. All real numbers between -2 and 4.

26. All real numbers between -7 and -1.

27. All real numbers less than -7 or greater than -2.

28. All real numbers greater than 5 or less than -2.

29. All real numbers between -4 and -2 or greater than 0

30. All real numbers between -1 and 1 or less than -6.

In Exercises 31–50, evaluate each expression.

31. $|8|$

32. $|9|$

33. $|-8|$

34. $|-9|$

35. $|0|$

36. $|-2|^2$

37. $|3|^2$

38. $|-5|^3$

39. $-|10|$

40. $-|-10|$

41. $-(|-2| + |-3|)$

42. $|-3| - |-2|$

43. $\left|\dfrac{5}{3} + \dfrac{3}{5}\right|$

44. $\left|\dfrac{7}{2} - \dfrac{3}{5}\right|$

45. $3|15 - 8| + 2|13 - 9|$

46. $5|3 - 1|^2 + 2$

47. $2|-3|^3 - |-4|^2$

48. $8|5 - 3| - |-4|^2$

49. $|5|^2 - |-1|^3 - |-2|^2$

50. $|5^2| - |1^3| - |-3|^2$

In Exercises 51–72, evaluate each expression if $x = 2$, $y = -3$, and $z = 0$.

51. $-x$

52. $-y$

53. $-(-y)$

54. $-(-x)$

55. $|x|$

56. $|y|$

57. $-|z|$

58. $|-z|$

59. $-|-y|$

60. $-|-x|$

61. $|x| + |y|$

62. $|-x| \cdot |y|$

63. $|x - z|^2$

64. $|x + 3z|^3$

65. $|2x + 3|^2$

66. $|x^2 - x|^3$

67. $\dfrac{|-y + z|}{|2x - 1|}$

68. $\dfrac{|x + z|}{|x - z|}$

69. $\dfrac{|3z + 6|}{|5x - 7|}$

70. $\dfrac{8 - |x + 2z|}{3|-x|}$

71. $\dfrac{|x^4| + |y|^2}{|y|^2 + |xz| - |4|}$

72. $\dfrac{(|x| + |y| + |z|)^3}{(|x| + |y|)^2}$

73. Explain why the set of even natural numbers is a subset of the rational numbers.

74. Explain why the set of prime numbers is not a subset of the odd natural numbers.

75. Explain why the absolute value of a number is equal to the absolute value of the negative of that number.

76. Find two numbers that have an absolute value of 2.

REVIEW EXERCISES

In Review Exercises 1–4, simplify each expression.

1. $3 + 7^2$

2. $(3 + 7)^2$

3. $4 \cdot 3^2$

4. $(4 \cdot 3)^2$

In Review Exercises 5–8, let x = 3, y = 5, and z = 1. Place the symbol <, =, or > in the box.

5. $x(y - z)$ ☐ $4x$

6. $xy - z$ ☐ y^2

7. $x^z + y$ ☐ y^z

8. $(x^2 - y)^x$ ☐ $(x + z)^3$

9. A concrete block weighs $37\frac{1}{2}$ pounds. How much will 30 of these blocks weigh?

10. Which has the greater area, an $8\frac{1}{2}$-by-14-inch sheet of legal paper, or a 12-inch-square floor tile?

11. The volume, V, of the right-circular cone shown in Illustration 1 is given by

$$V = \frac{1}{3} \pi r^2 h$$

If $r = 7$ and $h = 1$, find V.

ILLUSTRATION 1

12. How many terms are in the expression $5x^2y + 7xy^2 + 9x + 2$?

1.6 ADDING AND SUBTRACTING REAL NUMBERS

Because the positive direction of the number line is to the right, positive numbers can be represented by arrows that point to the right; negative numbers can be represented by arrows that point to the left. Using arrows can help us establish rules for adding signed numbers.

Suppose we wish to add the real numbers $+2$ and $+3$. We can represent $+2$ with an arrow of length 2, pointing to the right. We can represent $+3$ with an arrow of length 3, also pointing to the right. To find the sum $(+2) + (+3)$, we place the two arrows end to end as in Figure 1-16. The endpoint of the second arrow is the point with coordinate $+5$. Thus,

$$(+2) + (+3) = +5$$

The addition problem

$$(-2) + (-3)$$

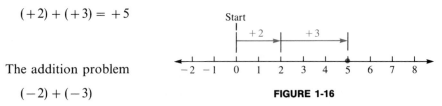

FIGURE 1-16

can be represented by two arrows on the number line, as shown in Figure 1-17. The number -2 can be represented with an arrow that begins at the origin, is 2 units long, and points to the *left*. The number -3 can be represented with an arrow that is 3 units long, continuing from point -2, and also pointing to the *left*. The endpoint of the final arrow is the point -5. Thus,

$$(-2) + (-3) = -5$$

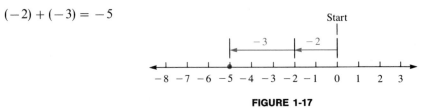

FIGURE 1-17

The addition of two real numbers with the same sign can always be represented with arrows pointing in the same direction and placed end to end. This fact is the basis for the following rule.

Adding Real Numbers with Like Signs.   If two real numbers have the same sign, we find their sum by adding their absolute values and keeping their common sign.

EXAMPLE 1

a. $(+4) + (+6) = 4 + 6$
$= 10$

b. $(-4) + (-6) = -(4 + 6)$
$= -10$

c. $+5 + (+10) = 5 + 10$
$= 15$

d. $-\dfrac{1}{2} + \left(-\dfrac{3}{2}\right) = -\left(\dfrac{1}{2} + \dfrac{3}{2}\right)$
$= -\dfrac{4}{2}$
$= -2$ ■

Two real numbers with *unlike* signs can be represented by arrows on a number line that point in *opposite* directions. For example, the addition problem

$$(-6) + (+2)$$

can be represented on a number line as shown in Figure 1-18. The first arrow starts at the origin, is 6 units long, and points to the *left*. The second arrow begins at -6, is 2 units long, and points to the *right*. The endpoint of this final arrow is the point with coordinate -4. Thus,

$$(-6) + (+2) = -4$$

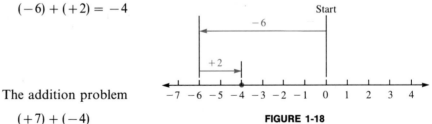

The addition problem

$$(+7) + (-4)$$

FIGURE 1-18

can be represented on a number line as in Figure 1-19. The first arrow begins at the origin, is 7 units long, and points to the *right*. The second arrow begins at 7, is 4 units long, and points to the *left*. The endpoint of the final arrow is the point with coordinate $+3$. Thus,

$$(+7) + (-4) = +3$$

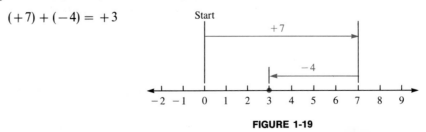

FIGURE 1-19

Two real numbers that have unlike signs can be represented by arrows pointing in opposite directions. This fact is the basis for the following rule.

> **Adding Real Numbers with Unlike Signs.**   If two real numbers have unlike signs, we find their sum by subtracting their absolute values (the smaller from the larger) and using the sign of the number with the greater absolute value.

EXAMPLE 2   **a.** $(+6) + (-5) = 6 - 5$ **b.** $(-2) + (+3) = +(3 - 2)$
$$= 1$$ $$= 1$$

c. $+6 + (-9) = -(9 - 6)$ **d.** $-10 + (+4) = -(10 - 4)$
$$= -3$$ $$= -6$$ ■

EXAMPLE 3   **a.** $[(+3) + (-7)] + (-4) = (-4) + (-4)$ Do the work in the brackets first.
$$= -8$$

b. $-3 + [(-2) + (-8)] = -3 + [-10]$ Do the work in the brackets first.
$$= -13$$ ■

EXAMPLE 4   If $x = -4$, $y = 5$, and $z = -13$, evaluate **a.** $x + y$ and   **b.** $2y + z$.

Solution   Substitute -4 for x, 5 for y, and -13 for z and then simplify.

a. $x + y = (-4) + (5)$ **b.** $2y + z = 2 \cdot 5 + (-13)$
$$= 1$$ $$= 10 + (-13)$$
$$= -3$$ ■

Sometimes the addition of real numbers is done vertically, as shown in the next example.

EXAMPLE 5   **a.** $\begin{array}{r} +5 \\ +2 \\ \hline +7 \end{array}$ **b.** $\begin{array}{r} +5 \\ -2 \\ \hline +3 \end{array}$ **c.** $\begin{array}{r} -5 \\ +2 \\ \hline -3 \end{array}$ **d.** $\begin{array}{r} -5 \\ -2 \\ \hline -7 \end{array}$ ■

Words and phrases such as *found, gain, credit, up, increase, forward, rises, in the future,* and *to the right* indicate a positive direction. Words and phrases such as *lost, loss, debit, down, backward, falls, in the past,* and *to the left* indicate a negative direction.

EXAMPLE 6   Laura opens a checking account by depositing $350 in the bank. The bank debits her account $9 for check printing, and Laura writes a check for $22. Find the balance in Laura's account after these transactions.

Solution   The initial deposit can be represented by the positive number 350. A debit of $9 can be represented by the negative number -9, and a check written for $22 can be represented by the negative number -22. The balance in Laura's account after these three transactions is the sum of 350, -9, and -22.

$$350 + (-9) + (-22) = 341 + (-22)$$
$$= 319$$

Laura's account balance is $319.   ■

Subtracting Real Numbers

In arithmetic, subtraction is often thought of as a take-away process. For example, the expression

$$7 - 4 = 3$$

can be thought of as taking four objects away from seven objects, leaving three objects. This take-away interpretation of subtraction is not very useful in algebra.

A better approach treats the subtraction problem

$$7 - 4$$

as an equivalent addition problem:

$$7 + (-4)$$

In either case, the answer is 3.

$$7 - 4 = 3 \quad \text{and} \quad 7 + (-4) = 3$$

This idea can be used to define the *difference* when b is subtracted from a.

Finding the Difference of Real Numbers. If a and b are two real numbers, then
$$a - b = a + (-b)$$

This rule points out that subtracting a number is equivalent to adding the negative (or opposite) of that number. Thus, to *subtract* one number from a second, we can add the opposite of the number to the second number.

EXAMPLE 7 Evaluate **a.** $12 - 4$, **b.** $-13 - 5$, and **c.** $-14 - (-6)$.

Solution Use the rule for finding the difference of two real numbers.

a. $12 - 4 = 12 + (-4)$ **b.** $-13 - 5 = -13 + (-5)$
$\qquad\quad = 8$ $\qquad\qquad\qquad\qquad\qquad = -18$

c. $-14 - (-6) = -14 + [-(-6)]$
$\qquad\qquad\quad = -14 + 6 \qquad\qquad$ Use the double negative rule.
$\qquad\qquad\quad = -8$ ■

EXAMPLE 8 If $x = -5$ and $y = -3$, evaluate **a.** $\dfrac{y - x}{7 + x}$ and **b.** $\dfrac{6 + x}{y - x} - \dfrac{y - 4}{7 + x}$.

Solution   Substitute -5 for x and -3 for y into each expression. Then simplify.

a. $\dfrac{y - x}{7 + x} = \dfrac{-3 - (-5)}{7 + (-5)}$

$= \dfrac{-3 + [-(-5)]}{2}$ Use the rule for subtracting two numbers.

$= \dfrac{-3 + 5}{2}$ $-(-5) = +5.$

$= \dfrac{2}{2}$

$= 1$

b. $\dfrac{6 + x}{y - x} - \dfrac{y - 4}{7 + x} = \dfrac{6 + (-5)}{-3 - (-5)} - \dfrac{-3 - 4}{7 + (-5)}$

$= \dfrac{1}{-3 + 5} - \dfrac{-3 + (-4)}{2}$

$= \dfrac{1}{2} - \dfrac{-7}{2}$

$= \dfrac{1 - (-7)}{2}$

$= \dfrac{1 + [-(-7)]}{2}$ Use the rule for subtracting two numbers.

$= \dfrac{1 + 7}{2}$ $-(-7) = +7.$

$= \dfrac{8}{2}$

$= 4$ ■

To use a vertical format for subtracting real numbers, we add the opposite of the number that is to be subtracted by changing the sign of the lower number (called the **subtrahend**) and proceeding as in addition.

EXAMPLE 9   Perform each indicated subtraction by performing an equivalent addition.

a. The subtraction $\begin{array}{r} 5 \\ - \ -4 \\ \hline \end{array}$ becomes the addition $\begin{array}{r} 5 \\ + \ +4 \\ \hline 9 \end{array}$

b. The subtraction $\begin{array}{r} -8 \\ - \ +3 \\ \hline \end{array}$ becomes the addition $\begin{array}{r} -8 \\ + \ -3 \\ \hline -11 \end{array}$ ■

EXAMPLE 10   Simplify **a.** $3 - [4 + (-6)]$ and **b.** $[-5 + (-3)] - [-2 - (+5)]$.

Solution **a.** $3 - [4 + (-6)] = 3 - (-2)$ Do the addition within the brackets first.

$= 3 + [-(-2)]$ Use the rule for subtracting two numbers.

$= 3 + 2$ Use the double negative rule.

$= 5$ Add.

b. $[-5 + (-3)] - [-2 - (+5)]$

$= [-5 + (-3)] - [-2 + (-5)]$ Use the rule for subtracting two numbers.

$= -8 - (-7)$ Do the work within the brackets.

$= -8 + [-(-7)]$ Use the rule for subtracting two numbers.

$= -8 + 7$ Use the double negative rule.

$= -1$ Add. ■

EXAMPLE 11 At noon the temperature was 7 degrees above zero. At midnight the temperature was 4 degrees below zero. Find the difference between these two temperatures.

Solution A temperature of 7 degrees above zero can be represented as $+7$. A temperature of 4 degrees below zero can be represented as -4. To find the difference between these two temperatures, set up a subtraction problem and simplify.

$$7 - (-4) = 7 + [-(-4)]$$
$$= 7 + 4 \quad \text{Use the double negative rule.}$$
$$= 11$$

The difference between the given temperatures is 11 degrees. Figure 1-20 shows this difference.

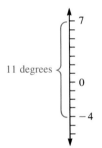

FIGURE 1-20 ■

EXERCISE 1.6

In Exercises 1–20, find each sum.

1. $4 + 8$

2. $(-4) + (-2)$

3. $(-3) + (-7)$

4. $(+4) + 11$

5. $6 + (-4)$

6. $5 + (-3)$

7. $9 + (-11)$

8. $10 + (-13)$

9. $(-5) + (-7)$

10. $(-6) + (-4)$

11. $(-0.4) + (-0.9)$

12. $(-1.2) + (-5.3)$

13. $\dfrac{1}{5} + \left(+\dfrac{1}{7}\right)$

14. $\dfrac{2}{3} + \left(-\dfrac{1}{4}\right)$

15. $\left(-\dfrac{3}{4}\right) + \left(+\dfrac{2}{3}\right)$

16. $\dfrac{3}{5} + \left(-\dfrac{2}{3}\right)$

17. $\begin{array}{r} 5 \\ + \ -4 \\ \hline \end{array}$

18. $\begin{array}{r} -20 \\ + \ -17 \\ \hline \end{array}$

19. $\begin{array}{r} -1.3 \\ + \ \ 3.5 \\ \hline \end{array}$

20. $\begin{array}{r} 1.3 \\ + \ -2.5 \\ \hline \end{array}$

In Exercises 21–34, evaluate each expression.

21. $5 + [4 + (-2)]$

22. $-6 + [(-3) + 8]$

23. $-2 + (-4 + 5)$

24. $5 + [-4 + (-6)]$

25. $[-4 + (-3)] + [2 + (-2)]$

26. $[3 + (-1)] + [-2 + (-3)]$

27. $-4 + [-3 + 2] + (-3)$

28. $5 + [2 + (-5)] + (-2)$

29. $-|-9 + (-3)| + (-6)$

30. $-|8 + (-4)| + 7$

31. $\left|\dfrac{3}{5} + \left(-\dfrac{4}{5}\right)\right|$

32. $\left|\dfrac{1}{6} + \left(-\dfrac{5}{6}\right)\right|$

33. $-5.2 + |-2.5 + (-4)|$

34. $6.8 + |8.6 + (-1.1)|$

In Exercises 35–50, let $x = 2$, $y = -3$, $z = -4$, and $u = 5$. Evaluate each expression.

35. $x + y$

36. $x + z$

37. $x + z + u$

38. $y + z + u$

39. $(x + u) + 3$

40. $(y + 5) + x$

41. $x + (-1 + z)$

42. $-7 + (z + x)$

43. $(x + z) + (u + z)$

44. $(z + u) + (x + y)$

45. $x + [5 + (y + u)]$

46. $y + \{u + [z + (-6)]\} + y$

47. $|2x + y|$

48. $3|x + y + z|$

49. $|x + z| + |x + y + z|$

50. $|z + z| + |y + y|$

In Exercises 51–70, find each difference.

51. $8 - 4$

52. $-8 - 4$

53. $8 - (-4)$

54. $-9 - (-5)$

55. $-12 - 5$

56. $11 - (+4)$

57. $0 - (-5)$

58. $0 - 75$

59. $\dfrac{5}{3} - \dfrac{7}{6}$

60. $-\dfrac{5}{9} - \dfrac{5}{3}$

61. $-5 - \left(-\dfrac{3}{5}\right)$

62. $\dfrac{7}{8} - (-3)$

63. $-3\dfrac{1}{2} - 5\dfrac{1}{4}$

64. $2\dfrac{1}{2} - \left(-3\dfrac{1}{2}\right)$

65. $-6.7 - (-2.5)$

66. $25.3 - 17.5$

67. $\begin{array}{r} 8 \\ -\ 4 \\ \hline \end{array}$

68. $\begin{array}{r} 8 \\ -\ -3 \\ \hline \end{array}$

69. $\begin{array}{r} -10 \\ -\ -3 \\ \hline \end{array}$

70. $\begin{array}{r} -13 \\ -\ \ 5 \\ \hline \end{array}$

In Exercises 71–84, evaluate each quantity.

71. $+3 - [(-4) - 3]$

72. $-5 - [4 - (-2)]$

73. $(5 - 3) + (3 - 5)$

74. $(3 - 5) - [5 - (-3)]$

75. $5 - [4 + (-2) - 5]$

76. $3 - [-(-2) + 5]$

77. $[5 - (-34)] - [-2 + (-23)]$

78. $-5 + \{-3 - [-2 - (+4)]\}$

79. $\left(\dfrac{5}{2} - 3\right) - \left(\dfrac{3}{2} - 5\right)$

80. $\left(\dfrac{7}{3} - \dfrac{5}{6}\right) - \left[\dfrac{5}{6} - \left(-\dfrac{7}{3}\right)\right]$

81. $(5.2 - 2.5) - (5.25 - 5)$

82. $\left(3\dfrac{1}{2} - 2\dfrac{1}{2}\right) - \left[5\dfrac{1}{3} - \left(-5\dfrac{2}{3}\right)\right]$

83. $-|-9 - (-7)| - (-3)$

84. $-|8 - (-4)| - 7$

In Exercises 85–100, let $x = -4$, $y = 5$, *and* $z = -6$. *Evaluate each quantity.*

85. $x + y$

86. $y - z$

87. $x - y - z$

88. $y + z - x$

89. $x - (y - z)$

90. $y + (z - x)$

91. $3 - \{[x + (-3)] - z\}$

92. $[(-2) + z - (x - y)] + 10$

93. $|x - y + z|$

94. $|y - z - x|$

95. $|x|^2 - |y|^2 - |z|^2$

96. $|z|^2 - |x|^2 + |y|^2$

97. $\dfrac{y - x}{3 - z}$

98. $\dfrac{y - z}{3y + x}$

99. $\dfrac{x - y}{y} - \dfrac{z}{y}$

100. $\dfrac{y}{x - z} - \dfrac{x}{8 + z}$

In Exercises 101–108, let $a = 2$, $b = -3$, *and* $c = -4$. *Evaluate each quantity.*

101. $a + b - c$

102. $a - b + c$

103. $b - (c + a)$

104. $c + (a - b)$

105. $\dfrac{a + b}{b - c}$

106. $\dfrac{c - a}{-(a + b)}$

107. $\dfrac{|b + c|}{a - c}$

108. $\dfrac{a - b - c}{|a + b|}$

In Exercises 109–130, solve each problem by finding the sum of two or more signed numbers.

109. Wendy has $25. If she pays Bruce the $17 she owes him, what amount will she have left?

110. Scott weighed 150 pounds. After losing 4 pounds during an illness, what does Scott weigh?

111. The temperature rose 7 degrees in 1 hour. It then dropped 3 degrees in the next hour. What signed number represents the net change in temperature?

112. Kris lost 12 pounds but then gained back 5 pounds. What signed number represents her net change in weight?

113. The temperature fell from zero to 14 below one night. By 5:00 P.M. the next day the temperature had risen 20 degrees. What was the temperature at 5:00 P.M.?

114. In 1897 Joseph Thompson discovered the electron. Fifty-four years later, the first fission reactor was built. Nineteen years before the reactor, James Chadwick discovered the neutron. In what year was the neutron discovered?

115. John deposited $212 in a new checking account, wrote a check for $173.30, and deposited another $312.50. Find the balance in John's account.

116. An army retreated 2300 meters. After regrouping, it moved forward 1750 meters. The next day it gained another 1875 meters. What was the army's net gain?

117. A football player gained and lost the following yards on six consecutive plays: $+5$, $+7$, -5, $+1$, -2, and -6. What was the net outcome?

118. On January 1, Sally had $437 in the bank. During the month, she had deposits of $25, $37, and $45, and she had withdrawals of $17, $83, and $22. How much was in her account at the end of the month?

119. At the opening bell on Monday, the Dow Jones Industrial Average was 2153. At the close of light trading, the Dow was down 12 points, but news of a half-point drop in interest rates on Tuesday sent prices up. With a volume of 181 million shares, Tuesday's Dow was up 21 points. What was the Dow average after the market closed on Tuesday?

120. On Monday morning, the Dow Jones average opened at 2917. For the week the Dow rose 29 points on Monday and 12 points on Wednesday. However, it fell 53 points on Tuesday and 27 points on both Thursday and Friday. Where did the Dow close on Friday?

121. Andy owned 500 shares of Transitronics Corporation before the company declared a two-for-one stock split. After the split, Andy sold 300 shares. How many shares does Andy now own?

122. Pat gained 12 yards on the first play of a football game. However, on the next play, he lost 15 yards. Find Pat's net gain or loss.

123. Tuesday's high and low prices for Transitronics stock were $37\frac{1}{8}$ and $31\frac{5}{8}$. Find the range of prices for this stock.

124. Find the difference between a temperature of 32 degrees above zero and a temperature of 27 degrees above zero.

125. Find the difference between a temperature of 3 degrees below zero and a temperature of 21 degrees below zero.

126. Maria earned $2532 in a part-time business. However, 25% of the earnings went for taxes. Find Maria's net earnings.

127. The Greek mathematician Euclid was alive in 300 B.C. The English mathematician Sir Isaac Newton was alive in A.D. 1700. How many years apart did they live?

128. Juan owed his mother $75 and his brother $32. However, his sister owed Juan $47. Use a signed number to express Juan's financial position.

129. Susan owed her mother $125. Her mother agreed to cancel $70 of this debt. Use a signed number to express Susan's current financial position.

130. Mike owed Paul $350. However, Paul canceled $43 of the debt when Mike tutored him in algebra. Another $75 of the debt was canceled when Mike gave Paul his golf clubs. How much does Mike still owe Paul?

REVIEW EXERCISES

1. Simplify the fraction $\dfrac{24}{27}$.

2. Add: $\dfrac{3}{5} + \dfrac{1}{3}$.

3. Simplify: $3(7 - 2)$.

4. Simplify: $3(7) - 2$.

In Review Exercises 5–6, let $x = -3$ and $y = 5$. Perform the indicated operations.

5. $|y||x| - 2$

6. $|y|(|x| - 2)$

In Review Exercises 7–10, write an algebraic expression that represents the given quantity.

7. The sum of x and twice y.

8. Twice the sum of x and y.

9. The absolute value of the sum of x and y.

10. Twice the sum of the absolute values of x and y.

11. Jim has one penny, seven nickels, x dimes, and y quarters. Find the total value of his coins.

12. On her first quiz, Jill earned a grade of k points. On each successive quiz, her grade improved by 2 points. Find Jill's score on the seventh quiz.

1.7 MULTIPLYING AND DIVIDING REAL NUMBERS

To develop rules for multiplying real numbers, we rely on the definition of multiplication. The expression $5 \cdot 4$ indicates that 4 is to be used as a term in a sum five times. That is,

$$5(4) = 4 + 4 + 4 + 4 + 4 = 20$$

Likewise, the expression $5(-4)$ indicates that -4 is to be used as a term in a sum five times. Thus,

$$5(-4) = (-4) + (-4) + (-4) + (-4) + (-4) = -20$$

If multiplying by a positive number indicates repeated addition, then it is reasonable to assume that multiplication by a negative number indicates repeated subtraction. The expression $(-5)4$, for example, means that 4 is to be used as a term in a repeated subtraction five times. That is,

$$\begin{aligned}
(-5)4 &= -(4) - (4) - (4) - (4) - (4) \\
&= (-4) + (-4) + (-4) + (-4) + (-4) \\
&= -20
\end{aligned}$$

Likewise, the expression $(-5)(-4)$ indicates that -4 is to be used as a term in a repeated subtraction five times. Thus,

$$\begin{aligned}
(-5)(-4) &= -(-4) - (-4) - (-4) - (-4) - (-4) \\
&= -(-4) + [-(-4)] + [-(-4)] + [-(-4)] + [-(-4)] \\
&= 4 + 4 + 4 + 4 + 4 \\
&= 20
\end{aligned}$$

The expression $0(-2)$ indicates that -2 is to be used zero times as a term in a repeated addition. Thus,

$$0(-2) = 0$$

This suggests that the product of any number and 0 is 0.

Finally, the expression $(-3)(1) = -3$ suggests that the product of any number and 1 is the number itself.

The previous results suggest the following rules.

Rules for Multiplying Real Numbers.

1. The product of two real numbers with like signs is the positive product of their absolute values.
2. The product of two real numbers with unlike signs is the negative of the product of their absolute values.
3. Any number multiplied by 0 is 0: $a \cdot 0 = 0 \cdot a = 0$.
4. Any number multiplied by 1 is that number itself: $a \cdot 1 = 1 \cdot a = a$.

EXAMPLE 1   Find each product: **a.** $4(-7)$, **b.** $(-5)(-4)$, **c.** $(-7)(6)$, **d.** $8(6)$,
e. $(-3)(5)(-4)$, and **f.** $(-4)(-2)(-3)$.

Solution   **a.** $4(-7) = -(4 \cdot 7)$ **b.** $(-5)(-4) = +(5 \cdot 4)$
$\qquad = -28$ $\qquad = +20$

c. $(-7)(6) = -(7 \cdot 6)$ **d.** $8(6) = +(8 \cdot 6)$
$\qquad = -42$ $\qquad = +48$

e. $(-3)(5)(-4) = (-15)(-4)$ **f.** $(-4)(-2)(-3) = (8)(-3)$
$\qquad = 60$ $\qquad = -24$ ∎

EXAMPLE 2   If $x = -3$, $y = 2$, and $z = 4$, evaluate **a.** $y + xz$ and **b.** $x(y - z)$.

Solution   Substitute -3 for x, 2 for y, and 4 for z in each expression and simplify.

a. $y + xz = 2 + (-3)(4)$ **b.** $x(y - z) = -3[2 - 4]$
$\qquad = 2 + (-12)$ $\qquad = -3[2 + (-4)]$
$\qquad = -10$ $\qquad = -3[-2]$
$\qquad\qquad\qquad\qquad\qquad\qquad\qquad = 6$ ∎

EXAMPLE 3   If $x = -2$ and $y = 3$, evaluate **a.** $x^2 - y^2$ and **b.** $-x^2$.

Solution   **a.** Substitute -2 for x and 3 for y and simplify.

$$x^2 - y^2 = (-2)^2 - 3^2$$
$$= 4 - 9 \qquad\qquad \text{Simplify the exponential expressions first.}$$
$$= -5 \qquad\qquad \text{Do the subtraction: } 4 - 9 = 4 + (-9) = -5.$$

b. Substitute -2 for x and simplify.

$$-x^2 = -(-2)^2$$
$$= -4 \qquad\qquad (-2)^2 = (-2)(-2) = 4.$$ ∎

EXAMPLE 4   Find each product: **a.** $\left(-\dfrac{2}{3}\right)\left(-\dfrac{6}{5}\right)$ and **b.** $\left(\dfrac{3}{10}\right)\left(-\dfrac{5}{9}\right)$.

Solution   **a.** Because both fractions are negative, their product is positive.

$$\left(-\frac{2}{3}\right)\left(-\frac{6}{5}\right) = +\left(\frac{2}{3} \cdot \frac{6}{5}\right)$$
$$= +\frac{2 \cdot 6}{3 \cdot 5}$$
$$= +\frac{12}{15}$$
$$= +\frac{4}{5}$$

b. Because the fractions are of opposite signs, their product is negative.

$$\left(\frac{3}{10}\right)\left(-\frac{5}{9}\right) = -\frac{3}{10} \cdot \frac{5}{9}$$
$$= -\frac{3 \cdot 5}{10 \cdot 9}$$
$$= -\frac{15}{90}$$
$$= -\frac{1}{6}$$ ∎

EXAMPLE 5   If the temperature is dropping 4 degrees each hour, how much warmer was it 3 hours ago?

Solution   A temperature drop of 4 degrees per hour can be represented by -4 degrees per hour. Three hours ago can be represented by -3. The temperature 3 hours ago can be represented by the product $(-3)(-4)$. Because

$$(-3)(-4) = +12$$

the temperature was 12 degrees warmer 3 hours ago.   ■

Dividing Real Numbers

We have seen that 8 divided by 4 is 2, because there are two 4's in 8. That is,

$$\frac{8}{4} = 2 \qquad \text{because } 2 \cdot 4 = 8$$

Likewise,

$$\frac{18}{6} = 3 \qquad \text{because } 3 \cdot 6 = 18$$

In general, the rule

$$\frac{a}{b} = c \qquad \text{if and only if} \qquad c \cdot b = a$$

is true for the division of *any* real number, a, by any nonzero real number, b. For example,

$$\frac{+10}{+2} = +5 \qquad \text{because } (+5)(+2) = +10$$

$$\frac{-10}{-2} = +5 \qquad \text{because } (+5)(-2) = -10$$

$$\frac{+10}{-2} = -5 \qquad \text{because } (-5)(-2) = +10$$

$$\frac{-10}{+2} = -5 \qquad \text{because } (-5)(+2) = -10$$

These four examples suggest that the rules for dividing real numbers are similar to the rules for multiplying real numbers.

Rules for Dividing Real Numbers.

1. The quotient of two real numbers with like signs is the positive quotient of their absolute values.
2. The quotient of two real numbers with unlike signs is the negative of the quotient of their absolute values.
3. Division by zero is undefined.
4. If $a \neq 0$, then $\dfrac{0}{a} = 0$.

We now see that a negative fraction can have the $-$ sign either in front of the fraction or in the numerator of the fraction. For example, because $-\frac{10}{5}$ and $\frac{-10}{5}$ are both equal to -2,

$$-\frac{10}{5} = \frac{-10}{5}$$

The negative fraction $-\frac{10}{5}$ is a rational number because it can be written as the quotient of two integers. Likewise, fractions such as $-\frac{2}{3}$ and $-\frac{7}{2}$ are rational numbers because they can be written as the quotient of two integers:

$$-\frac{2}{3} = \frac{-2}{3} \quad \text{and} \quad -\frac{7}{2} = \frac{-7}{2}$$

EXAMPLE 6   Find each quotient: **a.** $\dfrac{36}{18}$, **b.** $\dfrac{-44}{11}$, **c.** $\dfrac{27}{-9}$, and **d.** $\dfrac{-64}{-8}$.

Solution   **a.** $\dfrac{36}{18} = +\dfrac{|36|}{|18|} = \dfrac{36}{18} = 2$

The quotient of two real numbers with like signs is the positive quotient of their absolute values.

b. $\dfrac{-44}{11} = -\dfrac{|-44|}{|11|} = -\dfrac{44}{11} = -4$

The quotient of two real numbers with unlike signs is the negative of the quotient of their absolute values.

c. $\dfrac{27}{-9} = -\dfrac{|27|}{|-9|} = -\dfrac{27}{9} = -3$

The quotient of two real numbers with unlike signs is the negative of the quotient of their absolute values.

d. $\dfrac{-64}{-8} = +\dfrac{|-64|}{|-8|} = \dfrac{64}{8} = 8$

The quotient of two real numbers with like signs is the positive quotient of their absolute values.   ∎

EXAMPLE 7   If $x = -64$, $y = 16$, and $z = -4$, evaluate **a.** $\dfrac{yz}{-x}$, **b.** $\dfrac{z^3 y}{x}$, and **c.** $\dfrac{x + y}{-z^2}$.

Solution   Substitute -64 for x, 16 for y, and -4 for z in each expression and simplify.

a. $\dfrac{yz}{-x} = \dfrac{16(-4)}{-(-64)}$ **b.** $\dfrac{z^3 y}{x} = \dfrac{(-4)^3(16)}{-64}$ **c.** $\dfrac{x + y}{-z^2} = \dfrac{-64 + 16}{-(-4)^2}$

$\qquad = \dfrac{-64}{+64}$ $\qquad = \dfrac{(-64)(16)}{(-64)}$ $\qquad = \dfrac{-48}{-16}$

$\qquad = -1$ $\qquad = 16$ $\qquad = 3$ ∎

EXAMPLE 8   If $x = -50$, $y = 10$, and $z = -5$, evaluate **a.** $\dfrac{xyz}{x - 5z}$ and **b.** $\dfrac{3xy + 2yz}{2(x + y)}$.

Solution   Substitute -50 for x, 10 for y, and -5 for z in each expression and simplify.

a. $\dfrac{xyz}{x - 5z} = \dfrac{(-50)(10)(-5)}{-50 - 5(-5)}$

$\qquad = \dfrac{(-500)(-5)}{-50 + 25}$

$\qquad = \dfrac{2500}{-25}$

$\qquad = -100$

b. $\dfrac{3xy + 2yz}{2(x + y)} = \dfrac{3(-50)(10) + 2(10)(-5)}{2(-50 + 10)}$

$\qquad = \dfrac{-150(10) + (20)(-5)}{2(-40)}$

$\qquad = \dfrac{-1500 - 100}{-80}$

$\qquad = \dfrac{-1600}{-80}$

$\qquad = 20$ ∎

EXAMPLE 9 Twelve gamblers lost $336. If they lost equal amounts, how much did each gambler lose?

Solution A loss of $336 can be represented by -336. Because there are 12 gamblers, the amount lost by each gambler can be represented by the quotient $\frac{-336}{12}$.

$$\frac{-336}{12} = -28$$

Each gambler lost $28. ∎

Remember these facts about dividing real numbers.

---

1. Division by zero is undefined.

2. If 0 is divided by a nonzero number, the result is zero: $\dfrac{0}{a} = 0$.

3. Any number divided by 1 is that number itself: $\dfrac{a}{1} = a$.

4. Any nonzero number divided by itself is 1: $\dfrac{a}{a} = 1$.

EXERCISE 1.7

In Exercises 1–24, find each product.

1. $(+1)(+3)$

2. $(-2)(-5)$

3. $(-3)(-6)$

4. $(4)(-6)$

5. $(+8)(+4)$

6. $(+8)(-4)$

7. $(-8)(4)$

8. $(-8)(-4)$

9. $(+9)(-6)$

10. $(-9)(6)$

11. $\left(\dfrac{1}{2}\right)(-32)$

12. $\left(-\dfrac{3}{4}\right)(12)$

13. $\left(-\dfrac{3}{4}\right)\left(-\dfrac{8}{3}\right)$

14. $\left(-\dfrac{2}{5}\right)\left(\dfrac{15}{2}\right)$

15. $(-3)\left(-\dfrac{1}{3}\right)$

16. $(5)\left(-\dfrac{2}{5}\right)$

17. $(3)(-4)(-6)$

18. $(-1)(-3)(-6)$

19. $(-2)(3)(4)$

20. $(5)(0)(-3)$

21. $(2)(-5)(-6)(-7)$

22. $(-3)(-5)(-5)(-2)$

23. $(-2)(-2)(-2)(-3)(-4)$

24. $(-5)(4)(3)(-2)(-1)$

In Exercises 25–48, let $x = -1$, $y = 2$, and $z = -3$. Evaluate each expression.

25. y^2 **26.** x^2 **27.** $-z^2$ **28.** $-xz$

29. xy **30.** yz **31.** $y + xz$ **32.** $z - xy$

33. $(x + y)z$ **34.** $y(x - z)$ **35.** $(x - z)(x + z)$ **36.** $(y + z)(x - z)$

37. $xy + yz$ **38.** $zx - zy$ **39.** xyz **40.** x^2y

41. y^2z^2 **42.** z^3y **43.** $y(x - y)^2$ **44.** $z(y - x)^2$

45. $x^2(y - z)$ **46.** $y^2(x - z)$ **47.** $(-x)(-y) + z^2$ **48.** $(-x)(-z) - y^2$

In Exercises 49–60, simplify each expression.

49. $\dfrac{8}{-2}$ **50.** $\dfrac{-6}{3}$ **51.** $\dfrac{-10}{-5}$ **52.** $\dfrac{20}{4}$

53. $\dfrac{-16}{4}$ **54.** $\dfrac{-25}{-5}$ **55.** $\dfrac{32}{-16}$ **56.** $\dfrac{18}{-3}$

57. $\dfrac{8 - 12}{-2}$ **58.** $\dfrac{16 - 2}{2 - 9}$ **59.** $\dfrac{20 - 2(5)^2}{(-2)^3 - 2(11)}$ **60.** $\dfrac{2(15)^2 - 2}{-2^3 + 1}$

In Exercises 61–72, evaluate each expression if $x = -2$, $y = 3$, $z = 4$, $t = 5$, and $w = -18$.

61. $\dfrac{yz}{x}$ **62.** $\dfrac{zt}{x}$ **63.** $\dfrac{tw}{y}$ **64.** $\dfrac{w}{xy}$

65. $\dfrac{z + w}{x}$ **66.** $\dfrac{xyx}{y - 1}$ **67.** $\dfrac{xtz}{y + 1}$ **68.** $\dfrac{x + y + z}{t}$

69. $\dfrac{wz - xy}{x + y}$ **70.** $\dfrac{x^2y^3}{yz}$ **71.** $\dfrac{yw + xy}{xt}$ **72.** $\dfrac{tw}{xz - w}$

In Exercises 73–84, evaluate each expression if $x = 4$, $y = -6$, and $z = -3$. Use a calculator.

73. $\dfrac{2x^2 + 2y}{x + y}$ **74.** $\dfrac{y^2 + z^2}{y + z}$ **75.** $\dfrac{2x^2 - 2z^2}{x + z}$ **76.** $\dfrac{8x^3 - 8y^2}{x - z}$

77. $\dfrac{y^3 + 4z^3}{(x + y)^2}$ **78.** $\dfrac{x^2 - 2xz + z^2}{x - y + z}$ **79.** $\dfrac{xy^2z + x^2y}{2y - 2z}$ **80.** $\dfrac{(x^2 - 2y)z^2}{-xz}$

81. $\dfrac{xyz - y^2z}{y(x + z)^4}$ **82.** $\dfrac{-3x^2 - 2z^2 + x^2}{(x + y + z)^2}$

83. $\dfrac{2(x - y)(y - z)(x - z)}{2x - 3y + y}$ **84.** $\dfrac{x^3y - (yz)^2 - 10y - 2}{x^2 + y^2 + z^3}$

In Exercises 85–92, evaluate each expression if $x = \frac{1}{2}$, $y = -\frac{2}{3}$, and $z = -\frac{3}{4}$.

85. $x + y$ **86.** $y + z$ **87.** $x + y + z$ **88.** $y + x - z$

89. $(x + y)(x - y)$ **90.** $(x - z)(x + z)$ **91.** $(x + y + z)(xyz)$ **92.** $xyz(x - y - z)$

In Exercises 93–102, solve each problem by finding a product or quotient of two signed numbers.

93. If the temperature is increasing 2 degrees each hour, how much warmer will it be in 3 hours?

94. If the temperature is decreasing 3 degrees each hour, how much colder will it be in 2 hours?

95. In Las Vegas, Robert lost $30 per hour playing the slot machines. Use a signed number to express Robert's financial condition after he had gambled for 15 hours.

96. Rafael worked all day mowing lawns and was paid $8 per hour. If he had $94 at the end of an 8-hour day, how much did he have before he started working?

97. If a drain is emptying a pool at the rate of 12 gallons per minute, how much more water was in the pool 1 hour ago?

98. The flow of water from a pipe is filling a pool at the rate of 23 gallons per minute. How much less water was in the pool 2 hours ago?

99. At a carnival, three boys lost a total of $21. Use a signed number to express how much each lost, assuming that they lost equal amounts.

100. Suppose that the temperature is dropping at the rate of 3 degrees each hour. If the temperature has dropped 18 degrees, what signed number expresses how many hours the temperature has been falling?

101. Suppose that the temperature is dropping at the rate of 4 degrees each hour. Use a signed number to show when the temperature was 20 degrees warmer.

102. A man has lost 37.5 pounds. If he lost 2.5 pounds each week, how long has he been dieting?

REVIEW EXERCISES

In Review Exercises 1–10, place one of the symbols $=$, $<$, or $>$ in the box to make the statement true.

1. $5 + (-3) \;\boxed{}\; -5 - (-7)$

2. $-3 - 3 \;\boxed{}\; (-2)(-3)$

3. $2(7 - 9) \;\boxed{}\; \dfrac{-12}{4}$

4. $-(-7) \;\boxed{}\; -5 + [3 - (-2)]$

5. $|-2[9 - (-1)]| \;\boxed{}\; -(-5) + (-2)$

6. $\dfrac{-3 + 5}{8(5 - 3)} \;\boxed{}\; \dfrac{-1}{2} + \dfrac{3 - (-2)}{2}$

7. $(-2)^2 \;\boxed{}\; (-2)^3$

8. $(5 - 7)^4 \;\boxed{}\; (7 - 4)^4$

9. $\left(-\dfrac{24}{8}\right)^2 \;\boxed{}\; |-9|$

10. $3^2 + 2^2 \;\boxed{}\; (3 + 2)^2$

11. On the number line, graph the real numbers between -3 and $\dfrac{3}{2}$.

12. On the number line, graph the real numbers from x to $2x$ if x is the negative of -3.

1.8 PROPERTIES OF REAL NUMBERS

Basic to understanding algebra is knowledge of the properties governing the operations of addition, subtraction, multiplication, and division of real numbers. We will discuss these properties in this section.

The **closure properties** guarantee that the sum, difference, product, or quotient (except for division by zero) of any two real numbers is a real number.

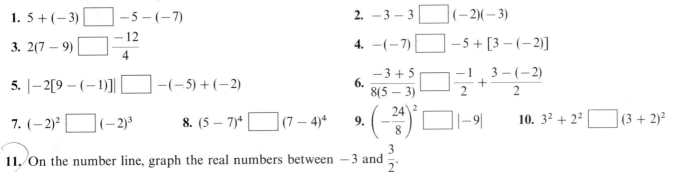

The Closure Properties. If a and b are real numbers, then

$a + b$ is a real number

$a - b$ is a real number

ab is a real number

$\dfrac{a}{b}$ is a real number, provided that $b \neq 0$

EXAMPLE 1   Assume that $x = 8$ and $y = -4$. Find the real-number answer to show that **a.** $x + y$, **b.** $x - y$ **c.** xy, and **d.** $\frac{x}{y}$ all represent real numbers.

Solution   Substitute 8 for x and -4 for y in each expression and simplify.

a. $x + y = 8 + (-4)$
$\qquad = 4$

b. $x - y = 8 - (-4)$
$\qquad = 8 + 4$
$\qquad = 12$

c. $xy = 8(-4)$
$\qquad = -32$

d. $\dfrac{x}{y} = \dfrac{8}{-4}$
$\qquad = -2$ ∎

The **commutative properties** assert that the addition or multiplication of two real numbers may be done in either order.

> **The Commutative Properties.**   If a and b are two real numbers, then
> $a + b = b + a$ $\qquad$ commutative property of addition
> $ab = ba$ $\qquad$ commutative property of multiplication

EXAMPLE 2   Assume that $x = -3$ and $y = 7$. Show that **a.** $x + y = y + x$ and **b.** $xy = yx$.

Solution   **a.** Show that the sum $x + y$ is the same number as the sum $y + x$ by substituting -3 for x and 7 for y in each expression and simplifying.

$$x + y = -3 + 7 = 4$$
$$y + x = 7 + (-3) = 4$$

b. Show that the product xy is the same number as the product yx by substituting -3 for x and 7 for y in each expression and simplifying.

$$xy = -3(7) = -21$$
$$yx = 7(-3) = -21$$ ∎

The **associative properties** are necessary to find sums and products if more than two numbers are involved.

> **The Associative Properties.**   If a, b, and c are real numbers, then
> $(a + b) + c = a + (b + c)$ $\qquad$ associative property of addition
> $(ab)c = a(bc)$ $\qquad$ associative property of multiplication

The associative property of addition permits us to group, or *associate,* the numbers in a sum in any way that we wish. For example,

$$(3 + 4) + 5 = 7 + 5 = 12$$

and

$$3 + (4 + 5) = 3 + 9 = 12$$

The answer is 12 regardless of how we group the three numbers.

The associative property of multiplication permits us to group, or associate, the numbers in a product in any way that we wish. For example,

$$(3 \cdot 4) \cdot 7 = 12 \cdot 7 = 84$$

and

$$3 \cdot (4 \cdot 7) = 3 \cdot 28 = 84$$

The answer is 84 regardless of how we group the three numbers.

The **distributive property** shows how to multiply the sum of two numbers by a third number. Because of this property, we can often add first and then multiply, or multiply first and then add. For example, $2(3 + 7)$ can be calculated in two different ways. One way is to perform the indicated addition and then the multiplication:

$$2(3 + 7) = 2(10) = 20$$

The other way is to distribute the multiplication by 2 over the addition by first multiplying each number within the parentheses by 2 and then adding:

$$2(3 + 7) = 2 \cdot 3 + 2 \cdot 7 = 6 + 14 = 20$$

Either way, the result is 20.

In general, we have

> **The Distributive Property.**   If a, b, and c are real numbers, then
> $$a(b + c) = ab + ac$$

Because multiplication is commutative, the distributive property can be written in the form

$$(b + c)a = ba + ca$$

EXAMPLE 3 Use the distributive property to evaluate each expression in two different ways: **a.** $3(5 + 9)$ and   **b.** $-2(-7 + 3)$.

Solution **a.** $3(5 + 9) = 3(14) = 42$

$3(5 + 9) = 3 \cdot 5 + 3 \cdot 9 = 15 + 27 = 42$

b. $-2(-7 + 3) = -2(-4) = 8$

$-2(-7 + 3) = -2(-7) + (-2)3 = 14 + (-6) = 8$ ∎

EXAMPLE 4 Use the distributive property to write the expressions **a.** $3(x + 2)$ and **b.** $-2(y - 3)$ without using parentheses.

Solution **a.** $3(x + 2) = 3x + 3 \cdot 2$ **b.** $-2(y - 3) = -2y - (-2)(3)$

$\qquad\qquad = 3x + 6$ $\qquad\qquad = -2y + 6$ ∎

The distributive property can be extended to three or more terms. For example, if a, b, c, and d are real numbers, then

$$a(b + c + d) = ab + ac + ad$$

The numbers 0 and 1 play special roles in the arithmetic of real numbers. The number 0 is the only number that can be added to another number, say a, and give an answer of that same number a:

$$0 + a = a + 0 = a$$

The number 1 is the only number that can be multiplied by another number, say a, and give an answer of that same number a:

$$1 \cdot a = a \cdot 1 = a$$

Because adding 0 to a number or multiplying a number by 1 leaves that number identically the same, the numbers 0 and 1 are called **identity elements.**

The Identity Elements. The number 0 is called the **identity element for addition.** The number 1 is called the **identity element for multiplication.**

If the sum of two numbers is 0, the numbers are called **negatives,** or **additive inverses,** of each other. For example, because $3 + (-3) = 0$, the numbers 3 and -3 are negatives or additive inverses of each other. In general, because

$$a + (-a) = 0$$

the numbers represented by a and $-a$ are negatives or additive inverses of each other.

If the product of two numbers is 1, the numbers are called **reciprocals,** or **multiplicative inverses,** of each other. For example, because $7(\frac{1}{7}) = 1$, the numbers 7 and $\frac{1}{7}$ are reciprocals. Because $(-0.25)(-4) = 1$, the numbers -0.25 and -4 are reciprocals. In general, because

$$a\left(\frac{1}{a}\right) = 1 \qquad \text{provided } a \neq 0$$

the numbers represented by a and $\frac{1}{a}$ are reciprocals or multiplicative inverses of each other.

The Additive and Multiplicative Inverse Properties. Because
$$a + (-a) = 0$$
the numbers represented by a and $-a$ are called **negatives** or **additive inverses** of each other. Because

$$a\left(\frac{1}{a}\right) = 1 \qquad \text{provided } a \neq 0$$

the numbers represented by a and $\frac{1}{a}$ are called **reciprocals** or **multiplicative inverses** of each other.

EXAMPLE 5   The statement in the right column justifies the statement in the left column.

$3 + 4$ is a real number	The closure property of addition.
$\dfrac{8}{3}$ is a real number	The closure property of division.
$3 + 4 = 4 + 3$	The commutative property of addition.
$-3 + (2 + 7) = (-3 + 2) + 7$	The associative property of addition.
$(5)(-4) = (-4)(5)$	The commutative property of multiplication.
$(ab)c = a(bc)$	The associative property of multiplication.
$3(a + 2) = 3a + 3 \cdot 2$	The distributive property.
$3 + 0 = 3$	The additive identity property.
$3(1) = 3$	The multiplicative identity property.
$2 + (-2) = 0$	The additive inverse property.
$\left(\dfrac{2}{3}\right)\left(\dfrac{3}{2}\right) = 1$	The multiplicative inverse property. ∎

The properties of the real numbers are summarized as follows:

For all real numbers a, b, and c,	Addition	Multiplication
Closure properties	$a + b$ is a real number	$a \cdot b$ is a real number
Commutative properties	$a + b = b + a$	$a \cdot b = b \cdot a$
Associative properties	$(a + b) + c = a + (b + c)$	$(ab)c = a(bc)$
Identity properties	$a + 0 = a$	$a \cdot 1 = a$
Inverse properties	$a + (-a) = 0$	$a \cdot \left(\dfrac{1}{a}\right) = 1 \qquad a \neq 0$
Distributive property	$a(b + c) = ab + ac$	

EXERCISE 1.8

In Exercises 1–8, assume that $x = 12$ and $y = -2$. Show that each expression represents a real number by finding the real-number answer.

1. $x + y$
2. $y - x$
3. xy
4. $\dfrac{x}{y}$

5. x^2
6. y^2
7. $\dfrac{x}{y^2}$
8. $\dfrac{2x}{3y}$

In Exercises 9–14, assume that x = 5 and y = 7. Show that both given expressions have the same value.

9. $x + y; y + x$

10. $xy; yx$

11. $3x + 2y; 2y + 3x$

12. $3xy; 3yx$

13. $x(x + y); (x + y)x$

14. $xy + y^2; y^2 + xy$

In Exercises 15–20, assume that x = 2, y = −3, and z = 1. Show that both given expressions have the same value.

15. $(x + y) + z; x + (y + z)$

16. $(xy)z; x(yz)$

17. $(xz)y; x(yz)$

18. $(x + y) + z; y + (x + z)$

19. $x^2(yz^2); (x^2y)z^2$

20. $x(y^2z^3); (xy^2)z^3$

In Exercises 21–32, use the distributive property to write each expression without parentheses. Simplify each result if possible.

21. $3(x + y)$

22. $4(a + b)$

23. $x(x + 3)$

24. $y(y + z)$

25. $-x(a + b)$

26. $a(x + y)$

27. $4(x^2 + x)$

28. $-2(a^2 + 3)$

29. $-5(t + 2)$

30. $2x(a - x)$

31. $-2a(x + a)$

32. $-p(p - q)$

In Exercises 33–44, give the additive and the multiplicative inverse of each number when possible.

33. 2

34. 3

35. $\dfrac{1}{3}$

36. $-\dfrac{1}{2}$

37. 0

38. -2

39. $-\dfrac{5}{2}$

40. 0.5

41. -0.2

42. 0.75

43. $\dfrac{4}{3}$

44. -1.25

In Exercises 45–56, state which property of real numbers justifies each statement.

45. $3 + x = x + 3$

46. $(3 + x) + y = 3 + (x + y)$

47. $xy = yx$

48. $(3 \cdot 2)x = 3(2x)$

49. $-2(x + 3) = -2x + (-2)(3)$

50. $x(y + z) = (y + z)x$

51. $(x + y) + z = z + (x + y)$

52. $3(x + y) = 3x + 3y$

53. $5 \cdot 1 = 5$

54. $x + 0 = x$

55. $3 + (-3) = 0$

56. $9 \cdot \dfrac{1}{9} = 1$

In Exercises 57–66, use the given property to rewrite the expression in a different form.

57. $3(x + 2)$; distributive property

58. $x + y$; commutative property of addition

59. y^2x; commutative property of multiplication

60. $x + (y + z)$; associative property of addition

61. $(x + y)z$; commutative property of addition

62. $x(y + z)$; distributive property

63. $(xy)z$; associative property of multiplication

64. $1x$; multiplicative identity property

65. $0 + x$; additive identity property

66. $5 \cdot \dfrac{1}{5}$; multiplicative inverse property

REVIEW EXERCISES

In Review Exercises 1–4, write each English phrase as a mathematical expression.

1. The sum of x and the square of y.

2. Three more than the square of x.

3. The square of three more than x.

4. The sum of x and y is greater than or equal to the cube of z.

In Review Exercises 5–8, fill each box with the appropriate response.

5. If $x \geq 0$, then $|x| = \boxed{}$.

6. If $x < \boxed{}$, then $|x| = -x$.

7. $x - y = x + (\boxed{})$.

8. The product of two negative numbers is a $\boxed{}$ number.

In Review Exercises 9–12, let $x = 10$ and $y = -5$. Evaluate each expression.

9. $\dfrac{1}{3}(x - y)$

10. $\dfrac{3}{5}(y^2 - x)$

11. $\dfrac{x}{y + 1}$

12. $|x + y|^2$

CHAPTER SUMMARY

Key Words

absolute value (1.5)

additive inverse (1.8)

algebraic expressions (1.3)

algebraic terms (1.3)

area (1.4)

associative properties (1.8)

base (1.4)

circumference (1.4)

closure properties (1.8)

coefficient (1.3)

commutative properties (1.8)

composite numbers (1.1)

constants (1.3)

coordinate (1.1)

denominator (1.2)

diameter (1.4)

difference (1.3)

distributive property (1.8)

element of a set (1.1)

even natural number (1.1)

exponential expression (1.4)

exponents (1.4)

factors (1.2)

fractions (1.2)

graph (1.1)

identity elements (1.8)

improper fractions (1.2)

integers (1.5)

irrational numbers (1.5)

lowest (or least) common denominator (1.2)

lowest terms (1.2)

mixed number (1.2)

multiplicative inverse (1.8)

natural numbers (1.1)

negative numbers (1.5)

number line (1.1)

numerator (1.2)

numerical coefficient (1.3)

odd natural number (1.1)

origin (1.1)

perimeter (1.4)

positive numbers (1.5)

prime-factored form (1.2)

prime numbers (1.1)

product (1.3)

proper fractions (1.2)

quotient (1.3)

radius (1.4)

rational numbers (1.5)

real numbers (1.5)

reciprocal (1.2)

set (1.1)

signed numbers (1.5)

subset (1.5)

sum (1.3)

variables (1.2)

volume (1.4)

whole numbers (1.1)

Key Ideas

(1.1) Sets of numbers can be graphed on the number line.

(1.2) To simplify a fraction, first factor the numerator and denominator and then divide out all common factors.

To multiply two fractions, multiply their numerators and multiply their denominators.

To divide one fraction by another, multiply the first fraction by the reciprocal of the second fraction.

To add (or subtract) two fractions with the same denominator, add (or subtract) their numerators and keep their common denominator.

To add (or subtract) fractions with unlike denominators, write the fractions as fractions with a common denominator, add (or subtract) their numerators, and use their common denominator.

(1.3) If a term has no written numerical coefficient, the numerical coefficient is 1.

(1.4) If n is a natural number, then

$$x^n = \overbrace{x \cdot x \cdot x \cdot x \cdots \cdot x}^{n \text{ factors of } x}$$

(1.5) If x represents a number, then $-(-x) = x$.

If $x \geq 0$, then $|x| = x$.

If $x < 0$, then $|x| = -x$.

(1.6) If two real numbers x and y have the same sign, find their sum by adding their absolute values and using their common sign.

If two real numbers x and y have unlike signs, find their sum by subtracting their absolute values (the smaller from the larger) and using the sign of the number with the greatest absolute value.

If x and y are two real numbers, then $x - y = x + (-y)$.

(1.7) The product of two real numbers with like signs is the positive product of their absolute values.

The product of two real numbers with unlike signs is the negative of the product of their absolute values.

The quotient of two real numbers with like signs is the positive quotient of their absolute values.

The quotient of two real numbers with unlike signs is the negative of the quotient of their absolute values.

Division by zero is undefined.

(1.8) If x, y, and z are real numbers, then:

The closure properties:

$x + y$ is a real number $x - y$ is a real number

xy is a real number $\dfrac{x}{y}$ is a real number ($y \neq 0$)

The commutative properties:

$$x + y = y + x \qquad xy = yx$$

The associative properties:

$$(x + y) + z = x + (y + z) \qquad (xy)z = x(yz)$$

The distributive property:

$$x(y + z) = xy + xz$$

The identity elements:

0 is the identity for addition   1 is the identity for multiplication

The additive and multiplicative inverse properties:

$$x + (-x) = 0 \qquad x\left(\frac{1}{x}\right) = 1 \qquad \text{provided } x \neq 0$$

CHAPTER 1 REVIEW EXERCISES

(1.1) *In Review Exercises 1–4, consider the set of numbers* $\{0, 1, 2, 3, 4, 5\}$.

1. Which numbers are natural numbers?

2. Which numbers are prime numbers?

3. Which numbers are odd natural numbers?

4. Which numbers are composite numbers?

5. Draw a number line and graph the composite numbers from 10 to 20.

6. Draw a number line and graph the whole numbers between 15 and 25.

In Review Exercises 7–8, insert one of the symbols =, <, *or* > *to make the statement true.*

7. $8 + 3 \boxed{} 13 - 3$

8. $\dfrac{15}{4} \boxed{} \dfrac{34}{7}$

(1.2) *In Review Exercises 9–10, simplify each fraction.*

9. $\dfrac{45}{27}$

10. $\dfrac{121}{11}$

In Review Exercises 11–14, perform each operation.

11. $\dfrac{31}{15} \cdot \dfrac{10}{62}$

12. $\dfrac{18}{21} \div \dfrac{6}{7}$

13. $\dfrac{1}{3} + \dfrac{1}{7}$

14. $\dfrac{2}{3} - \dfrac{1}{7}$

(1.3) *In Review Exercises 15–16, follow the given directions.*

15. Let x represent a certain number. Write an algebraic expression denoting a number that is three more than twice x.

16. Let x and y represent two numbers. Write an algebraic expression to denote the difference obtained when twice x is subtracted from the product of 3 and y.

In Review Exercises 17–18, let x represent 6 and y represent 8. Evaluate each expression.

17. $\dfrac{x + y}{x - 4}$

18. $\dfrac{xy - 12}{4 + y}$

19. How many terms does the expression $3x + 4y + 9$ have?

20. What is the numerical coefficient of the term $7xy$?

21. What is the numerical coefficient of the term xy?

22. Find the sum of the numerical coefficients in the expression $2x^3 + 4x^2 + 3x$.

(1.4) *In Review Exercises 23–26, assume that x = 2 and y = 3. Find the value of each expression.*

23. y^4

24. x^y

25. $x^2 + xy^2$

26. $\dfrac{x^2 + y}{x^3 - 1}$

In Review Exercises 27–28, find the area of each figure.

27.

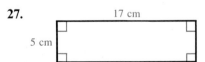

28.

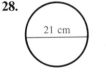

In Review Exercises 29–30, find the volume of each solid.

29.

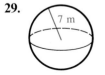

30.

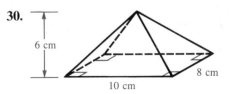

(1.5) *In Review Exercises 31–34, place one of the symbols* =, <, *or* > *in the box to make a true statement.*

31. $-5 \boxed{} 1$ **32.** $-7 \boxed{} -5$ **33.** $-1 \boxed{} -3$ **34.** $3 \boxed{} -3$

35. Division by a certain number is undefined. What is the number?

36. Draw a number line and graph all real numbers less than -2 or greater than 2.

37. Draw a number line and graph all real numbers between -4 and 3.

38. Evaluate $|6|$.

39. Evaluate $-|5|$.

40. Evaluate $|-8|$.

(1.6–1.7) *In Review Exercises 41–56, let* $x = 2$, $y = -3$, *and* $z = -1$. *Evaluate each expression.*

41. $y + z$ **42.** $x + y$ **43.** $x + (y + z)$ **44.** $x - y$

45. $x - (y - z)$ **46.** $(x - y) - z$ **47.** xy **48.** yz

49. $x(x + z)$ **50.** xyz **51.** $y^2 z + x$ **52.** $yz^3 + (xy)^2$

53. $\dfrac{xy}{z}$ **54.** $\dfrac{|xy|}{3z}$ **55.** $\dfrac{3y^2 - x^2 + 1}{y|z|}$ **56.** $\dfrac{2y^2 - xyz}{x^2|yz|}$

(1.8) *In Review Exercises 57–66, tell which property of real numbers justifies each statement. Assume that all variables represent real numbers.*

57. $x + y$ is a real number

58. $3 \cdot (4 \cdot 5) = (4 \cdot 5) \cdot 3$

59. $3 + (4 + 5) = (3 + 4) + 5$

60. $3(x + 2) = 3 \cdot x + 3 \cdot 2$

61. $a + x = x + a$

62. $3 \cdot (4 \cdot 5) = (3 \cdot 4) \cdot 5$

63. $3 + (x + 1) = (x + 1) + 3$

64. $x \cdot 1 = x$

65. $17 + (-17) = 0$

66. $x + 0 = x$

CHAPTER 1 TEST

1. List the prime numbers between 30 and 50.

2. Graph the composite numbers less than 10 on a number line.

3. Graph the real numbers from 5 to 15 on a number line.

4. Let x and y represent two real numbers. Write an algebraic expression to denote the quotient obtained when the product of the two numbers is divided by their sum.

5. Let x and y represent two real numbers. Write an algebraic expression to denote the difference obtained when the sum of x and y is subtracted from the product of 5 and y.

In Problems 6–7, which of the symbols =, <, *or* > *placed in the box will make a true statement?*

6. $3(4 - 2) \boxed{} -2(2 - 5)$

7. $1 + 4 \cdot 3 \boxed{} -2(-7)$

In Problems 8–12, simplify each expression.

_____ **8.** $\dfrac{26}{40}$

_____ **9.** $\dfrac{7}{8} \cdot \dfrac{24}{21}$

_____ **10.** $\dfrac{18}{35} \div \dfrac{9}{14}$

_____ **11.** $\dfrac{24}{16} + 3$

_____ **12.** $\dfrac{17 - 5}{36} - \dfrac{2(13 - 5)}{12}$

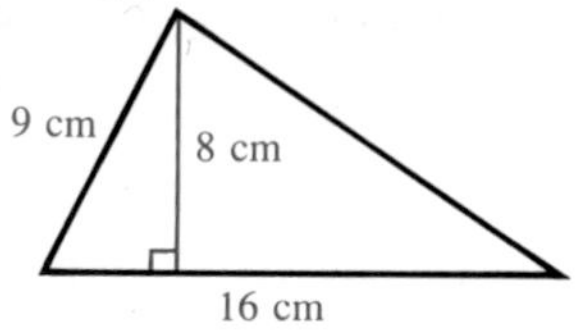

_____ **13.** What is the numerical coefficient of the term $3xy^2$?

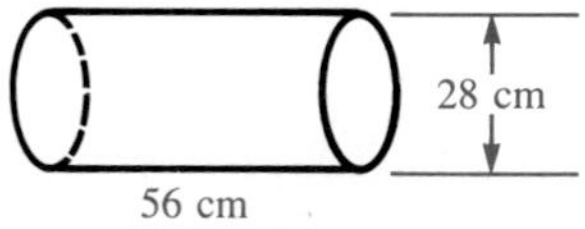

_____ **14.** Evaluate $|-16|$.

_____ **15.** Evaluate $-|23|$.

_____ **16.** Evaluate $-|7| + |-7|$.

_____ **17.** Find the area of the figure shown in Illustration 1.

_____ **18.** Find the volume of the solid shown in Illustration 2.

ILLUSTRATION 1

ILLUSTRATION 2

In Problems 19–24, let $x = -2$, $y = 3$, and $z = 4$. Evaluate each expression.

_____ **19.** $x + y + z$

_____ **20.** $\dfrac{z + 4y}{2x}$

_____ **21.** $|x^y - z|$

_____ **22.** $x^3 + y^2 + z$

_____ **23.** $|x - y| - |z|$

_____ **24.** $|x| - 3|y| - 4|z|$

_____ **25.** What is the identity element for addition?

_____ **26.** What is the multiplicative inverse of $\dfrac{1}{5}$?

In Problems 27–30, state which property of the real numbers justifies each statement.

_____ **27.** $(xy)z = z(xy)$

_____ **28.** $3(x + y) = 3x + 3y$

_____ **29.** $2 + x = x + 2$

_____ **30.** $7 \cdot \dfrac{1}{7} = 1$

_____ **31.** Jack lives 12 miles from work and 7 miles from the grocery store. If he made x round trips to work and y round trips to the store, how many miles did Jack drive?

_____ **32.** A baseball costs a dollars and a bat costs b dollars. How much will 12 baseballs and 8 bats cost?

2

Equations
and Inequalities

"How many?" "How far?" "How fast?" "How heavy?" To determine answers to such questions, we often make use of mathematical statements called **equations.** In this chapter we discuss this important idea.

2.1   THE ADDITION AND SUBTRACTION PROPERTIES OF EQUALITY

An **equation** is a mathematical statement indicating that two quantities are equal. Each of the following statements is an example of an equation.

$$x + 5 = 21$$
$$3(3x + 4) = 3 + x$$
$$3x^2 - 4x + 5 = 0$$

In the equation $x + 5 = 21$, the expression $x + 5$ is called the **left-hand side** of the equation and 21 is called the **right-hand side.** The letter x is called the **variable** (or the **unknown**) of the equation.

An equation can be true or false. For example, the equation $16 + 5 = 21$ is a **true equation,** whereas the equation $10 + 5 = 21$ is a **false equation.** An equation such as $2x - 5 = 11$ might be true or false, depending upon the value of x. If $x = 8$, the equation is true because

$$2(8) - 5 = 16 - 5 = 11$$

However, this equation is false for all other values of x. Any number that makes an equation true when substituted for its variable is said to **satisfy** the equation. All numbers that satisfy an equation are called its **solutions** or **roots.** Because 8 is the only number that satisfies the equation $2x - 5 = 11$, it is the only solution of the equation.

François Vieta (Viête) (1540–1603) By using letters in place of unknown numbers, Vieta simplified algebra and brought its notation closer to the notation that we use today. The one symbol he didn't use was the equal sign.

EXAMPLE 1   Verify that 16 is a solution of the equation $x + 5 = 21$.

Solution   Substitute 16 for the variable x in the equation $x + 5 = 21$, and verify that the left- and right-hand sides become equal numbers.

$$x + 5 = 21$$
$$16 + 5 \overset{?}{=} 21 \qquad \text{Replace } x \text{ with 16.}$$
$$21 = 21 \qquad \text{Simplify.}$$

Because the left- and right-hand sides both equal 21, the number 16 is a solution of the equation $x + 5 = 21$.   ∎

EXAMPLE 2   Is 6 a solution of the equation $3x - 5 = 2x$?

Solution   Substitute 6 for the variable x in the given equation and simplify.

$$3x - 5 = 2x$$
$$3 \cdot 6 - 5 \overset{?}{=} 2 \cdot 6 \qquad \text{Replace } x \text{ with 6.}$$
$$18 - 5 \overset{?}{=} 12 \qquad \text{Simplify.}$$
$$13 = 12 \qquad \text{Simplify.}$$

Because the left- and right-hand sides of the final equation are *not* equal, the number 6 is *not* a solution of the equation $3x - 5 = 2x$.   ∎

To *solve an equation* means to find all of its solutions. To do so, we will use certain properties of equality to get the variable by itself on one side of the equal sign and a single number on the other side. In this section we will consider the addition and subtraction properties of equality. When either of these properties is applied, the resulting equation will have the same solutions as the original equation.

The Addition Property of Equality.   If a, b, and c are real numbers, and if $a = b$, then
$$a + c = b + c$$

The **addition property of equality** can be stated in this way: *If the same quantity is added to equal quantities, the results will be equal quantities.*

EXAMPLE 3   Solve the equation $x - 5 = 2$.

Solution   To solve for x, isolate x on one side of the equal sign. Do this by using the addition property of equality and adding 5 to both sides of the equation.

$$x - 5 = 2$$
$$x - 5 + 5 = 2 + 5 \qquad \text{Add 5 to both sides of the equation.}$$
$$x = 7 \qquad \text{Simplify.}$$

The solution of the given equation is 7. Check this solution by substituting 7 for x in the original equation and simplifying.

$$x - 5 = 2$$
$$7 - 5 \stackrel{?}{=} 2 \qquad \text{Replace } x \text{ with 7.}$$
$$2 = 2 \qquad \text{Simplify.}$$

The solution checks.   ∎

The Subtraction Property of Equality.   If a, b, and c are real numbers, and if $a = b$, then
$$a - c = b - c$$

The **subtraction property of equality** can be stated in this way: *If the same quantity is subtracted from equal quantities, the results will be equal quantities.*

EXAMPLE 4   Solve the equation $x + 4 = 9$.

Solution   To isolate x on one side of the equal sign, undo the indicated addition of 4 by subtracting 4 from both sides of the equation.

$$x + 4 = 9$$
$$x + 4 - 4 = 9 - 4 \qquad \text{Subtract 4 from both sides.}$$
$$x = 5 \qquad \text{Simplify.}$$

The solution of the given equation is 5. Check this solution by substituting 5 for x in the original equation and simplifying.

$$x + 4 = 9$$
$$5 + 4 \overset{?}{=} 9 \qquad \text{Replace } x \text{ with 5.}$$
$$9 = 9 \qquad \text{Simplify.}$$

The solution checks. Note that instead of subtracting 4 from both sides, we could just as well have added -4 to both sides. ∎

EXAMPLE 5   A stockbroker lost 17 clients who went bankrupt last year. He now has 73 clients. How many did he have before?

Solution   Let c represent the original number of clients. Translate the words of the problem into an equation as follows:

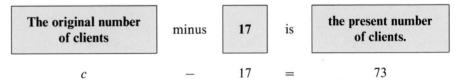

The original number of clients	minus	17	is	the present number of clients.
c	$-$	17	$=$	73

Solve the equation $c - 17 = 73$.

$$c - 17 = 73$$
$$c - 17 + 17 = 73 + 17 \qquad \text{Add 17 to both sides.}$$
$$c = 90 \qquad \text{Simplify.}$$

Originally the stockbroker had 90 clients. After losing 17, he has $90 - 17$, or 73 clients left. The solution checks. ∎

EXAMPLE 6   A 21-year-old girl is 5 years older than her brother. How old is her brother?

Solution   Let a represent the age of the brother. Translate the words of the problem into an equation as follows:

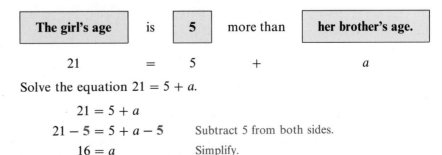

The girl's age	is	5	more than	her brother's age.
21	$=$	5	$+$	a

Solve the equation $21 = 5 + a$.

$$21 = 5 + a$$
$$21 - 5 = 5 + a - 5 \qquad \text{Subtract 5 from both sides.}$$
$$16 = a \qquad \text{Simplify.}$$

Because 16 and a represent the same number, the previous result can be written as

$$a = 16$$

The brother is 16 years old. The girl, at age 21, is 5 years older than her 16-year-old brother. The solution checks. ∎

EXAMPLE 7   If Mike had more money, he could buy a ten-speed bike that is on sale for $209. Mike has $192. How much more money does he need?

Solution   Let x represent the extra money that Mike needs to purchase the bike. Translate the words of the problem into an equation as follows:

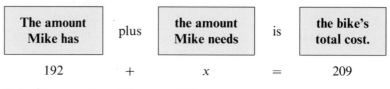

The amount Mike has	plus	the amount Mike needs	is	the bike's total cost.
192	+	x	=	209

Solve the equation $192 + x = 209$.

$$192 + x = 209$$
$$192 + x - \mathbf{192} = 209 - \mathbf{192} \qquad \text{Subtract 192 from both sides.}$$
$$x = 17 \qquad\qquad \text{Simplify.}$$

Mike needs $17 more. Then he will have $192 + $17, or $209, to buy the bike. The solution checks. ∎

EXERCISE 2.1

In Exercises 1–8, indicate whether each statement is an equation.

1. $x = 2$

2. $y - 3$

3. $7x < 8$

4. $7 + x = 2$

5. $x + y = 0$

6. $3 - 3y > 2$

7. $1 + 1 = 3$

8. $5 = a + 2$

In Exercises 9–26, indicate whether the indicated number is a solution of the given equation.

9. $x + 2 = 3$; 1

10. $x - 2 = 4$; 6

11. $a - 7 = 0$; 7

12. $x + 4 = 4$; 0

13. $8 - y = y$; -4

14. $10 - c = c$; 5

15. $2x + 32 = 0$; 16

16. $3x - 9 = 0$; -3

17. $z + 7 = z$; -7

18. $n - 9 = n$; 9

19. $2x = x$; 0

20. $3x = 2x$; 0

21. $3k + 5 = 5k - 1$; 3

22. $2s - 1 = s + 7$; 6

23. $\dfrac{x}{2} + 3 = 5$; 4

24. $x - \dfrac{x}{7} = 12$; 21

25. $\dfrac{x - 5}{6} + x = x + 1$; 11

26. $\dfrac{5 + x}{10} - x = \dfrac{1}{2}$; 0

In Exercises 27–46, use the addition property of equality to solve each equation. Check all solutions.

27. $x - 7 = 3$

28. $y - 3 = 7$

29. $a - 2 = -5$

30. $z - 3 = -9$

31. $2 = -5 + b$

32. $3 = -7 + t$

33. $x - 4 = 0$

34. $c - 3 = 0$

35. $y - 7 = 6$

36. $a - 2 = 4$

37. $17 = -15 + x$

38. $16 = -9 + b$

39. $312 = x - 428$

40. $x - 307 = -113$

41. $x - \dfrac{5}{2} = -\dfrac{5}{2}$

42. $y - \dfrac{3}{5} = 0$

43. $z - \dfrac{2}{3} = \dfrac{1}{2}$

44. $r - \dfrac{1}{5} = \dfrac{1}{3}$

45. $h - \dfrac{1}{3} = 2$

46. $w - 3 = -\dfrac{1}{3}$

In Exercises 47–66, use the subtraction property of equality to solve each equation. Check all solutions.

47. $x + 9 = 3$

48. $x + 3 = 9$

49. $y + 7 = 12$

50. $c + 11 = 22$

51. $t + 19 = 28$

52. $s + 34 = -45$

53. $23 + x = -13$

54. $34 + y = 34$

55. $3 = 4 + c$

56. $41 = 23 + x$

57. $-19 = r + 43$

58. $-92 = r + 37$

59. $\dfrac{5}{3} = \dfrac{5}{3} + x$

60. $\dfrac{29}{12} = \dfrac{29}{12} + x$

61. $d + \dfrac{2}{3} = \dfrac{3}{2}$

62. $s + \dfrac{2}{3} = \dfrac{1}{5}$

63. $w + \dfrac{1}{3} = 5$

64. $2 = x + \dfrac{3}{5}$

65. $r + 2 = -\dfrac{1}{2}$

66. $-\dfrac{1}{3} = s + \dfrac{2}{3}$

In Exercises 67–90, solve each equation. Check all solutions.

67. $3 + x = -7$

68. $4 + b = -8$

69. $y - 5 = 7$

70. $z - 9 = 3$

71. $-4 + a = 12$

72. $5 + x = 13$

73. $13 + x = 34$

74. $-23 + x = 19$

75. $-37 + z = 37$

76. $-43 + a = -43$

77. $-57 = b - 29$

78. $-93 = 67 + y$

79. $493 = -313 + x$

80. $347 = -4132 + y$

81. $137 = x - 323$

82. $-773 = -577 + y$

83. $\dfrac{13}{17} = x + \dfrac{13}{17}$

84. $\dfrac{23}{71} = \dfrac{23}{71} + x$

85. $\dfrac{5}{7} = x - \dfrac{3}{5}$

86. $\dfrac{21}{3} = x + \dfrac{7}{4}$

87. $x - \dfrac{13}{2} = \dfrac{21}{4}$

88. $\dfrac{2}{13} + x = 1$

89. $-\dfrac{5}{17} + y = -3$

90. $x + \dfrac{11}{17} = \dfrac{31}{34}$

In Exercises 91–106, translate the words into an equation and solve it. Check all solutions.

91. What number added to 6 gives 11?

92. What number decreased by 4 gives 8?

93. What number increased by 9 gives 27?

94. If a certain number is decreased by 41, the result is 31. Find the number.

95. Three of Jennifer's party invitations were lost in the mail, but 59 were delivered. How many invitations did Jennifer send?

96. Hank graduated from college 8 years ago. He is now 31 years old. How old was Hank at graduation?

97. The price of a condominium is $57,500 less than the cost of a house. The house costs $102,700. Find the price of the condominium.

98. Heidi ate 10 pancakes, 6 more than her sister Sarah. How many pancakes did Sarah eat?

99. Maria ate 4 hot dogs less than her brother Kurt. Kurt ate 7 hot dogs. How many did Maria eat?

100. Kim and Mary shared the costs of driving to Pittsburgh. Kim paid $12 more than Mary, who chipped in $68. How much did Kim pay?

101. A man needs $345 for a new set of golf clubs. How much more money does he need if he now has $317?

102. A student needs 42 cents more to buy a \$4.50 movie ticket. How much money does he have?

103. Heather paid \$48.50 for a sweater. Holly bought the same sweater on sale for \$9 less. How much did Holly pay for the sweater?

104. Last week, Molly made \$543.21 more than Max. How much money did Molly make if Max made \$123.45?

105. A woman paid \$29 less to have her car fixed at a muffler shop than she would have paid at a gas station. She would have paid \$219 at the gas station. How much did she pay to have her car fixed?

106. A man had to wait 20 minutes for a bus today. Three days ago, he had to wait 15 minutes longer than he did today because four buses passed by without stopping. How long did he wait three days ago?

REVIEW EXERCISES

In Review Exercises 1–8, perform the indicated operations.

1. $3[2 - (-3)]$

2. $3 - 3(-2)$

3. $5(-3) + 3^3$

4. $(2 - 4)^4$

5. $\dfrac{2^3 - 14}{3^2 - 3}$

6. $\dfrac{5^2 - 2 \cdot 9}{5 \cdot 7 - 6^2}$

7. $\dfrac{1}{2^2} + \dfrac{3}{6 - 2}$

8. $\dfrac{3 + 5}{3} - \dfrac{5}{7 - 4}$

In Review Exercises 9–12, let $x = 3$, $y = -5$, and $z = 1$. Evaluate each expression.

9. $x(y - z)$

10. $xy - z$

11. $y^x + z$

12. $(x^2 + y)^x$

2.2 THE DIVISION AND MULTIPLICATION PROPERTIES OF EQUALITY

To solve many equations, it is necessary to divide or multiply both sides of the equation by same nonzero number. When this is done, the resulting equation will have the same solutions as the original equation.

> **The Division Property of Equality.** Let a and b be real numbers and let c be any nonzero real number.
>
> If $a = b$, then $\dfrac{a}{c} = \dfrac{b}{c}$.

The **division property of equality** can be stated this way: *If equal quantities are divided by the same nonzero quantity, the results will be equal quantities.*

EXAMPLE 1 Solve the equation $3x = 6$.

Solution To isolate x on one side of the equal sign, undo the indicated multiplication by 3 by dividing both sides of the equation by 3.

$$3x = 6$$

$$\frac{3x}{3} = \frac{6}{3} \qquad \text{Divide both sides by 3.}$$

$$x = 2 \qquad \text{Simplify.}$$

The solution of the equation is 2. Check it as follows:

$$3x = 6$$

$$3 \cdot 2 \stackrel{?}{=} 6 \qquad \text{Replace } x \text{ with 2.}$$

$$6 = 6 \qquad \text{Simplify.}$$

Because the final equation is a true statement, the solution checks. ■

EXAMPLE 2 Solve the equation $-5x = 15$.

Solution To isolate x on one side of the equal sign, undo the indicated multiplication by -5 by dividing both sides of the equation by -5.

$$-5x = 15$$

$$\frac{-5x}{-5} = \frac{15}{-5} \qquad \text{Divide both sides by } -5.$$

$$x = -3 \qquad \text{Simplify.}$$

The solution of the equation is -3. Check it as follows:

$$-5x = 15$$

$$-5(-3) \stackrel{?}{=} 15 \qquad \text{Replace } x \text{ with } -3.$$

$$15 = 15 \qquad \text{Simplify.}$$

The solution checks. ■

We can also multiply both sides of an equation by the same nonzero number. The equation that results has the same solutions as the original equation.

The Multiplication Property of Equality.   Let a, b, and c be real numbers, with $c \neq 0$.

If $a = b$, then $ca = cb$.

The **multiplication property of equality** can be stated this way: *If equal quantities are multiplied by the same nonzero quantity, the results will be equal quantities.*

EXAMPLE 3 Solve the equation $\dfrac{x}{5} = 7$.

Solution To find x, undo the indicated divison by 5 by multiplying both sides of the equation by 5.

$$\frac{x}{5} = 7$$

$$5 \cdot \frac{x}{5} = 5 \cdot 7 \qquad \text{Multiply both sides by 5.}$$

$$x = 35 \qquad \text{Simplify.}$$

Check: $\dfrac{x}{5} = 7$

$\dfrac{35}{5} \overset{?}{=} 7$ Replace x with 35.

$7 = 7$ Simplify.

The solution checks.   ■

EXAMPLE 4   Solve the equation $-12x + 5 = 1$.

Solution   To solve this equation, we must use both the addition and division properties of equality.

$$-12x + 5 = 1$$
$$-12x + 5 + (-5) = 1 + (-5) \qquad \text{Add } -5 \text{ to both sides.}$$
$$-12x = -4 \qquad \text{Simplify.}$$
$$\dfrac{-12x}{-12} = \dfrac{-4}{-12} \qquad \text{Divide both sides by } -12.$$
$$x = \dfrac{1}{3} \qquad \text{Simplify.}$$

Check: $-12x + 5 = 1$

$-12\left(\dfrac{1}{3}\right) + 5 \overset{?}{=} 1$ Replace x with $\frac{1}{3}$.

$-4 + 5 \overset{?}{=} 1$ Simplify.

$1 = 1$ Simplify.

The solution checks.   ■

EXAMPLE 5   Solve the equation $5(x - 3) = -10$.

Solution   Use the distributive property to remove the parentheses and simplify.

$$5(x - 3) = -10$$
$$5x - 5 \cdot 3 = -10 \qquad \text{Remove parentheses.}$$
$$5x - 15 = -10 \qquad \text{Simplify.}$$
$$5x - 15 + 15 = -10 + 15 \qquad \text{Add 15 to both sides.}$$
$$5x = 5 \qquad \text{Simplify.}$$
$$\dfrac{5x}{5} = \dfrac{5}{5} \qquad \text{Divide both sides by 5.}$$
$$x = 1 \qquad \text{Simplify.}$$

Check: $5(x - 3) = -10$

$5(1 - 3) \overset{?}{=} -10$ Replace x with 1.

$5(-2) \overset{?}{=} -10$ Simplify.

$-10 = -10$ Simplify.

The solution checks.   ■

EXAMPLE 6 Solve the equation $\dfrac{3(x-2)}{4} = -6$.

Solution

$$\frac{3(x-2)}{4} = -6$$

$$4 \cdot \frac{3(x-2)}{4} = 4(-6) \qquad \text{Multiply both sides by 4.}$$

$$3(x-2) = -24 \qquad \text{Simplify.}$$

$$3x - 6 = -24 \qquad \text{Remove parentheses.}$$

$$3x - 6 + 6 = -24 + 6 \qquad \text{Add 6 to both sides.}$$

$$3x = -18 \qquad \text{Simplify.}$$

$$\frac{3x}{3} = \frac{-18}{3} \qquad \text{Divide both sides by 3.}$$

$$x = -6 \qquad \text{Simplify.}$$

Check: $\dfrac{3(x-2)}{4} = -6$

$$\frac{3(-6-2)}{4} \stackrel{?}{=} -6 \qquad \text{Replace } x \text{ with } -6.$$

$$\frac{3(-8)}{4} \stackrel{?}{=} -6 \qquad \text{Simplify.}$$

$$\frac{-24}{4} \stackrel{?}{=} -6 \qquad \text{Simplify.}$$

$$-6 = -6 \qquad \text{Simplify.}$$

The solution checks. ∎

EXAMPLE 7 Solve the equation $-\dfrac{2}{3}(4x-6) = 2$.

Solution

$$-\frac{2}{3}(4x-6) = 2$$

$$3\left[-\frac{2}{3}(4x-6)\right] = 3(2) \qquad \text{Multiply both sides by 3.}$$

$$-2(4x-6) = 6 \qquad \text{Simplify.}$$

$$-8x + 12 = 6 \qquad \text{Remove parentheses.}$$

$$-8x + 12 + (-12) = 6 + (-12) \qquad \text{Add } -12 \text{ to both sides.}$$

$$-8x = -6 \qquad \text{Simplify.}$$

$$\frac{-8x}{-8} = \frac{-6}{-8} \qquad \text{Divide both sides by } -8.$$

$$x = \frac{3}{4} \qquad \text{Simplify.}$$

$$Check: \quad -\frac{2}{3}(4x-6) = 2$$

$$-\frac{2}{3}\left[4\left(\frac{3}{4}\right)-6\right] \overset{?}{=} 2 \qquad \text{Replace } x \text{ with } \tfrac{3}{4}.$$

$$-\frac{2}{3}(3-6) \overset{?}{=} 2$$

$$-\frac{2}{3}(-3) \overset{?}{=} 2$$

$$2 = 2$$

The solution checks.   ■

EXAMPLE 8   If one-third of a number is decreased by 5, the result is 2. Find the number.

Solution   Let x represent the unknown number. Translate the statement into an equation, solve the equation, and check the solution.

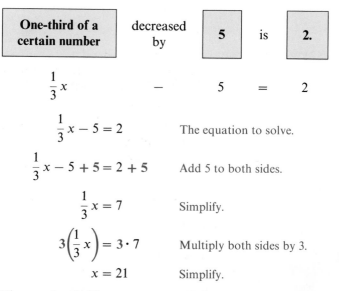

$$\frac{1}{3}x - 5 = 2 \qquad \text{The equation to solve.}$$

$$\frac{1}{3}x - 5 + 5 = 2 + 5 \qquad \text{Add 5 to both sides.}$$

$$\frac{1}{3}x = 7 \qquad \text{Simplify.}$$

$$3\left(\frac{1}{3}x\right) = 3 \cdot 7 \qquad \text{Multiply both sides by 3.}$$

$$x = 21 \qquad \text{Simplify.}$$

The number is 21.

Check: One-third of 21 is 7. If 7 is decreased by 5, the result is 2. The answer checks.   ■

EXAMPLE 9   Bill's sales commissions this year were $2500 less than three times his last year's commissions. This year, Bill earned $38,000 in commissions. How much did he earn in commissions last year?

Solution   Let c represent Bill's earnings last year. Then translate the words of the problem into an equation. His earnings this year (in dollars) can be expressed in two ways: as 2500 less than $3c$, and as 38,000.

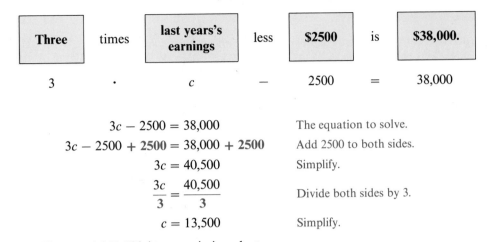

$$3 \cdot c - 2500 = 38{,}000$$

$3c - 2500 = 38{,}000$	The equation to solve.
$3c - 2500 + 2500 = 38{,}000 + 2500$	Add 2500 to both sides.
$3c = 40{,}500$	Simplify.
$\dfrac{3c}{3} = \dfrac{40{,}500}{3}$	Divide both sides by 3.
$c = 13{,}500$	Simplify.

Bill earned \$13,500 in commissions last year.

Check: If Bill's commissions of \$38,000 had been \$2500 greater, he would have had commissions of \$38,000 + \$2500, or \$40,500. Three times last year's commissions is 3(\$13,500), or \$40,500. The solution checks. ∎

EXAMPLE 10 The manager of a fast-food restaurant hires a woman to distribute advertising circulars door to door. The payment schedule is \$5 a day plus a nickel for every advertisement distributed. After one day's work, she receives \$42.50. How many circulars did she distribute?

Solution Let a represent the number of circulars that the woman distributed. Then translate the words of the problem into an equation. The total payment can be expressed in two ways: as \$5 more than the nickel-apiece cost of distributing the circulars, and as \$42.50.

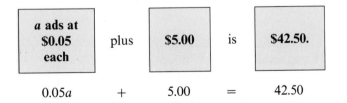

$$0.05a + 5.00 = 42.50$$

$0.05a + 5.00 = 42.50$	The equation to solve.
$0.05a + 5.00 + (-5.00) = 42.50 + (-5.00)$	Add -5.00 to both sides.
$0.05a = 37.50$	Simplify.
$\dfrac{0.05a}{0.05} = \dfrac{37.50}{0.05}$	Divide both sides by 0.05.
$a = 750$	Simplify.

The woman distributed 750 advertisements. Check this result. ∎

EXERCISE 2.2

In Exercises 1–64, solve each equation and check each solution.

1. $3x = 3$
2. $5x = 5$
3. $6x = 15$
4. $4x = 18$

5. $7y = -21$
6. $9y = -27$
7. $-4y = -20$
8. $-8y = -32$

9. $-32z = 64$
10. $-17z = -51$
11. $9z + 2 = 11$
12. $3z + 5 = 11$

13. $3x - 8 = 1$
14. $7x - 19 = 2$
15. $11x + 17 = -5$
16. $13x - 29 = -3$

17. $43t + 72 = 158$
18. $96t + 23 = -265$
19. $-47 - 21s = 58$
20. $-151 + 13s = -229$

21. $5(x - 3) = 0$
22. $3(x + 7) = 6$
23. $-9(3 + y) = -9$
24. $4(y + 3) = 40$

25. $2(x + 7) = 10$
26. $3(x - 9) = 15$
27. $-4(2 + a) = -8$
28. $-7(2 + b) = 14$

29. $\frac{1}{3}b + 5 = 2$
30. $\frac{1}{5}a - 3 = -4$
31. $\frac{s}{11} + 9 = 6$
32. $\frac{s}{12} + 2 = 4$

33. $\frac{b + 5}{3} = 11$
34. $\frac{2 + a}{11} = 3$
35. $\frac{r + 7}{3} = 4$
36. $\frac{t - 2}{7} = -3$

37. $\frac{3(u - 2)}{5} = 3$
38. $\frac{5(v - 7)}{3} = -5$
39. $\frac{7(x - 4)}{4} = -21$
40. $\frac{9(3 + y)}{5} = -27$

41. $\frac{3x - 12}{2} = 9$
42. $\frac{5x + 10}{7} = 0$
43. $\frac{1}{2}(a + 8) = 1$
44. $\frac{1}{3}(2x - 3) = \frac{-17}{3}$

45. $\frac{2}{3}(5x - 3) = 38$
46. $\frac{3}{2}(3x + 2) = 3$
47. $\frac{3}{4}(7k - 5) = 12$
48. $\frac{5}{3}(6 + 5b) = 10$

49. $\frac{1}{3}(4n - 3) = 5$
50. $\frac{7}{11}(4x + 1) = -7$
51. $\frac{3}{8}(5x + 2) = -3$
52. $\frac{2}{7}(3m - 7) = -2$

53. $\frac{1}{4}\left(\frac{q}{2} + 1\right) = 1$
54. $\frac{1}{2}\left(\frac{a}{3} - 4\right) = -3$
55. $\frac{2r - 10}{6} + 7 = 17$
56. $\frac{5a + 10}{10} + 7 = 0$

57. $\frac{3z + 2}{17} = 0$
58. $\frac{10t - 4}{2} = 1$
59. $\frac{17k - 28}{21} + \frac{4}{3} = 0$
60. $\frac{5a - 2}{3} = \frac{1}{6}$

61. $-\frac{x}{3} - \frac{1}{2} = -\frac{5}{2}$
62. $2\left(\frac{17 - a}{3}\right) = 4$
63. $\frac{1}{3}\left(\frac{9 - w}{5}\right) = \frac{2}{5}$
64. $\frac{3t - 5}{5} + \frac{1}{2} = -\frac{19}{2}$

In Exercises 65–90, translate the words into an equation and solve it. Check all solutions.

65. Six less than 3 times a certain number is 9. Find the number.

66. Four times a certain number is increased by 7. The result is 43. Find the number.

67. If 2 is added to twice a certain number, the result is 12. Find the number.

68. If 5 is subtracted from a certain number, the result is 11. Find the number.

69. Four more than 3 times a certain number is 25. Find the number.

70. If a certain number is decreased by 7, and that result is doubled, the number 6 is obtained. Find the original number.

71. When 4 more than a certain number is divided by 3, the result is 4. Find the number.

72. If 12 is decreased by a certain number, and that result is multiplied by 5, the result is 15. Find the number.

73. If 5 is added to a certain number, and that result is divided by 6, the value obtained is 2. Find the number.

74. If a certain number is tripled, increased by 7, and the result is then divided by 2, the number 5 is obtained. Find the original number.

75. If a certain number is tripled, decreased by 7, and the result is then divided by 2, the number 1 is obtained. Find the original number.

76. Seventeen more than one-third of a certain number is 18. Find the number.

77. The number 9 decreased by one-half of a certain number is 11. Find the number.

78. One-half of the sum of 3 and one-half of a certain number is 4. Find the number.

79. Two-thirds of the movie audience left the theater in disgust. If 78 angry patrons walked out, how many were there originally?

80. Mandy was disturbed to learn that only four-sevenths of her class passed the first exam. If 16 students passed the exam, how large is the class?

81. Susan's sales in February were $1000 less than three times her sales in January. If her February sales were $5000, how much did she sell in January?

82. Jennifer is 22, which is twice as old as her brother will be next year. How old is her brother now?

83. When Maggie is 51, she will be 3 times as old as she was last year. How old is she now?

84. If a girl had twice as much money as she has now, she would have enough money to buy a $3.75 movie ticket and an 85-cent box of popcorn. How much money does she have now?

85. In 5 years, Roger will be 3 times as old as Jack is now. In 5 years, Jack will be 27. How old is Roger now?

86. Jane won't move to that bigger apartment because its $400 monthly rent is $100 less than twice what she is now paying. What is her current rent?

87. A mechanic charged $20 an hour to repair the water pump on a car, plus $95 for parts. The total bill was $155. How many hours did the repair take?

88. Vacation boarding for a dog at the kennel is $16 plus $12 a day. The stay cost the owners $100. How many days were her owners gone?

89. Wanda's monthly water bill is $3, plus 20 cents for every 100 gallons of water used. Last month, Wanda paid $5.40 for water. How much water did she use?

90. A music group charges $1500 for each performance, plus one-fifth of the total ticket sales. After the concert, the promoter gave the group's agent a check for $2980. How much money did the ticket sales raise?

REVIEW EXERCISES

In Review Exercises 1–6, refer to the formulas given in Section 1.4.

1. Find the perimeter of a rectangle with sides measuring 8.5 cm and 16.5 cm.

2. Find the area of a rectangle with sides measuring 2.3 in. and 3.7 in.

3. Find the area of a triangle with a base of 9 cm and a height of 6 cm.

4. Find the area of a trapezoid with a height of 8.5 in. and bases measuring 6.7 in. and 12.2 in.

5. Find the volume of a rectangular solid with dimensions of 8.2 cm by 7.6 cm by 10.2 cm.

6. Find the volume of a sphere with a radius of 21 ft. (Use $\pi = \frac{22}{7}$.)

In Review Exercises 7–12, indicate the property of real numbers or the property of equality that justifies each statement.

7. $3 + 31$ is a real number

8. $3(x + y) = 3x + 3y$

9. $a + (3 + b) = (3 + b) + a$

10. $a + (3 + b) = (a + 3) + b$

11. $3xy + (-3xy) = 0$

12. If $a + b = c$, then $a + b - x = c - x$.

2.3   SIMPLIFYING EXPRESSIONS TO SOLVE EQUATIONS

Recall that a *term* is either a number or the product of numbers and variables. Some examples of terms are $7x$, $-3xy$, y^2, and 8. The **numerical coefficient** of the term $7x$ is 7, the numerical coefficient of the term $-3xy$ is -3, and the numerical coefficient of the term y^2 is the understood factor of 1. The number 8 is the numerical coefficient of the term 8.

> **DEFINITION.**   **Like terms,** or **similar terms,** are terms with exactly the same variables and exponents.

The terms $3x$ and $5x$ are **like terms,** as are $9x^2$ and $-3x^2$. The terms $4xy$ and $3x^2$ are **unlike terms** because they have different variables. The terms $4x$ and $5x^2$ are unlike terms because the variables have different exponents.

The distributive property can be used to combine terms of algebraic expressions that contain sums or differences of like terms. For example, the terms of the algebraic expression $3x + 5x$ can be combined as follows:

$$3x + 5x = (3 + 5)x = 8x$$

Likewise, the distributive property can be used to combine the like terms of the expression $9xy^2 - 11xy^2$:

$$9xy^2 - 11xy^2 = (9 - 11)xy^2 = -2xy^2$$

These examples suggest the following rule:

> To combine like terms, add their numerical coefficients and keep the same variables and exponents.

If the terms of an expression are not like terms, they cannot be combined. Because the terms of the expression $9xy^2 - 11x^2y$, for example, have variables with different exponents, they are unlike terms and cannot be combined.

EXAMPLE 1   Simplify the expression $3(x + 2) + 2(x - 8)$.

Solution

$$
\begin{aligned}
3(x + 2) + 2(x - 8) &= 3x + 3 \cdot 2 + 2x - 2 \cdot 8 && \text{Remove parentheses.}\\
&= 3x + 6 + 2x - 16 && \text{Simplify.}\\
&= 3x + 2x + 6 - 16 && \text{Use the associative and}\\
& && \text{commutative properties of}\\
& && \text{addition to rearrange terms.}\\
&= 5x - 10 && \text{Combine terms.}\quad\blacksquare
\end{aligned}
$$

EXAMPLE 2   Simplify the expression $3(x - 3) - 5(x + 4)$.

Solution

$$3(x - 3) - 5(x + 4) = 3(x - 3) + (-5)(x + 4)$$ $a - b = a + (-b)$.

$$= 3x - 3 \cdot 3 + (-5)x + (-5)4$$ Remove parentheses.

$$= 3x - 9 + (-5x) + (-20)$$ Multiply.

$$= -2x - 29$$ Combine terms. ■

Solving More Equations

We must combine like terms to solve the equations in the next examples.

EXAMPLE 3   Solve the equation $3(x + 2) - 5x = 0$.

Solution

$$3(x + 2) - 5x = 0$$

$$3x + 3 \cdot 2 - 5x = 0$$ Remove parentheses.

$$3x - 5x + 6 = 0$$ Rearrange terms and simplify.

$$-2x + 6 = 0$$ Combine terms.

$$-2x + 6 + (-6) = 0 + (-6)$$ Add -6 to both sides.

$$-2x = -6$$ Combine terms.

$$\frac{-2x}{-2} = \frac{-6}{-2}$$ Divide both sides by -2.

$$x = 3$$ Simplify.

Check: $3(x + 2) - 5x = 0$

$$3(3 + 2) - 5 \cdot 3 \overset{?}{=} 0$$ Replace x with 3.

$$3 \cdot 5 - 5 \cdot 3 \overset{?}{=} 0$$

$$15 - 15 \overset{?}{=} 0$$

$$0 = 0$$

The solution checks.   ■

EXAMPLE 4   Solve the equation $3(x - 5) = 4(x + 9)$.

Solution

$$3(x - 5) = 4(x + 9)$$

$$3x - 15 = 4x + 36$$ Remove parentheses.

$$3x - 15 + (-3x) = 4x + 36 + (-3x)$$ Add $-3x$ to both sides.

$$3x - 3x - 15 = 4x - 3x + 36$$ Rearrange terms.

$$-15 = x + 36$$ Combine terms.

$$-15 + (-36) = x + 36 + (-36)$$ Add -36 to both sides.

$$-51 = x$$ Combine terms.

or

$$x = -51$$

$$Check: \quad 3(x - 5) = 4(x + 9)$$

$$3(-51 - 5) \stackrel{?}{=} 4(-51 + 9) \qquad \text{Replace } x \text{ with } -51.$$

$$3(-56) \stackrel{?}{=} 4(-42)$$

$$-168 = -168$$

The solution checks.   ■

EXAMPLE 5   Solve the equation $\dfrac{3x + 11}{5} = x + 3$.

Solution   First clear the equation of fractions by multiplying both sides of the equation by 5. When multiplying the right-hand side of the equation by 5, be sure to multiply the *entire* right-hand side by 5.

$$\frac{3x + 11}{5} = x + 3$$

$$5 \cdot \frac{3x + 11}{5} = 5(x + 3) \qquad\qquad \text{Multiply both sides by 5.}$$

$$3x + 11 = 5(x + 3) \qquad\qquad \text{Simplify the left-hand side.}$$

$$3x + 11 = 5x + 15 \qquad\qquad \text{Remove parentheses.}$$

$$3x + 11 + (-11) = 5x + 15 + (-11) \qquad \text{Add } -11 \text{ to both sides.}$$

$$3x = 5x + 4 \qquad\qquad \text{Combine terms.}$$

$$3x + (-5x) = 5x + 4 + (-5x) \qquad \text{Add } -5x \text{ to both sides.}$$

$$-2x = 4 \qquad\qquad \text{Combine terms.}$$

$$\frac{-2x}{-2} = \frac{4}{-2} \qquad\qquad \text{Divide both sides by } -2.$$

$$x = -2 \qquad\qquad \text{Simplify.}$$

$$Check: \quad \frac{3x + 11}{5} = x + 3$$

$$\frac{3(-2) + 11}{5} \stackrel{?}{=} (-2) + 3 \qquad \text{Replace } x \text{ with } -2.$$

$$\frac{-6 + 11}{5} \stackrel{?}{=} 1 \qquad\qquad \text{Simplify.}$$

$$\frac{5}{5} \stackrel{?}{=} 1$$

$$1 = 1$$

The solution checks.   ■

EXAMPLE 6   Solve the equation $0.2x + 0.4(50 - x) = 19$.

Solution   Because $0.2 = \frac{2}{10}$ and $0.4 = \frac{4}{10}$, this equation contains fractions. To clear the equation of these fractions, multiply both sides by 10. Then solve the resulting equation.

$$0.2x + 0.4(50 - x) = 19$$

$$10[0.2x + 0.4(50 - x)] = 10(19) \qquad \text{Multiply both sides by 10.}$$

$$10(0.2x) + 10[0.4(50 - x)] = 10(19) \qquad \text{Use the distributive property on the left-hand side.}$$

$$2x + 4(50 - x) = 190 \qquad \text{Multiply.}$$

$$2x + 200 - 4x = 190 \qquad \text{Remove parentheses.}$$

$$-2x + 200 = 190 \qquad \text{Combine terms.}$$

$$-2x = -10 \qquad \text{Add } -200 \text{ to both sides.}$$

$$x = 5 \qquad \text{Divide both sides by } -2.$$

Verify that the solution checks. ∎

EXAMPLE 7 Solve the equation $x(x - 5) = x^2 + 15$.

Solution

$$x(x - 5) = x^2 + 15$$

$$x^2 - 5x = x^2 + 15 \qquad \text{Remove parentheses.}$$

$$x^2 - 5x + (-x^2) = x^2 + 15 + (-x^2) \qquad \text{Add } -x^2 \text{ to both sides.}$$

$$-5x = 15 \qquad \text{Combine terms.}$$

$$\frac{-5x}{-5} = \frac{15}{-5} \qquad \text{Divide both sides by } -5.$$

$$x = -3 \qquad \text{Simplify.}$$

Verify that the solution checks. ∎

Identities and Impossible Equations

An equation that is true for all values of its variable is called an **identity.** For example, the equation

$$x + x = 2x$$

is an identity because it is true for all values of x. For example, if $x = 8$, then

$$8 + 8 = 2 \cdot 8$$

$$16 = 16$$

Because no number can equal a number that is 1 larger than itself, the equation $x = x + 1$ is true for no values of x. Such equations are called **impossible equations** or **contradictions.**

EXAMPLE 8 Solve the equation $3(x + 8) + 5x = 2(12 + 4x)$.

Solution

$$3(x + 8) + 5x = 2(12 + 4x)$$

$$3x + 24 + 5x = 24 + 8x \qquad \text{Remove parentheses.}$$

$$8x + 24 = 24 + 8x \qquad \text{Combine terms.}$$

$$8x + 24 + (-8x) = 24 + 8x + (-8x) \qquad \text{Add } -8x \text{ to both sides.}$$

$$24 = 24 \qquad \text{Combine terms.}$$

The result $24 = 24$ is a true equation, and it is true for every number x. Thus, every value of x is a solution of the original equation. This equation is an identity. ∎

EXAMPLE 9   Solve the equation $3(x + 7) - x = 2(x + 10)$.

Solution

$$3(x + 7) - x = 2(x + 10)$$

$3x + 21 - x = 2x + 20$	Remove parentheses.
$2x + 21 = 2x + 20$	Combine terms.
$2x + 21 + (-2x) = 2x + 20 + (-2x)$	Add $-2x$ to both sides.
$21 = 20$	Combine terms.

The result $21 = 20$ is a false equation; no value of x can possibly make 21 equal to 20. Thus, the original equation has no solution. It is an impossible equation. ∎

EXERCISE 2.3

In Exercises 1–30, simplify each expression, if possible.

1. $3x + 17x$ **2.** $12y - 15y$ **3.** $8x^2 - 5x^2$ **4.** $17x^2 + 3x^2$

5. $9x + 3y$ **6.** $5x + 5y$ **7.** $3(x + 2) + 4x$ **8.** $9(y - 3) + 2y$

9. $5(z - 3) + 2z$ **10.** $4(y + 9) - 6y$ **11.** $12(x + 11) - 11$ **12.** $-3(3 + z) + 2z$

13. $-23(x^2 - 2) + x^2$ **14.** $5(17 + x) - 27$ **15.** $3x^3 - 2(x^3 + 5)$ **16.** $-7z^2 - 2z^3 + z$

17. $8x(x + 3) - 3x^2$ **18.** $2x + x(x + 3)$ **19.** $8(y + 7) - 2(y - 3)$ **20.** $9(z + 2) + 5(3 - z)$

21. $2x + 4(y - x) + 3y$ **22.** $3y - 6(y + z) + y$ **23.** $9(x + y) - 9(x - y)$ **24.** $2(x + z) + 3(x - z)$

25. $(x + 2) + (x - y)$ **26.** $3z + 2(y - z) + y$

27. $2\left(4x + \dfrac{9}{2}\right) - 3\left(x + \dfrac{2}{3}\right)$ **28.** $7\left(3x - \dfrac{2}{7}\right) - 5\left(2x - \dfrac{3}{5}\right) + x$

29. $2(7x + 2^2) - 3(3x - 3^2)$ **30.** $2^2\left(2x - \dfrac{3}{2}\right) + 3^3\left(\dfrac{x}{3} - \dfrac{2}{3}\right)$

In Exercises 31–88, solve each equation, if possible. Check all solutions.

31. $3x + 2 = 2x$ **32.** $5x + 7 = 4x$ **33.** $5x + 3 = 4x$ **34.** $4x + 3 = 5x$

35. $9y - 3 = 6y$ **36.** $8y + 4 = 4y$ **37.** $8y - 7 = y$ **38.** $9y - 8 = y$

39. $3(a + 2) = 4a$ **40.** $4(a - 5) = 3a$ **41.** $5(b + 7) = 6b$ **42.** $8(b + 2) = 9b$

43. $-9x + 3 = 8x + 20$ **44.** $-11x + 3 = 4x - 27$ **45.** $-7a + 5 = 11a - 22$ **46.** $-6a + 7 = 9a - 8$

47. $4x + 8 + 3(x - 2) = -12$ **48.** $7x + 2 + 4(x - 3) = 12$

49. $2 + 3(x - 5) = 4(x - 1)$ **50.** $2 - (5x + 7) = 7 + x$

51. $10x + 3(2 - x) = 5(x + 2) - 4$ **52.** $11x + 6(3 - x) = 3$

53. $3(a + 2) = 2(a - 7)$

54. $9(t - 1) = 6(t + 2) - t$

55. $9(x + 11) + 5(13 - x) = 0$

56. $3(x + 15) + 4(11 - x) = 0$

57. $5(x + 2) = 3x - 2$

58. $3(x + 7) = 2(2x + 4) + x - 1$

59. $2 + 3(x - 5) = 4(x - 1)$

60. $2 - (4x + 7) = 3 + 2(x + 2)$

61. $13(b - 9) - 2 = 2(3b - 4) + b$

62. $21(b - 1) + 3 = 4(5b - 6)$

63. $\dfrac{3(t - 7)}{2} = t - 6$

64. $\dfrac{2(t + 9)}{3} = t - 8$

65. $\dfrac{5(2 - s)}{3} = s + 6$

66. $\dfrac{8(5 - s)}{5} = -2s$

67. $\dfrac{4(2x - 10)}{3} = 2(x - 4)$

68. $\dfrac{11(x - 12)}{2} = 9 - 2x$

69. $\dfrac{5(2x - 23) - (6 - x)}{9} = -5\left(x + \dfrac{1}{5}\right)$

70. $\dfrac{2(4x + 40) - (x - 4)}{5} = 7(x - 2)$

71. $3.1(x - 2) = 1.3x + 2.8$

72. $0.6x - 0.8 = 0.8(2x - 1) - 0.7$

73. $2.7(y + 1) = 0.3(3y + 33)$

74. $1.5(5 - y) = 3y + 12$

75. $19.1x - 4(x + 0.3) = -46.5$

76. $18.6x + 7.2 = 1.5(48 - 2x)$

77. $14.3(x + 2) + 13.7(x - 3) = 15.5$

78. $1.25(x - 1) = 0.5(3x - 1) - 1$

79. $x(2x - 3) = 2x^2 + 15$

80. $2x(3x + 4) = 6x^2 + 32$

81. $a(a + 2) = a(a - 4) + 16$

82. $b(b - 1) + 18 = b(b + 5)$

83. $\dfrac{x(2x - 8)}{2} = x(x + 2)$

84. $\dfrac{3x(2x + 1)}{2} = 3x^2 - 6$

85. $2y^2 - 9 = y(y + 3) + y^2$

86. $y(3y - 4) - y^2 = 2y(y + 3) + 20$

87. $\dfrac{x(4x + 3) + 2(x^2 + 9)}{2} = 3x(x + 2)$

88. $x(x + 2) + 24 = \dfrac{x(x + 2) + x(2x - 8)}{3}$

In Exercises 89–100, solve each equation. If it is an identity or an impossible equation, so indicate.

89. $8x + 3(2 - x) = 5(x + 2) - 4$

90. $5(x + 2) = 5x - 2$

91. $s(s + 2) = s^2 + 2s + 1$

92. $21(b - 1) + 3 = 3(7b - 6)$

93. $\dfrac{2(t - 1)}{6} - 2 = \dfrac{t + 2}{6}$

94. $\dfrac{2(2r - 1)}{6} + 5 = \dfrac{3(r + 7)}{6}$

95. $2(3z + 4) = 2(3z - 2) + 13$

96. $x + 7 = \dfrac{2x + 6}{2} + 4$

97. $2(y - 3) - \dfrac{y}{2} = \dfrac{3}{2}(y - 4)$

98. $\dfrac{20 - a}{2} = \dfrac{3}{2}(a + 4)$

99. $\dfrac{3x + 14}{2} = x - 2 + \dfrac{x + 18}{2}$

100. $\dfrac{5(x + 3)}{3} - x = \dfrac{2(x + 8)}{3}$

REVIEW EXERCISES

In Review Exercises 1–6, let $x = -3$, $y = -5$, and $z = 0$. Evaluate each expression.

1. $x^2 z(y^3 - z)$

2. $z - x^2 y$

3. $y^3 - x$

4. $x^3(y - z^{23})$

5. $\dfrac{x - y^2}{2y - 1 + x}$

6. $\dfrac{2y + 1}{x} - x$

In Review Exercises 7–10, solve each equation. Check all solutions.

7. $3x + 5 = 11$ **8.** $7 - 8y = -9$ **9.** $\dfrac{3}{7}(w - 2) - 2 = 1$ **10.** $\dfrac{2}{13}(21 - z) = 4$

In Review Exercises 11–12, translate the words into an equation and solve it. Then check your solution.

11. What number added to 37 gives 20?

12. Mark needs $27 more to buy a $140 bicycle. How much money does Mark have?

2.4   LITERAL EQUATIONS

Equations that contain several variables are called **literal equations.** Often these equations are called **formulas**—such as $A = lw$, the formula for finding the area of a rectangle. Suppose we wish to use this formula to find the lengths of several rectangles whose areas and widths are known. It would be tedious to substitute values for A and w into the formula, and then repeatedly solve the formula for l. It would be better to solve the formula $A = lw$ for the variable l first, and then substitute values for A and w and compute l directly.

To *solve an equation for a variable* means to isolate that variable on one side of the equation, with all other quantities on the opposite side.

EXAMPLE 1   Solve $A = lw$ for the variable l.

Solution   To isolate l on the left-hand side of the equation, undo the multiplication by w. To do this, divide both sides of the equation by w and simplify.

$$A = lw$$

$$\frac{A}{w} = \frac{lw}{w} \qquad \text{Divide both sides by } w.$$

$$\frac{A}{w} = l \qquad \text{Simplify.}$$

or

$$l = \frac{A}{w}$$

■

EXAMPLE 2   The formula $A = \frac{1}{2}bh$ gives the area of a triangle with base b and height h. Solve this equation for b.

Solution

$$A = \frac{1}{2}bh$$

$$2 \cdot A = 2 \cdot \frac{1}{2}bh \qquad \text{Multiply both sides by 2.}$$

$$2A = bh \qquad \text{Simplify.}$$

$$\frac{2A}{h} = \frac{bh}{h} \qquad \text{Divide both sides by } h.$$

$$\frac{2A}{h} = b \qquad \text{Simplify.}$$

or

$$b = \frac{2A}{h}$$

If the area A and the height h of a triangle are known, the base b is given by the formula $b = \frac{2A}{h}$. ∎

EXAMPLE 3 The formula $C = \frac{5}{9}(F - 32)$ is used to convert Fahrenheit temperature readings into their Celsius equivalents. Solve this equation for F.

Solution

$$C = \frac{5}{9}(F - 32)$$

$$9C = 9 \cdot \frac{5}{9}(F - 32) \qquad \text{Multiply both sides by 9.}$$

$$9C = 5(F - 32) \qquad \text{Simplify.}$$

$$\frac{9C}{5} = \frac{5(F - 32)}{5} \qquad \text{Divide both sides by 5.}$$

$$\frac{9}{5}C = 1(F - 32) \qquad \text{Simplify.}$$

$$\frac{9}{5}C = F - 32 \qquad \text{Remove parentheses.}$$

$$\frac{9}{5}C + 32 = F - 32 + 32 \qquad \text{Add 32 to both sides.}$$

$$\frac{9}{5}C + 32 = F \qquad \text{Combine terms.}$$

$$F = \frac{9}{5}C + 32$$

The formula $F = \frac{9}{5}C + 32$ is used to convert degrees Celsius to degrees Fahrenheit. ∎

EXAMPLE 4 The area A of the trapezoid shown in Figure 2-1 is given by the formula

$$A = \frac{1}{2}(B + b)h$$

where B and b are its bases and h is its height. Solve the formula for b.

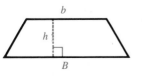

FIGURE 2-1

Solution   *Method 1:* $A = \dfrac{1}{2}(B + b)h$

$2 \cdot A = 2 \cdot \dfrac{1}{2}(B + b)h$ Multiply both sides by 2.

$2A = Bh + bh$ Simplify and
remove parentheses.

$2A + (-Bh) = Bh + bh + (-Bh)$ Add $-Bh$ to both sides.

$2A - Bh = bh$

$\dfrac{2A - Bh}{h} = \dfrac{bh}{h}$ Divide both sides by h.

$\dfrac{2A - Bh}{h} = b$ Simplify.

Method 2: $A = \dfrac{1}{2}(B + b)h$

$2 \cdot A = 2 \cdot \dfrac{1}{2}(B + b)h$ Multiply both sides by 2.

$2A = (B + b)h$ Simplify.

$\dfrac{2A}{h} = \dfrac{(B + b)h}{h}$ Divide both sides by h.

$\dfrac{2A}{h} = B + b$ Simplify.

$\dfrac{2A}{h} + (-B) = B + b + (-B)$ Add $-B$ to both sides.

$\dfrac{2A}{h} - B = b$ Combine terms.

Albert Einstein (1879–1955)
Einstein was a theoretical physicist
best known for his theory of
relativity. Although Einstein was
born in Germany, he became a
Swiss citizen and earned his
doctorate at the University of Zurich
in 1905. In 1910 he returned to
Germany to teach. He fled Germany
because of the Nazi government,
and became a United States citizen
in 1940. He is famous for his
equation, $E = mc^2$

Although they look different, the results of Methods 1 and 2 are equivalent.   ∎

EXAMPLE 5   Use the formula $P = 2l + 2w$ to find l if $P = 56$ and $w = 11$.

Solution   Solve the formula $P = 2l + 2w$ for l, and then substitute the given values of P
and w to determine the value of l.

$P = 2l + 2w$

$P + (-2w) = 2l + 2w + (-2w)$ Add $-2w$ to both sides.

$P - 2w = 2l$ Combine terms.

$\dfrac{P - 2w}{2} = \dfrac{2l}{2}$ Divide both sides by 2.

$\dfrac{P - 2w}{2} = l$ Simplify.

or

$l = \dfrac{P - 2w}{2}$

Then substitute 56 for P and 11 for w, and simplify.

$$l = \frac{P - 2w}{2} = \frac{56 - 2(11)}{2} = \frac{56 - 22}{2} = \frac{34}{2} = 17$$

Thus, $l = 17$. ∎

EXAMPLE 6 The volume V of the right-circular cone shown in Figure 2-2 is given by the formula

$$V = \frac{1}{3} Bh$$

where B is the area of its circular base and h is its height. Solve the formula for h and find the height of a right-circular cone with a volume of 64 cubic centimeters and a base area of 16 square centimeters.

Solution Solve the formula for h as follows:

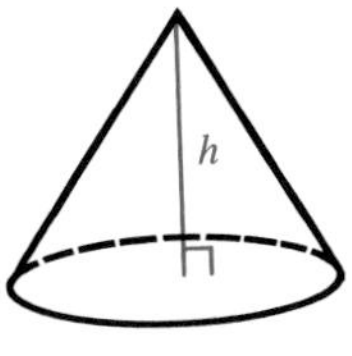

$$V = \frac{1}{3} Bh$$

$$3V = 3 \cdot \frac{1}{3} Bh \qquad \text{Multiply both sides by 3.}$$

$$3V = Bh \qquad \text{Simplify.}$$

$$\frac{3V}{B} = \frac{Bh}{B} \qquad \text{Divide both sides by } B.$$

$$\frac{3V}{B} = h \qquad \text{Simplify.}$$

or

$$h = \frac{3V}{B}$$

Then substitute 64 for V and 16 for B and simplify.

$$h = \frac{3V}{B} = \frac{3(64)}{16} = 3(4) = 12$$

The height of the right-circular cone is 12 centimeters. ∎

FIGURE 2-2

EXERCISE 2.4

In Exercises 1–24, solve each formula for the variable indicated.

1. $E = IR$; for I **2.** $i = prt$; for r **3.** $V = lwh$; for w **4.** $K = A + 32$; for A

5. $P = a + b + c$; for b **6.** $P = 4s$; for s **7.** $P = 2l + 2w$; for w **8.** $d = rt$; for t

9. $A = P + Prt$; for t **10.** $A = \frac{1}{2}(B + b)h$; for h **11.** $C = 2\pi r$; for r **12.** $I = \frac{E}{R}$; for R

13. $K = \frac{wv^2}{2g}$; for w **14.** $V = \pi r^2 h$; for h **15.** $P = I^2 R$; for R **16.** $V = \frac{1}{3}\pi r^2 h$; for h

17. $K = \frac{wv^2}{2g}$; for g **18.** $P = \frac{RT}{mV}$; for V **19.** $F = \frac{GMm}{d^2}$; for M **20.** $C = 1 - \frac{A}{a}$; for A

21. $F = \frac{GMm}{d^2}$; for d^2 **22.** $y = mx + b$; for x

23. $G = 2(r - 1)b$; for r **24.** $F = f(1 - M)$; for M

In Exercises 25–32, solve each formula for the variable indicated and then substitute numbers to find that variable's value.

25. $d = rt$ Find t if $d = 135$ and $r = 45$.

26. $d = rt$ Find r if $d = 275$ and $t = 5$.

27. $i = prt$ Find t if $i = 12$, $p = 100$, and $r = 0.06$.

28. $i = prt$ Find r if $i = 120$, $p = 500$, and $t = 6$.

29. $P = a + b + c$ Find c if $P = 47$, $a = 15$, and $b = 19$.

30. $y = mx + b$ Find x if $y = 30$, $m = 3$, and $b = 0$.

31. $K = \frac{1}{2}h(a + b)$ Find h if $K = 48$, $a = 7$, and $b = 5$.

32. $\frac{x}{2} + y = z^2$ Find x if $y = 3$ and $z = 3$.

33. The formula $E = IR$, called *Ohm's law,* is used in electronics. Solve for I and then calculate the current I if the voltage E is 48 volts and the resistance R is 12 ohms. Current has units of *amperes.*

34. The volume V of a cone is given by the formula $V = \frac{1}{3}\pi r^2 h$. Solve the formula for h and then calculate the height h if V is 36π cubic inches and the radius r is 6 inches.

35. The circumference C of a circle is given by $C = 2\pi r$, where r is the radius of the circle. Solve the formula for r and then calculate the radius of a circle with a circumference of 17 feet. Use $\pi \approx 3.14$ and give your answer to the nearest hundredth of a foot.

36. At a simple interest rate r, an amount of money P grows to an amount A in t years according to the formula $A = P(1 + rt)$. Solve the formula for P. After $t = 3$ years, a girl has an amount $A = \$4300$ on deposit. What amount P did she start with? Assume an interest rate of 9% ($r = 0.09$).

37. The power P lost when an electric current I passes through a resistance R is given by the formula $P = I^2 R$. Solve for R. If P is 2700 watts and I is 14 amperes, calculate R to the nearest hundredth of an ohm.

38. The perimeter P of a rectangle with length l and width w is given by the formula $P = 2l + 2w$. Solve this formula for w. If the perimeter of a certain rectangle is 58 meters and its length is 17 meters, find its width.

39. The force of gravitation F between two objects with masses m and M and separated by distance d is given by the formula

$$F = \frac{GmM}{d^2}$$

where G is a constant. Solve for m.

40. In thermodynamics, the Gibbs free-energy function is given by

$$G = U - TS + pV$$

Solve this equation for the pressure, p.

41. The approximate length L of a belt joining two pulleys of radii r and R feet with centers D feet apart is given by the formula $L = 2D + 3.25(r + R)$. See Illustration 1. Solve the formula for D. If a 25-foot belt joins pulleys of radius 1 foot and 3 feet, how far apart are the centers of the pulleys?

42. The measure a of an interior angle of a regular polygon with n sides is given by the formula $a = 180°(1 - \frac{2}{n})$. See Illustration 2. Solve the formula for n. How many sides does a regular polygon have if an interior angle is 108°? (*Hint:* Distribute first.)

ILLUSTRATION 1   **ILLUSTRATION 2**

REVIEW EXERCISES

In Review Exercises 1–8, simplify each expression, if possible.

1. $3x - 5x$

2. $2x - 5y + 3x$

3. $2x^2y + 5x^2y^2$

4. $8x^2 + 3(x^2 + 2)$

5. $2(y^3 - x^2) + 5(y^3 + 3) - 10$

6. $7(x - 5) - 7(5 - x)$

7. $\frac{3}{5}(x + 5) - \frac{8}{5}(10 + x)$

8. $\frac{2}{11}(22x - y^2) + \frac{9}{11}y^2$

In Review Exercises 9–12, solve each equation. Check all solutions.

9. $3(x - 5) = 2(x + 3)$

10. $7(3 - 2x) - 3x = 4$

11. $\frac{3x - 5}{13} = x - 5$

12. $3x - 2(5 - x) = 15x$

2.5 APPLICATIONS OF EQUATIONS

In this section we discuss the solutions to several types of word problems. We will follow these steps as we solve each problem.

> 1. Read the problem several times and analyze the facts. What information is given? What are we asked to find? Occasionally a sketch, chart, or diagram will help us visualize the facts of the problem.
> 2. Pick a variable to represent the quantity to be found and write a sentence stating what that variable represents. Express all other important quantities mentioned in the problem as expressions involving this single variable.
> 3. Organize the data and find a way to express a quantity in two different ways.
> 4. Write an equation showing that the two quantities found in step 3 are equal.
> 5. Solve the equation.
> 6. State the solution or solutions.
> 7. Check the answer in the words of the problem. Have all the questions been answered?

EXAMPLE 1   Six more than twice a certain number is equal to 1 less than 3 times that number. Find the number.

Analysis   We are to find a certain number. Although we do not know the number, we can pick a variable such as x to represent it. Then we can express the other information of the problem in terms of x.

Solution   Let x represent the certain number. Then

$2x + 6$ represents *6 more than twice the certain number*, and
$3x - 1$ represents *1 less than 3 times the number*.

We know that these two quantities are equal.

Six more than twice a number	is equal to	1 less than 3 times that number.
$2x + 6$	$=$	$3x - 1$

$$2x + 6 = 3x - 1 \qquad \text{The equation to solve.}$$
$$6 = x - 1 \qquad \text{Add } -2x \text{ to both sides.}$$
$$7 = x \qquad \text{Add 1 to both sides.}$$

The number is 7.

Check this solution in the words of the original problem. First calculate *6 more than twice the number*. The result is $2 \cdot 7 + 6$, or 20. Then calculate *1 less than 3 times the number*. This result is $3 \cdot 7 - 1$, or 20. Because the results agree, the solution checks.   ■

EXAMPLE 2   Jim wants to cut a 17-foot pipe into three sections. The longest section is to be 3 times as long as the shortest, and the middle-sized section is to be 2 feet longer than the shortest. How long should each section be?

Analysis   The information in this problem is given in terms of the length of the shortest of the three sections of the pipe. Therefore, let a variable represent the length of that shortest section. Then express the other facts in terms of that variable.

Solution   Let x represent the length of the shortest section. Then

$3x$ represents the length of the longest section, and
$x + 2$ represents the length of the middle-sized section.

Sketch the situation described in the problem, as shown in Figure 2-3.

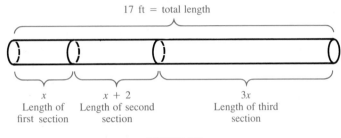

FIGURE 2-3

The sum of the lengths of these three sections must equal the total length of the pipe.

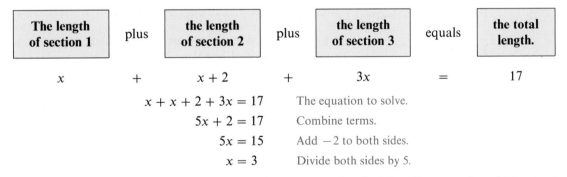

The length of section 1	plus	the length of section 2	plus	the length of section 3	equals	the total length.
x	$+$	$x + 2$	$+$	$3x$	$=$	17

$$x + x + 2 + 3x = 17 \quad \text{The equation to solve.}$$
$$5x + 2 = 17 \quad \text{Combine terms.}$$
$$5x = 15 \quad \text{Add } -2 \text{ to both sides.}$$
$$x = 3 \quad \text{Divide both sides by 5.}$$

The length of the pipe's shortest section is 3 feet. Because the middle-sized section is 2 feet longer than the shortest, it is 5 feet long. Because the longest section is 3 times the shortest, it is 9 feet long.

Check: Because 3 feet, 5 feet, and 9 feet total 17 feet, the solution checks. ∎

EXAMPLE 3   A truck made five trips to the dump, and a larger truck made eight trips. When fully loaded, the larger truck carries 2 tons more garbage than the smaller truck. These two trucks hauled a total of 55 tons. How much garbage can the smaller truck carry in one load?

Analysis   Pick a variable such as x to represent the number of tons that the smaller truck can carry when it is fully loaded. Because the larger truck can carry 2 tons more than the smaller truck, it can carry $(x + 2)$ tons when fully loaded. Since the smaller truck can carry x tons in each load, it can carry $5x$ tons in five trips. Since the larger truck can carry $(x + 2)$ tons in each load, it can carry $8(x + 2)$ tons in eight trips. The total number of tons of garbage hauled can be expressed two ways: as the sum of $5x$ and $8(x + 2)$ and as the number 55.

Solution   Let x represent the capacity of the smaller truck in tons. Then

$x + 2$ represents the capacity of the larger truck in tons.

Total tonnage carried by small truck	plus	total tonnage carried by large truck	equals	total tonnage carried by both trucks.
$5x$	$+$	$8(x + 2)$	$=$	55

$$5x + 8(x + 2) = 55 \qquad \text{The equation to solve.}$$
$$5x + 8x + 16 = 55 \qquad \text{Remove parentheses.}$$
$$13x + 16 = 55 \qquad \text{Combine terms.}$$
$$13x = 39 \qquad \text{Add } -16 \text{ to both sides.}$$
$$x = 3 \qquad \text{Divide both sides by 13.}$$

The smaller truck can carry 3 tons of garbage in one trip.

Check: Each load on the small truck is 3 tons. Because the larger truck carries 2 tons more than that, it carries 5 tons. The small truck made five trips carrying 3 tons on each trip, and the larger truck made eight trips carrying 5 tons on each trip. The small truck carried a total of $5 \cdot 3$, or 15 tons, while the larger truck hauled $8 \cdot 5$, or 40 tons. In total they carried $15 + 40$, or 55 tons. The solution checks.   ■

EXAMPLE 4   In a fourth-grade class of 32 students, only 29 will be promoted. There are 12 more girls in the class than there are boys, and only 1 of the girls will have to repeat fourth grade. How many boys are in the class?

Analysis   We are asked to find how many boys are presently in the fourth-grade class. It is not important to know how many of them will pass, or how many will not. The important information is the total number in the class (32) and the fact that there are 12 more girls than boys. Let b represent the number of boys in the class. Since there are 12 more girls than boys, the expression $b + 12$ represents the number of girls. Thus, the total number of students can be expressed as $b + (b + 12)$ and as 32.

Solution   Let b represent the number of boys in the class. Then

$b + 12$ represents the number of girls.

The number of boys in the class	plus	the number of girls in the class	equals	the total number in the class.
b	$+$	$b + 12$	$=$	32

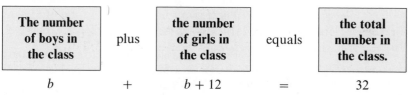

$$b + b + 12 = 32 \qquad \text{The equation to solve.}$$
$$2b + 12 = 32 \qquad \text{Combine terms.}$$
$$2b = 20 \qquad \text{Add } -12 \text{ to both sides.}$$
$$b = 10 \qquad \text{Divide both sides by 2.}$$

There are 10 boys in the class.

Check: There are 10 boys in the class and 22 girls (12 more than 10). The total number is $10 + 22$, or 32. The solution checks.   ∎

EXERCISE 2.5

In Exercises 1–30, pick a variable to represent the unknown quantity, set up an equation involving the variable, solve the equation, and check it.

 1. If a number is doubled, the result is 3 less than three times the number. Find the number.

 2. If a number is multiplied by 4, the result is 7 greater than 3 times the number. Find the number.

 3. If 6 is added to a number, the result is equal to 4 times the number. Find the number.

 4. If 9 is subtracted from 3 times a number, the result is 1 less than the original number. Find the number.

 5. Ten less than 7 times a number is 12 more than 5 times the number. Find the number.

 6. Thirteen minus 2 times a number is the same as 4 more than the number. Find the number.

 7. One-fifth of a number added to the number itself gives 42. Find the number.

 8. One-fourth of a number added to twice the number equals 27. Find the number.

 9. Six more than 3 times a certain number is 5 times the number. Find the number.

10. The number 5 decreased by a certain number becomes 13 greater than the number. Find the number.

11. What number is equal to its own double?

12. What number is equal to one-half of itself?

13. A 12-foot board has been cut into two sections, one twice as long as the other. How long is each section?

14. A 20-foot pipe has been cut into two sections, one 3 times as long as the other. How long is each section?

15. A 45-meter length of wire has been cut into three pieces. The longest of the three is 5 times as long as the shortest. The middle-sized section is 3 times as long as the shortest. How long is each piece?

16. A 30-foot steel beam must be cut into two pieces so that the longer piece is 2 feet more than 3 times as long as the shorter piece. How long will each piece be?

17. A 35-foot beam, 1 foot wide and 2 inches thick, is cut into three sections. One section is 14 feet long. Of the remaining two sections, one is twice as long as the other. How long is each section?

18. Two tanks hold a total of 45 gallons of water. One tank holds 6 gallons more than twice the amount in the other. How much water is in each tank?

19. The price of a chair is reduced by one-fourth of its retail price, making it one-half of the cost of a matching sofa. The reduced price is $237. What was the chair's retail price?

20. If you buy one bottle of vitamins, you can get a second bottle for half price. Two bottles cost $2.25. What is the usual price for a single bottle of vitamins?

21. Thirty-six children received science fair awards. There were three first-place awards, and there were twice as many third-place awards as second-place awards. How many of each award were given?

22. A glazed donut has 50 fewer calories than 5 times the number of calories in a slice of whole wheat bread. The donut and the bread together contain 280 calories. How many calories are in the slice of bread?

23. A slice of pie with a scoop of ice cream contains 850 calories. The calories in the pie alone are 100 greater than double the calories in the ice cream alone. How many calories are in the ice cream?

24. Juanita buys a biology text and its lab manual for $69.50. The text costs $2 more than four times the cost of the manual. Find the cost of the text.

25. A hat and a coat together cost $100. The coat costs $80 more than the hat. How much does the hat cost?

26. A novel can be purchased in a hardcover edition for $15.95 or in paperback for $4.95. The publisher printed 11 times as many paperbacks as hardcover books, a total of 114,000 copies. How many hardcover books were printed?

27. Concrete contains 3 times as much gravel as cement. How much cement is in 500 pounds of dry concrete mix?

28. Sam gave 3 times as much to a charity as Bill did, and Bob donated $4 more than Bill. If their contributions totaled $109, how much did each contribute?

29. Water is made up of hydrogen and oxygen. The mass of the oxygen in water is 8 times the mass of the hydrogen. How much hydrogen is in 2700 grams of water?

30. A man can buy several 20-cent stamps, using all of the money in his pocket. If he buys 50-cent stamps instead, he can buy six fewer stamps, but he would then receive 30 cents change. How many 20-cent stamps can he buy?

REVIEW EXERCISES

In Review Exercises 1–6, give an example that illustrates the given property of real numbers.

1. The commutative property of addition.

2. The distributive property.

3. The associative property of multiplication.

4. The closure property of addition.

5. The double negative rule.

6. The multiplicative identity property.

In Review Exercises 7–12, find the requested number.

7. The absolute value of the negative of 12.

8. The negative of the absolute value of 12.

9. The negative of the negative of 12.

10. The negative of the reciprocal of 12.

11. The sum of 12 and its additive inverse.

12. The product of 12 and its multiplicative inverse.

2.6 MORE APPLICATIONS OF EQUATIONS

In this section we consider more types of word problems.

Integer Problems

EXAMPLE 1   The sum of two consecutive even integers is 22. Find the integers.

Analysis Recall that consecutive even integers differ by 2; numbers such as 2, 4, 6, 8, and 10 are consecutive even integers. Hence, if x is an even integer, then the expressions x and $x + 2$ represent two consecutive even integers. The sum of these two integers can be written in two different ways: as $x + (x + 2)$ and as 22.

Solution Let x represent the first even integer. Then

$x + 2$ represents the next consecutive even integer.

We can form the equation

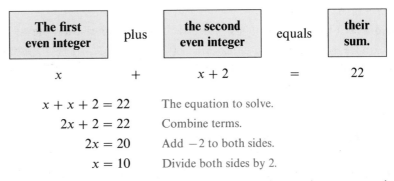

| The first even integer | plus | the second even integer | equals | their sum. |

$$x \qquad + \qquad x + 2 \qquad = \qquad 22$$

$x + x + 2 = 22$	The equation to solve.
$2x + 2 = 22$	Combine terms.
$2x = 20$	Add -2 to both sides.
$x = 10$	Divide both sides by 2.

The first of the two consecutive even integers is 10. The next even integer is 12. Hence the two consecutive even integers are 10 and 12.

Check: The numbers 10 and 12 are indeed consecutive even integers, and their sum is 22. The answers check. ∎

Geometric Problems

EXAMPLE 2 The length of a rectangle is 4 meters longer than twice its width. If the perimeter of the rectangle is 26 meters, find its dimensions.

Analysis Recall that the perimeter of a rectangle is given by the formula $P = 2l + 2w$, where P represents the perimeter, l represents the length, and w represents the width of the rectangle. Since the length of the rectangle shown in Figure 2-4 is 4 meters longer than twice its width, the length of the rectangle can be represented by the expression $2w + 4$. Thus, its perimeter is $P = 2(2w + 4) + 2w$. The perimeter is also 26.

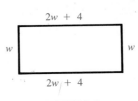

FIGURE 2-4

Solution Let w represent the width of the rectangle. Then

$2w + 4$ represents the length of the rectangle.

We can form the equation

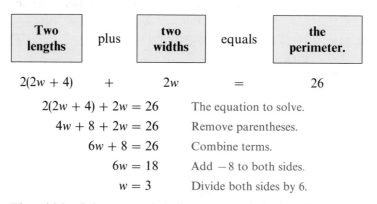

$$2(2w + 4) \quad + \quad 2w \quad = \quad 26$$

$2(2w + 4) + 2w = 26$	The equation to solve.
$4w + 8 + 2w = 26$	Remove parentheses.
$6w + 8 = 26$	Combine terms.
$6w = 18$	Add -8 to both sides.
$w = 3$	Divide both sides by 6.

The width of the rectangle is 3 meters and the length, $4 + 2w$, is 10 meters.

Check: If a rectangle has a width of 3 meters and a length of 10 meters, then the length is 4 meters longer than twice the width ($2 \cdot 3 + 4 = 10$). Furthermore, the perimeter is $2(10) + 2(3)$ or 26 meters. The solution checks.   ∎

Another Geometric Problem

EXAMPLE 3   The vertex angle of an isosceles triangle is 56°. Find the measure of each base angle.

Analysis   An **isosceles triangle** has two equal sides, which meet to form the **vertex angle.** The angles opposite those sides, called **base angles,** are also equal (see Figure 2-5). If we let x represent the measure of one base angle, then the measure of the other base angle is also x. In any triangle, the sum of the three angles is 180°.

Solution   Let x represent the measure of one base angle. Then

x also represents the measure of the other base angle.

We can form the equation

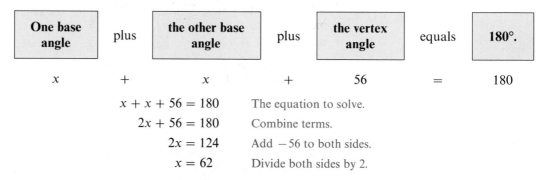

$$x \quad + \quad x \quad + \quad 56 \quad = \quad 180$$

$x + x + 56 = 180$	The equation to solve.
$2x + 56 = 180$	Combine terms.
$2x = 124$	Add -56 to both sides.
$x = 62$	Divide both sides by 2.

The measure of each base angle is 62°.

FIGURE 2-5

Check: The measure of each base angle is 62°, and the vertex angle measures 56°. These three angles total 180°. The solution checks.   ■

Coin Problems

EXAMPLE 4   Jill has several pennies, nickels, dimes, and quarters. She has twice as many nickels as she has dimes, and four times as many quarters as nickels. She has 36 coins in all, and three of them are pennies. How many nickels, dimes, and quarters does Jill have?

Analysis   Suppose d represents the number of dimes. Because Jill has twice as many nickels as dimes, she has $2d$ nickels. Because she has four times as many quarters as nickels, she has $4(2d)$, or $8d$, quarters. Jill also has 3 pennies. The total number of pennies, nickels, dimes, and quarters is 36. The number of coins can be expressed in two ways:

as $(3 + 2d + d + 8d)$ coins   and as   36 coins

Solution   Let d represent the number of dimes. Then

$2d$ represents the number of nickels and
$8d$ represents the number of quarters.

We can form the equation

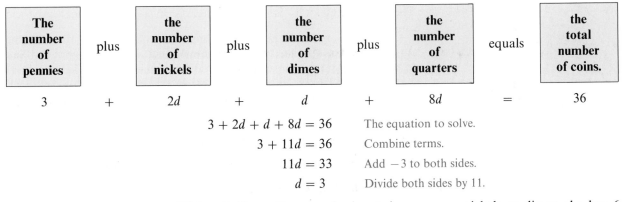

The number of pennies	plus	the number of nickels	plus	the number of dimes	plus	the number of quarters	equals	the total number of coins.
3	+	2d	+	d	+	8d	=	36

$$3 + 2d + d + 8d = 36 \qquad \text{The equation to solve.}$$
$$3 + 11d = 36 \qquad \text{Combine terms.}$$
$$11d = 33 \qquad \text{Add } -3 \text{ to both sides.}$$
$$d = 3 \qquad \text{Divide both sides by 11.}$$

Jill has 3 dimes. Because she has twice as many nickels as dimes, she has 6 nickels. Because she has four times as many quarters as nickels, she has 24 quarters.

Check: Because 3 dimes, 6 nickels, 24 quarters, and 3 pennies total 36 coins, the solution checks.   ■

Another Coin Problem

EXAMPLE 5   George has $2 in nickels, dimes, and quarters. He has 5 times as many quarters as he has nickels and two more dimes than he has quarters. How many of each type of coin does he have?

Analysis It is important to distinguish between the *number* of coins and the *value* of those coins. If n represents the number of nickels, then those nickels are worth $5n$ cents. Because there are $5n$ quarters, their value is $25(5n)$ cents. Finally, there are $(5n + 2)$ dimes (there are two more dimes than quarters, and there are $5n$ quarters). The value of these dimes is $10(5n + 2)$ cents. The total value can be expressed in two ways:

as $5n + 25(5n) + 10(5n + 2)$ cents, and as 200 cents

Solution Let n represent the number of nickels. Then

$5n$ represents the number of quarters, and
$5n + 2$ represents the number of dimes.

The value of the nickels (in cents)	plus	the value of the quarters (in cents)	plus	the value of the dimes (in cents)	equals	the total value (in cents).
$5n$	$+$	$25(5n)$	$+$	$10(5n + 2)$	$=$	200

$$5n + 25(5n) + 10(5n + 2) = 200 \qquad \text{The equation to solve.}$$
$$5n + 125n + 50n + 20 = 200 \qquad \text{Remove parentheses.}$$
$$180n + 20 = 200 \qquad \text{Combine terms.}$$
$$180n = 180 \qquad \text{Add } -20 \text{ to both sides.}$$
$$n = 1 \qquad \text{Divide both sides by 18.}$$

George has 1 nickel. Because he has five times as many quarters, he has 5 quarters. Because he has two more dimes than quarters, he has 7 dimes.

Check: Because the values of 1 nickel, 5 quarters, and 7 dimes add up to $2, the solution checks. ∎

Break-Even Analysis

An example from business that requires solving equations is called **break-even analysis.**

In any type of manufacturing business, there are two types of costs—**fixed costs** and **unit costs.** Fixed costs do not depend on the amount of product manufactured. Fixed costs would include the cost of plant rental, insurance, and machinery. Unit costs do depend on the amount of product manufactured. Unit costs would include the cost of raw materials and labor.

Break-even analysis is used to find a production level at which revenue will just offset the cost of production. When production exceeds the break-even point, the company will make a profit.

EXAMPLE 6 An electronics company has fixed costs of $6405 a week and a unit cost of $75 for each compact disk player it manufactures. The company can sell all the CD players it can make at a wholesale price of $90. Find the company's break-even point.

Analysis   Suppose the company manufacturers x CD players each week. The cost of manufacturing these players is the sum of the fixed and unit costs. We are given that the fixed costs are $6405 each week. The unit cost is the product of x, the number of CD players manufactured each week, and $75, the cost of manufacturing a single player. Thus, the weekly cost is

$$cost = 75x + 6405$$

Since the company can sell all the machines it can make, the weekly revenue is the product of x, the number of players manufactured (and sold) each week, and $90, the wholesale price of each CD player. Thus, the weekly revenue is

$$revenue = 90x$$

The break-even point is the value of x for which the weekly revenue is equal to the weekly cost.

Solution   Let x represent the number of CD players manufactured each week. Then

$90x$ represents the weekly revenue, and
$75x + 6405$ represents the weekly cost.

Because the break-even point occurs when revenue equals cost, set up and solve the following equation:

The weekly revenue	=	the weekly cost.

$$90x = 75x + 6405$$
$$15x = 6405$$
$$x = 427$$

If the company manufactures 427 CD players, its revenue will equal its costs, and the company will break even. Verify that this is true.   ∎

EXERCISE 2.6

In Exercises 1–28, pick a variable to represent the unknown quantity, set up an equation involving the variable, solve the equation, and check it.

Integer Problems

1. The sum of two consecutive even integers is 54. Find the integers.
2. The sum of two consecutive odd integers is 88. Find the integers.
3. The sum of three consecutive integers is 120. Find the integers.
4. The sum of three consecutive even integers is 72. Find the integers. $x + (x + 2) + (x + 4) = 72$
5. The sum of an integer and twice the next integer is 23. Find the smaller integer.
6. If 4 times the smallest of three consecutive integers is added to the largest, the result is 112. Find the three integers.

7. The larger of two integers is 10 greater than the smaller. The larger is 3 less than twice the smaller. Find the smaller integer.

8. The smaller of two integers is one-half of the larger, and 11 greater than one-third of the larger. Find the smaller integer.

Geometric Problems

9. The perimeter of a triangle is 57 feet. If all three sides are equal, find the length of each side.

10. The length of a rectangle is 7 centimeters longer than the width. The perimeter is 90 centimeters. Find the dimensions of the rectangle.

11. The width of a rectangle is 11 meters less than the length. The perimeter is 94 meters. Find the dimensions of the rectangle.

12. One of the two equal sides of an isosceles triangle is 4 feet less than the third side. The perimeter is 25 feet. Find the lengths of the sides.

13. The length of a rectangle is 5 inches longer than twice the width. The perimeter is 112 inches. Find the dimensions of the rectangle.

14. One of the two equal angles of an isosceles triangle is 4 times the third angle (the vertex angle). What is the measure of the vertex angle?

15. The three angles of an equilateral triangle are all equal. What is the measure of each?

16. The perimeter of a certain square is twice the perimeter of a certain equilateral (equal-sided) triangle. If one side of the square is 6 inches, what is the length of a side of the triangle?

Coin Problems

17. Liz has some nickels, dimes, and quarters—nine coins in all. She has twice as many dimes as nickels, and one less quarter than dimes. How many nickels does she have?

18. A man has twice as many pennies as he has dimes, and four less quarters than pennies. He has 41 coins in all. How many coins of each type does he have?

19. A many has $3.15 in nickels, dimes, and quarters. He has as many nickels as dimes, and three more quarters than dimes. How many of each coin does he have?

20. A woman has $6.75 in pennies, dimes, and quarters. She has twice as many dimes as pennies, and just six quarters. How many pennies and how many dimes does she have?

21. Margaret has twice as many dimes as Fern has quarters. When they pool their resources, they have $4.05. How many quarters did Fern have?

22. Jeanne has three times as many dimes as quarters, and half as many nickels as dimes. The coins are worth $3.75. How many quarters does Jeanne have?

23. If all of Kim's nickels were dimes, she would be 35 cents richer. How many nickels does she have?

24. If all of Bob's dimes were quarters, he would be $1.05 richer. How many dimes does he have?

Break-Even Analysis

25. A shoe company has fixed costs of $9600 per month and a unit cost of $20 per pair of shoes. The company can sell all the shoes it can make at a wholesale price of $30 per pair. Find the break-even point.

26. A belt company has fixed costs of $5400 per month and a unit cost of $12 per belt. The company can sell all the belts it can make at a wholesale price of $15. Find the break-even point.

27. A machine shop has two machines that can mill a certain brass plate. One machine has a setup cost of $500 and a cost per plate of $2, and the other machine has a setup cost of $800 and a cost per plate of $1. How many plates should be manufactured if the cost is to be the same using either machine?

28. A rug manufacturer has two looms for weaving Oriental-style rugs. One loom has a setup cost of $750 and can produce a rug for $115. The other loom has a setup cost of $950 and can produce a rug for $95. How many rugs can be manufactured if the costs are the same on each loom?

In Exercises 29–32, a paint manufacturer can choose between two processes for manufacturing house paint, with monthly costs shown in Table 1. The paint can be sold for $21 per gallon.

TABLE 1	Process	Fixed costs	Unit cost (per gallon)
	A	$75,000	$11
	B	$128,000	$5

29. Find the break-even point for process A.   **30.** Find the break-even point for process B.

31. If expected sales will be 8800 gallons per month, which process should the company choose?

32. If expected sales will be 9000 gallons per month, which process should the company choose?

REVIEW EXERCISES

In Review Exercises 1–6, refer to the formulas in Section 1.4.

1. Find the volume of a pyramid that has a height of 6 centimeters and a square base, 10 centimeters on each side.

2. Find the volume of a cone with a height of 6 centimeters and a circular base with radius 6 centimeters.

3. Find the volume of a sphere whose diameter is 14 meters.

4. Find the volume of a cylinder with a height of 8 centimeters and a circular base with diameter of 20 centimeters.

5. Find the number of cubic feet in one cubic yard.

6. Find the number of cubic inches in one cubic foot.

In Review Exercises 7–12, simplify each expression.

7. $3(x + 2) + 4(x - 3)$

8. $4(x - 2) - 3(x + 1)$

9. $\dfrac{1}{2}(x + 1) - \dfrac{1}{2}(x + 4)$

10. $\dfrac{3}{2}\left(x + \dfrac{2}{3}\right) + \dfrac{1}{2}(x + 8)$

11. $\dfrac{2}{3}(x^2 + x) - \dfrac{1}{3}(2x^2 - x)$

12. $\dfrac{3}{4}(x^2 + 4) + \dfrac{1}{4}(x^2 - 12)$

2.7 EVEN MORE APPLICATIONS OF EQUATIONS

In this section we consider more types of word problems.

Investment Problems

EXAMPLE 1   Kristy invested part of $12,000 at 9% annual interest, and the rest at 11%. The annual income from these investments was $1230. How much did she invest at each rate?

Analysis   The interest i earned by an amount p invested at an annual rate r for t years is given by the formula $i = prt$. In this example, $t = 1$ year. Hence, if x dollars were invested at 9%, the interest earned would be $0.09x$ dollars. If x dollars were invested at 9%, then the rest of the money $(12,000 - x)$ would be invested at 11%. The interest earned on that money would be $0.11(12,000 - x)$ dollars. The total interest earned in dollars can be expressed in two ways: as 1230 and as the sum $0.09x + 0.11(12,000 - x)$.

Solution   Let x represent the amount of money invested at 9%. Then

$12,000 - x$ represents the amount of money invested at 11%.

Form an equation as follows:

The interest earned at 9%	plus	the interest earned at 11%	equals	the total interest.
$0.09x$	$+$	$0.11(12,000 - x)$	$=$	1230

$$0.09x + 0.11(12,000 - x) = 1230 \qquad \text{The equation to solve.}$$

$$9x + 11(12,000 - x) = 123,000 \qquad \text{Multiply both sides by 100 to clear the equation of decimals.}$$

$$9x + 132,000 - 11x = 123,000 \qquad \text{Remove parentheses.}$$

$$-2x + 132,000 = 123,000 \qquad \text{Combine terms.}$$

$$-2x = -9000 \qquad \text{Add } -132,000 \text{ to both sides.}$$

$$x = 4500 \qquad \text{Divide both sides by } -2.$$

Kristy invested $4500 at 9% and $12,000 − $4500, or $7500, at 11%.

Check: The first investment yielded 9% of $4500, or $405. The second investment yielded 11% of $7500, or $825. Because the total return was $405 + $825, or $1230, the answers check.   ■

Uniform Motion Problems

EXAMPLE 2   Chicago, Illinois, and Green Bay, Wisconsin, are about 200 miles apart. A car leaves Chicago traveling toward Green Bay at 55 miles per hour. At the same time, a truck leaves Green Bay bound for Chicago at 45 miles per hour. How long will it take them to meet?

Analysis   Uniform motion problems are based on the formula $d = rt$, where d is the distance traveled, r is the rate, and t is the time. Organize the information of this problem in chart form as in Figure 2-6.

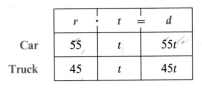

	r	$\cdot$	t	$=$	d
Car	55		t		$55t$
Truck	45		t		$45t$

FIGURE 2-6

We know that the two vehicles travel for the same amount of time, say t hours. The faster car travels $55t$ miles, and the slower truck travels $45t$ miles. The total distance can be expressed in two ways: as the sum $55t + 45t$, and as 200 miles.

Solution   Let t represent the time that each vehicle travels until they meet. Then

$55t$ represents the distance traveled by the car, and
$45t$ represents the distance traveled by the truck.

We can form the equation

The distance the car goes	plus	the distance the truck goes	equals	the total distance.
$55t$	$+$	$45t$	$=$	200

$$55t + 45t = 200 \quad \text{The equation to solve.}$$
$$100t = 200 \quad \text{Combine terms.}$$
$$t = 2 \quad \text{Divide both sides by 100.}$$

The vehicles meet after 2 hours.

Check: During those 2 hours, the car travels $55 \cdot 2$, or 110 miles, while the truck travels $45 \cdot 2$, or 90 miles. The total distance traveled is $110 + 90$, or 200 miles. This is the total distance between Chicago and Green Bay. The answer checks.   ■

Liquid Mixture Problems

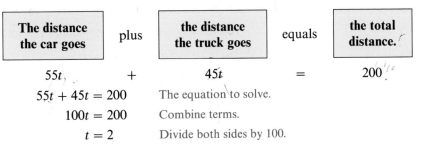

EXAMPLE 3   A chemistry instructor has one solution that is 50% sulfuric acid and another that is 20% sulfuric acid. How much of each should she use to make 12 liters of a solution that is 30% acid?

Analysis   The sulfuric acid present in the final mixture comes from the two solutions to be mixed. If x represents the number of liters of the 50% solution required for the mixture, then the rest of the mixture ($(12 - x)$ liters) must be the 20% solution. Only 50% of the x liters, and only 20% of the $(12 - x)$ liters, is pure sulfuric acid. The total of these amounts is also the amount of acid in the final mixture, which is 30% of 12 liters.

Solution   Let x represent the required number of liters of the 50% solution. Then $12 - x$ represents the required number of liters of the 20% solution.

We can form the equation

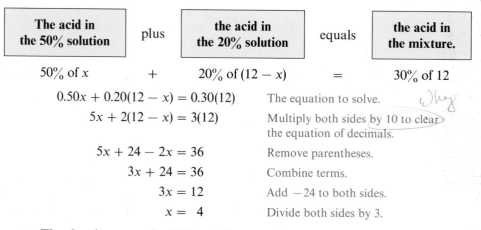

| The acid in the 50% solution | plus | the acid in the 20% solution | equals | the acid in the mixture. |

$$50\% \text{ of } x \qquad + \qquad 20\% \text{ of } (12 - x) \qquad = \qquad 30\% \text{ of } 12$$

$0.50x + 0.20(12 - x) = 0.30(12)$	The equation to solve.
$5x + 2(12 - x) = 3(12)$	Multiply both sides by 10 to clear the equation of decimals.
$5x + 24 - 2x = 36$	Remove parentheses.
$3x + 24 = 36$	Combine terms.
$3x = 12$	Add -24 to both sides.
$x = 4$	Divide both sides by 3.

The chemist must mix 4 liters of the 50% solution and 8 liters ($(12 - 4)$ liters) of the 20% solution.

Check these results.   ∎

Dry Mixture Problems

EXAMPLE 4   Cashews can be sold at $6 per pound, and peanuts are selling at $3 per pound. How many pounds of peanuts should be combined with 50 pounds of cashews to obtain a mixture that can be sold at $4 per pound?

Analysis   The money the storekeeper could make selling the peanuts and cashews separately should equal the amount received by selling the mixture. Suppose x pounds of peanuts are used in the mixture. At $3 per pound, they are worth $3x$ dollars. At $6 per pound, the 50 pounds of cashews are worth $6 \cdot 50$, or 300 dollars. The mixture will weigh $(50 + x)$ pounds, and at $4 per pound, its value will be $4(50 + x)$ dollars. The value of the peanuts and cashews, $(3x + 300)$ dollars, is equal to the value of the mixture, $4(50 + x)$ dollars.

Solution   Let x represent the number of pounds of peanuts in the mixture.

We can form the equation

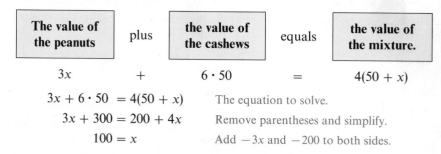

| The value of the peanuts | plus | the value of the cashews | equals | the value of the mixture. |

$$3x \qquad + \qquad 6 \cdot 50 \qquad = \qquad 4(50 + x)$$

$3x + 6 \cdot 50 = 4(50 + x)$	The equation to solve.
$3x + 300 = 200 + 4x$	Remove parentheses and simplify.
$100 = x$	Add $-3x$ and -200 to both sides.

The storekeeper should use 100 pounds of peanuts in the mixture.

Check:

The value of 100 pounds of peanuts at $3 per pound is	$300
The value of 50 pounds of cashews at $6 per pound is	$300
The value of the mixture is	$600

The value of 150 pounds of mixture at $4 per pound is $600.
The answer checks. ■

EXERCISE 2.7

In Exercises 1–28, pick a variable to represent the unknown quantity, set up an equation involving the variable, solve the equation, and check it.

Investment Problems

1. Steve's $24,000 is invested in two accounts, one earning 9% annual interest and the other earning 6%. After 1 year, his combined interest is $1575. How much was invested at each rate?

2. Martha's $18,750 is invested in two accounts, one earning 12% interest and the other earning 10%. After 1 year, her combined interest income is $2117. How much has she invested at each rate?

3. Carol invested equal amounts in each of two investments. One investment pays 8% and the other pays 11%. Her combined interest income for 1 year is $712.50. How much did she invest at each rate?

4. Craig invested equal amounts in each of three accounts, paying 7%, 8%, and 10.5%. His year's combined interest income is $1249.50. How much did he invest in each account?

5. Marilu earned $587.50 by investing in two accounts, one paying 5% annual interest and the other paying 10%. The amount invested at the lower rate was half that invested at the higher rate. How much did she invest at each rate?

6. Equal amounts are invested in accounts paying 11% and 13%. The difference in the annual interests is $150. How much is invested in each account?

7. Twice as much money is invested at 12% annual interest than at 10%. The difference in the annual incomes is $350. How much is invested at each rate?

8. The amount of annual interest earned by $8000 invested at a certain rate is $200 less than $12,000 would earn at a 1% lower rate. At what rate is the $8000 invested? (*Hint:* 1% = 0.01.)

Uniform Motion Problems

9. Two cities, A and B, are 315 miles apart. A car leaves A bound for B at 50 miles per hour. At the same time, another car leaves B and heads toward A at 55 miles per hour. In how many hours will the two cars meet?

10. Granville and Preston are 535 miles apart. A car leaves Preston bound for Granville at 47 miles per hour. At the same time, another car leaves Granville and heads toward Preston at 60 miles per hour. How long will it take them to meet?

11. Two cars leave Peoria at the same time, one heading east at 60 miles per hour and the other west at 50 miles per hour. How long will it take them to be 715 miles apart?

12. Two boats steam out of port, one heading north at 35 knots (nautical miles per hour), the other south at 47 knots. If they leave port at the same time, how long will it take them to be 738 nautical miles apart?

13. Two cars start together and head north, one at 42 miles per hour and the other at 53 miles per hour. In how many hours will the cars be 82.5 miles apart?

14. Two trains are 330 miles apart, and their speeds differ by 20 miles per hour. They travel toward each other and meet in 3 hours. Find the speed of each train.

15. Two planes are 6000 miles apart, and their speeds differ by 200 miles per hour. They travel toward each other and meet in 5 hours. Find the speed of the slower plane.

16. An automobile averaged 40 miles per hour for part of a trip and 50 miles per hour for the remainder. If the 5-hour trip covered 210 miles, for how long did the car average 40 miles per hour?

Mixture Problems

17. How many gallons of fuel costing $1.15 per gallon must be mixed with 20 gallons of a fuel costing $.85 per gallon to obtain a mixture costing $1 per gallon?

18. Paint costing $19 per gallon is to be mixed with 5 gallons of a $3 per gallon thinner to make a paint that can be sold for $14 per gallon. How much paint will be produced?

19. How many gallons of a 30% salt solution must be mixed with 50 gallons of a 70% solution to obtain a 50% solution?

20. How many gallons of milk containing 4% butterfat should be mixed with 10 gallons of milk containing 1% butterfat to obtain a mixture containing 2% butterfat?

21. A nurse wishes to add water to 30 ounces of a 10% alcohol solution to dilute it to an 8% solution. How much water must the nurse add?

22. A chemist wishes to mix 2 liters of a 5% silver iodide solution with a 10% solution to get a 7% solution. How many liters of 10% solution must the chemist add?

23. Lemon drops worth $1.90 per pound are to be mixed with jelly beans that cost $1.20 per pound to make 100 pounds of a mixture worth $1.48 per pound. How many pounds of each candy should be used?

24. One grade of tea, worth $3.20 per pound, is to be mixed with another grade worth $2 per pound to make 20 pounds that will sell for $2.72 per pound. How much of each grade of tea must be used?

25. A pound of hard candy is worth $.30 less than a pound of soft candy. Equal amounts of each are used to make 40 pounds of a mixture that sells for $1.05 per pound. How much is a pound of soft candy worth?

26. Twenty pounds of a candy that sells for $1.70 per pound is mixed with candy that sells for $2 per pound. If the mixture is to sell for $1.80 per pound, how much of the more expensive candy should be used?

27. A store sells regular coffee for $4 a pound and a gourmet coffee for $7 a pound. To get rid of 40 pounds of the gourmet coffee, the shopkeeper makes a gourmet blend to sell for $5 a pound. How many pounds of regular coffee should be used?

28. A garden store sells Kentucky bluegrass seed for $6 per pound and ryegrass seed for $3 per pound. How much rye must be mixed with 100 pounds of bluegrass to obtain a blend that will sell for $5 per pound?

REVIEW EXERCISES

In Review Exercises 1–8, graph each set on a number line.

1. The integers between −2 and 4.

2. The even integers between 3 and 10.

3. All real numbers between -2 and 4.

4. All real numbers less than -2 or greater than or equal to 4.

5. All negative real numbers greater than or equal to -6.

6. All positive numbers less than 3.

7. All real numbers less than $\frac{3}{2}$.

8. All real numbers greater than $-\frac{5}{3}$.

9. The amount, A, on deposit in a bank account bearing simple interest is given by the formula

$$A = P + Prt$$

Determine A when $P = \$1200$, $r = 0.08$, and $t = 3$.

10. The distance, s, that a certain object falls in t seconds is given by the formula

$$s = -16t^2 + vt + 350$$

Determine s when $t = 4$ and $v = -3$.

11. David needs \$3.50 more and Heidi needs \$1.50 more to split equally the \$13 cost of a pizza. How much money does David have?

12. Find a number that is 1 less than three times itself.

2.8 SOLVING INEQUALITIES

Recall the meaning of the following symbols.

> $<$ means "is less than"
> $>$ means "is greater than"
> $\leq$ means "is less than or equal to"
> $\geq$ means "is greater than or equal to"

An **inequality** is a mathematical expression that indicates that two quantities are not necessarily equal. A **solution of an inequality** is any number that makes the inequality a true statement. The number 2 is a solution of the inequality

$$x \leq 3$$

because $2 \leq 3$. The inequality $x \leq 3$ has many more solutions because *any* real number that is less than or equal to 3 will satisfy the inequality. We can use a graph on the number line to exhibit the solutions of the inequality $x \leq 3$. The colored arrow in Figure 2-7 indicates all those points with coordinates that satisfy the inequality $x \leq 3$.

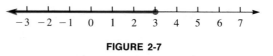

FIGURE 2-7

The solid circle at the point with coordinate 3 indicates that the number 3 is a solution of the inequality $x \leq 3$.

The graph of the inequality $x > 1$ appears in Figure 2-8.

FIGURE 2-8

The colored arrow indicates all those points whose coordinates satisfy the inequality $x > 1$. The open circle at the point with coordinate 1 indicates that 1 is not a solution of the inequality $x > 1$.

To solve more complicated inequalities, we need to use the addition, subtraction, multiplication, and division properties of inequalities. When we use any of these properties, the resulting inequality will have the same solutions as the original inequality.

The Addition Property of Inequality.   If a, b, and c are real numbers, and

If $a < b$,   then $a + c < b + c$.

Similar statements can be made for the symbols $>$, $\leq$, and $\geq$.

The **addition property of inequality** can be stated this way: *If any quantity is added to both sides of an inequality, the resulting inequality has the same direction as the original inequality.*

The Subtraction Property of Inequality.   If a, b, and c are real numbers, and

If $a < b$,   then $a - c < b - c$.

Similar statements can be made for the symbols $>$, $\leq$, and $\geq$.

The **subtraction property of inequality** can be stated this way: *If any quantity is subtracted from both sides of an inequality, the resulting inequality has the same direction as the original inequality.*

The subtraction property of inequality is included in the addition property: To *subtract* a number c from both sides of an inequality, we could instead *add* the *negative* of c to both sides.

EXAMPLE 1   Solve the inequality $2x + 5 > x - 4$ and graph its solution on a number line.

Solution   To isolate the x on the left-hand side of the $>$ sign, we proceed as we would when solving equations.

$$2x + 5 > x - 4$$
$2x + 5 - 5 > x - 4 - 5$ Subtract 5 from both sides.
$2x > x - 9$ Combine terms.
$2x - x > x - 9 - x$ Subtract x from both sides.
$x > -9$ Combine terms.

The graph of this solution (see Figure 2-9) includes all points to the right of -9 but does not include -9 itself. For that reason, use an open circle at -9.

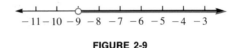

FIGURE 2-9   ■

If both sides of the true inequality $2 < 5$ are multiplied by a *positive* number, such as 3, another true inequality results.

$$2 < 5$$
$3 \cdot 2 < 3 \cdot 5$ Multiply both sides by 3.
$6 < 15$ Simplify.

The inequality $6 < 15$ is a true inequality. However, if both sides of the inequality $2 < 5$ are multiplied by a negative number, such as -3, the direction of the inequality symbol must be reversed to produce another true inequality.

$$2 < 5$$
$-3 \cdot 2 > -3 \cdot 5$ Multiply both sides by the *negative* number -3 and reverse the direction of the inequality.
$-6 > -15$ Simplify.

The inequality $-6 > -15$ is a true inequality because -6 lies to the right of -15 on the number line.

Thus, if both sides of an inequality are multiplied by a *positive* number, the direction of the resulting inequality remains the same. However, if both sides of an inequality are multiplied by a *negative* number, the direction of the resulting inequality must be reversed. This is stated formally as follows:

The Multiplication Property of Inequality.   If a, b, and c are real numbers, and

If $a < b$ and $c > 0$,   then $ac < bc$
If $a < b$ and $c < 0$,   then $ac > bc$

There is a similar property for division.

> **The Division Property of Inequality.** If a, b, and c are real numbers, and
>
> If $a < b$ and $c > 0$,   then $\dfrac{a}{c} < \dfrac{b}{c}$
>
> If $a < b$ and $c < 0$,   then $\dfrac{a}{c} > \dfrac{b}{c}$

To *divide* both sides of an inequality by a nonzero number c, we could instead *multiply* both sides by $\frac{1}{c}$.

The multiplication and division properties of inequality are also true for $\leq$, $>$, and $\geq$.

EXAMPLE 2   Solve the inequality $3x + 7 \leq -5$ and graph the solution.

Solution

$$3x + 7 \leq -5$$
$$3x + 7 + (-7) \leq -5 + (-7) \qquad \text{Add } -7 \text{ to both sides.}$$
$$3x \leq -12 \qquad \text{Combine terms.}$$
$$\frac{3x}{3} \leq \frac{-12}{3} \qquad \text{Divide both sides by 3.}$$
$$x \leq -4$$

The solution of the inequality $3x + 7 \leq -5$ consists of all real numbers less than, and also including, -4. The solid circle at -4 in the graph of the solution in Figure 2-10 indicates that -4 is one of the solutions of the given inequality.

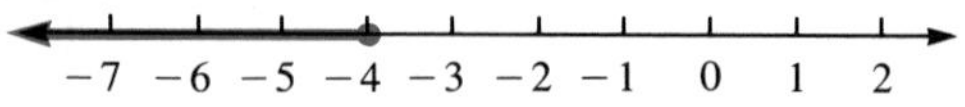

FIGURE 2-10   ■

EXAMPLE 3   Solve the inequality $5 - 3x \leq 14$ and graph the solution.

Solution

$$5 - 3x \leq 14$$
$$5 - 3x + (-5) \leq 14 + (-5) \qquad \text{Add } -5 \text{ to both sides.}$$
$$-3x \leq 9 \qquad \text{Combine terms.}$$

1. $\dfrac{-3x}{-3} \geq \dfrac{9}{-3}$ Divide both sides by -3 and reverse the direction of the $\leq$ symbol.

$$x \geq -3$$

In Statement 1, both sides of the inequality were divided by -3. Because -3 is negative, the direction of the inequality was *reversed*. The graph of the solution appears in Figure 2-11. The solid circle at -3 indicates that -3 is one of the solutions.

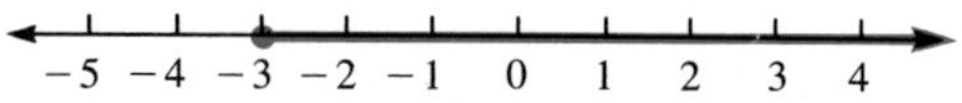

FIGURE 2-11   ■

EXAMPLE 4   Solve the inequality $x(x - 8) < x^2 + 3(x + 11)$ and graph the solution.

Solution

$$x(x - 8) < x^2 + 3(x + 11)$$
$$x^2 - 8x < x^2 + 3x + 33 \qquad \text{Remove parentheses.}$$
$$-8x < 3x + 33 \qquad \text{Add } -x^2 \text{ to both sides.}$$
$$-11x < 33 \qquad \text{Add } -3x \text{ to both sides.}$$
$$x > -3 \qquad \text{Divide both sides by } -11 \text{ and reverse the direction of the inequality sign.}$$

The graph of the solution appears in Figure 2-12. The open circle at -3 indicates that -3 is not a solution.

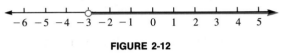

FIGURE 2-12

Double or Compound Inequalities

Two inequalities can be combined into a **double** or **compound inequality** to indicate that numbers lie *between* two fixed values. The inequality $2 < x < 5$, for example, indicates that x is greater than 2 and that x is *also* less than 5. The solution of the double inequality $2 < x < 5$ consists of all numbers that lie *between* 2 and 5. The graph of this set appears in Figure 2-13. Sets of numbers such as this are called **intervals.**

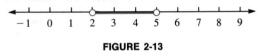

FIGURE 2-13

EXAMPLE 5   Solve the inequality $-4 < 2(x - 1) \leq 4$ and graph the solution.

Solution   To isolate x between the inequality symbols, proceed as follows:

$$-4 < 2(x - 1) \leq 4$$
$$-4 < 2x - 2 \leq 4 \qquad \text{Remove parentheses.}$$
$$-2 < 2x \leq 6 \qquad \text{Add 2 to all three parts.}$$
$$-1 < x \leq 3 \qquad \text{Divide all three parts by 2.}$$

The graph of the solution appears in Figure 2-14.

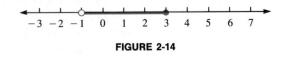

FIGURE 2-14

EXAMPLE 6   A labor union requires that its members work no more than 45 hours per week. How many minutes may a member work in a week?

Solution   Let x represent the number of minutes a union member may work each week. Because there are 60 minutes in 1 hour, there are $60 \cdot 45$ or 2700 minutes in a 45-hour workweek. Thus,

$$x \leq 2700$$

Of course no one can work less than 0 minutes per week, so the solution is given by the double inequality

$$0 \leq x \leq 2700$$

A union member may work from 0 to 2700 minutes each week. The graph of this solution appears in Figure 2-15.

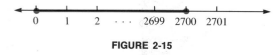

FIGURE 2-15   ■

EXAMPLE 7   The perimeter of a certain equilateral triangle can be no less than 15 feet. How long could one side be?

Solution   Recall that each side of an equilateral triangle is the same length and that the perimeter of a triangle is the sum of the lengths of its three sides.

Let x represent the length of one side of the triangle.

Then $x + x + x$ represents the perimeter.

Because the perimeter can be no less than 15 feet, we have the following inequality:

$$x + x + x \geq 15$$
$$3x \geq 15 \qquad \text{Combine terms.}$$
$$x \geq 5 \qquad \text{Divide both sides by 3.}$$

Each side of the triangle must be at least 5 feet long.   ■

EXERCISE 2.8

In Exercises 1–40, solve each inequality and graph the solution.

1. $x + 2 > 5$ **2.** $x + 5 \geq 2$ **3.** $-x - 3 \leq 7$ **4.** $-x - 9 > 3$

5. $3 + x < 2$ **6.** $5 + x \geq 3$ **7.** $2x - 3 \leq 5$ **8.** $-3x - 5 < 4$

9. $-3x - 7 > -1$ **10.** $-5x + 7 \leq 12$ **11.** $-4x + 1 > 17$ **12.** $7x - 9 > 5$

13. $2x + 9 \leq x + 8$

14. $3x + 7 \leq 4x - 2$

15. $9x + 13 \geq 8x$

16. $7x - 16 < 6x$

17. $8x + 4 > 6x - 2$

18. $7x + 6 \geq 4x$

19. $5x + 7 < 2x + 1$

20. $7x + 2 > 4x - 1$

21. $7 - x \leq 3x - 1$

22. $2 - 3x \geq 6 + x$

23. $9 - 2x > 24 - 7x$

24. $13 - 17x < 34 - 10x$

25. $3(x - 8) < 5x + 6$

26. $9(x - 11) > 13 + 7x$

27. $8(5 - x) \leq 10(8 - x)$

28. $17(3 - x) \geq 3 - 13x$

29. $x(5x - 5) > 5x^2 + 15$

30. $x(x - 7) < x^2 + 3x + 20$

31. $x(x + 8) \leq x^2 + 24$

32. $x(5 + 3x) \leq 15 + 3x^2$

33. $89x^2 - 178 > 89x(x - 1)$

34. $31x^2 + 124 > 3.1x(10x + 20)$

35. $\dfrac{5}{2}(7x - 15) + x \geq \dfrac{13}{2}x - \dfrac{3}{2}$

36. $\dfrac{5}{3}(x + 1) \leq -x + \dfrac{2}{3}$

37. $\dfrac{3x - 3}{2} < 2x + 2$

38. $\dfrac{x + 7}{3} \geq x - 3$

39. $\dfrac{2(x + 5)}{3} \leq 3x - 6$

40. $\dfrac{3(x - 1)}{4} > x + 1$

In Exercises 41–60, solve each inequality and graph the solution.

41. $2 < x - 5 < 5$

42. $3 < x - 2 < 7$

43. $-5 < x + 4 \leq 7$

44. $-9 \leq x + 8 < 1$

45. $0 \leq x + 10 \leq 10$

46. $-8 < x - 8 < 8$

47. $4 < -2x < 10$

48. $-4 \leq -4x < 12$

49. $-3 \leq \dfrac{x}{2} \leq 5$

50. $-12 \leq \dfrac{x}{3} < 0$

51. $3 \leq 2x - 1 < 5$

52. $4 < 3x - 5 \leq 7$

53. $0 < 10 - 5x \leq 15$

54. $1 \leq -7x + 8 \leq 15$

55. $-6 < 3(x + 2) < 9$

56. $-18 \leq 9(x - 5) < 27$

57. $x^2 + 3 \leq x(x - 3) \leq x^2 + 9$

58. $x^2 < x(x + 4) \leq x^2 + 8$

59. $3 - x < 5 < 7 - x$

60. $x + 1 < 2x + 3 < x + 5$

In Exercises 61–76, express each solution as an inequality.

61. The perimeter of an equilateral triangle is at most 57 feet. What could be the length of a side? (*Hint:* All three sides of an equilateral triangle are equal.)

62. The perimeter of a square is no less than 68 centimeters. How long can a side be?

63. The land elevations in Nevada range from the 13,143-foot height of Boundary Peak to the Colorado River at 470 feet. To the nearest tenth, what is the range of these elevations in miles? (*Hint:* 1 mile is 5280 feet.)

64. A teacher requires that students do homework at least 2 hours a day, seven days a week. How many minutes is a student required to work each week?

65. A pilot plans to fly at an altitude of between 17,500 and 21,700 feet. To the nearest tenth, what will be the range of altitudes in miles? (*Hint:* There are 5280 feet in 1 mile.)

66. A doctor advises Jake to exercise at least 15 minutes but less than 30 minutes per day. How many hours of exercise will Jake get each week?

67. To hold the temperature of a room between 19° and 22° Celsius, what Fahrenheit temperatures must be maintained? (*Hint:* Fahrenheit temperature (F) and Celsius temperature (C) are related by the formula $C = \frac{5}{9}(F - 32)$.)

68. To melt iron, the temperature of a furnace must be at least 1540°C but no more than 1650°C. What range of Fahrenheit temperatures must be maintained? See Exercise 67.

69. The radii of phonograph records must lie between 5.9 and 6.1 inches. What variation in circumference can occur? (*Hint:* The circumference of a circle is given by the formula $C = 2\pi r$, where r is the radius. $\pi \approx 3.14$.)

70. The heights of children in three fourth-grade classes range from 42 to 57 inches. What is the range of their heights in centimeters? (*Hint:* There are 2.54 centimeters in 1 inch.)

71. The normal weight for a 6-foot 2-inch man is between 150 and 190 pounds. To the nearest hundredth, what would such a person weigh in kilograms? (*Hint:* There are 2.2 pounds in 1 kilogram.)

72. Jim's car averages between 19 and 23 miles per gallon. At $1.20 per gallon, what might Jim expect to pay for fuel on a 350-mile trip?

73. Chen weighs 40 pounds more than his mother and 10 pounds less than his father. All together, they weigh no more than 420 pounds and no less than 390 pounds. What might Chen weigh?

74. The time required to assemble a television set at the factory is 2 hours. A stereo receiver requires only 1 hour. The labor force at the factory can supply at least 640 and at most 810 hours of assembly time per week. When the factory is producing 3 times as many television sets as stereos, how many stereos could be manufactured in 1 week?

75. A rectangle's length is 3 feet less than twice its width, and its perimeter is between 24 and 48 feet. What might be its width?

76. A rectangle's width is 8 feet less than 3 times its length, and its perimeter is between 8 and 16 feet. What might be its length?

REVIEW EXERCISES

In Review Exercises 1–6, simplify each expression, if possible.

1. $3x^2 - 2(y^2 - x^2)$

2. $5(xy + 2) - 3xy - 8$

3. $x(2x - y) - 3(x^2 - xy)$

4. $7x(x - 2) + x^2(2 + y)$

5. $\frac{1}{3}(x + 6) - \frac{4}{3}(x - 9)$

6. $\frac{4}{5}x(y + 1) - \frac{9}{5}y(x - 1)$

In Review Exercises 7–10, let $x = -\frac{2}{3}$ and $z = \frac{1}{3}$. Evaluate each expression.

7. $x^2 - z^2$

8. $(x + z)^2$

9. $\dfrac{x - z}{x + z}$

10. $\dfrac{x + z}{z - x}$

11. A man spent $16,500 for a car and a boat. The car cost $500 more than three times the cost of the boat. Find the cost of the car.

12. On the last exam, one-half of a class received a grade of C, one-third a grade of B, and one-twelfth a grade of D. No one failed, and three students received an A. How many students are in the class?

CHAPTER SUMMARY

Key Words

*addition property of
    equality* (2.1)
*addition property of
    inequality* (2.8)
*base angles of an isosceles
    triangle* (2.6)
compound inequality (2.8)
*division property of
    equality* (2.2)
*division property of
    inequality* (2.8)
double inequality (2.8)
equation (2.1)

formula (2.4)
identity (2.3)
impossible equation (2.3)
inequality (2.8)
isosceles triangle (2.6)
like terms (2.3)
literal equation (2.4)
*multiplication property of
    equality* (2.2)
*multiplication property of
    inequality* (2.8)
numerical coefficient (2.3)
perimeter (2.6)

root of an equation (2.1)
similar terms (2.3)
solution of an equation (2.1)
solution of an inequality (2.8)
*subtraction property of
    equality* (2.1)
*subtraction property of
    inequality* (2.8)
unknown (2.1)
unlike terms (2.3)
variable (2.1)
*vertex angle of an isosceles
    triangle* (2.6)

Key Ideas

(2.1) Any real number can be added to (or subtracted from) both sides of an equation to form another equation with the same solutions as the original equation.

(2.2) Both sides of an equation can be multiplied (or divided) by any *nonzero* real number to form another equation with the same solutions as the original equation.

(2.3) Like terms can be combined by adding their numerical coefficients and using the same variables and exponents.

(2.4) A literal equation, or formula, can often be solved for any of its variables.

(2.5–2.7) Equations are useful in solving many applied problems.

(2.8) Inequalities are solved by techniques similar to those used to solve equations, with this important exception: *If both sides of an inequality are multiplied or divided by a negative number, the direction of the inequality must be reversed.*

The solution of an inequality can be graphed on the number line.

CHAPTER 2 REVIEW EXERCISES

(2.1) *In Review Exercises 1–8, indicate whether the indicated number is a solution of the given equation.*

1. $3x + 7 = 1$; -2

2. $5 - 2x = 3$; -1

3. $2(x + 3) = x$; -3

4. $5(3 - x) = 2 - 4x$; 13

5. $3\left(\dfrac{x}{2} - 1\right) = 9$; 8

6. $5\left(\dfrac{x}{3} + 3\right) = 40$; 15

7. $3(x + 5) = 2(x - 3)$; -21

8. $2(x - 7) = x + 14$; 0

(2.1–2.2) **In Review Exercises 9–32, solve each equation. Check all solutions.**

9. $x + 7 = 3$

10. $5 - x = 2$

11. $2x - 5 = 13$

12. $3x + 4 = -8$

13. $5y + 6 = 21$

14. $5y - 9 = 1$

15. $12z + 4 = -8$

16. $17z + 3 = 20$

17. $13 - 13t = 0$

18. $10 + 7t = -4$

19. $23a - 43 = 3$

20. $84 - 21a = -63$

21. $3x + 7 = 1$

22. $7 - 9x = 16$

23. $\dfrac{b + 3}{4} = 2$

24. $\dfrac{b - 7}{2} = -2$

25. $\dfrac{x - 8}{5} = 1$

26. $\dfrac{x + 10}{2} = -1$

27. $\dfrac{2(y - 1)}{4} = 2$

28. $\dfrac{3(y + 4)}{11} = 3$

29. $\dfrac{1}{2}(x + 7) = 11$

30. $\dfrac{1}{3}(x - 9) = 7$

31. $\dfrac{3}{4}\left(\dfrac{a}{3} + 3\right) = 6$

32. $\dfrac{3}{4}\left(\dfrac{b}{7} - 2\right) = 0$

(2.3) **In Review Exercises 33–44, simplify each expression, if possible.**

33. $5x + 9x$

34. $7a + 12a$

35. $18b - 13b$

36. $21x - 23x$

37. $5x + 7y$

38. $19x - 19$

39. $y^2 + 3(y^2 - 2)$

40. $2x^2 - 2(x^2 - 2)$

41. $7(x + 2) + 2(x - 7)$

42. $2(3 - x) + x - 6y$

43. $2^3(x + 9) - 8(x + 3^2)$

44. $2^3 + 3k + 3(k - 2)$

In Review Exercises 45–56, solve each equation. Check all solutions.

45. $5x + 7 = 4x$

46. $7x - 9 = 8x$

47. $8a + 2 = 2a + 8$

48. $12x + 3 = 5x - 11$

49. $2x - 19 = 2 - x$

50. $5b - 19 = 2b + 20$

51. $3x + 20 = 5 - 2x$

52. $9x + 100 = 7x + 18$

53. $10(t - 3) = 3(t + 11)$

54. $2(5x - 7) = 2(x - 35)$

55. $x(x + 6) = x^2 + 6$

56. $3y(y + 2) = 2(y^2 + y) + y^2$

(2.4) **In Review Exercises 57–68, solve each equation for the variable indicated.**

57. $E = IR$; for R

58. $i = prt$; for t

59. $P = I^2R$; for R

60. $d = rt$; for r

61. $V = lwh$; for h

62. $y = mx + b$; for m

63. $V = \pi r^2 h$; for h

64. $A = 2\pi rh$; for r

65. $F = \dfrac{GMm}{d^2}$; for G

66. $P = \dfrac{RT}{mV}$; for m

67. $T = n(V - 3)$; for V

68. $T = n(V - 3)$; for n

(2.5–2.7) **In Review Exercises 69–84, translate each problem into an equation, solve it, and check the solution.**

69. What number added to 19 gives 11?

70. If twice a number is decreased by 7, the result is 9. Find the original number.

71. If Mark had 10 cents more than twice the amount he has, he would have enough to pay $9.80 for a pizza. How much does Mark have?

72. Sue's electric rate is $17.50 per month, plus 18 cents for every kilowatt-hour of energy used. Her bill in July was $43.96. How many kilowatt-hours did she use that month?

73. The installation of rain gutters on Bert's house will cost $35, plus $1.50 per foot. He expects to pay $162.50 to replace the gutters. How many feet of gutter does he need?

74. The sum of two consecutive odd integers is 44. Find the two integers.

75. If it were 10 degrees warmer today, it would be 5 degrees colder than twice today's temperature. Find today's temperature.

76. A 45-foot rope is to be cut into three sections. One section is to be 15 feet long. Of the remaining sections, one is to be 2 feet less than 3 times the length of the other. Find the length of the shortest section.

77. The perimeter of a rectangle is 84 inches. If the length is 3 inches more than twice the width, how wide is the rectangle?

78. Costs for machining an automotive part run $668.50 per week, plus variable costs of $1.25 per unit. The company can sell as many as it can make for $3 each. Find the break-even point.

79. A company can manufacture baseball caps on either of two machines, with costs as shown in the following table. At the projected sales level, the company finds that costs on the two machines are equal. What is the expected production?

Machine	Startup cost	Unit cost (per cap)
1	$85	$3
2	$105	$2.50

80. Juanita invests $27,000 for 1 year. She invests part of it in a certificate of deposit paying 7% interest and the remaining amount in a cash management fund paying 9%. The total interest on her two investments is $2110. How much did she invest at each rate?

81. A bicycle path is 5 miles long. Jim starts at one end, walking at the rate of 3 miles per hour. At the same time, his friend Jerry bicycles from the other end, traveling 12 miles per hour. In how many minutes will they meet?

82. A store manager mixes candy worth 90 cents per pound with gumdrops worth $1.50 per pound to make 20 pounds of a mixture worth $1.20 per pound. How many pounds of each kind of candy does the manager use?

83. After a 6% raise, Julie makes $18,550. Find her salary before the raise.

84. After sales tax, a $35 book will cost $37.45. What is the sales tax rate?

(2.8) In Review Exercises 85–94, solve each inequality and graph its solution.

85. $3x + 2 < 5$

86. $-5x - 8 > 7$

87. $5x - 3 \geq 2x + 9$

88. $7x + 1 \leq 8x - 5$

89. $5(3 - x) \leq 3(x - 3)$

90. $3(5 - x) \geq 2x$

91. $8 < x + 2 < 13$

92. $0 \leq 2 - 2x < 4$

93. $x^2 < x(x + 1) \leq x^2 + 9$

94. $x^2 + 4 \geq x^2 + x > x^2 + 3$

CHAPTER 2 TEST

In Problems 1–4, state whether the indicated number is a solution of the given equation.

_____ **1.** $5x + 3 = -2$; -1

_____ **2.** $3(x + 2) = 2x$; -6

_____ **3.** $-3(2 - x) = 0$; -2

_____ **4.** $x(x + 1) = x^2 + 1$; 1

In Problems 5–10, solve each equation.

_____   **5.** $8x + 2 = -14$ _____   **6.** $3 = 5 - 2x$

_____   **7.** $23 - 5(x + 10) = -12$ _____   **8.** $x(x + 5) = x^2 + 3x - 8$

_____   **9.** $\dfrac{3(x - 6)}{2} = 6x$ _____   **10.** $\dfrac{5}{3}(x - 7) = 15(x + 1)$

In Problems 11–16, simplify each expression.

_____   **11.** $x + 5(x - 3)$ _____   **12.** $3x - 5(2 - x)$

_____   **13.** $x(x - 3) + 2x^2 - 3x$ _____   **14.** $2^2(x^2 - 3^2) - x(x + 36)$

_____   **15.** $-3x(x + 3) + 3x(x - 3)$ _____   **16.** $-4x(2x - 5) - 7x(4x + 1)$

In Problems 17–22, solve each equation for the variable indicated.

_____   **17.** $d = rt$; for t _____   **18.** $P = 2l + 2w$; for l

_____   **19.** $A = 2\pi rh$; for h _____   **20.** $A = P + Prt$; for r

_____   **21.** $P = \dfrac{RT}{v}$; for v _____   **22.** $A = \dfrac{1}{3}\pi r^2 h$; for h

_____   **23.** A chocolate chip cookie has two-thirds of the calories of a glass of milk. Jim has a glass of milk and 2 cookies and consumes 385 calories. How many calories are in a glass of milk?

_____   **24.** The sum of two consecutive odd integers is 36. Find the integers.

_____   **25.** Karen invested part of $13,750 at 9% annual interest and the rest at 8%. After 1 year, she received $1185 in interest. How much did she invest at the lower rate?

_____   **26.** A car leaves Rockford, Illinois at the rate of 65 miles per hour bound for Madison, Wisconsin. At the same time, a truck leaves Madison at the rate of 55 miles per hour bound for Rockford. If the cities are 72 miles apart, how long will it take for the car and the truck to meet?

_____   **27.** How many liters of water must be added to 30 liters of a 10% brine solution to dilute it to an 8% solution?

In Problems 28–30, solve each inequality and graph its solution.

28. $8x - 20 \geq 4$

29. $x^2 - x(x + 7) > 14$

30. $-4 \leq 2(x + 1) < 10$

note that the expression x^2 means that x is to be used as a factor two times and the expression x^3 means that x is to be used as a factor three times. Thus,

$$x^2 x^3 = \overbrace{x \cdot x}^{2 \text{ factors of } x} \cdot \overbrace{x \cdot x \cdot x}^{3 \text{ factors of } x} = \overbrace{x \cdot x \cdot x \cdot x \cdot x}^{5 \text{ factors of } x} = x^5$$

In general,

$$x^m x^n = \overbrace{x \cdot x \cdot x \cdots\cdots x}^{m \text{ factors of } x} \overbrace{x \cdot x \cdot x \cdots\cdots x}^{n \text{ factors of } x}$$

$$= \overbrace{x \cdot x \cdot x \cdot x \cdot x \cdot x \cdots\cdots x \cdot x \cdot x}^{m + n \text{ factors of } x}$$

$$= x^{m+n}$$

This discussion justifies the rule for multiplying exponential expressions: *To multiply two exponential expressions with the same base, we keep the base and add the exponents.*

The Product Rule for Exponents.   If m and n are natural numbers, then

$$x^m x^n = x^{m+n}$$

EXAMPLE 3   **a.** $x^3 x^4 = x^{3+4}$ **b.** $y^2 y^4 = y^{2+4}$
$\qquad\qquad\qquad = x^7$ $\qquad\qquad\qquad\qquad = y^6$

c. $zz^3 = z^1 z^3$ $\qquad\qquad\quad$ **d.** $x^2 x^3 x^6 = (x^2 x^3)x^6$
$\qquad = z^{1+3}$ $\qquad\qquad\qquad\qquad\quad = (x^{2+3})x^6$
$\qquad = z^4$ $\qquad\qquad\qquad\qquad\qquad = x^5 x^6$
$\qquad\qquad\qquad\qquad\qquad\qquad\qquad = x^{5+6}$
$\qquad\qquad\qquad\qquad\qquad\qquad\qquad = x^{11}$

e. $(2y^3)(3y^2) = 2(3)y^3 y^2$ $\qquad$ **f.** $(4x)(-3x^2) = 4(-3)xx^2$
$\qquad\qquad\quad = 6y^{3+2}$ $\qquad\qquad\qquad\qquad = -12x^{1+2}$
$\qquad\qquad\quad = 6y^5$ $\qquad\qquad\qquad\qquad = -12x^3$ ■

The product rule for exponents applies only to exponential expressions with the same base. An expression such as $x^2 y^3$ cannot be simplified because x^2 and y^3 have different bases.

To find another rule of exponents, we consider the expression $(x^3)^4$. The expression $(x^3)^4$ can be written as $x^3 \cdot x^3 \cdot x^3 \cdot x^3$. Because each of the four factors of x^3 contains three factors of x, there are $4 \cdot 3$, or 12, factors of x. Thus, the product can be written as x^{12}.

$$(x^3)^4 = x^3 \cdot x^3 \cdot x^3 \cdot x^3$$

$$\overbrace{= x \cdot x \cdot x \cdot x \cdot x \cdot x \cdot x \cdot x \cdot x \cdot x \cdot x \cdot x}^{\text{12 factors of } x}$$

$$\underbrace{x^3} \quad \underbrace{x^3} \quad \underbrace{x^3} \quad \underbrace{x^3}$$

$$= x^{12}$$

In general,

$$(x^m)^n = \overbrace{x^m \cdot x^m \cdot x^m \cdot \ \cdots \ \cdot x^m}^{n \text{ factors of } x^m}$$

$$= \overbrace{x \cdot x \cdot x \cdot x \cdot x \cdot x \cdot x \cdot \ \cdots \ \cdot x}^{mn \text{ factors of } x}$$

$$= x^{mn}$$

This discussion justifies a rule for raising an exponential expression to a power: *To raise an exponential expression to a power, we keep the same base and multiply the exponents.*

The First Power Rule for Exponents.   If m and n are natural numbers, then

$$(x^m)^n = x^{mn}$$

EXAMPLE 4   **a.** $(2^3)^7 = 2^{3 \cdot 7}$ **b.** $(y^5)^2 = y^{5 \cdot 2}$ **c.** $(z^7)^7 = z^{7 \cdot 7}$ **d.** $(u^x)^y = u^{x \cdot y}$

$\qquad\qquad\qquad = 2^{21} \qquad\qquad\qquad = y^{10} \qquad\qquad\qquad = z^{49} \qquad\qquad\qquad = u^{xy}$ ■

In Example 5, both the product and power rules for exponents are applied.

EXAMPLE 5   **a.** $(x^2 x^5)^2 = (x^7)^2$ **b.** $(yy^6 y^2)^3 = (y^9)^3$

$\qquad\qquad\qquad\quad = x^{14} \qquad\qquad\qquad\qquad = y^{27}$

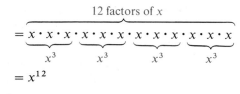

c. $(z^2)^4 (z^3)^3 = z^8 z^9$ **d.** $(x^3)^2 (x^5 x^2)^3 = x^6 (x^7)^3$

$\qquad\qquad\quad = z^{17} \qquad\qquad\qquad\qquad\quad = x^6 x^{21}$

$\qquad\qquad\qquad\qquad\qquad\qquad\qquad\qquad = x^{27}$ ■

To find two more power rules for exponents, we consider the expressions $(2x)^3$ and $\left(\frac{2}{x}\right)^3$.

$$(2x)^3 = (2x)(2x)(2x) \qquad\qquad \left(\frac{2}{x}\right)^3 = \left(\frac{2}{x}\right)\left(\frac{2}{x}\right)\left(\frac{2}{x}\right) \qquad (x \neq 0)$$

$$= (2 \cdot 2 \cdot 2)(x \cdot x \cdot x) \qquad\qquad\qquad = \frac{2 \cdot 2 \cdot 2}{x \cdot x \cdot x}$$

$$= 2^3 x^3$$

$$= 8x^3 \qquad\qquad\qquad\qquad\qquad\qquad = \frac{2^3}{x^3}$$

$$= \frac{8}{x^3}$$

These examples suggest that *to raise a product to a power, we raise each factor of the product to the power,* and *to raise a fraction to a power, we raise both the numerator and the denominator to the power.*

More Power Rules for Exponents. If n is a natural number, then

$$(xy)^n = x^n y^n \quad \text{and} \quad \left(\frac{x}{y}\right)^n = \frac{x^n}{y^n} \quad (y \neq 0)$$

EXAMPLE 6

a. $(ab)^4 = a^4 b^4$

b. $(3c)^3 = 3^3 c^3$
$= 27c^3$

c. $(x^2 y^3)^5 = (x^2)^5 (y^3)^5$
$= x^{10} y^{15}$

d. $(-2x^3 y)^2 = (-2)^2 (x^3)^2 y^2$
$= 4x^6 y^2$

e. $\left(\frac{4}{k}\right)^3 = \frac{4^3}{k^3}$
$= \frac{64}{k^3}$

f. $\left(\frac{3x^2}{2y^3}\right)^5 = \frac{3^5 (x^2)^5}{2^5 (y^3)^5}$
$= \frac{243 x^{10}}{32 y^{15}}$ ∎

To find a rule for dividing exponential expressions, we consider the fraction

$$\frac{4^5}{4^2}$$

where the exponent in the numerator is greater than the exponent in the denominator. We can simplify the fraction as follows:

$$\frac{4^5}{4^2} = \frac{4 \cdot 4 \cdot 4 \cdot 4 \cdot 4}{4 \cdot 4} = \frac{\overset{1}{\cancel{4}} \cdot \overset{1}{\cancel{4}} \cdot 4 \cdot 4 \cdot 4}{\underset{1}{\cancel{4}} \cdot \underset{1}{\cancel{4}}} = 4^3$$

The result of 4^3 has a base of 4 and an exponent of $5 - 2$, or 3. This suggests that *to divide exponential expressions with the same base, we keep the base and subtract the exponents.*

The Quotient Rule for Exponents. If m and n are natural numbers, $m > n$ and $x \neq 0$, then

$$\frac{x^m}{x^n} = x^{m-n}$$

EXAMPLE 7 If there are no divisions by 0, then

a. $\dfrac{x^4}{x^3} = x^{4-3} = x^1 = x$

b. $\dfrac{8y^2 y^6}{4y^3} = \dfrac{8y^8}{4y^3} = \dfrac{8}{4} y^{8-3} = 2y^5$

c. $\dfrac{a^3 a^5 a^7}{a^4 a} = \dfrac{a^{15}}{a^5} = a^{15-5} = a^{10}$

d. $\dfrac{(a^3 b^4)^2}{ab^5} = \dfrac{a^6 b^8}{ab^5} = a^{6-1} b^{8-5} = a^5 b^3$ ∎

The rules for positive exponents are summarized as follows:

Properties of Exponents.   If n is a natural number, then

$$x^n = \overbrace{x \cdot x \cdot x \cdot \cdots \cdot x}^{n \text{ factors of } x}$$

If m and n are natural numbers and there are no divisions by 0, then

$$x^m x^n = x^{m+n} \qquad (x^m)^n = x^{mn} \qquad (xy)^n = x^n y^n \qquad \left(\frac{x}{y}\right)^n = \frac{x^n}{y^n}$$

$$\frac{x^m}{x^n} = x^{m-n} \qquad \text{provided } m > n$$

EXERCISE 3.1

In Exercises 1–12, identify the base and the exponent in each expression.

1. 4^3

2. $(-5)^2$

3. x^5

4. y^8

5. $(2y)^3$

6. $(-3x)^2$

7. $-x^4$

8. $(-x)^4$

9. x

10. (xy)

11. $2x^3$

12. $-3y^6$

In Exercises 13–20, write each expression without using exponents.

13. 5^3

14. -4^5

15. x^7

16. $3x^3$

17. $-4x^5$

18. $(-2y)^4$

19. $(3t)^5$

20. $a^3 b^2$

In Exercises 21–28, write each expression using exponents.

21. $2 \cdot 2 \cdot 2$

22. $5 \cdot 5$

23. $x \cdot x \cdot x \cdot x$

24. $y \cdot y \cdot y \cdot y \cdot y \cdot y$

25. $(2x)(2x)(2x)$

26. $(-4y)(-4y)$

27. $-4t \cdot t \cdot t \cdot t$

28. $5 \cdot u \cdot u$

In Exercises 29–36, evaluate each expression.

29. 5^4

30. $(-3)^3$

31. $2^2 + 3^2$

32. $2^3 - 2^2$

33. $5^4 - 4^3$

34. $2(4^3 + 3^2)$

35. $-5(3^4 + 4^3)$

36. $-5^2(4^3 - 2^6)$

In Exercises 37–54, write each expression as an expression involving only one exponent.

37. $x^4 x^3$

38. $y^5 y^2$

39. $x^5 x^2$

40. yy^3

41. tt^2

42. $w^3 w^4$

43. $a^3 a^4 a^5$

44. $b^2 b^3 b^5$

45. $y^3(y^2 y^4)$

46. $(y^4 y)y^6$

47. $4x^2(3x^5)$

48. $-2y(y^3)$

49. $(-y^2)(4y^3)$

50. $(-4x^3)(-5x)$

51. $6x^3(2x^2)(x^4)$

52. $-2x(x^2)(3x)$

53. $(-2a^3)(3a^2)(-a)$

54. $(-a^3)(-a^2)(-a)$

In Exercises 55–70, write each expression as an expression involving only one exponent.

55. $(3^2)^4$

56. $(4^3)^3$

57. $(y^5)^3$

58. $(b^3)^6$

59. $(a^3)^7$

60. $(b^2)^3$

61. $(x^2x^3)^5$

62. $(y^3y^4)^4$

63. $(3zz^2z^3)^5$

64. $(4t^3t^6t^2)^2$

65. $(x^5)^2(x^7)^3$

66. $(y^3y)^2(y^2)^2$

67. $(r^3r^2)^4(r^3r^5)^2$

68. $(s^2)^3(s^3)^2(s^4)^4$

69. $(s^3)^3(s^2)^2(s^5)^4$

70. $(yy^3)^3(y^2y^3)^4(y^3y^3)^2$

In Exercises 71–90, write each expression without using parentheses.

71. $(xy)^3$

72. $(uv^2)^4$

73. $(r^3s^2)^2$

74. $(a^3b^2)^3$

75. $(4ab^2)^2$

76. $(3x^2y)^3$

77. $(-2r^2s^3t)^3$

78. $(-3x^2y^4z)^2$

79. $\left(\dfrac{a}{b}\right)^3$

80. $\left(\dfrac{r^2}{s}\right)^4$

81. $\left(\dfrac{x^2}{y^3}\right)^5$

82. $\left(\dfrac{u^4}{v^2}\right)^6$

83. $\left(\dfrac{-2a}{b}\right)^5$

84. $\left(\dfrac{2t}{3}\right)^4$

85. $\dfrac{(a^3a^5)^3}{(a^2)^4}$

86. $\dfrac{(x^2x^7)^3}{(x^3x^2)^2}$

87. $\left(\dfrac{y^3y}{2yy^2}\right)^2$

88. $\left(\dfrac{3t^3t^4t^5}{4t^2t^6}\right)^3$

89. $\left(\dfrac{-2r^3r^3}{3r^4r}\right)^3$

90. $\left(\dfrac{-6y^4y^5}{5y^3y^5}\right)^2$

In Exercises 91–102, simplify each expression.

91. $\dfrac{x^5}{x^3}$

92. $\dfrac{a^6}{a^3}$

93. $\dfrac{y^3y^4}{yy^2}$

94. $\dfrac{b^4b^5}{b^2b^3}$

95. $\dfrac{12a^2a^3a^4}{4(a^4)^2}$

96. $\dfrac{16(aa^2)^3}{2a^2a^3}$

97. $\dfrac{(ab^2)^3}{(ab)^2}$

98. $\dfrac{(m^3n^4)^3}{(mn^2)^3}$

99. $\dfrac{20(r^4s^3)^4}{6(rs^3)^3}$

100. $\dfrac{15(x^2y^5)^5}{21(x^3y)^2}$

101. $\dfrac{17(x^4y^3)^8}{34(x^5y^2)^4}$

102. $\dfrac{35(r^3s^2)^2}{49r^2s^3}$

REVIEW EXERCISES

1. Graph the real numbers 3, 0, -4, and $-\frac{5}{2}$ on a number line.

2. Graph the *negatives* of 5 and -3 on a number line.

In Review Exercises 3–6, perform the indicated operation.

3. $\dfrac{5}{6} + \dfrac{2}{9}$

4. $\dfrac{1}{2} - \dfrac{1}{5}$

5. $\dfrac{2}{7} \cdot \dfrac{21}{6}$

6. $\dfrac{1}{5} \div \dfrac{3}{15}$

In Review Exercises 7–10, translate each algebraic expression into an English phrase.

7. $3(x + y)$

8. $(x + y)(x - y)$

9. $|x - y|$

10. $\dfrac{x + y}{xy}$

In Review Exercises 11–12, translate each English phrase into an algebraic expression.

11. Three greater than the absolute value of twice x.

12. The sum of the numbers y and z, decreased by the sum of their squares.

3.2   ZERO AND NEGATIVE INTEGRAL EXPONENTS

In Section 3.1 we developed the quotient rule for exponents where the exponent in the numerator was greater than the exponent in the denominator. We now consider what happens when the exponents are equal, and when the denominator exponent is greater than the numerator exponent.

If we apply the quotient rule to the fraction $\frac{5^3}{5^3}$, where the exponents in the numerator and denominator are equal, we obtain

$$\frac{5^3}{5^3} = 5^{3-3} = 5^0$$

However, because any nonzero number divided by itself is equal to 1, we have

$$\frac{5^3}{5^3} = 1$$

To make the results of 5^0 and 1 consistent, we shall define 5^0 to be equal to 1. In general, we have

---

DEFINITION.   If x is any nonzero real number, then
$$x^0 = 1$$

EXAMPLE 1   **a.** $\left(\frac{1}{13}\right)^0 = 1$ **b.** $(-0.115)^0 = 1$ **c.** $\dfrac{4^2}{4^2} = 4^{2-2}$
$$= 4^0$$
$$= 1$$

d. $\dfrac{x^5}{x^5} = x^{5-5} \quad (x \neq 0)$ **e.** $3x^0 = 3(1)$ **f.** $(3x)^0 = 1$
$$= x^0 \qquad\qquad\qquad = 3$$
$$= 1$$

g. $\dfrac{6^n}{6^n} = 6^{n-n}$ **h.** $\dfrac{y^m}{y^m} = y^{m-m} \quad (y \neq 0)$
$$= 6^0 \qquad\qquad\qquad\quad = y^0$$
$$= 1 \qquad\qquad\qquad\quad = 1$$

Parts **e** and **f** show that $3x^0 \neq (3x)^0$.   ■

If we apply the quotient rule to the fraction $\frac{6^2}{6^5}$, where the exponent in the numerator is less than the exponent in the denominator, we obtain

$$\frac{6^2}{6^5} = 6^{2-5} = 6^{-3}$$

However, we know that

$$\frac{6^2}{6^5} = \frac{\overset{1}{\cancel{6}} \cdot \overset{1}{\cancel{6}}}{\underset{1}{\cancel{6}} \cdot \underset{1}{\cancel{6}} \cdot 6 \cdot 6 \cdot 6} = \frac{1}{6^3}$$

To make the results of 6^{-3} and $\frac{1}{6^3}$ consistent, we shall define 6^{-3} to be equal to $\frac{1}{6^3}$. In general,

---

DEFINITION. If x is any nonzero number and n is a natural number, then

$$x^{-n} = \frac{1}{x^n}$$

EXAMPLE 2   Write each quantity without using negative exponents or parentheses. Assume that no denominators are zero.

a. $3^{-5} = \dfrac{1}{3^5}$ **b.** $x^{-4} = \dfrac{1}{x^4}$ **c.** $(2x)^{-2} = \dfrac{1}{(2x)^2}$

$$= \frac{1}{4x^2}$$

d. $2x^{-2} = 2\left(\dfrac{1}{x^2}\right)$ **e.** $(-3a)^{-4} = \dfrac{1}{(-3a)^4}$ **f.** $\left(x^3x^2\right)^{-3} = \left(x^5\right)^{-3}$

$$= \frac{2}{x^2} \qquad\qquad = \frac{1}{81a^4} \qquad\qquad = \frac{1}{(x^5)^3}$$

$$= \frac{1}{x^{15}} \qquad \blacksquare$$

Because of the definition of negative and zero exponents, the product, power, and quotient rules are also true for all integral exponents. We restate the properties of exponents for integral exponents.

Properties of Exponents.   If m and n are natural numbers and there are no divisions by 0, then

$$x^m x^n = x^{m+n} \qquad (x^m)^n = x^{mn} \qquad (xy)^n = x^n y^n \qquad \left(\frac{x}{y}\right)^n = \frac{x^n}{y^n}$$

$$x^0 = 1 \quad (x \neq 0) \qquad x^{-n} = \frac{1}{x^n} \qquad \frac{x^m}{x^n} = x^{m-n}$$

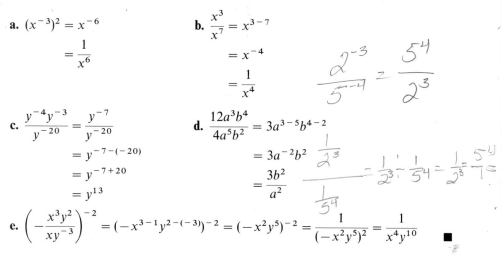

EXAMPLE 3 Simplify each quantity and write the result without using negative exponents. Assume that no denominators are zero.

a. $(x^{-3})^2 = x^{-6}$

$\qquad = \dfrac{1}{x^6}$

b. $\dfrac{x^3}{x^7} = x^{3-7}$

$\qquad = x^{-4}$

$\qquad = \dfrac{1}{x^4}$

c. $\dfrac{y^{-4}y^{-3}}{y^{-20}} = \dfrac{y^{-7}}{y^{-20}}$

$\qquad = y^{-7-(-20)}$

$\qquad = y^{-7+20}$

$\qquad = y^{13}$

d. $\dfrac{12a^3b^4}{4a^5b^2} = 3a^{3-5}b^{4-2}$

$\qquad = 3a^{-2}b^2$

$\qquad = \dfrac{3b^2}{a^2}$

e. $\left(-\dfrac{x^3y^2}{xy^{-3}}\right)^{-2} = (-x^{3-1}y^{2-(-3)})^{-2} = (-x^2y^5)^{-2} = \dfrac{1}{(-x^2y^5)^2} = \dfrac{1}{x^4y^{10}}$ ∎

The properties of exponents discussed thus far are also true for exponential expressions with variables in an exponent.

EXAMPLE 4

a. $x^m x^{3m} = x^{m+3m}$

$\qquad = x^{4m}$

b. $\dfrac{y^{2m}}{y^{4m}} = y^{2m-4m} \qquad (y \neq 0)$

$\qquad = y^{-2m}$

$\qquad = \dfrac{1}{y^{2m}}$

c. $a^{2m-1}a^{2m} = a^{2m-1+2m}$

$\qquad = a^{4m-1}$

d. $(b^{m+1})^2 = b^{(m+1)2}$

$\qquad = b^{2m+2}$ ∎

EXERCISE 3.2

In Exercises 1–64, simplify each expression. Write each answer without using parentheses or negative exponents.

1. $2^5 \cdot 2^{-2}$

2. $10^2 \cdot 10^{-4} \cdot 10^5$

3. $4^{-3} \cdot 4^{-2} \cdot 4^5$

4. $3^{-4} \cdot 3^5 \cdot 3^{-3}$

5. $\dfrac{3^5 \cdot 3^{-2}}{3^3}$

6. $\dfrac{6^2 \cdot 6^{-3}}{6^{-2}}$

7. $\dfrac{2^5 \cdot 2^7}{2^6 \cdot 2^{-3}}$

8. $\dfrac{5^{-2} \cdot 5^{-4}}{5^{-6}}$

9. $2x^0$

10. $(2x)^0$

11. $(-x)^0$

12. $-y^0$

13. $\left(\dfrac{a^2b^3}{ab^4}\right)^0$

14. $\dfrac{2}{3}\left(\dfrac{xyz}{x^2y}\right)^0$

15. $\dfrac{x^0 - 5x^0}{2x^0}$

16. $\dfrac{4a^0 + 2a^0}{3a^0}$

17. x^{-2}

18. y^{-3}

19. b^{-5}

20. c^{-4}

21. $(2y)^{-4}$

22. $(-3x)^{-1}$

23. $(ab^2)^{-3}$

24. $(m^2n^3)^{-2}$

$a^{-3} = a^{2-3} = \dfrac{a^2}{a^5} = \dfrac{1}{a^3}$

25. $\dfrac{y^4}{y^5}$

26. $\dfrac{t^7}{t^{10}}$

27. $\dfrac{(r^2)^3}{(r^3)^4}$

28. $\dfrac{(b^3)^4}{(b^5)^4}$

29. $\dfrac{y^4 y^3}{y^4 y^{-2}}$

30. $\dfrac{x^{12}x^{-7}}{x^3 x^4}$

31. $\dfrac{a^4 a^{-2}}{a^2 a^0}$

32. $\dfrac{b^0 b^3}{b^{-3} b^4}$

33. $(ab^2)^{-2}$

34. $(c^2 d^3)^{-2}$

35. $(x^2 y)^{-3}$

36. $(-xy^2)^{-4}$

37. $(x^{-4} x^3)^3$

38. $(y^{-2} y)^3$

39. $(y^3 y^{-2})^{-2}$

40. $(x^{-3} x^{-2})^2$

41. $(a^{-2} b^{-3})^{-4}$

42. $(y^{-3} z^5)^{-6}$

43. $(-2x^3 y^{-2})^{-5}$

44. $(-3u^{-2} v^3)^{-3}$

45. $\left(\dfrac{a^3}{a^{-4}}\right)^2$

46. $\left(\dfrac{a^4}{a^{-3}}\right)^3$

47. $\left(\dfrac{b^5}{b^{-2}}\right)^{-2}$

48. $\left(\dfrac{b^{-2}}{b^3}\right)^{-3}$

49. $\left(\dfrac{4x^2}{3x^{-5}}\right)^4$

50. $\left(\dfrac{-3r^4 r^{-3}}{r^{-3} r^7}\right)^3$

51. $\left(\dfrac{12 y^3 z^{-2}}{3 y^{-4} z^3}\right)^2$

52. $\left(\dfrac{6xy^3}{3x^{-1} y}\right)^3$

53. $\left(\dfrac{2x^3 y^{-2}}{4xy^2}\right)^7$

54. $\left(\dfrac{9u^2 v^3}{18u^{-3} v}\right)^4$

55. $\left(\dfrac{14u^{-2} v^3}{21u^{-3} v}\right)^4$

56. $\left(\dfrac{-27u^{-5} v^{-3} w}{18u^3 v^{-2}}\right)^4$

57. $\left(\dfrac{6a^2 b^3}{2ab^2}\right)^{-2}$

58. $\left(\dfrac{15r^2 s^{-2} t}{3r^{-3} s^3}\right)^{-3}$

59. $\left(\dfrac{18a^2 b^3 c^{-4}}{3a^{-1} b^2 c}\right)^{-3}$

60. $\left(\dfrac{21x^{-2} y^2 z^{-2}}{7x^3 y^{-1}}\right)^{-2}$

61. $\dfrac{(2x^{-2} y)^{-3}}{(4x^2 y^{-1})^3}$

62. $\dfrac{(ab^{-2} c)^2}{(a^{-2} b)^{-3}}$

63. $\dfrac{(17x^5 y^{-5} z)^{-3}}{(17x^{-5} y^3 z^2)^{-4}}$

64. $\dfrac{16(x^{-2} yz)^{-2}}{(2x^{-3} z^0)^4}$

In Exercises 65–80, write each expression with a single exponent.

65. $x^{2m} x^m$

66. $y^{3m} y^{2m}$

67. $u^{2m} v^{3n} u^{3m} v^{-3n}$

68. $r^{2m} s^{-3} r^{3m} s^3$

69. $y^{3m+2} y^{-m}$

70. $x^{m+1} x^m$

71. $\dfrac{y^{3m}}{y^{2m}}$

72. $\dfrac{z^{4m}}{z^{2m}}$

73. $\dfrac{x^{3n}}{x^{6n}}$

74. $\dfrac{x^m}{x^{5m}}$

75. $(x^{m+1})^2$

76. $(y^2)^{m+1}$

77. $(x^{3-2n})^{-4}$

78. $(y^{1-n})^{-3}$

79. $(y^{2-n})^{-4}$

80. $(x^{3-4n})^{-2}$

REVIEW EXERCISES

1. If $a = -2$ and $b = 3$, evaluate $\dfrac{3a^2 + 4b + 8}{a + 2b^2}$.

2. Evaluate $|-3 + 5 \cdot 2|$.

In Review Exercises 3–4, solve each equation.

3. $3(x - 4) = 12$

4. $\dfrac{2}{3}(4x - 3) = 6$ $3-2 = 1m$

5. If 5 times a certain number is increased by 6, the result is 51. Find the number.

6. Solve the equation $\dfrac{5(2 - x)}{6} = \dfrac{x + 6}{2}$.

7. Solve the equation $P = L + \dfrac{s}{f} i$ for s.

8. A man wants to cut a 33-foot rope into three pieces. He wants the shortest piece to be 8 feet shorter than the middle-sized piece and 13 feet shorter than the longest piece. After the rope is cut, how long is each piece?

In Review Exercises 9–10, solve each inequality and graph the solution set.

9. $-4(r + 1) > 2(r - 3)$

10. $2 < -2(x + 3) < 10$

11. A woman buys two cars for a total of $41,400. One of them costs $2400 more than twice the other. Find the cost of each.

12. The sum of two consecutive even numbers is 54. Find the numbers.

3.3   SCIENTIFIC NOTATION

Scientists often deal with extremely large and extremely small numbers. For example, the distance from the Earth to the sun is approximately 150,000,000 kilometers, and ultraviolet light emitted from a mercury arc has a wavelength of approximately 0.000025 centimeter. The large number of zeros in these numbers makes them difficult to read and hard to remember. To make such numbers easier to work with, scientists use a compact form of notation called **scientific notation.**

> **DEFINITION.**   A number is written in **scientific notation** if it is written as the product of a number between 1 (including 1) and 10 and a power of 10.

EXAMPLE 1   Change 150,000,000 to scientific notation.

Solution   To write the number 150,000,000 in scientific notation, we must express it as a product of a number between 1 and 10 and some power of 10. Note that the number 1.5 lies between 1 and 10. To obtain the number 150,000,000, the decimal point in the number 1.5 must be moved eight places to the right. Because multiplying a number by 10 moves the decimal place one place to the right, we can accomplish this by multiplying 1.5 by 10 eight times. To see this, count from the decimal point in 1.5 to where the decimal point should be in 150,000,000:

$$1.\underset{\smile}{5\,0\,0\,0\,0\,0\,0\,0}\,_{\wedge} \qquad \text{8 places to the right}$$

Thus, the number 150,000,000 written in scientific notation is 1.5×10^8.   ∎

EXAMPLE 2   Change 0.000025 to scientific notation.

Solution   To write the number 0.000025 in scientific notation, we must express it as a product of a number between 1 and 10 and some power of 10. To obtain the number 0.000025, the decimal point in the number 2.5 must be moved five places to the left. We can accomplish this by dividing 2.5 by 10^5, which is equivalent to multiplying 2.5 by $\frac{1}{10^5}$, or by 10^{-5}. To see this, count from the decimal point in 2.5 to where the decimal point should be in 0.000025:

$$0\,0\,0\,0\,2\,.5 \qquad \text{5 places to the left}$$

Thus, the number 0.000025 written in scientific notation is 2.5×10^{-5}.   ∎

EXAMPLE 3   Write **a.** 235,000 and **b.** 0.00000235 in scientific notation.

Solution   **a.** $235{,}000 = 2.35 \times 10^5$ because $2.35 \times 10^5 = 235{,}000$ and 2.35 is between 1 and 10.

b. $0.00000235 = 2.35 \times 10^{-6}$ because $2.35 \times 10^{-6} = 0.00000235$ and 2.35 is between 1 and 10.   ∎

If a number is written in scientific notation, it can be changed to **standard notation.** For example, to write the number 9.3×10^7 in standard notation, we multiply 9.3 by 10^7.

$$9.3 \times 10^7 = 9.3 \times 10{,}000{,}000 = 93{,}000{,}000$$

EXAMPLE 4   Write **a.** 3.4×10^5 and **b.** 2.1×10^{-4} in standard notation.

Solution   **a.** $3.4 \times 10^5 = 3.4 \times 100{,}000$
$= 340{,}000$

b. $2.1 \times 10^{-4} = 2.1 \times \dfrac{1}{10^4}$
$= 2.1 \times \dfrac{1}{10{,}000}$
$= 0.00021$ ∎

Each of the following numbers is written in both scientific and standard notation. In each case, the exponent gives the number of places that the decimal point moves, and the sign of the exponent indicates the direction that it moves.

$$5.32 \times 10^5 = 5\,3\,2\,0\,0\,0. \qquad \text{5 places to the right}$$
$$2.37 \times 10^6 = 2\,3\,7\,0\,0\,0\,0. \qquad \text{6 places to the right}$$
$$8.95 \times 10^{-4} = 0.0\,0\,0\,8\,9\,5 \qquad \text{4 places to the left}$$
$$8.375 \times 10^{-3} = 0.0\,0\,8\,3\,7\,5 \qquad \text{3 places to the left}$$
$$9.77 \times 10^0 = 9.77 \qquad \text{no movement of the decimal point}$$

EXAMPLE 5   Write 432.0×10^5 in scientific notation.

Solution   The number 432.0×10^5 is not written in scientific notation because 432.0 is not a number between 1 and 10. To write the number in scientific notation, proceed as follows:

$$432.0 \times 10^5 = 4.32 \times 10^2 \times 10^5 \qquad \text{Write 432.0 in scientific notation.}$$
$$= 4.32 \times 10^7 \qquad \text{Simplify.} \qquad \blacksquare$$

Another advantage of scientific notation becomes apparent when we must simplify fractions such as

$$\frac{(0.0032)(25,000)}{0.00040}$$

that contain very large or very small numbers. Although we can simplify this fraction by using ordinary arithmetic, scientific notation provides an easier way. First we write each number in scientific notation and then do the arithmetic on the numbers and the exponential expressions separately. Then we write the answer in standard form, if desired.

$$\frac{(0.0032)(25,000)}{0.00040} = \frac{(3.2 \times 10^{-3})(2.5 \times 10^4)}{4.0 \times 10^{-4}}$$
$$= \frac{(3.2)(2.5)}{4.0} \times \frac{10^{-3}10^4}{10^{-4}}$$
$$= \frac{8.0}{4.0} \times 10^{-3+4-(-4)}$$
$$= 2.0 \times 10^5$$
$$= 200,000$$

EXAMPLE 6   In a vacuum, light travels 1 meter in approximately 0.000000003 second. How long does it take for light to travel 500 kilometers?

Solution   Because 1 kilometer is equal to 1000 meters, the length of time for light to travel 500 kilometers (500 · 1000 meters) is given by

$$(0.000000003)(500)(1000) = (3 \times 10^{-9})(5 \times 10^2)(1 \times 10^3)$$
$$= 3(5) \times 10^{-9+2+3}$$
$$= 15 \times 10^{-4}$$
$$= 1.5 \times 10^1 \times 10^{-4}$$
$$= 1.5 \times 10^{-3}$$
$$= 0.0015$$

Light travels 500 kilometers in approximately 0.0015 second. $\blacksquare$

EXERCISE 3.3

In Exercises 1–12, write each number in scientific notation.

1. 23,000   **2.** 4750   **3.** 1,700,000   **4.** 290,000

5. 0.062   **6.** 0.00073   **7.** 0.0000051   **8.** 0.04

9. 42.5×10^2 **10.** 0.3×10^3 **11.** 0.25×10^{-2} **12.** 25.2×10^{-3}

In Exercises 13–24, write each number in standard notation.

13. 2.3×10^2 **14.** 3.75×10^4 **15.** 8.12×10^5 **16.** 1.2×10^3

17. 1.15×10^{-3} **18.** 4.9×10^{-2} **19.** 9.76×10^{-4} **20.** 7.63×10^{-5}

21. 25×10^6 **22.** 0.07×10^3 **23.** 0.51×10^{-3} **24.** 617×10^{-2}

25. The distance from the Earth to the nearest star outside our solar system is approximately 25,200,000,000,000 miles. Express this number in scientific notation.

26. The speed of sound in air is 33,100 centimeters per second. Express this number in scientific notation.

27. The distance from Mars to the sun is approximately 1.14×10^8 miles. Express this number in standard notation.

28. The distance from Venus to the sun is approximately 6.7×10^7 miles. Express this number in standard notation.

29. One meter is approximately 0.00622 mile. Use scientific notation to express this number.

30. One angstrom is 1×10^{-7} millimeter. Express this number in standard notation.

In Exercises 31–36, use scientific notation to simplify each expression. Give all answers in standard notation.

31. $(3.4 \times 10^2)(2.1 \times 10^3)$ **32.** $(4.1 \times 10^{-3})(3.4 \times 10^4)$

33. $\dfrac{9.3 \times 10^2}{3.1 \times 10^{-2}}$ **34.** $\dfrac{7.2 \times 10^6}{1.2 \times 10^8}$

35. $\dfrac{96,000}{(12,000)(0.00004)}$ **36.** $\dfrac{(0.48)(14,400,000)}{96,000,000}$

37. The distance from Mercury to the sun is approximately 3.6×10^7 miles. Use scientific notation to express this distance in feet. (*Hint:* 5280 feet = 1 mile.)

38. The mass of one proton is approximately 1.7×10^{-24} gram. Use scientific notation to express the mass of 1 million protons.

39. The speed of sound in air is approximately 3.3×10^4 centimeters per second. Use scientific notation to express this speed in kilometers per second. (*Hint:* 100 centimeters = 1 meter and 1000 meters = 1 kilometer.)

40. One light year is approximately 5.87×10^{12} miles. Use scientific notation to express this distance in feet. (*Hint:* 5280 feet = 1 mile.)

REVIEW EXERCISES

1. List the prime numbers between 10 and 30.

2. If $y = -3$, find the value of $-5y^2$.

3. Evaluate $(3 + 4 \cdot 3) \div 5$.

4. Evaluate $\dfrac{3a^2 - 2b}{2a + 2b}$ if $a = 4$ and $b = 3$.

In Review Exercises 5–6, tell which property of real numbers justifies each statement

5. $5 + z = z + 5$ **6.** $7(u + 3) = 7u + 7 \cdot 3$

In Review Exercises 7–8, solve each equation.

7. $3(x - 4) - 6 = 0$ **8.** $8(3x - 5) - 4(2x + 3) = 12$

In Review Exercises 9–10, solve each inequality and graph the solution set.

9. $5(x + 4) \geq 3(x + 2)$ **10.** $3 < 3x + 4 \leq 10$

11. The perimeter of a square is not less than 44 feet and not greater than 60 feet. What are the possible lengths of the square's edge?

12. The average acceleration of a moving object is given by the formula

$$a = \frac{v - v_0}{t}$$

Solve the formula for v_0.

3.4   POLYNOMIALS

Expressions such as

$$3x \qquad 4y^2 \qquad -8x^2y^3 \qquad \text{and} \qquad 25$$

that contain constant and/or variable factors are called **algebraic terms.** The numerical coefficients of the first three of these terms are 3, 4, and -8, respectively. Because $25 = 25x^0$, the number 25 is considered to be the numerical coefficient of the term 25.

An algebraic expression that is the sum of one or more terms containing whole number exponents on its variables is called a **polynomial.** The expressions

$$8xy^2t \qquad 3x + 2 \qquad 4y^2 - 2y + 3 \qquad \text{and} \qquad 3a - 4b - 4c + 8d$$

are all examples of polynomials. The expression $2x^3 - 3y^{-2}$, however, is not a polynomial because the second term contains a negative exponent on a variable base.

A polynomial with exactly one term is called a **monomial.** A polynomial with exactly two terms is called a **binomial.** A polynomial with exactly three terms is called a **trinomial.** Here are some examples of each:

Monomials	**Binomials**	**Trinomials**
$5x^2y$	$3u^3 - 4u^2$	$-5t^2 + 4t + 3$
$-6x$	$18a^2b + 4ab$	$27x^3 - 6x - 2$
29	$-29z^{17} - 1$	$-32r^6 + 7y^3 - z$

The monomial $7x^6$ is called a **monomial of sixth degree** or a **monomial of degree 6** because the variable x occurs as a factor six times. The monomial $3x^3y^4$ is a monomial of the seventh degree because the variables x and y occur as factors a total of seven times. Other examples are:

$-2x^3$ is a monomial of degree 3.
$47x^2y^3$ is a monomial of degree 5.
$18x^4y^2z^8$ is a monomial of degree 14.
8 is a monomial of degree 0, because $8 = 8x^0$.

Amalie Noether (1882–1935)
Albert Einstein described Noether
as the most creative female
mathematical genius since the
beginning of higher education for
women. Her work was in the area
of abstract algebra. Although she
received a doctoral degree in
mathematics, she was denied a
mathematics position in Germany
because she was a woman.

These examples illustrate the following definition.

> **DEFINITION.** If a is a nonzero constant, the **degree of the monomial** ax^n is n. The **degree of a monomial** containing several variables is the sum of the exponents of those variables.
> The constant 0 has no defined degree.

Because each term of a polynomial is a monomial, we define the degree of a polynomial by considering the degrees of each of its terms.

> **DEFINITION.** The **degree of a polynomial** is the same as the degree of its term with largest degree.

For example,

$x^2 + 2x$ is a binomial of degree 2 because the degree of its first term is 2, and the degree of its other term is less than 2.

$3x^3y^2 + 4x^4y^4 - 3x^3$ is a trinomial of degree 8 because the degree of its second term is 8, and the degrees of its other terms are less than 8.

$25x^4y^3z^7 - 15xy^8z^{10} - 32x^8y^8z^3 + 4$ is a polynomial of degree 19 because its second and third terms are of degree 19. Its other terms have degrees less than 19.

Polynomials that contain a single variable are often denoted by symbols such as

$P(x)$ Read as "P of x."

$Q(t)$ Read as "Q of t."

or

$R(z)$ Read as "R of z."

where the letter within the parentheses represents the variable of the polynomial. The symbol $P(x)$ does not indicate the product of P and x. Instead, it represents a polynomial with the variable x. The symbols $P(x)$, $Q(t)$, and $R(z)$ could represent the polynomials

$$P(x) = 3x + 4$$
$$Q(t) = 3t^2 + 4t - 5$$
$$R(z) = -z^3 - 2z + 3$$

The symbol $P(x)$ is convenient to use because it provides a way to indicate the value of a polynomial in x at different values of x. If $P(x) = 3x + 4$, for example, then $P(1)$ represents the value of the polynomial $P(x) = 3x + 4$ when $x = 1$.

$$P(x) = 3x + 4$$
$$P(1) = 3(1) + 4 = 7$$

Likewise, if $Q(t) = 3t^2 + 4t - 5$, then $Q(-2)$ represents the value of the polynomial $Q(t) = 3t^2 + 4t - 5$ when $t = -2$.

$$Q(t) = 3t^2 + 4t - 5$$
$$Q(-2) = 3(-2)^2 + 4(-2) - 5$$
$$= 3(4) - 8 - 5$$
$$= 12 - 8 - 5$$
$$= -1$$

EXAMPLE 1   Consider the polynomial $P(z)$ where $P(z) = 3z^2 + 2$. Find **a.** $P(0)$, **b.** $P(2)$, **c.** $P(-3)$, and **d.** $P(s)$.

Solution   **a.** $P(z) = 3z^2 + 2$ **b.** $P(z) = 3z^2 + 2$
$\qquad\quad P(0) = 3(0)^2 + 2$ $\qquad\qquad\quad P(2) = 3(2)^2 + 2$
$\qquad\qquad\quad = 2$ $\qquad\qquad\qquad\qquad = 3(4) + 2$
$\qquad\qquad\qquad\qquad\qquad\qquad\qquad = 12 + 2$
$\qquad\qquad\qquad\qquad\qquad\qquad\qquad = 14$

c. $\quad P(z) = 3z^2 + 2$ **d.** $P(z) = 3z^2 + 2$
$\qquad P(-3) = 3(-3)^2 + 2$ $\qquad\quad P(s) = 3s^2 + 2$
$\qquad\qquad\quad = 3(9) + 2$
$\qquad\qquad\quad = 27 + 2$
$\qquad\qquad\quad = 29$ ■

EXAMPLE 2   Consider the polynomial $Q(y)$ where $Q(y) = -2y^2 - y - 7$. Find **a.** $Q(0)$, **b.** $Q(2)$, **c.** $Q(-1)$, and **d.** $Q(-x)$.

Solution   **a.** $Q(y) = -2y^2 - y - 7$ **b.** $Q(y) = -2y^2 - y - 7$
$\qquad\quad Q(0) = -2(0)^2 - (0) - 7$ $\qquad\quad Q(2) = -2(2)^2 - (2) - 7$
$\qquad\qquad\quad = -7$ $\qquad\qquad\qquad\quad = -8 - 2 - 7$
$\qquad\qquad\qquad\qquad\qquad\qquad\qquad = -17$

c. $\quad Q(y) = -2y^2 - y - 7$ **d.** $\quad Q(y) = -2y^2 - y - 7$
$\qquad Q(-1) = -2(-1)^2 - (-1) - 7$ $\qquad\quad Q(-x) = -2(-x)^2 - (-x) - 7$
$\qquad\qquad\quad = -2 + 1 - 7$ $\qquad\qquad\qquad = -2x^2 + x - 7$
$\qquad\qquad\quad = -8$ ■

EXAMPLE 3   Consider the polynomial $P(x)$ where $P(x) = x^3 + 1$. Find **a.** $P(2t)$, **b.** $P(-3y)$, **c.** $P(s^4)$, and **d.** $P(x) + P(a)$.

Solution   **a.** $P(x) = x^3 + 1$ **b.** $\qquad P(x) = x^3 + 1$
$\qquad\quad P(2t) = (2t)^3 + 1$ $\qquad\qquad P(-3y) = (-3y)^3 + 1$
$\qquad\qquad\quad = 8t^3 + 1$ $\qquad\qquad\qquad\quad = -27y^3 + 1$

c. $P(x) = x^3 + 1$ **d.** $P(x) + P(a) = x^3 + 1 + a^3 + 1$
$\qquad P(s^4) = (s^4)^3 + 1$ $\qquad\qquad\qquad\qquad = x^3 + a^3 + 2$
$\qquad\qquad\quad = s^{12} + 1$ ■

EXERCISE 3.4

In Exercises 1–12, classify each polynomial as a monomial, a binomial, or a trinomial, if possible.

1. $3x + 7$

2. $3y - 5$

3. $3y^2 + 4y + 3$

4. $3xy$

5. $3z^2$

6. $3x^4 - 2x^3 + 3x - 1$

7. $5t - 32$

8. $9x^2y^3z^4$

9. $s^2 - 23s + 31$

10. $12x^3 - 12x^2 + 36x - 3$

11. $3x^5 - 2x^4 - 3x^3 + 17$

12. x^3

In Exercises 13–24, give the degree of each polynomial.

13. $3x^4$

14. $3x^5 - 4x^2$

15. $-2x^2 + 3x^3$

16. $-5x^5 + 3x^2 - 3x$

17. $3x^2y^3 + 5x^3y^5$

18. $-2x^2y^3 + 4x^3y^2z$

19. $-5r^2s^2t - 3r^3st^2 + 3$

20. $4r^2s^3t^3 - 5r^2s^8$

21. $x^{12} + 3x^2y^3z^4$

22. 17^2x

23. 38

24. -25

In Exercises 25–32, let $P(x) = 5x - 3$. Find each value.

25. $P(2)$

26. $P(0)$

27. $P(-1)$

28. $P(-2)$

29. $P(w)$

30. $P(t)$

31. $P(-y)$

32. $P(2t)$

In Exercises 33–40, let $Q(z) = -z^2 - 4$. Find each value.

33. $Q(0)$

34. $Q(1)$

35. $Q(-1)$

36. $Q(-2)$

37. $Q(r)$

38. $Q(-u)$

39. $Q(3s)$

40. $Q(-2x)$

In Exercises 41–48, let $R(y) = y^2 - 2y + 3$. Find each value.

41. $R(0)$

42. $R(3)$

43. $R(-2)$

44. $R(-1)$

45. $R(-b)$

46. $R(t)$

47. $R\left(-\dfrac{1}{4}w\right)$

48. $R\left(\dfrac{1}{2}u\right)$

In Exercises 49–64, let $P(x) = 5x - 2$. Find each value.

49. $P\left(\dfrac{1}{5}\right)$

50. $P\left(\dfrac{1}{10}\right)$

51. $P(u^2)$

52. $P(-v^4)$

53. $P(-4z^6)$

54. $P(10x^7)$

55. $P(x^2y^2)$

56. $P(x^3y^3)$

57. $P(x + h)$

58. $P(x - h)$

59. $P(x) + P(h)$

60. $P(x) - P(h)$

61. $P(2y + z)$

62. $P(-3r + 2s)$

63. $P(2y) + P(z)$

64. $P(-3r) + P(2s)$

REVIEW EXERCISES

In Review Exercises 1–2, solve each equation.

1. $5(u - 5) + 9 = 2(u + 4)$

2. $8(3a - 5) - 12 = 4(2a + 3)$

In Review Exercises 3–4, solve each inequality and graph the solution set.

3. $-4(3y + 2) \le 20$

4. $-5 < 3t + 4 \le 13$

5. The monthly cost of electricity in a certain city is $7 plus 11 cents per kilowatt-hour used. How many kilowatt-hours are used in a month when the bill is $68.60?

6. A rectangle with a perimeter of 28 inches is 6 inches longer than it is wide. Find its dimensions.

In Review Exercises 7–10, write each expression without using parentheses or negative exponents.

7. $(x^2 x^4)^3$

8. $(a^2)^3 (a^3)^2$

9. $\left(\dfrac{y^2 y^5}{y^4} \right)^3$

10. $\left(\dfrac{2t^3}{t} \right)^{-4}$

11. The estimated mass of the planet Jupiter is 1.9×10^{27} kilograms. Express this number in standard notation.

12. Solve the equation $y - 3 = m(x - 2)$ for x.

3.5   ADDING AND SUBTRACTING POLYNOMIALS

Recall that **like terms** are terms that contain the same variables with the same exponents. For example,

$$3xyz^2 \quad \text{and} \quad -2xyz^2$$

are like terms, and

$$\frac{1}{2} ab^2 c \quad \text{and} \quad \frac{1}{3} a^2 bd^2$$

are unlike terms.

Because of the distributive property, we can combine like terms by adding their coefficients and using the same variables and exponents. For example,

$$2y + 5y = (2 + 5)y = 7y$$
$$-3x^2 + 7x^2 = (-3 + 7)x^2 = 4x^2$$

Likewise,

$$4x^3 y^2 + 9x^3 y^2 = 13x^3 y^2$$
$$4r^2 s^3 t^4 + 7r^2 s^3 t^4 = 11r^2 s^3 t^4$$

Thus, to add like monomials together, we simply combine like terms.

EXAMPLE 1   a. $5xy^3 + 7xy^3 = 12xy^3$

b. $-7x^2 y^2 + 6x^2 y^2 + 3x^2 y^2 = -x^2 y^2 + 3x^2 y^2$
$$= 2x^2 y^2$$

c. $(2x^2)^2 + (3x)^4 = 4x^4 + 81x^4$
$$= 85x^4$$

d. $2(x + y) + 3(x + y) = 5(x + y)$
$$= 5x + 5y \qquad \blacksquare$$

Recall from Section 1.6 that

$$a - b = a + (-b)$$

Thus, to subtract one monomial from another, we can add the negative of the monomial that is to be subtracted.

EXAMPLE 2   **a.** $8x^2 - 3x^2 = 8x^2 + (-3x^2)$
$$= 5x^2$$

b. $6x^3y^2 - 9x^3y^2 = 6x^3y^2 + (-9x^3y^2)$
$$= -3x^3y^2$$

c. $-3r^2st^3 - 5r^2st^3 = -3r^2st^3 + (-5r^2st^3)$
$$= -8r^2st^3 \qquad \blacksquare$$

Because of the distributive property, we can remove parentheses enclosing several terms when the sign preceding the parentheses is a $+$ sign. We simply drop the parentheses.

$$+(3x^2 + 3x - 2) = +1(3x^2 + 3x - 2)$$
$$= 1(3x^2) + 1(3x) + 1(-2)$$
$$= 3x^2 + 3x + (-2)$$
$$= 3x^2 + 3x - 2$$

Polynomials are added by removing parentheses, if necessary, and then combining any like terms that are contained within the polynomials.

EXAMPLE 3 $(3x^2 - 3x + 2) + (2x^2 + 7x - 4) = 3x^2 - 3x + 2 + 2x^2 + 7x - 4$
$$= 3x^2 + 2x^2 - 3x + 7x + 2 + (-4)$$
$$= 5x^2 + 4x - 2 \qquad \blacksquare$$

To make the addition easier, problems such as Example 3 are often written with the terms aligned vertically.

$$3x^2 - 3x + 2$$
$$\underline{2x^2 + 7x - 4}$$
$$5x^2 + 4x - 2$$

EXAMPLE 4   Add:

$$4x^2y + 8x^2y^2 - 3x^2y^3$$
$$\underline{3x^2y - 8x^2y^2 + 8x^2y^3}$$
$$7x^2y \qquad\quad + 5x^2y^3 \qquad \blacksquare$$

Because of the distributive property, we can also remove parentheses enclosing several terms when the sign preceding the parentheses is a $-$ sign. We simply drop the $-$ sign and the parentheses, and *change the sign of every term within the parentheses*.

$$-(3x^2 + 3x - 2) = -1(3x^2 + 3x - 2)$$
$$= -1(3x^2) + (-1)(3x) + (-1)(-2)$$
$$= -3x^2 + (-3x) + 2$$
$$= -3x^2 - 3x + 2$$

This suggests that the way to subtract polynomials is to remove parentheses and combine like terms.

EXAMPLE 5   **a.** $(3x - 4) - (5x + 7) = 3x - 4 - 5x - 7 = -2x - 11$

b. $(3x^2 - 4x - 6) - (2x^2 - 6x + 12) = 3x^2 - 4x - 6 - 2x^2 + 6x - 12$
$$= x^2 + 2x - 18$$

c. $(-4rt^3 + 2r^2t^2) - (-3rt^3 + 2r^2t^2) = -4rt^3 + 2r^2t^2 + 3rt^3 - 2r^2t^2 = -rt^3$
∎

To subtract polynomials in vertical form, we add the negative of the **subtrahend** (the bottom polynomial) to the **minuend** (the top polynomial).

EXAMPLE 6   Subtract $3x^2y - 2xy^2$ from $2x^2y + 4xy^2$.

Solution   Write the problem in vertical form, change the signs of the terms of the subtrahend, and add:

$$\begin{array}{r} 2x^2y + 4xy^2 \\ - \underline{\ 3x^2y - 2xy^2} \end{array} \longrightarrow \begin{array}{r} 2x^2y + 4xy^2 \\ + \underline{-3x^2y + 2xy^2} \\ - \ x^2y + 6xy^2 \end{array}$$

In horizontal form, the solution is

$$2x^2y + 4xy^2 - (3x^2y - 2xy^2) = 2x^2y + 4xy^2 - 3x^2y + 2xy^2$$
$$= -x^2y + 6xy^2$$
∎

EXAMPLE 7   Subtract $6xy^2 + 4x^2y^2 - x^3y^2$ from $-2xy^2 - 3x^3y^2$.

Solution
$$\begin{array}{r} -2xy^2 \qquad\ - 3x^3y^2 \\ - \underline{\ \ 6xy^2 + 4x^2y^2 - \ x^3y^2} \end{array} \longrightarrow \begin{array}{r} -2xy^2 \qquad\ - 3x^3y^2 \\ + \underline{-6xy^2 - 4x^2y^2 + \ x^3y^2} \\ -8xy^2 - 4x^2y^2 - 2x^3y^2 \end{array}$$

In horizontal form, the solution is

$$-2xy^2 - 3x^3y^2 - (6xy^2 + 4x^2y^2 - x^3y^2)$$
$$= -2xy^2 - 3x^3y^2 - 6xy^2 - 4x^2y^2 + x^3y^2$$
$$= -8xy^2 - 4x^2y^2 - 2x^3y^2$$
∎

Because of the distributive property, we can remove parentheses enclosing several terms when a monomial precedes the parentheses. We simply multiply every term within the parentheses by that monomial. For example, to add $3(2x + 5)$ and $2(4x - 3)$, we proceed as follows:

$$3(2x + 5) + 2(4x - 3) = 6x + 15 + 8x - 6$$
$$= 6x + 8x + 15 - 6$$
$$= 14x + 9$$

EXAMPLE 8   **a.** $3(x^2 + 4x) + 2(x^2 - 4) = 3x^2 + 12x + 2x^2 - 8$
$$= 5x^2 + 12x - 8$$

b. $8(y^2 - 2y + 3) - 4(2y^2 + y - 3) = 8y^2 - 16y + 24 - 8y^2 - 4y + 12$
$$= -20y + 36$$

c. $-4x(xy^2 - xy + 3) - x(xy^2 - 2) + 3(x^2y^2 + 2x^2y)$
$$= -4x^2y^2 + 4x^2y - 12x - x^2y^2 + 2x + 3x^2y^2 + 6x^2y$$
$$= -2x^2y^2 + 10x^2y - 10x$$

■

EXERCISE 3.5

In Exercises 1–12, tell whether the terms are like or unlike terms. If they are like terms, add them.

1. $3y, 4y$ **2.** $3x^2, 5x^2$ **3.** $3x, 3y$ **4.** $3x^2, 6x$

5. $3x^3, 4x^3, 6x^3$ **6.** $-2y^4, -6y^4, 10y^4$ **7.** $-5x^3y^2, 13x^3y^2$ **8.** $23, 12x$

9. $-23t^6, 32t^6, 56t^6$ **10.** $32x^5y^3, -21x^5y^3, -11x^5y^3$

11. $-x^2y, xy, 3xy^2$ **12.** $4x^3y^2z, -6x^3y^2z, 2x^3y^2z$

In Exercises 13–30, simplify each expression if possible.

13. $4y + 5y$ **14.** $-2x + 3x$ **15.** $-8t^2 - 4t^2$ **16.** $15x^2 + 10x^2$

17. $32u^3 - 16u^3$ **18.** $25xy^2 - 7xy^2$ **19.** $18x^5y^2 - 11x^5y^2$ **20.** $17x^6y - 22x^6y$

21. $3rst + 4rst + 7rst$ **22.** $-2ab + 7ab - 3ab$

23. $-4a^2bc + 5a^2bc - 7a^2bc$ **24.** $(xy)^2 + 4x^2y^2 - 2x^2y^2$

25. $(3x)^2 - 4x^2 + 10x^2$ **26.** $(2x)^4 - (3x^2)^2$

27. $5x^2y^2 + 2(xy)^2 - (3x^2)y^2$ **28.** $-3x^3y^6 + 2(xy^2)^3 - (3x)^3y^6$

29. $(-3x^2y)^4 + (4x^4y^2)^2 - 2x^8y^4$ **30.** $5x^5y^{10} - (2xy^2)^5 + (3x)^5y^{10}$

In Exercises 31–62, perform the indicated operations and simplify.

31. $(3x + 7) + (4x - 3)$ **32.** $(2y - 3) + (4y + 7)$

33. $(4a + 3) + (2a + 4)$ **34.** $(5b - 7) - (3b + 5)$

35. $(2x + 3y) + (5x - 10y)$ **36.** $(5x - 8y) - (2x + 5y)$

37. $(-8x - 3y) + (11x + y)$ **38.** $(-4a + b) + (5a - b)$

39. $(3x^2 - 3x - 2) + (3x^2 + 4x - 3)$ **40.** $(3a^2 - 2a + 4) - (a^2 - 3a + 7)$

41. $(2b^2 + 3b - 5) - (2b^2 - 4b - 9)$ **42.** $(4c^2 + 3c - 2) + (3c^2 + 4c + 2)$

43. $(2x^2 - 3x + 1) - (4x^2 - 3x + 2) + (2x^2 + 3x + 2)$

44. $(-3z^2 - 4z + 7) + (2z^2 + 2z - 1) - (2z^2 - 3z + 7)$

45. $2(x + 3) + 3(x + 3)$

46. $5(x + y) + 7(x + y)$

47. $-8(x - y) + 11(x - y)$

48. $-4(a - b) - 5(a - b)$

49. $2(x^2 - 5x - 4) - 3(x^2 - 5x - 4) + 6(x^2 - 5x - 4)$

50. $7(x^2 + 3x + 1) + 9(x^2 + 3x + 1) - 5(x^2 + 3x + 1)$

51. Add: $3x^2 + 4x + 5$
$2x^2 - 3x + 6$

52. Add: $2x^3 + 2x^2 - 3x + 5$
$3x^3 - 4x^2 - x - 7$

53. Add: $2x^3 - 3x^2 + 4x - 7$
$-9x^3 - 4x^2 - 5x + 6$

54. Add: $-3x^3 + 4x^2 - 4x + 9$
$2x^3 + 9x - 3$

55. Add: $-3x^2y + 4xy + 25y^2$
$5x^2y - 3xy - 12y^2$

56. Add: $-6x^3z - 4x^2z^2 + 7z^3$
$-7x^3z + 9x^2z^2 - 21z^3$

57. Subtract: $3x^2 + 4x - 5$
$-2x^2 - 2x + 3$

58. Subtract: $3y^2 - 4y + 7$
$6y^2 - 6y - 13$

59. Subtract: $4x^3 + 4x^2 - 3x + 10$
$5x^3 - 2x^2 - 4x - 4$

60. Subtract: $3x^3 + 4x^2 + 7x + 12$
$-4x^3 + 6x^2 + 9x - 3$

61. Subtract: $-2x^2y^2 - 4xy + 12y^2$
$10x^2y^2 + 9xy - 24y^2$

62. Subtract: $25x^3 - 45x^2z + 31xz^2$
$12x^3 + 27x^2z - 17xz^2$

63. Find the sum when $x^2 + x - 3$ is added to the sum of $2x^2 - 3x + 4$ and $3x^2 - 2$.

64. Find the sum when $3y^2 - 5y + 7$ is added to the sum of $-3y^2 - 7y + 4$ and $5y^2 + 5y - 7$.

65. Find the difference when $t^3 - 2t^2 + 2$ is subtracted from the sum of $3t^3 + t^2$ and $-t^3 + 6t - 3$.

66. Find the difference when $-3z^3 - 4z + 7$ is subtracted from the sum of $2z^2 + 3z - 7$ and $-4z^3 - 2z - 3$.

67. Find the sum when $3x^2 + 4x - 7$ is added to the sum of $-2x^2 - 7x + 1$ and $-4x^2 + 8x - 1$.

68. Find the difference when $32x^2 - 17x + 45$ is subtracted from the sum of $23x^2 - 12x - 7$ and $-11x^2 + 12x + 7$.

In Exercises 69–78, simplify each expression.

69. $2(x + 3) + 4(x - 2)$

70. $3(y - 4) - 5(y + 3)$

71. $-2(x^2 + 7x - 1) - 3(x^2 - 2x + 7)$

72. $-5(y^2 - 2y - 2) + 6(2y^2 + 2y + 4)$

73. $2(2y^2 - 2y + 2) - 4(3y^2 - 4y - 1) + 4y(y^2 - y - 1)$

74. $-4(z^2 - 5z) - 5(4z^2 - 1) + 6(2z - 3)$

75. $2a(ab^2 - b) - 3b(a + 2ab) + b(b - a + a^2b)$

76. $3y(xy + y) - 2y^2(x - 4 + y) + 2(y^3 + y^2)$

77. $-4xy^2(x + y + z) - 2x(xy^2 - 4y^2z) - 2y(8xy^2 - 1)$

78. $-3uv(u - v^2 + w) + 4w(uv + w) - 3w(w + uv)$

In Exercises 79–80, let $P(x) = 3x - 5$. Find each value.

79. $P(x + h) + P(x)$

80. $P(x + h) - P(x)$

REVIEW EXERCISES

1. On the number line, graph the set of even integers between -5 and 5.

2. Evaluate $-|3 - 5|$.

In Review Exercises 3–6, let $a = 3$, $b = -2$, $c = -1$, and $d = 2$. Evaluate each expression.

3. $ab + cd$ **4.** $ad + bc$ **5.** $a(b + c)$ **6.** $d(b + a)$

In Review Exercises 7–8, solve each equation.

7. $3(a + 2) - (2 - a) = a - 5$

8. $\dfrac{3}{2}x = 7(x + 11)$

9. A rectangle with a perimeter of 72 feet is 6 feet longer than it is wide. Find its area.

10. Solve the inequality $-4(2x - 9) \geq 12$ and graph the solution set.

11. The **kinetic energy** of a moving object is given by

$$K = \frac{mv^2}{2}$$

Solve the formula for m.

12. A rectangular garden is 3 feet longer than twice its width. Its perimeter is 48 feet. Find its area.

3.6 MULTIPLYING POLYNOMIALS

In Section 3.1, we multiplied certain monomials by other monomials. To multiply $4x^2$ by $-2x^3$, for example, we use the commutative and associative properties of multiplication to group the numerical factors and the variable factors together and multiply.

$$4x^2(-2x^3) = 4(-2)x^2x^3 = -8x^5$$

EXAMPLE 1   **a.** $3x^5(2x^5) = 3(2)x^5x^5$ **b.** $-2a^2b^3(5ab^2) = -2(5)a^2ab^3b^2$

$\qquad\qquad\qquad\qquad = 6x^{10}$ $\qquad\qquad\qquad\qquad\qquad\qquad = -10a^3b^5$

c. $-4y^5z^2(2y^3z^3)(3yz) = -4(2)(3)y^5y^3yz^2z^3z$

$\qquad\qquad\qquad\qquad\quad = -24y^9z^6$ ∎

The previous examples suggest the following rule:

> To multiply two monomials, first multiply the numerical factors and then multiply the variable factors.

To find the product of a monomial and a polynomial with more than one term, we use the distributive property. To multiply $x + 4$ by $3x$, for example, we proceed as follows:

$$3x(x + 4) = 3x \cdot x + 3x \cdot 4 = 3x^2 + 12x$$

EXAMPLE 2   **a.** $2a^2(3a^2 - 4a) = 2a^2 \cdot 3a^2 - 2a^2 \cdot 4a$
$$= 6a^4 - 8a^3$$

b. $-2xz^2(2x - 3z + 2x^2z^2) = -2xz^2(2x) - (-2xz^2)(3z) + (-2xz^2)(2x^2z^2)$
$$= -4x^2z^2 + 6xz^3 - 4x^3z^4 \qquad \blacksquare$$

The results of Example 2 suggest the following rule:

> To multiply a polynomial with more than one term by a monomial, use the distributive property to remove parentheses, and simplify.

We must use the distributive property more than once to multiply a polynomial by a binomial. For example, to multiply $3x^2 + 3x - 5$ by $2x + 3$, we proceed as follows:

$$(2x + 3)(3x^2 + 3x - 5) = (2x + 3)3x^2 + (2x + 3)3x - (2x + 3)5$$
$$= 3x^2(2x + 3) + 3x(2x + 3) - 5(2x + 3)$$
$$= 6x^3 + 9x^2 + 6x^2 + 9x - 10x - 15$$
$$= 6x^3 + 15x^2 - x - 15$$

EXAMPLE 3   **a.** $(2x - 4)(3x + 5) = (2x - 4)3x + (2x - 4)5$
$$= 2x \cdot 3x - 4 \cdot 3x + 2x \cdot 5 - 4 \cdot 5$$
$$= 6x^2 - 12x + 10x - 20$$
$$= 6x^2 - 2x - 20$$

b. $(3x - 2y)(2x + 3y) = (3x - 2y)2x + (3x - 2y)3y$
$$= 3x \cdot 2x - 2y \cdot 2x + 3x \cdot 3y - 2y \cdot 3y$$
$$= 6x^2 - 4xy + 9xy - 6y^2$$
$$= 6x^2 + 5xy - 6y^2$$

c. $(3y + 1)(3y^2 + 2y + 2)$

$$= (3y + 1)3y^2 + (3y + 1)2y + (3y + 1)2$$
$$= 3y \cdot 3y^2 + 1 \cdot 3y^2 + 3y \cdot 2y + 1 \cdot 2y + 3y \cdot 2 + 1 \cdot 2$$
$$= 9y^3 + 3y^2 + 6y^2 + 2y + 6y + 2$$
$$= 9y^3 + 9y^2 + 8y + 2 \qquad\blacksquare$$

The results of Example 3 suggest the following rule:

> To multiply one polynomial by another, multiply every term of one polynomial by every term of the other polynomial and combine like terms.

It is often convenient to organize the work vertically, as shown in the next example.

EXAMPLE 4   **a.** Multiply:

$$
\begin{array}{r}
2x - 4 \\
3x + 2 \\
\hline
6x^2 - 12x \\
+ 4x - 8 \\
\hline
6x^2 - 8x - 8
\end{array}
$$

$3x(2x - 4) \longrightarrow$
$2(2x - 4) \longrightarrow$

b. Multiply:

$$
\begin{array}{r}
3a^2 - 4a + 7 \\
2a + 5 \\
\hline
6a^3 - 8a^2 + 14a \\
+ 15a^2 - 20a + 35 \\
\hline
6a^3 + 7a^2 - 6a + 35
\end{array}
$$

$2a(3a^2 - 4a + 7) \longrightarrow$
$5(3a^2 - 4a + 7) \longrightarrow$

c. Multiply:

$$
\begin{array}{r}
3y^2 - 5y + 4 \\
- 4y^2 - 3 \\
\hline
-12y^4 + 20y^3 - 16y^2 \\
- 9y^2 + 15y - 12 \\
\hline
-12y^4 + 20y^3 - 25y^2 + 15y - 12
\end{array}
$$

$-4y^2(3y^2 - 5y + 4) \longrightarrow$
$-3(3y^2 - 5y + 4) \longrightarrow$

$\qquad\blacksquare$

We can use a method called the **FOIL** method to multiply one binomial by another. In this method we multiply every term of one binomial by every term of the other binomial. **FOIL** is an acronym for First terms, Outer terms, Inner terms, and Last terms. To use the **FOIL** method to multiply $2a - 4$ by $3a + 5$, we

1. multiply the First terms $2a$ and $3a$ to obtain $6a^2$,
2. multiply the Outer terms $2a$ and 5 to obtain $10a$,
3. multiply the Inner terms -4 and $3a$ to obtain $-12a$, and
4. multiply the Last terms -4 and 5 to obtain -20.

Then, we simplify the resulting polynomial, if possible.

$$(2a - 4)(3a + 5) = 2a(3a) + 2a(5) + (-4)(3a) + (-4)(5)$$
$$= 6a^2 + 10a - 12a - 20 \qquad \text{Simplify.}$$
$$= 6a^2 - 2a - 20 \qquad \text{Combine terms.}$$

First terms, Last terms, Inner terms, Outer terms

EXAMPLE 5   Use the **FOIL** method to find each product.

a. $(3x + 4)(2x - 3) = 3x(2x) + 3x(-3) + 4(2x) + 4(-3)$
$$= 6x^2 - 9x + 8x - 12$$
$$= 6x^2 - x - 12$$

b. $(2y - 7)(5y - 4) = 2y(5y) + 2y(-4) + (-7)(5y) + (-7)(-4)$
$$= 10y^2 - 8y - 35y + 28$$
$$= 10y^2 - 43y + 28$$

c. $(2r - 3s)(2r + t) = 2r(2r) + 2r(t) - 3s(2r) - 3s(t)$
$$= 4r^2 + 2rt - 6rs - 3st$$

■

EXAMPLE 6   Simplify each expression.

a. $3(2x - 3)(x + 1) = 3(2x^2 + 2x - 3x - 3)$ ⎢ Use **FOIL** to multiply the binomials.
$$= 3(2x^2 - x - 3) \qquad \text{Combine terms.}$$
$$= 6x^2 - 3x - 9 \qquad \text{Use the distributive property to remove parentheses.}$$

b. $(x + 1)(x - 2) - 3x(x + 3) = x^2 - 2x + x - 2 - 3x^2 - 9x$
$$= -2x^2 - 10x - 2 \qquad \text{Combine like terms.}$$

■

The products discussed in Example 7 are called **special products.**

EXAMPLE 7   Use the **FOIL** method to find each special product.

a. $(x + y)^2 = (x + y)(x + y)$
$$= x^2 + xy + xy + y^2$$
$$= x^2 + 2xy + y^2$$

The square of the sum of two quantities such as $x + y$ has three terms: the square of the first quantity, plus twice the product of the quantities, plus the square of the second quantity.

b. $(x - y)^2 = (x - y)(x - y)$
$$= x^2 - xy - xy + y^2$$
$$= x^2 - 2xy + y^2$$

The square of the difference of two quantities such as $x - y$ has three terms: the square of the first quantity, minus twice the product of the quantities, plus the square of the second quantity.

c. $(x + y)(x - y) = x^2 - xy + xy - y^2$
$$= x^2 - y^2$$

The product of a sum and a difference of two quantities such as $x + y$ and $x - y$ is a binomial. It is the product of the first quantities minus the product of the second quantities. Binomials that have the same terms but different signs between them are often called **conjugate binomials.** For example, the *conjugate* of $2x + 12$ is $2x - 12$, and the conjugate of $ab - c$ is $ab + c$. ■

Because the special products discussed in Example 7 occur so often, you should learn their forms.

$$(x + y)^2 = x^2 + 2xy + y^2$$
$$(x - y)^2 = x^2 - 2xy + y^2$$
$$(x + y)(x - y) = x^2 - y^2$$

Multiplying Binomials to Solve Equations

To solve an equation such as $(x + 2)(x + 3) = x(x + 7)$, we must first remove the parentheses. To do so, we use the **FOIL** method on the left-hand side of the equation, use the distributive property on the right-hand side, and proceed as follows:

$$(x + 2)(x + 3) = x(x + 7)$$

$$x^2 + 3x + 2x + 6 = x^2 + 7x$$

$5x + 6 = 7x$	Add $-x^2$ to both sides and combine terms.
$6 = 2x$	Add $-5x$ to both sides.
$3 = x$	Divide both sides by 2.

Check: $(x + 2)(x + 3) = x(x + 7)$	
$(3 + 2)(3 + 3) \stackrel{?}{=} 3(3 + 7)$	Replace x with 3.
$5(6) \stackrel{?}{=} 3(10)$	Perform the additions within parentheses.
$30 = 30$	

EXAMPLE 8   Solve the equation $(x + 5)(x + 4) = (x + 9)(x + 10)$.

Solution   Use the **FOIL** method to remove parentheses on both sides of the equation. Then proceed as follows:

$$(x + 5)(x + 4) = (x + 9)(x + 10)$$

$$x^2 + 4x + 5x + 20 = x^2 + 10x + 9x + 90$$

$$9x + 20 = 19x + 90 \qquad \text{Add } -x^2 \text{ to both sides and combine terms.}$$

$$20 = 10x + 90 \qquad \text{Add } -9x \text{ to both sides.}$$

$$-70 = 10x \qquad \text{Add } -90 \text{ to both sides.}$$

$$-7 = x \qquad \text{Divide both sides by 10.}$$

Check: $(x + 5)(x + 4) = (x + 9)(x + 10)$

$$(-7 + 5)(-7 + 4) \overset{?}{=} (-7 + 9)(-7 + 10) \qquad \text{Replace } x \text{ with } -7.$$

$$(-2)(-3) \overset{?}{=} (2)(3) \qquad \text{Perform the additions within parentheses.}$$

$$6 = 6$$

■

EXAMPLE 9   The square painting in Figure 3-1 is surrounded by a border 2 inches wide. If the area of the border is 96 square inches, find the dimensions of the painting.

Solution   Let x represent the length of each side of the square painting. Then the outer rectangle is also a square, and its dimensions are $(x + 4)$ by $(x + 4)$ inches. If we subtract the area of the painting from the area of the large square, the difference is 96.

$$(x + 4)^2 - x^2 = 96$$

$$(x + 4)(x + 4) - x^2 = 96$$

$$x^2 + 8x + 16 - x^2 = 96$$

$$8x + 16 = 96$$

$$8x = 80$$

$$x = 10$$

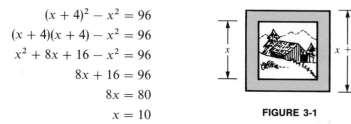

FIGURE 3-1

The dimensions of the painting are 10 inches by 10 inches.

■

EXERCISE 3.6

In Exercises 1–12, find each product.

1. $(3x^2)(4x^3)$

2. $(-2a^3)(3a^2)$

3. $(3b^2)(-2b)(4b^3)$

4. $(3y)(2y^2)(-y^4)$

5. $(2x^2y^3)(3x^3y^2)$

6. $(-x^3y^6z)(x^2y^2z^7)$

7. $(x^2y^5)(x^2z^5)(-3y^2z^3)$

8. $(-r^4st^2)(2r^2st)(rst)$

9. $(x^2y^3)^5$

10. $(a^3b^2c)^4$

11. $(a^3b^2c)(abc^3)^2$

12. $(xyz^3)(xy^2z^2)^3$

In Exercises 13–34, find each product.

13. $3(x + 4)$

14. $-3(a - 2)$

15. $-4(t + 7)$

16. $6(s^2 - 3)$

17. $3x(x - 2)$

18. $4y(y + 5)$

19. $-2x^2(3x^2 - x)$

20. $4b^3(2b^2 - 2b)$

21. $3xy(x + y)$

22. $-4x^2(3x^2 - x)$

23. $2x^2(3x^2 + 4x - 7)$

24. $3y^3(2y^2 - 7y - 8)$

25. $\frac{1}{4} x^2(8x^5 - 4)$

26. $\frac{4}{3} a^2b(6a - 5b)$

27. $-\frac{2}{3} r^2t^2(9r - 3t)$

28. $-\frac{4}{5} p^2q(10p + 15q)$

29. $-3x^3y^3(x^2y^3 - x^3y^2 + xy^2)$

30. $7rst(r^2 + s^2 - t^2)$

31. $-abc^3(a^3 + b^3 - c^3)$

32. $4x^5y^3z^2(3x^2 - 4y^2z - 4)$

33. $(3xy)(-2x^2y^3)(x + y)$

34. $(-2a^2b)(-3a^3b^2)(3a - 2b)$

*In Exercises 35–58, use the **FOIL** method to find each product.*

35. $(a + 4)(a + 5)$

36. $(y - 3)(y + 5)$

37. $(3x - 2)(x + 4)$

38. $(t + 4)(2t - 3)$

39. $(2a + 4)(3a - 5)$

40. $(2b - 1)(3b + 4)$

41. $(3x - 5)(2x + 1)$

42. $(2y - 5)(3y + 7)$

43. $(x + 3)(2x - 3)$

44. $(2x + 3)(2x - 5)$

45. $(2t + 3s)(3t - s)$

46. $(3a - 2b)(4a + b)$

47. $(x + y)(x - 2y)$

48. $(a - b)(2a - b)$

49. $(-2r - 3s)(2r + 7s)$

50. $(-4a + 3)(-2a - 3)$

51. $(-2a - 4b)(3a + 2b)$

52. $(3u + 2v)(-2u + 5v)$

53. $(4t - u)(-3t + u)$

54. $(-3t + 2s)(2t - 3s)$

55. $(x + y)(x + z)$

56. $(a - b)(x + y)$

57. $(u + v)(u + 2t)$

58. $(x - 5y)(a + 2y)$

In Exercises 59–64, find each product.

59. $\begin{array}{r} 4x + 3 \\ x + 2 \\ \hline \end{array}$

60. $\begin{array}{r} 5r + 6 \\ 2r - 1 \\ \hline \end{array}$

61. $\begin{array}{r} 4x - 2y \\ 3x + 5y \\ \hline \end{array}$

62. $\begin{array}{r} 5r + 6s \\ 2r - s \\ \hline \end{array}$

63. $\begin{array}{r} x^2 + x + 1 \\ x - 1 \\ \hline \end{array}$

64. $\begin{array}{r} 4x^2 - 2x + 1 \\ 2x + 1 \\ \hline \end{array}$

In Exercises 65–84, find each special product.

65. $(x + 4)(x + 4)$

66. $(a + 3)(a + 3)$

67. $(t - 3)(t - 3)$

68. $(z - 5)(z - 5)$

69. $(r + 4)(r - 4)$ **70.** $(b + 2)(b - 2)$ **71.** $(x + 5)^2$ **72.** $(y - 6)^2$

73. $(2s + 1)(2s + 1)$ **74.** $(3t - 2)(3t - 2)$ **75.** $(4x + 5)(4x - 5)$ **76.** $(5z + 1)(5z - 1)$

77. $(3r + 4s)(3r + 4s)$ **78.** $(2u - 3v)(2u - 3v)$ **79.** $(x - 2y)^2$ **80.** $(3a + 2b)^2$

81. $(2a - 3b)^2$ **82.** $(2x + 3y)^2$ **83.** $(4x + 5y)(4x - 5y)$ **84.** $(6p + 5q)(6p - 5q)$

In Exercises 85–96, find each product.

85. $2(x - 4)(x + 1)$

86. $-3(2x + 3y)(3x - 4y)$

87. $3a(a + b)(a - b)$

88. $-2r(r + s)(r + s)$

89. $-3y^2z(y + 2z)(2y - z)$

90. $4t^3u^2(3t - 2u)(t + 2u)$

91. $(4t + 3)(t^2 + 2t + 3)$

92. $(3x + y)(2x^2 - 3xy + y^2)$

93. $(-3x + y)(x^2 - 8xy + 16y^2)$

94. $(3x - y)(x^2 + 3xy - y^2)$

95. $(x - 2y)(x^2 + 2xy + 4y^2)$

96. $(2m + n)(4m^2 - 2mn + n^2)$

In Exercises 97–106, simplify each expression.

97. $2t(t + 2) + 3t(t - 5)$

98. $3y(y + 2) + (y + 1)(y - 1)$

99. $3xy(x + y) - 2x(xy - x)$

100. $(a + b)(a - b) - (a + b)(a + b)$

101. $(x + y)(x - y) + x(x + y)$

102. $(2x - 1)(2x + 1) + x(2x + 1)$

103. $(x + 2)^2 - (x - 2)^2$

104. $(x - 3)^2 - (x + 3)^2$

105. $(2s - 3)(s + 2) + (3s + 1)(s - 3)$

106. $(3x + 4)(2x - 2) - (2x + 1)(x + 3)$

In Exercises 107–116, solve each equation.

107. $(s - 4)(s + 1) = s^2 + 5$

108. $(y - 5)(y - 2) = y^2 - 4$

109. $z(z + 2) = (z + 4)(z - 4)$

110. $(z + 3)(z - 3) = z(z - 3)$

111. $(x + 4)(x - 4) = (x - 2)(x + 6)$

112. $(y - 1)(y + 6) = (y - 3)(y - 2) + 8$

113. $(a - 3)^2 = (a + 3)^2$

114. $(b + 2)^2 = (b - 1)^2$

115. $4 + (2y - 3)^2 = (2y - 1)(2y + 3)$

116. $7s^2 + (s - 3)(2s + 1) = (3s - 1)^2$

117. If 3 less than a certain number is multiplied by 4 more than the number, the product is 6 less than the square of the number. Find the number.

118. The difference between the squares of two consecutive positive integers is 11. Find the integers.

119. The difference between the squares of two consecutive odd positive integers is 32. Find the integers.

120. The sum of the squares of three consecutive odd integers is 112 less than 3 times the square of the largest integer. Find the integers.

121. In major league baseball, the distance between bases is 30 feet greater than it is in softball. The bases in major league baseball mark the corners of a square that has an area 4500 square feet greater than for softball. Find the distance between the bases in baseball.

122. Two square sheets of cardboard differ in area by 44 square inches. An edge of the larger square is 2 inches greater than an edge of the smaller square. Find the length of an edge of the smaller square.

123. The radius of one circle is 1 inch greater than the radius of another circle, and their areas differ by 4π square inches. Find the radius of the smaller circle.

124. The radius of one circle is 3 meters greater than the radius of another circle, and their areas differ by 15π square meters. Find the radius of the larger circle.

REVIEW EXERCISES

In Review Exercises 1–4, list the numbers in the set $\{-\frac{1}{2}, 2, 4, \frac{7}{8}\}$ *that satisfy the given condition.*

1. negative number
2. prime number
3. composite number
4. integer

In Review Exercises 5–8, tell which property of real numbers justifies each statement.

5. $3(x + 5) = 3x + 3 \cdot 5$

6. $(x + 3) + y = x + (3 + y)$

7. $3(ab) = (ab)3$

8. $a + 0 = a$

9. Solve the equation $\dfrac{5}{3}(5y + 6) - 10 = 0$.

10. Solve the equation $F = \dfrac{GMm}{d^2}$ for m.

11. The **parsec,** a unit of distance used in astronomy, is 3×10^{16} meters. The distance to Betelgeuse, a star in the constellation *Orion,* is 160 parsecs. Use scientific notation to express this distance in meters.

12. What number is one greater than one-half of itself?

3.7 DIVIDING POLYNOMIALS BY MONOMIALS

Dividing by a number is equivalent to multiplying by its reciprocal. For example, dividing the number 8 by 2 gives the same answer as multiplying 8 by $\frac{1}{2}$.

$$\frac{8}{2} = 4 \qquad \text{and} \qquad \frac{1}{2} \cdot 8 = 4$$

In general, if $b \neq 0$, then

$$\frac{a}{b} = \frac{1}{b} \cdot a$$

Recall that to simplify a fraction, we write both its numerator and denominator as the product of several factors and then divide out all common factors. For example, to simplify $\frac{4}{6}$, we can write

$$\frac{4}{6} = \frac{2 \cdot 2}{2 \cdot 3} = \frac{\overset{1}{\cancel{2}} \cdot 2}{\underset{1}{\cancel{2}} \cdot 3} = \frac{2}{3}$$

To simplify the fraction $\frac{20}{25}$, we can write

$$\frac{20}{25} = \frac{4 \cdot 5}{5 \cdot 5} = \frac{4 \cdot \overset{1}{\cancel{5}}}{\underset{1}{\cancel{5}} \cdot 5} = \frac{4}{5}$$

To simplify algebraic fractions, we can either use the previous method for simplifying arithmetic fractions or use the rules of exponents.

EXAMPLE 1   Simplify **a.** $\dfrac{x^2 y}{xy^2}$ and   **b.** $\dfrac{-8a^3 b^2}{4ab^3}$.

Solution   **Method for arithmetic fractions**

a. $\dfrac{x^2 y}{xy^2} = \dfrac{xxy}{xyy}$

$\quad = \dfrac{\overset{1}{\cancel{x}}\overset{}{x}\overset{1}{\cancel{y}}}{\underset{1}{\cancel{x}}y\underset{1}{\cancel{y}}}$

$\quad = \dfrac{x}{y}$

b. $\dfrac{-8a^3 b^2}{4ab^3} = \dfrac{-2 \cdot 4aaabb}{4abbb}$

$\quad = \dfrac{-2 \cdot \overset{11}{\cancel{4}}aa\overset{11}{\cancel{bb}}}{\underset{1111}{\cancel{4}\cancel{a}\cancel{b}\cancel{b}}}$

$\quad = \dfrac{-2a^2}{b}$

Using the rules of exponents

a. $\dfrac{x^2 y}{xy^2} = x^{2-1} y^{1-2}$

$\quad = x^1 y^{-1}$

$\quad = \dfrac{x}{y}$

b. $\dfrac{-8a^3 b^2}{4ab^3} = \dfrac{-2^3 a^3 b^2}{2^2 ab^3}$

$\quad = -2^{3-2} a^{3-1} b^{2-3}$

$\quad = -2^1 a^2 b^{-1}$

$\quad = \dfrac{-2a^2}{b}$

∎

To divide a polynomial with more than one term by a monomial, we rewrite the division as a product, use the distributive law to remove parentheses, and simplify each resulting fraction.

EXAMPLE 2   Simplify $\dfrac{9x + 6y}{3xy}$.

Solution $\dfrac{9x + 6y}{3xy} = \dfrac{1}{3xy}(9x + 6y)$ $\dfrac{a}{b} = \dfrac{1}{b} \cdot a$

$\quad\quad\quad = \dfrac{9x}{3xy} + \dfrac{6y}{3xy}$ Remove parentheses.

$\quad\quad\quad = \dfrac{3}{y} + \dfrac{2}{x}$ Simplify each fraction.

∎

EXAMPLE 3   Simplify $\dfrac{6x^2y^2 + 4x^2y - 2xy}{2xy}$.

Solution $\dfrac{6x^2y^2 + 4x^2y - 2xy}{2xy} = \dfrac{1}{2xy}(6x^2y^2 + 4x^2y - 2xy)$

$= \dfrac{6x^2y^2}{2xy} + \dfrac{4x^2y}{2xy} - \dfrac{2xy}{2xy}$ Remove parentheses.

$= 3xy + 2x - 1$ Simplify each fraction. ∎

EXAMPLE 4   Simplify $\dfrac{12a^3b^2 - 4a^2b + a}{6a^2b^2}$.

Solution $\dfrac{12a^3b^2 - 4a^2b + a}{6a^2b^2} = \dfrac{1}{6a^2b^2}(12a^3b^2 - 4a^2b + a)$

$= \dfrac{12a^3b^2}{6a^2b^2} - \dfrac{4a^2b}{6a^2b^2} + \dfrac{a}{6a^2b^2}$ Remove parentheses.

$= 2a - \dfrac{2}{3b} + \dfrac{1}{6ab^2}$ Simplify each fraction. ∎

EXAMPLE 5   Simplify $\dfrac{(x - y)^2 - (x + y)^2}{xy}$.

Solution $\dfrac{(x - y)^2 - (x + y)^2}{xy} = \dfrac{x^2 - 2xy + y^2 - (x^2 + 2xy + y^2)}{xy}$ Square the binomials in the numerator.

$= \dfrac{x^2 - 2xy + y^2 - x^2 - 2xy - y^2}{xy}$ Remove parentheses.

$= \dfrac{-4xy}{xy}$ Combine terms.

$= -4$ Divide out xy. ∎

EXERCISE 3.7

In Exercises 1–12, simplify each fraction.

1. $\dfrac{5}{15}$

2. $\dfrac{64}{128}$

3. $\dfrac{-125}{75}$

4. $\dfrac{-98}{21}$

5. $\dfrac{120}{160}$

6. $\dfrac{70}{420}$

7. $\dfrac{-3612}{-3612}$

8. $\dfrac{-288}{-112}$

9. $\dfrac{-90}{360}$

10. $\dfrac{8423}{-8423}$

11. $\dfrac{5880}{2660}$

12. $\dfrac{-762}{366}$

In Exercises 13–40, perform each division by simplifying each fraction. Write all answers without using negative or zero exponents.

13. $\dfrac{xy}{yz}$

14. $\dfrac{a^2b}{ab^2}$

15. $\dfrac{r^3s^2}{rs^3}$

16. $\dfrac{y^4z^3}{y^2z^2}$

17. $\dfrac{8x^3y^2}{4xy^3}$

18. $\dfrac{-3y^3z}{6yz^2}$

19. $\dfrac{12u^5v}{-4u^2v^3}$

20. $\dfrac{16rst^2}{-8rst^3}$

21. $\dfrac{-16r^3y^2}{-4r^2y^4}$

22. $\dfrac{35xyz^2}{-7x^2yz}$

23. $\dfrac{-65rs^2t}{15r^2s^3t}$

24. $\dfrac{112u^3z^6}{-42u^3z^6}$

25. $\dfrac{x^2x^3}{xy^6}$

26. $\dfrac{(xy)^2}{x^2y^3}$

27. $\dfrac{(a^3b^4)^3}{ab^4}$

28. $\dfrac{(a^2b^3)^3}{a^6b^6}$

29. $\dfrac{15(r^2s^3)^2}{-5(rs^5)^3}$

30. $\dfrac{-5(a^2b)^3}{10(ab^2)^3}$

31. $\dfrac{-32(x^3y)^3}{128(x^2y^2)^3}$

32. $\dfrac{68(a^6b^7)^2}{-96(abc^2)^3}$

33. $\dfrac{(5a^2b)^3}{(2a^2b^2)^3}$

34. $\dfrac{-(4x^3y^3)^2}{(x^2y^4)^8}$

35. $\dfrac{-(3x^3y^4)^3}{-(9x^4y^5)^2}$

36. $\dfrac{(2r^3s^2t)^2}{-(4r^2s^2t^2)^2}$

37. $\dfrac{(a^2a^3)^4}{(a^4)^3}$

38. $\dfrac{(b^3b^4)^5}{(bb^2)^2}$

39. $\dfrac{(z^3z^{-4})^3}{(z^{-3})^2}$

40. $\dfrac{(t^{-3}t^5)}{(t^2)^{-3}}$

In Exercises 41–54, perform each indicated division.

41. $\dfrac{6x+9y}{3xy}$

42. $\dfrac{8x+12y}{4xy}$

43. $\dfrac{5x-10y}{25xy}$

44. $\dfrac{2x-32}{16x}$

45. $\dfrac{3x^2+6y^3}{3x^2y^2}$

46. $\dfrac{4a^2-9b^2}{12ab}$

47. $\dfrac{15a^3b^2-10a^2b^3}{5a^2b^2}$

48. $\dfrac{9a^4b^3-16a^3b^4}{12a^2b}$

49. $\dfrac{4x-2y+8z}{4xy}$

50. $\dfrac{5a^2+10b^2-15ab}{5ab}$

51. $\dfrac{12x^3y^2-8x^2y-4x}{4xy}$

52. $\dfrac{12a^2b^2-8a^2b-4ab}{4ab}$

53. $\dfrac{-25x^2y+30xy^2-5xy}{-5xy}$

54. $\dfrac{-30a^2b^2-15a^2b-10ab^2}{-10ab}$

In Exercises 55–64, simplify each numerator and perform the indicated division.

55. $\dfrac{5x(4x-2y)}{2y}$

56. $\dfrac{9y^2(x^2-3xy)}{3x^2}$

57. $\dfrac{(-2x)^3+(3x^2)^2}{6x^2}$

58. $\dfrac{(-3x^2y)^3+(3xy^2)^3}{27x^3y^4}$

59. $\dfrac{4x^2y^2-2(x^2y^2+xy)}{2xy}$

60. $\dfrac{-5a^3b-5a(ab^2-a^2b)}{10a^2b^2}$

61. $\dfrac{(3x-y)(2x-3y)}{6xy}$

62. $\dfrac{(2m-n)(3m-2n)}{-3m^2n^2}$

63. $\dfrac{(a+b)^2 - (a-b)^2}{2ab}$

64. $\dfrac{(x-y)^2 + (x+y)^2}{2x^2y^2}$

REVIEW EXERCISES

In Review Exercises 1–2, let $P(x) = 3x^2 + x$.

1. Find $P(4)$.

2. Find $P(-2)$.

3. Write 0.000265 in scientific notation.

4. Write 5.67×10^3 in standard notation.

In Review Exercises 5–8, simplify each expression.

5. $(3x^2)^0$

6. $(a^2b^3a^4b^5)^3$

7. $\dfrac{8x^4y^3}{4x^5y}$

8. $\left(\dfrac{9r^2st^3}{3r^4st^2}\right)^2$

9. Solve $x(x+2) = (x+3)^2 - 1$.

10. Solve $y = mx + b$ for m.

11. The product of two consecutive integers is 19 greater than the square of the smaller integer. Find the integers.

12. The total surface area, A, of a box with dimensions l, w, and d (see Illustration 1) is given by the formula

$$A = 2lw + 2wd + 2ld$$

If $A = 202$ square inches, $l = 9$ inches, and $w = 5$ inches, find d.

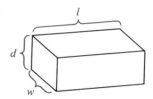

ILLUSTRATION 1

3.8 DIVIDING POLYNOMIALS BY POLYNOMIALS

If we use pencil and paper to divide 156 by 12, the work would probably look like this:

$$
\begin{array}{r}
13 \\
12\overline{)156} \\
12 \\
\hline
36 \\
36 \\
\hline
0
\end{array}
$$

To find a process for dividing a polynomial by a polynomial with more than one term, we will use a slightly different format for the division. We begin by noting that the **dividend** (the number 156) can be written as

$$156 = 100 + 50 + 6$$

and that the **divisor** (the number 12) can be written as

$$12 = 10 + 2$$

The division can then be accomplished as follows:

Step 1

$$\begin{array}{r} 10 \\ 10 + 2 \overline{)\ 100 + 50 + 6} \end{array}$$

How many times does 10 divide 100? $100/10 = 10$. Place the 10 above the division symbol.

Step 2

$$\begin{array}{r} 10 \\ 10 + 2 \overline{)\ 100 + 50 + 6} \\ 100 + 20 \end{array}$$

Multiply each term in the divisor by 10. Place the product under $100 + 50$ as indicated and draw a line.

Step 3

$$\begin{array}{r} 10 \\ 10 + 2 \overline{)\ 100 + 50 + 6} \\ 100 + 20 \\ \hline 30 + 6 \end{array}$$

Subtract $100 + 20$ from $100 + 50$ by adding the negative of $100 + 20$ to $100 + 50$. Bring down the next term.

Step 4

$$\begin{array}{r} 10 + \ \ 3 \\ 10 + 2 \overline{)\ 100 + 50 + 6} \\ 100 + 20 \\ \hline 30 + 6 \end{array}$$

How many times does 10 divide 30? $30/10 = +3$. Place the $+3$ above the division symbol.

Step 5

$$\begin{array}{r} 10 + \ \ 3 \\ 10 + 2 \overline{)\ 100 + 50 + 6} \\ 100 + 20 \\ \hline 30 + 6 \\ 30 + 6 \end{array}$$

Multiply each term in the divisor by 3. Place the product under the $30 + 6$ as indicated and draw a line.

Step 6

$$\begin{array}{r} 10 + \ \ 3 \\ 10 + 2 \overline{)\ 100 + 50 + 6} \\ 100 + 20 \\ \hline 30 + 6 \\ 30 + 6 \\ \hline 0 \end{array}$$

Subtract $30 + 6$ from $30 + 6$ by adding the negative of $30 + 6$.

Because the remainder is 0, the division "comes out even." The **quotient** is $10 + 3$, or 13.

Step 7

We check the answer by verifying that the product of the divisor and the quotient is equal to the dividend

$$12 \cdot 13 = 156$$

The answer checks.

We will now see how to use this method to divide one polynomial by another. The first example closely parallels the previous discussion.

EXAMPLE 1   Divide $x^2 + 5x + 6$ by $x + 2$.

Solution   *Step 1*

$$x + 2 \overline{) \; x^2 + 5x + 6} $$
with x above

How many times does x divide x^2? $x^2/x = x$.
Place the x above the division symbol.

Step 2

$$x + 2 \overline{) \; x^2 + 5x + 6}$$
$$\underline{x^2 + 2x}$$
with x above

Multiply each term in the divisor by x.
Place the product under $x^2 + 5x$ as indicated and draw a line.

Step 3

$$x + 2 \overline{) \; x^2 + 5x + 6}$$
$$\underline{x^2 + 2x}$$
$$3x + 6$$
with x above

Subtract $x^2 + 2x$ from $x^2 + 5x$ by adding the negative of $x^2 + 2x$ to $x^2 + 5x$.

Bring down the next term.

Step 4

$$x + 2 \overline{) \; x^2 + 5x + 6}$$
$$\underline{x^2 + 2x}$$
$$3x + 6$$
with $x + 3$ above

How many times does x divide $3x$? $3x/x = +3$.
Place the $+3$ above the division symbol.

Step 5

$$x + 2 \overline{) \; x^2 + 5x + 6}$$
$$\underline{x^2 + 2x}$$
$$3x + 6$$
$$\underline{3x + 6}$$
with $x + 3$ above

Multiply each term in the divisor by 3.
Place the product under the $3x + 6$ as indicated and draw a line.

Step 6

$$x + 2 \overline{) \; x^2 + 5x + 6}$$
$$\underline{x^2 + 2x}$$
$$3x + 6$$
$$\underline{3x + 6}$$
$$0$$
with $x + 3$ above

Subtract $3x + 6$ from $3x + 6$ by adding the negative of $3x + 6$.

The quotient is $x + 3$ and the remainder is 0.

Step 7

Check the work by verifying that the product of $x + 2$ and $x + 3$ is $x^2 + 5x + 6$.

$$(x + 2)(x + 3) = x^2 + 3x + 2x + 6$$
$$= x^2 + 5x + 6$$

The answer checks.   ■

EXAMPLE 2   Divide: $\dfrac{6x^2 - 7x - 2}{2x - 1}$.

Solution   **Step 1**

$$
\begin{array}{r}
3x \\
2x - 1 \overline{)6x^2 - 7x - 2}
\end{array}
$$

How many times does $2x$ divide $6x^2$?
$6x^2/2x = 3x$. Place the $3x$ above the division symbol.

Step 2

$$
\begin{array}{r}
3x \\
2x - 1 \overline{)6x^2 - 7x - 2} \\
6x^2 - 3x
\end{array}
$$

Multiply each term in the divisor by $3x$. Place the product under $6x^2 - 7x$ as indicated and draw a line.

Step 3

$$
\begin{array}{r}
3x \\
2x - 1 \overline{)6x^2 - 7x - 2} \\
6x^2 - 3x \\
\hline
-4x - 2
\end{array}
$$

Subtract $6x^2 - 3x$ from $6x^2 - 7x$ by adding the negative of $6x^2 - 3x$ to $6x^2 - 7x$.

Bring down the next term.

Step 4

$$
\begin{array}{r}
3x - 2 \\
2x - 1 \overline{)6x^2 - 7x - 2} \\
6x^2 - 3x \\
\hline
-4x - 2
\end{array}
$$

How many times does $2x$ divide $-4x$?
$-4x/2x = -2$. Place the -2 above the division symbol.

Step 5

$$
\begin{array}{r}
3x - 2 \\
2x - 1 \overline{)6x^2 - 7x - 2} \\
6x^2 - 3x \\
\hline
-4x - 2 \\
-4x + 2
\end{array}
$$

Multiply each term in the divisor by -2. Place the product under the $-4x - 2$ as indicated and draw a line.

Step 6

$$
\begin{array}{r}
3x - 2 \\
2x - 1 \overline{)6x^2 - 7x - 2} \\
6x^2 - 3x \\
\hline
-4x - 2 \\
-4x + 2 \\
\hline
-4
\end{array}
$$

Subtract $-4x + 2$ from $-4x - 2$ by adding the negative of $-4x + 2$.

In this example, the quotient is $3x - 2$ and the remainder is -4. It is common to write the answer in the following way:

$$3x - 2 + \frac{-4}{2x - 1}$$

where the fraction $\dfrac{-4}{2x - 1}$ is formed by dividing the remainder by the divisor.

Step 7

To check the answer, multiply $3x - 2 + \dfrac{-4}{2x - 1}$ by $2x - 1$. The product should be the dividend.

$$(2x - 1)\left(3x - 2 + \frac{-4}{2x - 1}\right) = (2x - 1)(3x - 2) + (2x - 1)\left(\frac{-4}{2x - 1}\right)$$
$$= (2x - 1)(3x - 2) - 4$$
$$= 6x^2 - 4x - 3x + 2 - 4$$
$$= 6x^2 - 7x - 2$$

Because the result is the dividend, the answer checks. ∎

EXAMPLE 3   Divide $4x^2 + 2x^3 + 12 - 2x$ by $x + 3$.

Solution   The division process works most efficiently if both the divisor and the dividend are written in descending powers of x. This means that the term involving the highest power of x appears first, the term involving the second highest power of x appears second, and so on. Use the commutative property to rearrange the terms of the dividend in descending order and divide as follows:

$$
\begin{array}{r}
2x^2 - 2x\ + 4 \\
x + 3 \overline{\smash{\big)}\ 2x^3 + 4x^2 - 2x + 12} \\
\underline{2x^3 + 6x^2} \\
-2x^2 - 2x \\
\underline{-2x^2 - 6x} \\
+ 4x + 12 \\
\underline{+ 4x + 12} \\
0
\end{array}
$$

Check: $(x + 3)(2x^2 - 2x + 4) = 2x^3 - 2x^2 + 4x + 6x^2 - 6x + 12$
$$= 2x^3 + 4x^2 - 2x + 12$$

The answer checks. ∎

EXAMPLE 4   Divide: $\dfrac{x^2 - 4}{x + 2}$.

Solution   Note that the binomial $x^2 - 4$ does not have a term involving x. To perform this division, you must either include the term $0x$ or leave a space for it. After this adjustment, the division is routine.

$$
\begin{array}{r}
x + 2 \\
x + 2 \overline{) x^2 + 0x - 4} \\
\underline{x^2 + 2x} \\
-2x - 4 \\
\underline{-2x - 4} \\
0
\end{array}
$$

Check: $(x + 2)(x - 2) = x^2 - 2x + 2x - 4 = x^2 - 4$

The answer checks.   ∎

EXAMPLE 5   Divide $x^3 + y^3$ by $x + y$.

Solution   Write $x^3 + y^3$ leaving spaces for any missing terms, and proceed as follows.

$$
\begin{array}{r}
x^2 - xy + y^2 \\
x + y \overline{) x^3 \qquad\qquad + y^3} \\
\underline{x^3 + x^2 y} \\
-x^2 y \\
\underline{-x^2 y - xy^2} \\
+xy^2 + y^3 \\
\underline{xy^2 + y^3} \\
0
\end{array}
$$

Check: $(x + y)(x^2 - xy + y^2) = x^3 - x^2 y + xy^2 + x^2 y - xy^2 + y^3 = x^3 + y^3$

The answer checks.   ∎

EXERCISE 3.8

1. Divide $x^2 + 4x + 4$ by $x + 2$.

2. Divide $x^2 - 5x + 6$ by $x - 2$.

3. Divide $y^2 + 13y + 12$ by $y + 1$.

4. Divide $z^2 - 7z + 12$ by $z - 3$.

5. Divide $a^2 + 2ab + b^2$ by $a + b$.

6. Divide $a^2 - 2ab + b^2$ by $a - b$.

In Exercises 7–12, perform each division.

7. $\dfrac{6a^2 + 5a - 6}{2a + 3}$

8. $\dfrac{8a^2 + 2a - 3}{2a - 1}$

9. $\dfrac{3b^2 + 11b + 6}{3b + 2}$

10. $\dfrac{3b^2 - 5b + 2}{3b - 2}$

11. $\dfrac{2x^2 - 7xy + 3y^2}{2x - y}$

12. $\dfrac{3x^2 + 5xy - 2y^2}{x + 2y}$

In Exercises 13–24, rearrange the terms so that the powers of x are in descending order. Then perform each division.

13. $5x + 3 \overline{) 11x + 10x^2 + 3}$

14. $2x - 7 \overline{) -x - 21 + 2x^2}$

15. $4 + 2x \overline{) -10x - 28 + 2x^2}$

16. $1 + 3x \overline{) 9x^2 + 1 + 6x}$

17. $2x - y \overline{) xy - 2y^2 + 6x^2}$

18. $2y + x \overline{) 3xy + 2x^2 - 2y^2}$

19. $x + 3y \overline{) 2x^2 - 3y^2 + 5xy}$

20. $2x - 3y \overline{) 2x^2 - 3y^2 - xy}$

21. $3x - 2y \overline{) -10y^2 + 13xy + 3x^2}$

22. $2x + 3y \overline{) -12y^2 + 10x^2 + 7xy}$

23. $4x + y \overline{) -19xy + 4x^2 - 5y^2}$

24. $x - 4y \overline{) 5x^2 - 4y^2 - 19xy}$

In Exercises 25–30, perform each division.

25. $2x + 3 \overline{) 2x^3 + 7x^2 + 4x - 3}$

26. $2x - 1 \overline{) 2x^3 - 3x^2 + 5x - 2}$

27. $3x + 2 \overline{) 6x^3 + 10x^2 + 7x + 2}$

28. $4x + 3 \overline{) 4x^3 - 5x^2 - 2x + 3}$

29. $2x + y \overline{) 2x^3 + 3x^2y + 3xy^2 + y^3}$

30. $3x - 2y \overline{) 6x^3 - x^2y + 4xy^2 - 4y^3}$

In Exercises 31–40, perform each division. If there is a remainder, leave the answer in quotient $+ \dfrac{remainder}{divisor}$ form.

31. $\dfrac{2x^2 + 5x + 2}{2x + 3}$

32. $\dfrac{3x^2 - 8x + 3}{3x - 2}$

33. $\dfrac{4x^2 + 6x - 1}{2x + 1}$

34. $\dfrac{6x^2 - 11x + 2}{3x - 1}$

35. $\dfrac{x^3 + 3x^2 + 3x + 1}{x + 1}$

36. $\dfrac{x^3 + 6x^2 + 12x + 8}{x + 2}$

37. $\dfrac{2x^3 + 7x^2 + 4x + 3}{2x + 3}$

38. $\dfrac{6x^3 + x^2 + 2x + 1}{3x - 1}$

39. $\dfrac{2x^3 + 4x^2 - 2x + 3}{x - 2}$

40. $\dfrac{3y^3 - 4y^2 + 2y + 3}{y + 3}$

In Exercises 41–50, perform each division.

41. $\dfrac{x^2 - 1}{x - 1}$

42. $\dfrac{x^2 - 9}{x + 3}$

43. $\dfrac{4x^2 - 9}{2x + 3}$

44. $\dfrac{25x^2 - 16}{5x - 4}$

45. $\dfrac{x^3 + 1}{x + 1}$

46. $\dfrac{x^3 - 8}{x - 2}$

47. $\dfrac{x^3 + x}{x + 3}$

48. $\dfrac{x^3 - 50}{x - 5}$

49. $3x - 4 \overline{) 15x^3 - 23x^2 + 16x}$

50. $2y + 3 \overline{) 21y^2 + 6y^3 - 20}$

REVIEW EXERCISES

1. List the composite numbers between 20 and 30.

2. Graph the set of prime numbers between 10 and 18 on a number line.

In Review Exercises 3–4, let $a = -2$ and $b = -3$. Evaluate each expression.

3. $4a - 2b$

4. $\dfrac{3a^2 + 2b^2}{3(a - b)}$

In Review Exercises 5–8, write each expression as an equivalent expression without absolute value symbols.

5. $|25|$

6. $|-32 - 5|$

7. $-|3 - 5|$

8. $-|16(-3)|$

In Review Exercises 9–10, simplify each expression.

9. $3(2x^2 - 4x + 5) + 2(x^2 + 3x - 7)$

10. $-2(y^3 + 2y^2 - y) - 3(3y^3 + y)$

11. The approximate area of the ring between two circles of radius r and R (see Illustration 1) is given by

$$A = \frac{22}{7}(R + r)(R - r)$$

If $r = 3$ inches and $R = 17$ inches, find A.

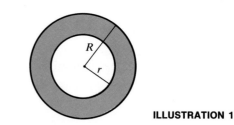

ILLUSTRATION 1

12. Laura bought a color TV set for $502.90, which included a 7% sales tax. Find the selling price of the TV before the tax was added.

CHAPTER SUMMARY

Key Words

algebraic terms (3.4)	*exponent* (3.1)	*quotient* (3.8)
base (3.1)	*FOIL* (3.6)	*scientific notation* (3.3)
binomial (3.4)	*like terms* (3.5)	*special products* (3.6)
degree of a monomial (3.4)	*minuend* (3.5)	*standard notation* (3.3)
degree of a polynomial (3.4)	*monomial* (3.4)	*subtrahend* (3.5)
dividend (3.8)	*polynomial* (3.4)	*trinomial* (3.4)
divisor (3.8)	*power* (3.1)	

Key Ideas

(3.1)–(3.2) **Properties of exponents.** If n is a natural number, then

$$x^n = \overbrace{x \cdot x \cdot x \cdot x \cdots \cdot x}^{n \text{ factors of } x}$$

If m and n are integers, then

$$x^m x^n = x^{m+n} \qquad (x^m)^n = x^{mn} \qquad (xy)^n = x^n y^n$$

$$\left(\frac{x}{y}\right)^n = \frac{x^n}{y^n} \qquad \text{provided } y \neq 0$$

$$\frac{x^m}{x^n} = x^{m-n} \qquad \text{provided } x \neq 0$$

$$x^0 = 1 \qquad \text{provided } x \neq 0$$

$$x^{-n} = \frac{1}{x^n} \qquad \text{provided } x \neq 0$$

(3.3) A number is written in scientific notation if it is written as the product of a number between 1 (including 1) and 10 and an integer power of 10.

(3.4) If $P(x)$ is a polynomial in x, then $P(r)$ is the value of the polynomial when $x = r$.

(3.5) When adding or subtracting polynomials, combine like terms by adding or subtracting the numerical coefficients and using the same variables and the same exponents.

(3.6) To multiply two monomials, first multiply the numerical factors and then multiply the variable factors.

To multiply a polynomial with more than one term by a monomial, multiply each term of the polynomial by the monomial, and simplify.

To multiply one polynomial by another, multiply each term of one polynomial by each term of the other polynomial, and simplify.

To multiply two binomials, use the **FOIL** method.

Special products:

$$(x + y)^2 = x^2 + 2xy + y^2$$
$$(x - y)^2 = x^2 - 2xy + y^2$$
$$(x + y)(x - y) = x^2 - y^2$$

(3.7) To simplify a fraction, divide out all factors common to the numerator and the denominator of the fraction.

To divide a polynomial by a monomial, rewrite the division as a product, use the distributive law to remove parentheses, and simplify each resulting fraction.

(3.8) Use long division to divide one polynomial by another.

CHAPTER 3 REVIEW EXERCISES

(3.1–3.2) *In Review Exercises 1–8, evaluate each expression.*

1. 5^3
2. 3^5
3. $(-8)^2$
4. -8^2

5. $3^2 + 2^2$
6. $(3 + 2)^2$
7. $3(3^3 + 3^3)$
8. $1^{17} + 17^1$

In Review Exercises 9–24, perform the indicated operation and simplify.

9. x^3x^2 $\times\ ^5$
10. $x(x^2y)$
11. y^7y^3
12. x^0y^5

13. $2b^3b^4b^5$
14. $(-z^2)(z^3y^2)$
15. $(4^4s)s^2$
16. $-3y(y^5)$

17. $(x^2x^3)^3$
18. $(2x^2y)^2$
19. $(3x^0)^2$
20. $(3x^2y^2)^0$

21. $\dfrac{x^7}{x^3}$
22. $\left(\dfrac{x^2y}{xy^2}\right)^2$
23. $\dfrac{8(y^2x)^2}{2^3(yx^2)^2}$
24. $\dfrac{(5x^0y^2z^3)^3}{25(yz)^5}$

In Review Exercises 25–32, write each expression without using negative exponents or parentheses.

25. $x^{-2}x^3$
26. y^4y^{-3}
27. $\dfrac{x^3}{x^{-7}}$
28. $(x^{-3}x^{-4})^{-2}$

29. $\dfrac{x^3}{x^7}$
30. $\left(\dfrac{x^2}{x}\right)^{-5}$
31. $\left(\dfrac{3s}{6s^2}\right)^3$
32. $\left(\dfrac{15z^4}{5z^3}\right)^{-2}$

(3.3) *In Review Exercises 33–40, write each number in scientific notation.*

33. 728 $7.28 \cdot 10^2$
34. 9370
35. 0.0136
36. 0.00942

37. 7.61
38. 795×10^3
39. 0.012×10^{-2}
40. 600×10^2

In Review Exercises 41–46, write each number in standard notation.

41. 7.26×10^5
42. 3.91×10^{-4}
43. 2.68×10^0
44. 5.76×10^1

45. 731×10^{-2}
46. 0.498×10^3

(3.4) *In Review Exercises 47–50, give the degree of each polynomial.*

47. $13x^7$
48. $5^3x + x^2$
49. $-3x^5 + x - 1$
50. $9x + 21x^3$

In Review Exercises 51–54, let $P(x) = 3x + 2$. Find each value.

51. $P(3)$
52. $P(0)$
53. $P(-2)$
54. $P(2t)$

In Review Exercises 55–58, let $P(x) = 5x^4 - x$. Find each value.

55. $P(3)$

56. $P(0)$

57. $P(-2)$

58. $P(2t)$

(3.5) *In Review Exercises 59–68, simplify each expression, if possible.*

59. $3x + 5x - x$

60. $2x + 3y$

61. $(xy)^2 + 3x^2y^2$

62. $-2x^2yz + 3yx^2z$

63. $3x^2y^0 + 2x^2$

64. $2(x + 7) + 3(x + 7)$

65. $(3x^2 + 2x) + (5x^2 - 8x)$

66. $(7a^2 + 2a - 5) - (3a^2 - 2a + 1)$

67. $3(9x^2 + 3x + 7) + 2(2x^2 - 8x + 3) - 2(11x^2 - 5x + 9)$

68. $4(4x^3 + 2x^2 - 3x - 8) - 5(2x^3 - 3x + 8)$

(3.6) *In Review Exercises 69–90, find each product.*

69. $(2x^2y^3)(5xy^2)$

70. $(xyz^3)(x^3z)^2$

71. $5(x + 3)$

72. $3(2x + 4)$

73. $x^2(3x^2 - 5)$

74. $2y^2(y^2 + 5y)$

75. $-x^2y(y^2 - xy)$

76. $-3xy(xy - x)$

77. $(x + 3)(x + 2)$

78. $(2x + 1)(x - 1)$

79. $(3a - 3)(2a + 2)$

80. $6(a - 1)(a + 1)$

81. $(a - b)(2a + b)$

82. $(3x - y)(2x + y)$

83. $(-3a - b)(3a - b)$

84. $(x + 5)(x - 5)$

85. $(y - 2)(y + 2)$

86. $(x + 4)^2$

87. $(y - 3)^2$

88. $y(y + 1)^2$

89. $(3x + 1)(x^2 + 2x + 1)$

90. $(2a - 3)(4a^2 + 6a + 9)$

In Review Exercises 91–96, solve each equation.

91. $x^2 + 3 = x(x + 3)$

92. $x^2 + x = (x + 1)(x + 2)$

93. $(x + 2)(x - 5) = (x - 4)(x - 1)$

94. $(x - 1)(x - 2) = (x - 3)(x + 1)$

95. $x^2 + x(x + 2) = x(2x + 1) + 1$

96. $(x + 5)(3x + 1) = x^2 + (2x - 1)(x - 5)$

(3.7–3.8) *In Review Exercises 97–106, perform each division. If there is a remainder, leave the answer in quotient $+$ $\dfrac{\text{remainder}}{\text{divisor}}$ form.*

97. $\dfrac{3x + 6y}{2xy}$

98. $\dfrac{14xy - 21x}{7xy}$

99. $\dfrac{15a^2bc + 20ab^2c - 25abc^2}{-5abc}$

100. $\dfrac{(x + y)^2 + (x - y)^2}{-2xy}$

101. $x + 2 \overline{)\, x^2 + 3x + 5}$

102. $x - 1 \overline{)\, x^2 - 6x + 5}$

103. $x + 3 \overline{)\, 2x^2 + 7x + 3}$

104. $3x - 1 \overline{)\, 3x^2 + 14x - 2}$

105. $2x - 1 \overline{)\, 6x^3 + x^2 + 1}$

106. $3x + 1 \overline{)\, -13x - 4 + 9x^3}$

CHAPTER 3 TEST

_____ **1.** Use exponents to rewrite $2xxxyyyy$.

_____ **2.** Evaluate $3^2 + 5^3$.

In Problems 3–6, write each expression as an expression containing only one exponent.

_____ **3.** $y^2(yy^3)$ _____ **4.** $(-3b^2)(2b^3)(-b^2)$

_____ **5.** $(2x^3)^5(x^2)^3$ _____ **6.** $(2rr^2r^3)^3$

In Problems 7–10, simplify each expression. Write answers without using parentheses or negative exponents.

_____ **7.** $3x^0$ _____ **8.** $2y^{-5}y^2$

_____ **9.** $\dfrac{y^2}{yy^{-2}}$ _____ **10.** $\left(\dfrac{a^2b^{-1}}{4a^3b^{-2}}\right)^{-3}$

_____ **11.** Express 28,000 in scientific notation.

_____ **12.** Express 0.0025 in scientific notation.

_____ **13.** Express 7.4×10^3 in standard notation.

_____ **14.** Express 9.3×10^{-5} in standard notation.

_____ **15.** Identify $3x^2 + 2$ as a monomial, binomial, or trinomial.

_____ **16.** Find the degree of the polynomial $3x^2y^3z^4 + 2x^3y^2z - 5x^2y^3z^5$.

_____ **17.** If $P(x) = x^2 + x - 2$, find $P(-2)$.

_____ **18.** Simplify $(xy)^2 + 5x^2y^2 - 3x^2y^2$.

_____ **19.** Simplify: $-6(x - y) + 2(x + y) - 3(x + 2y)$

_____ **20.** Simplify: $-2(x^2 + 3x - 1) - 3(x^2 - x + 2) + 5(x^2 + 2)$

_____ **21.** Add: $\begin{array}{r} 3x^3 + 4x^2 - x - 7 \\ 2x^3 - 2x^2 + 3x + 2 \\ \hline \end{array}$ _____ **22.** Subtract: $\begin{array}{r} 2x^2 - 7x + 3 \\ 3x^2 - 2x - 1 \\ \hline \end{array}$

In Problems 23–27, find each product.

_____ **23.** $(-2x^3)(2x^2y)$ _____ **24.** $3y^2(y^2 - 2y + 3)$

_____ **25.** $(2x - 5)(3x + 4)$ _____ **26.** $(3x - y)(3x + y)$

_____ **27.** $(2x - 3)(x^2 - 2x + 4)$

_____ **28.** Solve the equation $(a + 2)^2 = (a - 3)^2$.

_____ **29.** Simplify the fraction $\dfrac{8x^2y^3z^4}{16x^3y^2z^4}$. _____ **30.** Perform the division $\dfrac{6a^2 - 12b^2}{24ab}$.

_____ **31.** Divide: $2x + 3 \overline{)\, 2x^2 - x - 6}$ _____ **32.** Divide: $x - y \overline{)\, x^3 - y^3}$

4

Factoring Polynomials

In this chapter we shall reverse the operation of multiplication and show how to find the factors of a known product. The process of finding the individual factors of a product is called **factoring.** In this chapter we will use factoring to solve a new type of equation, called a **quadratic equation,** and in the next chapter we will use factoring to simplify algebraic fractions.

4.1 FACTORING OUT THE GREATEST COMMON FACTOR

Because the natural number 4 divides the natural number 12 exactly, 4 is called a **factor** of 12. The natural-number factors of 8, for example, are 1, 2, 4, and 8 because each of these numbers divides 8. The only natural-number factors of 17 are 1 and 17. Recall that a natural number greater than 1 whose only factors are 1 and the number itself is called a **prime number.** Thus, 17 is a prime number.

To **factor** a natural number means to write the number as the product of other natural numbers.

> **DEFINITION.** A natural number is said to be in **prime-factored form** if it is written as the product of factors that are prime numbers.

The right-hand sides of the equations

$$42 = 2 \cdot 3 \cdot 7 \qquad 60 = 2^2 \cdot 3 \cdot 5 \qquad 90 = 2 \cdot 3^2 \cdot 5$$

show the prime-factored forms or **prime factorizations** of 42, 60, and 90. A theorem called the *fundamental theorem of arithmetic* points out that there is *exactly one* prime factorization for every natural number greater than 1.

The largest natural number that divides each of several natural numbers is called the **greatest common factor** or the **greatest common divisor** of these numbers. For example, 6 is the greatest common factor of 42, 60, and 90 because

$$\frac{42}{6} = 7 \qquad \frac{60}{6} = 10 \qquad \text{and} \qquad \frac{90}{6} = 15$$

and no natural number greater than 6 divides 42, 60, and 90.

Algebraic monomials also have greatest common factors. The right-hand sides of the following equations

$$6a^2b^3 = 2 \cdot 3 \cdot a \cdot a \cdot b \cdot b \cdot b$$
$$4a^3b^2 = 2 \cdot 2 \cdot a \cdot a \cdot a \cdot b \cdot b$$
$$18a^2b = 2 \cdot 3 \cdot 3 \cdot a \cdot a \cdot b$$

show the prime factorizations of $6a^2b^3$, $4a^3b^2$, and $18a^2b$. Because all three of these monomials have one factor of 2, two factors of a and one factor of b in common, their greatest common factor is

$$2 \cdot a \cdot a \cdot b \qquad \text{or} \qquad 2a^2b$$

To find the greatest common factor of several monomials, we follow these steps:

> **1.** Find the prime factorization of each monomial.
> **2.** Use each common factor the least number of times it appears in any one monomial.
> **3.** Find the product of the factors found in step 2 to obtain the greatest common factor.

Recall that the distributive property provides the way to multiply a polynomial by a monomial. For example,

$$3x^2(2x - 3y) = 3x^2 \cdot 2x - 3x^2 \cdot 3y = 6x^3 - 9x^2y$$

Given a polynomial such as $6x^3 - 9x^2y$, we can factor each monomial and use the distributive property in reverse to factor the polynomial.

$$6x^3 - 9x^2y = 3x^2 \cdot 2x - 3x^2 \cdot 3y = 3x^2(2x - 3y)$$

Because $3x^2$ is the greatest common factor of the terms $6x^3$ and $-9x^2y$, this process is called **factoring out the greatest common factor.**

EXAMPLE 1   Factor $6x + 9$.

Solution   To find the greatest factor common, find the prime factorization of $6x$ and 9.

$$6x = 3 \cdot 2 \cdot x$$
$$9 = 3 \cdot 3$$

The greatest factor common in $6x$ and 9 is 3, and we can use the distributive property to factor it out.

$$6x + 9 = 3 \cdot 2x + 3 \cdot 3$$
$$= 3(2x + 3)$$

Check this result by verifying that $3(2x + 3) = 6x + 9$.   ∎

EXAMPLE 2   Factor $12y^2 + 20y$.

Solution   To find the greatest common factor, find the prime factorization of $12y^2$ and $20y$.

$$12y^2 = 2 \cdot 2 \cdot 3 \cdot y \cdot y$$
$$20y = 2 \cdot 2 \cdot 5 \cdot y$$

The greatest factor common in $12y^2$ and $20y$ is $2 \cdot 2 \cdot y$, or $4y$, and we can use the distributive property to factor it out.

$$12y^2 + 20y = 4y \cdot 3y + 4y \cdot 5 = 4y(3y + 5)$$

Check this result by verifying that $4y(3y + 5) = 12y^2 + 20y$.   ∎

EXAMPLE 3   Factor $35a^3b^2 - 14a^2b^3$.

Solution   To find the greatest common factor, find the prime factorization of $35a^3b^2$ and $-14a^2b^3$.

$$35a^3b^2 = 5 \cdot 7 \cdot a \cdot a \cdot a \cdot b \cdot b$$
$$-14a^2b^3 = -2 \cdot 7 \cdot a \cdot a \cdot b \cdot b \cdot b$$

The greatest factor common in $35a^2b^2$ and $-14a^2b^3$ is $7 \cdot a \cdot a \cdot b \cdot b$, or $7a^2b^2$, and we can use the distributive property to factor it out.

$$35a^3b^2 - 14a^2b^3 = 7a^2b^2 \cdot 5a - 7a^2b^2 \cdot 2b = 7a^2b^2(5a - 2b)$$

Check this result by verifying that $7a^2b^2(5a - 2b) = 35a^3b^2 - 14a^2b^3$. ∎

EXAMPLE 4   Factor $a^2b^2 - ab$.

Solution   Factor out the greatest common factor, which is ab.

$$a^2b^2 - ab = ab \cdot ab - ab \cdot 1 = ab(ab - 1)$$

It is important to understand where the 1 comes from. The last term of the binomial $a^2b^2 - ab$ has an implied coefficient of 1. When the ab is factored out, this coefficient of 1 must be written.

Check the result by verifying that $ab(ab - 1) = a^2b^2 - ab$. ∎

EXAMPLE 5   Factor $12x^3y^2z + 6x^2yz - 3xz$.

Solution   Factor out the greatest common factor, which is $3xz$.

$$12x^3y^2z + 6x^2yz - 3xz = 3xz \cdot 4x^2y^2 + 3xz \cdot 2xy - 3xz \cdot 1$$
$$= 3xz(4x^2y^2 + 2xy - 1)$$

Check the result by verifying that

$$3xz(4x^2y^2 + 2xy - 1) = 12x^3y^2z + 6x^2yz - 3xz.$$ ∎

EXAMPLE 6   Factor -1 out of the trinomial $-a^3 + 2a^2 - 4$.

Solution $$-a^3 + 2a^2 - 4 = (-1)a^3 + (-1)(-2a^2) + (-1)4 \quad \text{Note that}$$
$$\quad (-1)(-2a^2) = +2a^2.$$

$$= -1(a^3 - 2a^2 + 4)$$
$$= -(a^3 - 2a^2 + 4)$$

Check the result by verifying that $-(a^3 - 2a^2 + 4) = -a^3 + 2a^2 - 4$. ∎

EXAMPLE 7   Factor out the negative of the greatest common factor in $-18a^2b + 6ab^2 - 12a^2b^2$.

Solution   The greatest common factor in the trinomial is $6ab$. Since we want to factor out the negative of this greatest common factor, factor out $-6ab$ as follows:

$$-18a^2b + 6ab^2 - 12a^2b^2 = (-6ab)3a - (-6ab)b + (-6ab)2ab$$
$$= -6ab(3a - b + 2ab)$$

Check the result by verifying that

$$-6ab(3a - b + 2ab) = -18a^2b + 6ab^2 - 12a^2b^2.$$ ∎

EXERCISE 4.1

In Exercises 1–12, find the prime factorization of each number.

1. 12   **2.** 24   **3.** 15   **4.** 20

5. 40   **6.** 62   **7.** 98   **8.** 112

9. 225   **10.** 144   **11.** 288   **12.** 968

In Exercises 13–40, factor out the greatest common factor.

13. $3x + 6$ **14.** $2y - 10$ **15.** $xy - xz$ **16.** $uv + ut$

17. $t^2 + 2t$ **18.** $b^3 - 3b$ **19.** $2r^4 - 4r^2$ **20.** $a^3 + 3a^2$

21. $a^3b^3z^3 - a^2b^3z^2$ **22.** $r^3s^6t^9 + r^2s^2t^2$ **23.** $24x^2y^3z^4 + 8xy^2z^3$ **24.** $3x^2y^3 - 9x^4y^3z$

25. $12uvw^3 - 18uv^2w^2$ **26.** $14xyz - 16x^2y^2z$ **27.** $3x + 3y - 6z$ **28.** $2x - 4y + 8z$

29. $ab + ac - ad$ **30.** $rs - rt + ru$ **31.** $4y^2 + 8y - 2xy$ **32.** $3x^2 - 6xy + 9xy^2$

33. $12r^2 - 3rs + 9r^2s^2$ **34.** $6a^2 - 12a^3b + 36ab$

35. $abx - ab^2x + abx^2$ **36.** $a^2b^2x^2 + a^3b^2x^2 - a^3b^3x^3$

37. $4x^2y^2z^2 - 6xy^2z^2 + 12xyz^2$ **38.** $32xyz + 48x^2yz + 36xy^2z$

39. $70a^3b^2c^2 + 49a^2b^3c^3 - 21a^2b^2c^2$ **40.** $8a^2b^2 - 24ab^2c + 9b^2c^2$

In Exercises 41–52, factor out -1 from each polynomial.

41. $-a - b$ **42.** $-x - 2y$ **43.** $-2x + 5y$ **44.** $-3x + 8z$

45. $-2a + 3b$ **46.** $-2x + 5y$ **47.** $-3m - 4n + 1$ **48.** $-3r + 2s - 3$

49. $-3xy + 2z + 5w$ **50.** $-4ab + 3c - 5d$

51. $-3ab - 5ac + 9bc$ **52.** $-6yz + 12xz - 5xy$

In Exercises 53–62, factor each polynomial by factoring out the greatest common factor, including -1.

53. $-3x^2y - 6xy^2$ **54.** $-4a^2b^2 + 6ab^2$

55. $-4a^2b^3 + 12a^3b^2$ **56.** $-25x^4y^3z^2 + 30x^2y^3z^4$

57. $-4a^2b^2c^2 + 14a^2b^2c - 10ab^2c^2$ **58.** $-10x^4y^3z^2 + 8x^3y^2z - 20x^2y$

59. $-14a^6b^6 + 49a^2b^3 - 21ab$ **60.** $-35r^9s^9t^9 + 25r^6s^6t^6 + 75r^3s^3t^3$

61. $-5a^2b^3c + 15a^3b^4c^2 - 25a^4b^3c$ **62.** $-7x^5y^4z^3 + 49x^5y^5z^4 - 21x^6y^4z^3$

REVIEW EXERCISES

In Review Exercises 1–6, multiply the binomials.

1. $(3x + 2)(2y + 1)$ **2.** $(2r - 3)(r + t)$

3. $(2a + 1)(b - 1)$

5. $(3p - q)(2r - q)$

4. $(m - 4n)(m + 2)$

6. $(2u + v)(3m - v)$

In Review Exercises 7–12, remove parentheses.

7. $3(x + y) + a(x + y)$

9. $(x + 3)(x + 1) - y(x + 1)$

11. $(3x - y)(x^2 - 2) + 3(x^2 - 2)$

8. $x(y + 1) + 5(y + 1)$

10. $x(x^2 + 2) - y(x^2 + 2)$

12. $(x - 5y)(a + 2) - (x - 5y)b$

4.2 FACTORING BY GROUPING

Sometimes we can factor out a polynomial as the greatest common factor. For example, we can use the techniques of the previous section to factor $a + b$ out of the expression $(a + b)x + (a + b)y$. To do so, we note that the binomial $a + b$ is a common factor of both $(a + b)x$ and $(a + b)y$, and it can be factored out as follows:

$$(a + b)x + (a + b)y = (a + b) \cdot x + (a + b) \cdot y$$
$$= (a + b)(x + y)$$

We can check the result by verifying that $(a + b)(x + y) = (a + b)x + (a + b)y$.

EXAMPLE 1 Factor $a + 3$ out of the expression $(a + 3) + (a + 3)^2$.

Solution Recognize that $a + 3$ is equal to $(a + 3)1$ and that $(a + 3)^2$ is equal to $(a + 3)(a + 3)$. Then factor out $a + 3$ and simplify.

$$(a + 3) + (a + 3)^2 = (a + 3)1 + (a + 3)(a + 3)$$
$$= (a + 3)[1 + (a + 3)]$$
$$= (a + 3)(a + 4) \qquad \blacksquare$$

EXAMPLE 2 Factor $6a^2b^2(x + 2y) - 9ab(x + 2y)$.

Solution $6a^2b^2(x + 2y) - 9ab(x + 2y) = (x + 2y)(6a^2b^2 - 9ab)$ Factor out $(x + 2y)$.

$$= (x + 2y)3ab(2ab - 3) \qquad \text{Factor out } 3ab \text{ from } 6a^2b^2 - 9ab.$$

$$= 3ab(x + 2y)(2ab - 3) \qquad \blacksquare$$

Suppose we wish to factor the expression

$$ax + ay + cx + cy$$

Although no factor is common to all four terms, there is a common factor of a in $ax + ay$ and a common factor of c in $cx + cy$. We can factor out the a and the c to obtain

$$ax + ay + cx + cy = a(x + y) + c(x + y) = (x + y)(a + c)$$

This result can be checked by multiplication.

$$(x + y)(a + c) = ax + cx + ay + cy = ax + ay + cx + cy$$

Thus, $ax + ay + cx + cy$ factors as $(x + y)(a + c)$. This type of factoring is called **factoring by grouping.**

EXAMPLE 3   Factor $2c + 2d - cd - d^2$.

Solution $2c + 2d - cd - d^2 = 2(c + d) - d(c + d)$ Factor out 2 from $2c + 2d$ and factor out $-d$ from $-cd - d^2$.

                              $= (c + d)(2 - d)$ Factor out $c + d$.

Check by multiplication.

$$(c + d)(2 - d) = 2c - cd + 2d - d^2 = 2c + 2d - cd - d^2$$ ∎

EXAMPLE 4   Factor $x^2y - ax - xy + a$.

Solution $x^2y - ax - xy + a = x(xy - a) - 1(xy - a)$ Factor out x from $x^2y - ax$ and factor out -1 from $-xy + a$.

                              $= (xy - a)(x - 1)$ Factor out $xy - a$.

Check by multiplication.   ∎

The method of factoring by grouping often works on polynomials that contain more than four terms.

EXAMPLE 5   Factor $6am - 6bm + 6cm + 3an - 3bn + 3cn$.

Solution   Begin by factoring out the common factor of 3.

$$6am - 6bm + 6cm + 3an - 3bn + 3cn$$
$$= 3(2am - 2bm + 2cm + an - bn + cn)$$

Then factor $2m$ from $2am - 2bm + 2cm$ and n from $an - bn + cn$ to obtain

$$6am - 6bm + 6cm + 3an - 3bn + 3cn = 3[2m(a - b + c) + n(a - b + c)]$$

Then factor out the common factor of $(a - b + c)$.

$$6am - 6bm + 6cm + 3an - 3bn + 3cn = 3(a - b + c)(2m + n)$$

Check by multiplication.   ∎

EXAMPLE 6   Factor **a.** $a(c - d) + b(d - c)$ and   **b.** $ac + bd - ad - bc$.

Solution   **a.** $a(c - d) + b(d - c) = a(c - d) - b(-d + c)$ Factor -1 from $d - c$.

                     $= a(c - d) - b(c - d)$ $-d + c = c - d$.

                     $= (c - d)(a - b)$ Factor out $c - d$.

b. In this example, we cannot factor anything from the first two terms or the last two terms. However, if we rearrange the terms, the factoring is routine:

$$ac + bd - ad - bc = ac - ad + bd - bc \qquad bd - ad = -ad + bd.$$

$$= a(c - d) + b(d - c) \qquad \text{Factor } a \text{ from } ac - ad$$
$$\text{and } b \text{ from } bd - bc.$$

$$= (c - d)(a - b) \qquad \text{See part } \mathbf{a}. \qquad \blacksquare$$

EXERCISE 4.2

In Exercises 1–20, factor each expression.

1. $(x + y)2 + (x + y)b$

2. $(a - b)c + (a - b)d$

3. $3(x + y) - a(x + y)$

4. $x(y + 1) - 5(y + 1)$

5. $3(r - 2s) - x(r - 2s)$

7. $(x - 3)^2 + (x - 3)$

9. $2x(a^2 + b) + 2y(a^2 + b)$

11. $3x^2(r + 3s) - 6y^2(r + 3s)$

6. $x(a + 2b) + y(a + 2b)$

8. $(2t + 4) - (2t + 4)^2$

10. $3x(c - 3d) + 6y(c - 3d)$

12. $9a^2b^2(3x - 2y) - 6ab(3x - 2y)$

13. $3x(a + b + c) - 2y(a + b + c)$

14. $2m(a - 2b + 3c) - 21xy(a - 2b + 3c)$

15. $14x^2y(r + 2s - t) - 21xy(r + 2s - t)$

16. $15xy^3(2x - y + 3z) + 25xy^2(2x - y + 3z)$

17. $(x + 3)(x + 1) - y(x + 1)$

19. $(3x - y)(x^2 - 2) + (x^2 - 2)$

18. $x(x^2 + 2) - y(x^2 + 2)$

20. $(x - 5y)(a + 2) - (x - 5y)$

In Exercises 21–40, factor each expression.

21. $2x + 2y + ax + ay$

23. $7r + 7s - kr - ks$

25. $xr + xs + yr + ys$

27. $2ax + 2bx + 3a + 3b$

29. $2ab + 2ac + 3b + 3c$

31. $2x^2 + 2xy - 3x - 3y$

33. $3tv - 9tw + uv - 3uw$

35. $9mp + 3mq - 3np - nq$

37. $mp - np - m + n$

39. $x(a - b) + y(b - a)$

22. $bx + bz + 5x + 5z$

24. $9p - 9q + mp - mq$

26. $pm - pn + qm - qn$

28. $3xy + 3xz - 5y - 5z$

30. $3ac + a + 3bc + b$

32. $3ab + 9a - 2b - 6$

34. $ce - 2cf + 3de - 6df$

36. $ax + bx - a - b$

38. $6x^2u - 3x^2v + 2yu - yv$

40. $p(m - n) - q(n - m)$

In Exercises 41–48, factor each expression. Factor out all common factors first if they exist.

41. $ax^3 + bx^3 + 2ax^2y + 2bx^2y$

43. $4a^2b + 12a^2 - 8ab - 24a$

42. $x^3y^2 - 2x^2y^2 + 3xy^2 - 6y^2$

44. $-4abc - 4ac^2 + 2bc + 2c^2$

45. $x^3 + 2x^2 + x + 2$

47. $x^3y - x^2y - xy^2 + y^2$

46. $y^3 - 3y^2 - 5y + 15$

48. $2x^3z - 4x^2z + 32xz - 64z$

In Exercises 49–56, factor each expression completely.

49. $x^2 + xy + x + 2x + 2y + 2$

51. $am + bm + cm - an - bn - cn$

53. $ad - bd - cd + 3a - 3b - 3c$

55. $ax^2 - ay + bx^2 - by + cx^2 - cy$

50. $ax + ay + az + bx + by + bz$

52. $x^2 + xz - x - xy - yz + y$

54. $ab + ac - ad - b - c + d$

56. $a^2x - bx - a^2y + by + a^2z - bz$

In Exercises 57–68, factor each expression completely. You may have to rearrange some terms first.

57. $2r - bs - 2s + br$

59. $ax + by + bx + ay$

61. $ac + bd - ad - bc$

63. $ar^2 - brs + ars - br^2$

65. $ba + 3 + a + 3b$

67. $pr + qs - ps - qr$

58. $5x + ry + rx + 5y$

60. $mr + ns + ms + nr$

62. $sx - ry + rx - sy$

64. $a^2bc + a^2c + abc + ac$

66. $xy + 7 + y + 7x$

68. $ac - bd - ad + bc$

REVIEW EXERCISES

In Review Exercises 1–4, simplify each expression. Write all results without using negative exponents.

1. $u^3u^2u^4$

2. $\dfrac{y^6}{y^8}$

3. $\dfrac{a^3b^4}{a^2b^5}$

4. $(3x^5)^0$

5. Write the number 0.00045 in scientific notation.

6. Write the number 6.28×10^4 in standard notation.

In Review Exercises 7–10, multiply the binomials.

7. $(a + b)(a - b)$

8. $(2r + s)(2r - s)$

9. $(3x + 2y)(3x - 2y)$

10. $(4x^2 + 3)(4x^2 - 3)$

In Review Exercises 11–12, write each statement as an algebraic expression.

11. The quotient obtained when the sum of the numbers x and y is divided by their product.

12. The product of the numbers r and s, decreased by twice their sum.

4.3   FACTORING THE DIFFERENCE OF TWO SQUARES

Whenever we multiply a binomial of the form $x + y$ by a binomial of the form $x - y$, we obtain another binomial:

$$(x + y)(x - y) = x^2 - y^2$$

The binomial $x^2 - y^2$ is called the **difference of two squares** because x^2 is the square of x and y^2 is the square of y. The difference of the squares of two quantities such as x and y always factors into the sum of the quantities multiplied by their difference. In formula form, we have

Factoring the Difference of Two Squares.

$$x^2 - y^2 = (x + y)(x - y)$$

To factor the quantity $x^2 - 9$, for example, we note that $x^2 - 9$ can be written in the form $x^2 - 3^2$ and that $x^2 - 3^2$ is the difference of the squares of x and 3. Thus, it factors into the product of (x plus 3) and (x minus 3).

$$x^2 - 9 = x^2 - 3^2 = (x + 3)(x - 3)$$

We can check this result by verifying that $(x + 3)(x - 3) = x^2 - 9$.

EXAMPLE 1   Factor $4y^4 - 25z^2$.

Solution   Because the binomial $4y^4 - 25z^2$ can be written in the form $(2y^2)^2 - (5z)^2$, it represents the difference of the squares of $2y^2$ and $5z$. Thus, it factors into the sum of these quantities times their difference.

$$4y^4 - 25z^2 = (2y^2)^2 - (5z)^2 = (2y^2 + 5z)(2y^2 - 5z)$$

Check this result by multiplication.   ∎

We can often factor out a greatest common factor before factoring the difference of two squares. For example, to factor $8x^2 - 72$, we begin by factoring out the greatest common factor of 8 and then factor the resulting difference of two squares.

$$
\begin{aligned}
8x^2 - 72 &= 8(x^2 - 9) && \text{Factor out 8.}\\
&= 8(x^2 - 3^2) && \text{Write 9 as } 3^2.\\
&= 8(x + 3)(x - 3) && \text{Factor the difference of two squares.}
\end{aligned}
$$

We can verify this result by multiplication:

$$8(x + 3)(x - 3) = 8(x^2 - 9) = 8x^2 - 72$$

EXAMPLE 2   Factor $2a^2x^3y - 8b^2xy$.

Solution
$$
\begin{aligned}
2a^2x^3y - 8b^2xy &= 2xy(a^2x^2 - 4b^2) && \text{Factor out } 2xy.\\
&= 2xy[(ax)^2 - (2b)^2]\\
&= 2xy(ax + 2b)(ax - 2b) && \text{Factor the difference of two squares.}
\end{aligned}
$$

Use multiplication to verify this result.   ∎

Sometimes we must factor a difference of two squares more than once to factor a polynomial completely. For example, the binomial $625a^4 - 81b^4$ can be written in the form $(25a^2)^2 - (9b^2)^2$. Thus, it is the difference of the squares of $25a^2$ and $9b^2$ and factors as

$$625a^4 - 81b^4 = (25a^2)^2 - (9b^2)^2 = (25a^2 + 9b^2)(25a^2 - 9b^2)$$

The factor $25a^2 - 9b^2$, however, can be written in the form $(5a)^2 - (3b)^2$ and can be factored as $(5a + 3b)(5a - 3b)$. Thus, the complete factorization of $625a^4 - 81b^4$ is

$$625a^4 - 81b^4 = (25a^2 + 9b^2)(5a + 3b)(5a - 3b)$$

The binomial $25a^2 + 9b^2$ is called the **sum of two squares** because it can be written in the form $(5a)^2 + (3b)^2$. Such binomials cannot be factored if we are limited to integral coefficients. Polynomials that do not factor over the integers are called **irreducible** or **prime polynomials.**

EXAMPLE 3   Factor $2x^4y - 32y$.

Solution
$$2x^4y - 32y = 2y \cdot x^4 - 2y \cdot 16$$
$$= 2y(x^4 - 16) \qquad \text{Factor out } 2y.$$
$$= 2y(x^2 + 4)(x^2 - 4) \qquad \text{Factor } x^4 - 16.$$
$$= 2y(x^2 + 4)(x + 2)(x - 2) \qquad \text{Factor } x^2 - 4. \text{ Note that}$$
$$\qquad\qquad\qquad\qquad\qquad\qquad\qquad x^2 + 4 \text{ does not factor.} \quad \blacksquare$$

Example 4 requires the techniques of factoring out a common factor, factoring by grouping, and factoring the difference of two squares.

EXAMPLE 4   Factor $2x^3 - 8x + 2x^2y - 8y$.

Solution
$$2x^3 - 8x + 2x^2y - 8y = 2(x^3 - 4x + x^2y - 4y) \qquad \text{Factor out 2.}$$
$$= 2[x(x^2 - 4) + y(x^2 - 4)] \qquad \text{Factor out } x \text{ from}$$
$$\qquad\qquad\qquad\qquad\qquad\qquad\qquad x^3 - 4x \text{ and factor}$$
$$\qquad\qquad\qquad\qquad\qquad\qquad\qquad \text{out } y \text{ from } x^2y - 4y.$$
$$= 2[(x^2 - 4)(x + y)] \qquad \text{Factor out } x^2 - 4.$$
$$= 2(x + 2)(x - 2)(x + y) \qquad \text{Factor } x^2 - 4.$$

Check by multiplication. $\blacksquare$

EXERCISE 4.3

In Exercises 1–20, factor each expression, if possible. If a polynomial is prime, so indicate.

1. $x^2 - 16$ 　　　　**2.** $x^2 - 25$ 　　　　**3.** $y^2 - 49$ 　　　　**4.** $y^2 - 81$

5. $4y^2 - 49$ 　　　　**6.** $9z^2 - 4$ 　　　　**7.** $9x^2 - y^2$ 　　　　**8.** $4x^2 - z^2$

9. $25t^2 - 36u^2$ 　　　　**10.** $49u^2 - 64v^2$ 　　　　**11.** $16a^2 - 25b^2$ 　　　　**12.** $36a^2 - 121b^2$

13. $a^2 + b^2$ 　　　　**14.** $121a^2 - 144b^2$ 　　　　**15.** $a^4 - 4b^2$ 　　　　**16.** $9y^2 + 16z^2$

17. $49y^2 - 225z^4$ **18.** $25x^2 + 36y^2$ **19.** $196x^4 - 169y^2$ **20.** $144a^4 + 169b^4$

In Exercises 21–36, factor each expression completely. Factor out any common monomial factors first.

21. $8x^2 - 32y^2$

22. $2a^2 - 200b^2$

23. $2a^2 - 8y^2$

24. $32x^2 - 8y^2$

25. $3r^2 - 12s^2$

26. $45u^2 - 20v^2$

27. $x^3 - xy^2$

28. $a^2b - b^3$

29. $4a^2x - 9b^2x$

30. $4b^2y - 16c^2y$

31. $3m^3 - 3mn^2$

32. $2p^2q - 2q^3$

33. $4x^4 - x^2y^2$

34. $9xy^2 - 4xy^4$

35. $2a^3b - 242ab^3$

36. $50c^4d^2 - 8c^2d^4$

In Exercises 37–48, factor each expression completely.

37. $x^4 - 81$

38. $y^4 - 625$

39. $a^4 - 16$

40. $b^4 - 256$

41. $a^4 - b^4$

42. $m^4 - 16n^4$

43. $81r^4 - 256s^4$

44. $x^8 - y^4$

45. $a^4 - b^8$

46. $16y^8 - 81z^4$

47. $x^8 - y^8$

48. $x^8y^8 - 1$

In Exercises 49–68, factor each expression completely.

49. $2x^4 - 2y^4$

50. $a^5 - ab^4$

51. $a^4b - b^5$

52. $m^5 - 16mn^4$

53. $48m^4n - 243n^5$

54. $2x^4y - 512y^5$

55. $3a^5y + 6ay^5$

56. $2p^{10}q - 32p^2q^5$

57. $3a^{10} - 3a^2b^4$

58. $2x^9y + 2xy^9$

59. $2x^8y^2 - 32y^6$

60. $3a^8 - 243a^4b^8$

61. $a^6b^2 - a^2b^6c^4$

62. $a^2b^3c^4 - a^2b^3d^4$

63. $a^2b^7 - 625a^2b^3$

64. $16x^3y^4z - 81x^3y^4z^5$

65. $243r^5s - 48rs^5$

66. $1024m^5n - 324mn^5$

67. $16(x - y)^2 - 9$

68. $9(x + 1)^2 - y^2$

In Exercises 69–78, factor each expression completely.

69. $a^3 - 9a + 3a^2 - 27$

70. $b^3 - 25b - 2b^2 + 50$

71. $y^3 - 16y - 3y^2 + 48$

72. $a^3 - 49a + 2a^2 - 98$

73. $3x^3 - 12x + 3x^2 - 12$

74. $2x^3 - 18x - 6x^2 + 54$

75. $3m^3 - 3mn^2 + 3am^2 - 3an^2$

76. $ax^3 - axy^2 - bx^3 + bxy^2$

77. $2m^3n^2 - 32mn^2 + 8m^2 - 128$

78. $2x^3y + 4x^2y - 98xy - 196y$

REVIEW EXERCISES

In Review Exercises 1–10, multiply the binomials.

1. $(x + 6)(x + 6)$ **2.** $(y - 7)(y - 7)$ **3.** $(a - 3)(a - 3)$ **4.** $(r + 8)(r + 8)$

5. $(x + 4y)(x + 5y)$ **6.** $(r - 2s)(r - 3s)$ **7.** $(m + 3n)(m - 2n)$ **8.** $(a - 3b)(a + 4b)$

9. $(u - 3v)(u - 5v)$ **10.** $(x + 4y)(x - 6y)$

11. In the study of the flow of fluids, Bernoulli's Law is given by the equation

$$\frac{p}{w} + \frac{v^2}{2g} + h = k$$

Solve the equation for p.

12. Solve Bernoulli's Law for h. (See Review Exercise 11.)

4.4   FACTORING TRINOMIALS WITH LEAD COEFFICIENTS OF 1

The product of two binomials is often a trinomial. For example,

$$(x + 3)(x + 3) = x^2 + 6x + 9$$
$$(x - 4y)(x - 4y) = x^2 - 8xy + 16y^2$$

and

$$(3x - 4)(2x + 3) = 6x^2 + x - 12$$

Because the product of two binomials can be a trinomial, we should not be surprised that many trinomials factor into the product of two binomials.

Many trinomials can be factored by using the following two special product formulas, first discussed in Section 3.6.

$$(x + y)(x + y) = x^2 + 2xy + y^2$$
$$(x - y)(x - y) = x^2 - 2xy + y^2$$

The trinomials $x^2 + 2xy + y^2$ and $x^2 - 2xy + y^2$ are called **perfect square trinomials** because each one can be written as the square of a binomial.

1. $x^2 + 2xy + y^2 = (x + y)(x + y) = (x + y)^2$
2. $x^2 - 2xy + y^2 = (x - y)(x - y) = (x - y)^2$

To factor a perfect square trinomial such as $x^2 + 8x + 16$, we note that the trinomial can be written in the form

$$x^2 + 2(x)(4) + 4^2$$

and that if $y = 4$, this form matches the left-hand side of Equation 1. Thus, $x^2 + 8x + 16$ factors as

$$x^2 + 8x + 16 = x^2 + 2(x)(4) + 4^2 = (x + 4)(x + 4)$$

Carl Friedrich Gauss (1777–1855)
Many people consider Gauss to
be the greatest mathematician of
all time. He made contributions
in the areas of number theory,
solutions of equations, geometry of
curved surfaces, and statistics. For
his efforts, he has earned the
title "Prince of the Mathematicians."

This result can be verified by multiplication by using the **FOIL** method.

$$(x + 4)(x + 4) = x^2 + 4x + 4x + 16$$
$$= x^2 + 8x + 16 \qquad \text{Combine like terms.}$$

Likewise, the perfect square trinomial $a^2 - 4ab + 4b^2$ can be written in the form

$$a^2 - 2(a)(2b) + (2b)^2$$

If $x = a$ and $y = 2b$, this form matches the left-hand side of Equation 2. Thus, $a^2 - 4ab + 4b^2$ factors as

$$a^2 - 4ab + 4b^2 = a^2 - 2(a)(2b) + (2b)^2 = (a - 2b)(a - 2b)$$

This result can also be verified by multiplication.

Because the trinomial $x^2 + 5x + 6$ is not a perfect square trinomial, it cannot be factored by using a special product formula. However, it can be factored into the product of two binomials. To find those binomial factors, we note that the product of their first terms must be x^2. Thus, the first term of each binomial must be x.

$$\overbrace{(x \qquad)(x \qquad)}^{x^2}$$

Because the product of their last terms must be 6, and the sum of the products of the outer and inner terms must be $5x$, we must find two numbers whose product is 6 and whose sum is 5.

Two such numbers are $+3$ and $+2$. Thus, we have

3. $x^2 + 5x + 6 = (x + 3)(x + 2)$

This factorization can be verified by multiplying $x + 3$ and $x + 2$ and observing that the product is $x^2 + 5x + 6$.

$$(x + 3)(x + 2) = x^2 + 2x + 3x + 6$$
$$= x^2 + 5x + 6 \qquad \text{Combine like terms.}$$

Because of the commutative property of multiplication, the order of the factors listed in Equation 3 is not important. Equation 3 can be written as

$$x^2 + 5x + 6 = (x + 2)(x + 3)$$

EXAMPLE 1   Factor $y^2 - 7y + 12$.

Solution   If this trinomial is to be the product of two binomials, the product of their first terms must be y^2. Thus, the first term of each binomial must be y.

$$y^2$$
$$(y \quad)(y \quad)$$

Because the product of the last terms must be $+12$, and the sum of the products of the outer and inner terms must be $-7y$, we must find two negative numbers whose product is $+12$ and whose sum is -7.

$$
\underset{O + I = -7y}{(y - ?)(y - ?)}
$$

Two such numbers are -4 and -3. Hence,

$$y^2 - 7y + 12 = (y - 4)(y - 3)$$

Check by verifying that the product of $y - 4$ and $y - 3$ is $y^2 - 7y + 12$.

$$(y - 4)(y - 3) = y^2 - 3y - 4y + 12 = y^2 - 7y + 12 \qquad \blacksquare$$

EXAMPLE 2   Factor $a^2 + 2a - 15$.

Solution   Because the first term of this trinomial is a^2, the first term of each binomial factor must be a.

$$a^2$$
$$(a \quad)(a \quad)$$

Because the product of the last terms must be -15, and the sum of the products of the outer terms and inner terms must be $+2a$, we must find two numbers whose product is -15 and whose sum is $+2$.

$$
\underset{O + I = 2a}{(a \quad ?)(a \quad ?)}
$$

Two such numbers are $+5$ and -3. Hence,

$$a^2 + 2a - 15 = (a + 5)(a - 3)$$

Check by verifying that the product of $a + 5$ and $a - 3$ is $a^2 + 2a - 15$.

$$(a + 5)(a - 3) = a^2 - 3a + 5a - 15 = a^2 + 2a - 15 \qquad \blacksquare$$

EXAMPLE 3   Factor $z^2 - 4z - 21$.

Solution   Because the first term of this trinomial is z^2, the first term of each binomial factor must be z.

$$z^2$$
$$(z \quad)(z \quad)$$

Because the product of the last terms must be -21, and the sum of the products of the outer terms and inner terms must be $-4z$, we must find two numbers whose product is -21 and whose sum is -4.

$$-21$$
$$(z \quad ?)(z \quad ?)$$
$$O + I = -4z$$

Two such numbers are -7 and $+3$. Hence,

$$z^2 - 4z - 21 = (z - 7)(z + 3)$$

Check by verifying that the product of $z - 7$ and $z + 3$ is $z^2 - 4z - 21$.

$$(z - 7)(z + 3) = z^2 + 3z - 7z - 21 = z^2 - 4z - 21 \qquad \blacksquare$$

EXAMPLE 4 Factor $x^2 + xy - 6y^2$.

Solution Because the first term of this trinomial is x^2, the first term of each binomial factor must be x.

$$x^2$$
$$(x \quad)(x \quad)$$

Because the product of the last terms must be $-6y^2$, and the sum of the products of the outer terms and inner terms must be xy, we must find two numbers whose product is $-6y^2$ that will give a middle term of xy.

$$-6y^2$$
$$(x \quad ?)(x \quad ?)$$
$$O + I = xy$$

Two such numbers are $3y$ and $-2y$. Hence,

$$x^2 + xy - 6y^2 = (x + 3y)(x - 2y)$$

Check by verifying that the product of $x + 3y$ and $x - 2y$ is $x^2 + xy - 6y^2$.

$$(x + 3y)(x - 2y) = x^2 - 2xy + 3xy - 6y^2 = x^2 + xy - 6y^2 \qquad \blacksquare$$

When the coefficient of the first term of a trinomial is -1, begin by factoring out -1.

EXAMPLE 5 Factor $-x^2 + 11x - 18$.

Solution
$$-x^2 + 11x - 18 = -(x^2 - 11x + 18) \qquad \text{Factor out } -1.$$
$$= -(x - 9)(x - 2) \qquad \text{Factor } x^2 - 11x + 18.$$

Check the work by verifying that the product of -1, $x - 9$, and $x - 2$ is $-x^2 + 11x - 18$.

$$-(x - 9)(x - 2) = -(x^2 - 2x - 9x + 18)$$
$$= -(x^2 - 11x + 18)$$
$$= -x^2 + 11x - 18 \qquad \blacksquare$$

EXAMPLE 6   Factor $-x^2 + 2x + 15$.

Solution

$$-x^2 + 2x + 15 = -(x^2 - 2x - 15) \qquad \text{Factor out } -1.$$
$$= -(x - 5)(x + 3) \qquad \text{Factor } x^2 - 2x - 15.$$

Check the work by verifying that the product of -1, $x - 5$, and $x + 3$ is $-x^2 + 2x + 15$.

$$-(x - 5)(x + 3) = -(x^2 + 3x - 5x - 15)$$
$$= -(x^2 - 2x - 15)$$
$$= -x^2 + 2x + 15 \qquad \blacksquare$$

Not all trinomials are factorable. To attempt to factor the trinomial $x^2 + 2x + 3$, for example, we would begin by noting that the product of the first terms of the binomial factors is x^2. Thus, the first term of each binomial must be x.

$$x^2$$
$$(x \quad)(x \quad)$$

Because the last term of the trinomial is 3 and the middle term is $2x$, we must find two factors of 3 whose sum is 2 such that

$$3$$
$$(x \quad ?)(x \quad ?)$$
$$O + I = 2x$$

Because 3 factors only as $(1)(3)$ and $(-1)(-3)$, it has no factors whose sum is 2. Thus, $x^2 + 2x + 3$ cannot be factored. It is a prime polynomial.

EXAMPLE 7   Factor $-3ax^2 + 9a - 6ax$.

Solution   Write the trinomial in descending powers of x and factor out the common factor of $-3a$.

$$-3ax^2 + 9a - 6ax = -3ax^2 - 6ax + 9a = -3a(x^2 + 2x - 3)$$

Finally, factor the trinomial $x^2 + 2x - 3$.

$$-3ax^2 + 9a - 6ax = -3a(x + 3)(x - 1)$$

Check the result by multiplying.

$$-3a(x + 3)(x - 1) = -3a(x^2 + 2x - 3)$$
$$= -3ax^2 - 6ax + 9a$$
$$= -3ax^2 + 9a - 6ax \qquad \blacksquare$$

The next example requires factoring a trinomial and factoring a difference of two squares.

EXAMPLE 8   Factor $m^2 - 2mn + n^2 - 64a^2$.

Solution   Group the first three terms together and factor the resulting trinomial to obtain:

$$m^2 - 2mn + n^2 - 64a^2 = (m - n)(m - n) - 64a^2 = (m - n)^2 - (8a)^2$$

Then factor the resulting difference of two squares:

$$m^2 - 2mn + n^2 - 64a^2 = (m - n)^2 - 64a^2$$
$$= (m - n + 8a)(m - n - 8a) \quad \blacksquare$$

EXERCISE 4.4

In Exercises 1–12, factor each perfect square trinomial.

1. $x^2 + 6x + 9$ **2.** $x^2 + 10x + 25$ **3.** $y^2 - 8y + 16$ **4.** $z^2 - 2z + 1$

5. $t^2 + 20t + 100$ **6.** $r^2 + 24r + 144$ **7.** $u^2 - 18u + 81$ **8.** $v^2 - 14v + 49$

9. $x^2 + 4xy + 4y^2$ **10.** $a^2 + 6ab + 9b^2$ **11.** $r^2 - 10rs + 25s^2$ **12.** $m^2 - 12mn + 36n^2$

*In Exercises 13–40, factor each trinomial, if possible. If the trinomial is prime, so indicate. Use the **FOIL** method to check each result.*

13. $x^2 + 3x + 2$ **14.** $y^2 + 4y + 3$ **15.** $a^2 - 4a - 5$ **16.** $b^2 + 6b - 7$

17. $z^2 + 12z + 11$ **18.** $x^2 + 7x + 10$ **19.** $t^2 - 9t + 14$ **20.** $c^2 - 9c + 8$

21. $u^2 + 10u + 15$ **22.** $v^2 + 9v + 15$ **23.** $y^2 - y - 30$ **24.** $x^2 - 3x - 40$

25. $a^2 + 6a - 16$ **26.** $x^2 + 5x - 24$ **27.** $t^2 - 5t - 50$ **28.** $a^2 - 10a - 39$

29. $r^2 - 9r - 12$ **30.** $s^2 + 11s - 26$ **31.** $y^2 + 2yz + z^2$ **32.** $r^2 - 2rs + 4s^2$

33. $x^2 + 4xy + 4y^2$ **34.** $a^2 + 10ab + 9b^2$ **35.** $m^2 + 3mn - 10n^2$ **36.** $m^2 - mn - 12n^2$

37. $a^2 - 4ab - 12b^2$ **38.** $p^2 + pq - 6q^2$ **39.** $u^2 + 2uv - 15v^2$ **40.** $m^2 + 3mn - 10n^2$

In Exercises 41–52, factor each trinomial. Factor out -1 first.

41. $-x^2 - 7x - 10$ **42.** $-x^2 + 9x - 20$ **43.** $-y^2 - 2y + 15$ **44.** $-y^2 - 3y + 18$

45. $-t^2 - 15t + 34$ **46.** $-t^2 - t + 30$ **47.** $-r^2 + 14r - 40$ **48.** $-r^2 + 14r - 45$

49. $-a^2 - 4ab - 3b^2$ **50.** $-a^2 - 6ab - 5b^2$ **51.** $-x^2 + 6xy + 7y^2$ **52.** $-x^2 - 10xy + 11y^2$

In Exercises 53–64, write each trinomial in descending powers of one variable and then factor.

53. $4 - 5x + x^2$ **54.** $y^2 + 5 + 6y$ **55.** $10y + 9 + y^2$ **56.** $x^2 - 13 - 12x$

how do you choose order

57. $c^2 - 5 + 4c$ **58.** $b^2 - 6 - 5b$ **59.** $-r^2 + 2s^2 + rs$ **60.** $u^2 - 3v^2 + 2uv$

61. $4rx + r^2 + 3x^2$ **62.** $-a^2 + 5b^2 + 4ab$ **63.** $-3ab + a^2 + 2b^2$ **64.** $-13yz + y^2 - 14z^2$

In Exercises 65–76, completely factor each trinomial. Factor out any common monomials first (including -1 if necessary).

65. $2x^2 + 10x + 12$ **66.** $3y^2 - 21y + 18$ **67.** $3y^3 + 6y^2 + 3y$ **68.** $4x^4 + 16x^3 + 16x^2$

69. $-5a^2 + 25a - 30$ **70.** $-2b^2 + 20b - 18$ **71.** $3z^2 - 15tz + 12t^2$ **72.** $5m^2 + 45mn - 50n^2$

73. $12xy + 4x^2y - 72y$ **74.** $48xy + 6xy^2 + 96x$

75. $-4x^2y - 4x^3 + 24xy^2$ **76.** $3x^2y^3 + 3x^3y^2 - 6xy^4$

In Exercises 77–84, completely factor each expression.

77. $ax^2 + 4ax + 4a + bx + 2b$ **78.** $mx^2 + mx - 6m + nx - 2n$

79. $a^2 + 8a + 15 + ab + 5b$ **80.** $x^2 + 2xy + y^2 + 2x + 2y$

81. $a^2 + 2ab + b^2 - 4$ **82.** $a^2 + 6a + 9 - b^2$

83. $b^2 - y^2 - 4y - 4$ **84.** $c^2 - a^2 + 8a - 16$

REVIEW EXERCISES

Multiply the binomials.

1. $(2x + 1)(3x + 2)$ **2.** $(3y - 2)(2y - 5)$ **3.** $(4t - 3)(2t + 3)$ **4.** $(3r + 5)(2r - 5)$

5. $(2m - 3n)(3m - 2n)$ **6.** $(4a + 3b)(4a + b)$ **7.** $(4u - 3v)(5u + 2v)$ **8.** $(5c + 2d)(5c - 3d)$

9. $(5x^2 + 2y)(3x^2 + y)$ **10.** $(a - 2b^2)(a + 3b^2)$ **11.** $(3x^2 + 5y)(x^2 - 3y)$ **12.** $(z + 3t^2)(z + 3t^2)$

4.5 FACTORING GENERAL TRINOMIALS

There are more combinations of factors to consider when we factor trinomials with lead coefficients other than 1. To factor $2x^2 - 7x + 3$, for example, we must find binomials of the form $ax + b$ and $cx + d$ such that

$$2x^2 - 7x + 3 = (ax + b)(cx + d)$$

Because the first term of the trinomial $2x^2 - 7x + 3$ is $2x^2$, the first terms of the binomial factors must be $2x$ and x.

$$\overbrace{(2x \quad ?)(x \quad ?)}^{2x^2}$$

Because the product of the last terms is 3, and the sum of the products of the outer terms and inner terms is $-7x$, we must find two numbers with a product of 3 that will give a middle term of $-7x$.

$$+3$$
$$(2x \quad ?)(x \quad ?)$$
$$O + I = -7x$$

Because both $(3)(1)$ and $(-3)(-1)$ give a product of 3, there are four possible combinations to consider:

$$(2x + 3)(x + 1) \qquad (2x + 1)(x + 3)$$
$$(2x - 3)(x - 1) \qquad (2x - 1)(x - 3)$$

Of these possibilities, only the last one gives the required middle term of $-7x$. Thus,

$$2x^2 - 7x + 3 = (2x - 1)(x - 3)$$

We can check this result by multiplication.

$$(2x - 1)(x - 3) = 2x^2 - 6x - x + 3 = 2x^2 - 7x + 3$$

EXAMPLE 1 Factor $3y^2 - 4y - 4$.

Solution Because the first term of this trinomial is $3y^2$, the first terms of the binomial factors must be $3y$ and y.

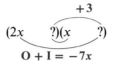

$$3y^2$$
$$(3y \quad)(y \quad)$$

The product of the last terms must be -4, and the sum of the products of the outer terms and inner terms must be $-4y$.

$$-4$$
$$(3y \quad ?)(y \quad ?)$$
$$O + I = -4y$$

Because $(1)(-4)$, $(-1)(4)$, and $(-2)(2)$ all give a product of -4, there are six possible combinations to consider:

$$(3y + 1)(y - 4) \qquad (3y - 4)(y + 1)$$
$$(3y - 1)(y + 4) \qquad (3y + 4)(y - 1)$$
$$(3y - 2)(y + 2) \qquad (3y + 2)(y - 2)$$

Again, only the last possibility gives the required middle term of $-4y$. Hence,

$$3y^2 - 4y - 4 = (3y + 2)(y - 2)$$

Check by multiplication.

$$(3y + 2)(y - 2) = 3y^2 - 6y + 2y - 4 = 3y^2 - 4y - 4 \qquad \blacksquare$$

EXAMPLE 2   Factor $2x^2 + 7xy + 6y^2$.

Solution   The first terms of the two binomial factors must be $2x$ and x.

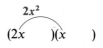

The product of the last terms must be $+6y^2$, and the sum of the products of the outer terms and inner terms must be $+7xy$.

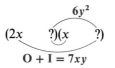

The products $(6y)(y)$, $(3y)(2y)$, $(-6y)(-y)$, and $(-3y)(-2y)$ all give a last term of $6y^2$. However, only the products $(6y)(y)$ and $(3y)(2y)$ can lead to a middle term that is preceded by a $+$ sign. Thus, there are only four possibilities to consider:

$$(2x + 6y)(x + y) \qquad (2x + y)(x + 6y)$$
$$(2x + 3y)(x + 2y) \qquad (2x + 2y)(x + 3y)$$

Of these possibilities, only $(2x + 3y)(x + 2y)$ gives the correct middle term of $7xy$. Hence,

$$2x^2 + 7xy + 6y^2 = (2x + 3y)(x + 2y)$$

Check by multiplication.

$$(2x + 3y)(x + 2y) = 2x^2 + 4xy + 3xy + 6y^2 = 2x^2 + 7xy + 6y^2 \qquad \blacksquare$$

EXAMPLE 3   Factor $6b^2 + 7b - 20$.

Solution   This time there are many possible combinations for the first terms of the binomial factors. They are

$$(b \quad)(6b \quad) \qquad (6b \quad)(b \quad)$$
$$(3b \quad)(2b \quad) \qquad (2b \quad)(3b \quad)$$

There are also many combinations for the last terms of the binomial factors. We must try to find one that will (in combination with our choice of first terms) give a last term of -20 and a sum of the products of the outer terms and inner terms of $+7b$.

Begin, for example, by picking factors of b and $6b$ for the first terms and $+4$ and -5 for the last terms. The possible factorization

$$\underbrace{(b + 4)(6b - 5)}_{O + I = 19b}$$

gives a middle term of $19b$, so it is incorrect.

Then try, for example, factors of $3b$ and $2b$ for the first terms and $+4$ and -5 for the last terms. The possible factorization

$$(3b + 4)(2b - 5)$$
$$O + I = -7b$$

gives a middle term of $-7b$, so it is incorrect.

The possible factorization

$$(3b - 4)(2b + 5)$$

does give a middle term of $+7b$ and a last term of -20, so it is correct. Hence, we have

$$6b^2 + 7b - 20 = (3b - 4)(2b + 5)$$

Check by multiplication.

$$(3b - 4)(2b + 5) = 6b^2 + 15b - 8b - 20 = 6b^2 + 7b - 20 \qquad \blacksquare$$

EXAMPLE 4   Factor $4x^2 + 4xy - 3y^2$.

Solution   Again, there are many combinations for the first terms of the binomial factors:

$$(4x \quad)(x \quad) \quad (x \quad)(4x \quad) \quad (2x \quad)(2x \quad)$$

We must try to find last terms that will give a third term of $-3y^2$ and a middle term of $+4xy$.

Begin by trying, for example, factors of $4x$ and x for the first terms and factors of $3y$ and $-y$ for the last terms. The possible factorization

$$(4x + 3y)(x - y)$$
$$O + I = -xy$$

gives a middle term of $-xy$, so it is incorrect.

Then try, for example, factors of $2x$ and $2x$ for the first terms and factors of $3y$ and $-y$ for the last terms. The possible factorization

$$(2x + 3y)(2x - y)$$
$$O + I = +4xy$$

gives a middle term of $+4xy$ and a last term of $-3y^2$, so it is correct. Hence,

$$4x^2 + 4xy - 3y^2 = (2x + 3y)(2x - y)$$

Check by multiplication.

$$(2x + 3y)(2x - y) = 4x^2 - 2xy + 6xy - 3y^2 = 4x^2 + 4xy - 3y^2 \qquad \blacksquare$$

EXAMPLE 5   Factor $-8x^3 + 22x^2 - 12x$.

Solution   Factor out the common monomial factor of $-2x$.

$$-8x^3 + 22x^2 - 12x = -2x(4x^2 - 11x + 6)$$

Then find the binomial factors of $4x^2 - 11x + 6$.

$$-8x^3 + 22x^2 - 12x = -2x(x - 2)(4x - 3)$$

Check by multiplication.

$$-2x(x - 2)(4x - 3) = -2x(4x^2 - 3x - 8x + 6)$$
$$= -2x(4x^2 - 11x + 6)$$
$$= -8x^3 + 22x^2 - 12x \qquad \blacksquare$$

It is not easy to give specific rules for factoring trinomials, because some guesswork is often necessary. However, the following hints are often helpful:

To factor a general trinomial, follow these steps:

1. Write the trinomial in descending powers of one variable.

2. Factor out any greatest common factor (including -1 if that is necessary to make the coefficient of the first term positive).

3. If the sign of the first term of the trinomial is $+$ and the sign of the third term is $+$, the signs between the terms of the binomial factors are the same as the sign of the middle term of the trinomial. If the sign of the third term is $-$, the signs between the terms of the binomial factors are opposite.

4. Mentally, try various combinations of first terms and last terms until you find one that works, or until you exhaust all the possibilities. In that case, the trinomial does not factor using only integer coefficients.

5. Check the factorization by multiplication.

EXAMPLE 6   Factor $2x^2y - 8x^3 + 3xy^2$.

Solution   *Step 1*
Rewrite the trinomial in descending powers of x.

$$-8x^3 + 2x^2y + 3xy^2$$

Step 2
Factor out $-x$.

$$-8x^3 + 2x^2y + 3xy^2 = -x(8x^2 - 2xy - 3y^2)$$

Step 3
Because the sign of the third term of the trinomial factor is $-$, the signs within its binomial factors must be different. Thus the sign between the terms in one binomial must be $+$, and the sign between the terms of the other binomial must be $-$.

Step 4
Find the binomial factors of the trinomial.

$$-8x^3 + 2x^2y + 3xy^2 = -x(8x^2 - 2xy - 3y^2)$$
$$= -x(2x + y)(4x - 3y)$$

Step 5

Check by multiplication.

$$-x(2x + y)(4x - 3y) = -x(8x^2 - 6xy + 4xy - 3y^2)$$
$$= -x(8x^2 - 2xy - 3y^2)$$
$$= -8x^3 + 2x^2y + 3xy^2$$
$$= 2x^2y - 8x^3 + 3xy^2 \qquad ■$$

The next examples combine the techniques of factoring by grouping, factoring the difference of two squares, and factoring trinomials.

EXAMPLE 7   Factor $4x^2 - 4xy + y^2 - 9$.

Solution $4x^2 - 4xy + y^2 - 9 = (2x - y)^2 - 9$ | Factor the first three terms as a perfect square.

$$= [(2x - y) + 3][(2x - y) - 3]$$ | Factor the difference of two squares.

$$= (2x - y + 3)(2x - y - 3)$$ | Remove parentheses.

Check by multiplication. $\qquad ■$

EXAMPLE 8   Factor $9 - 4x^2 - 4xy - y^2$.

Solution $9 - 4x^2 - 4xy - y^2 = 9 - (4x^2 + 4xy + y^2)$ | Factor -1 from $-4x^2 - 4xy - y^2$.

$$= 9 - (2x + y)(2x + y)$$ | Factor the trinomial.

$$= 9 - (2x + y)^2$$ | $(2x + y)(2x + y) = (2x + y)^2$.

$$= [3 + (2x + y)][3 - (2x + y)]$$ | Factor the difference of two squares.

$$= (3 + 2x + y)(3 - 2x - y)$$ | Remove parentheses.

Check by multiplication. $\qquad ■$

The Key Number Method

The method of factoring by grouping can be used to help factor trinomials of the form $ax^2 + bx + c$, where $a > 0$. For example, to factor the trinomial $4x^2 - 4x - 3$, where $a = 4$, $b = -4$, and $c = -3$, we proceed as follows:

1. Find the product of a and c. This is the **key number**: $ac = 4(-3) = -12$.

2. Find two factors of the key number whose sum is b. In this example, the key number is -12 and $b = -4$.

$$2(-6) = -12 \quad \text{and} \quad 2 + (-6) = -4$$

3. Use the factors 2 and -6 as coefficients of terms to be placed between $4x^2$ and -3:

$$4x^2 + 2x - 6x - 3$$

4. Factor by grouping:

$$4x^2 + 2x - 6x - 3 = 2x(2x + 1) - 3(2x + 1) = (2x + 1)(2x - 3)$$

We can verify this result by multiplication.

EXERCISE 4.5

In Exercises 1–24, factor each trinomial. Check each result.

1. $2x^2 - 3x + 1$
2. $2y^2 - 7y + 3$
3. $3a^2 + 13a + 4$
4. $2b^2 + 7b + 6$

5. $4z^2 + 13z + 3$
6. $4t^2 - 4t + 1$
7. $6y^2 + 7y + 2$
8. $4x^2 + 8x + 3$

9. $6x^2 - 7x + 2$
10. $4z^2 - 9z + 2$
11. $3a^2 - 4a - 4$
12. $8u^2 - 2u - 15$

13. $2x^2 - 3x - 2$
14. $12y^2 - y - 1$
15. $2m^2 + 5m - 12$
16. $10u^2 - 13u - 3$

17. $10y^2 - 3y - 1$
18. $6m^2 + 19m + 3$
19. $12y^2 - 5y - 2$
20. $10x^2 + 21x - 10$

21. $5t^2 + 13t + 6$
22. $16y^2 + 10y + 1$
23. $16m^2 - 14m + 3$
24. $16x^2 + 16x + 3$

In Exercises 25–36, factor each trinomial.

25. $3x^2 - 4xy + y^2$
26. $2x^2 + 3xy + y^2$
27. $2u^2 + uv - 3v^2$
28. $2u^2 + 3uv - 2v^2$

29. $4a^2 - 4ab + b^2$
30. $2b^2 - 5bc + 2c^2$
31. $6r^2 + rs - 2s^2$
32. $3m^2 + 5mn + 2n^2$

33. $4x^2 + 8xy + 3y^2$
34. $4b^2 + 15bc - 4c^2$
35. $4a^2 - 15ab + 9b^2$
36. $12x^2 + 5xy - 3y^2$

In Exercises 37–56, write the terms of each trinomial in descending powers of one variable. Then factor the trinomial, if possible. If a trinomial is prime, so indicate.

37. $-13x + 3x^2 - 10$
38. $-14 + 3a^2 - a$
39. $15 + 8a^2 - 26a$
40. $16 - 40a + 25a^2$

41. $12y^2 + 12 - 25y$
42. $12t^2 - 1 - 4t$
43. $3x^2 + 6 + x$
44. $25 + 2u^2 + 3u$

45. $2a^2 + 3b^2 + 5ab$
46. $11uv + 3u^2 + 6v^2$
47. $pq + 6p^2 - q^2$
48. $-11mn + 12m^2 + 2n^2$

49. $b^2 + 4a^2 + 16ab$

50. $3b^2 + 3a^2 - ab$

51. $12x^2 + 10y^2 - 23xy$

52. $5ab + 25a^2 - 2b^2$

53. $-19xy + 6x^2 + 15y^2$

54. $35r^2 - 6s^2 + rs$

55. $25a^2 - 16b^2 + 30ab$

56. $-10uv + 8u^2 - 7v^2$

In Exercises 57–80, factor completely. Remember to factor out any common factors first.

57. $4x^2 + 10x - 6$

58. $9x^2 + 21x - 18$

59. $y^3 + 13y^2 + 12y$

60. $2xy^2 + 8xy - 24x$

61. $6x^3 - 15x^2 - 9x$

62. $9y^3 + 3y^2 - 6y$

63. $2m^3 - m^2 - 3m$

64. $2a^3 + 8a^2 - 42a$

65. $6a^4 + 14a^3 - 40a^2$

66. $6b^5 - b^4 - 12b^3$

67. $30r^5 + 63r^4 - 30r^3$

68. $6s^5 - 26s^4 - 20s^3$

69. $4a^2 - 4ab - 8b^2$

70. $6x^2 + 3xy - 18y^2$

71. $8x^2 - 12xy - 8y^2$

72. $24a^2 + 14ab + 2b^2$

73. $2x^4y^2 + x^3y^3 - x^2y^4$

74. $2a^5b^2 + 7a^4b^3 + 6a^3b^4$

75. $-16m^3n - 20m^2n^2 - 6mn^3$

76. $-84x^4 - 100x^3y - 24x^2y^2$

77. $-28u^3v^3 + 26u^2v^4 - 6uv^5$

78. $-16x^4y^3 + 30x^3y^4 + 4x^2y^5$

79. $105x^3 - 3x^2 - 36x$

80. $30x^4 + 5x^3 - 200x^2$

In Exercises 81–90, factor each expression completely.

81. $4x^2 + 4xy + y^2 - 16$

82. $9x^2 - 6x + 1 - d^2$

83. $9 - a^2 - 4ab - 4b^2$

84. $25 - 9a^2 + 6ac - c^2$

85. $4x^2 + 4xy + y^2 - a^2 - 2ab - b^2$

86. $a^2 - 2ab + b^2 - x^2 + 2x - 1$

87. $2x^2z - 4xyz + 2y^2z - 18z^3$

88. $9s - r^2s + 2rs^2 - s^3$

89. $4x^2 + 4xy + y^2 + 6x + 3y$

90. $25 - y^2 - 2y^2 + 9y + 5$

In Exercises 91–100, use factoring by grouping to factor each trinomial.

91. $x^2 + 9x + 20$

92. $y^2 - 8y + 15$

93. $2r^2 + 9r + 10$

94. $2v^2 + 5v - 12$

95. $6x^2 - 7x - 5$

96. $2y^2 + 5y - 12$

97. $12t^2 + 13t - 4$

98. $2m^2 + 7m - 15$

99. $2x^2 - xy - 6y^2$

100. $2r^2 - 5rs - 3s^2$

REVIEW EXERCISES

In Review Exercises 1–10, find each product.

1. $(x - 3)(x^2 + 3x + 9)$

2. $(x + 2)(x^2 - 2x + 4)$

3. $(y + 4)(y^2 - 4y + 16)$

4. $(r - 5)(r^2 + 5r + 25)$

5. $(a - b)(a^2 + ab + b^2)$

6. $(a + b)(a^2 - ab + b^2)$

7. $(x + 2y)(x^2 - 2xy + 4y^2)$

9. $(r + s)(r^2 - rs + s^2)$

8. $(x - 2y)(x^2 + 2xy + 4y^2)$

10. $(2y - z)(4y^2 + 2yz + z^2)$

11. The nth term, l, of an arithmetic progression is

$$l = f + (n - 1)d$$

where f is the first term and d is the common difference. Remove the parentheses and solve the equation for n.

12. The sum, S, of n consecutive terms of an arithmetic progression is

$$S = \frac{n}{2}(f + l)$$

where f is the first term of the progression and l is the nth term. Solve for f.

4.6   FACTORING THE SUM AND DIFFERENCE OF TWO CUBES

There are formulas for factoring the sum of the cubes of two quantities and the difference of the cubes of two quantities. To discover these formulas, we need to find the following two products:

$$
\begin{aligned}
(x + y)(x^2 - xy + y^2) &= (x + y)x^2 - (x + y)xy + (x + y)y^2 \\
&= x^3 + x^2y - x^2y - xy^2 + xy^2 + y^3 \\
&= x^3 + y^3
\end{aligned}
$$

$$
\begin{aligned}
(x - y)(x^2 + xy + y^2) &= (x - y)x^2 + (x - y)xy + (x - y)y^2 \\
&= x^3 - x^2y + x^2y - xy^2 + xy^2 - y^3 \\
&= x^3 - y^3
\end{aligned}
$$

These results justify the formulas for factoring the **sum and difference of two cubes.**

Factoring the Sum and the Difference of Two Cubes.

$$x^3 + y^3 = (x + y)(x^2 - xy + y^2)$$
$$x^3 - y^3 = (x - y)(x^2 + xy + y^2)$$

The factorization of $x^3 + y^3$ has a first factor of $x + y$. The second factor has three terms: the square of x, the *negative* of the product of x and y, and the square of y.

The factorization of $x^3 - y^3$ has a first factor of $x - y$. The second factor has three terms: the square of x, the product of x and y, and the square of y.

EXAMPLE 1   Factor $x^3 + 8$.

Solution   The binomial $x^3 + 8$ is the sum of two cubes: the cube of x and the cube of 2.

$$x^3 + 8 = x^3 + 2^3$$

Hence, $x^3 + 8$ factors as the product of the *sum* of x and 2 and the trinomial $x^2 - 2x + 2^2$.

$$x^3 + 8 = x^3 + 2^3$$
$$= (x + 2)(x^2 - 2x + 2^2)$$
$$= (x + 2)(x^2 - 2x + 4)$$

Check by multiplication.

$$(x + 2)(x^2 - 2x + 4) = (x + 2)x^2 - (x + 2)2x + (x + 2)4$$
$$= x^3 + 2x^2 - 2x^2 - 4x + 4x + 8$$
$$= x^3 + 8 \qquad \blacksquare$$

EXAMPLE 2   Factor $a^3 - 64b^3$.

Solution   The binomial $a^3 - 64b^3$ is the difference of two cubes: the cube of a and the cube of $4b$.

$$a^3 - 64b^3 = a^3 - (4b)^3$$

Hence, its factors are the difference $a - 4b$ and the trinomial $a^2 + a(4b) + (4b)^2$.

$$a^3 - 64b^3 = a^3 - (4b)^3$$
$$= (a - 4b)[a^2 + a(4b) + (4b)^2]$$
$$= (a - 4b)(a^2 + 4ab + 16b^2)$$

Check by multiplication.

$$(a - 4b)(a^2 + 4ab + 16b^2) = (a - 4b)a^2 + (a - 4b)4ab + (a - 4b)16b^2$$
$$= a^3 - 4a^2b + 4a^2b - 16ab^2 + 16ab^2 - 64b^3$$
$$= a^3 - 64b^3 \qquad \blacksquare$$

Sometimes we must factor out a greatest common factor before factoring a sum or difference of two cubes.

EXAMPLE 3   Factor $-2t^5 + 128t^2$.

Solution $\qquad -2t^5 + 128t^2 = -2t^2(t^3 - 64)$ $\qquad$ Factor out $-2t^2$.
$\qquad\qquad\qquad\quad = -2t^2(t - 4)(t^2 + 4t + 16)$ $\qquad$ Factor $t^3 - 64$.

Verify this factorization by multiplication. $\qquad\qquad\qquad\qquad\qquad\qquad\qquad\quad \blacksquare$

EXAMPLE 4   Factor $x^6 - 64$.

Solution   The binomial $x^6 - 64$ is the difference of two squares and factors into the product of a sum and a difference.

$$x^6 - 64 = (x^3)^2 - 8^2 = (x^3 + 8)(x^3 - 8)$$

Because $x^3 + 8$ is the sum of two cubes and $x^3 - 8$ is the difference of two cubes, each of these binomials can be factored.

$$x^6 - 64 = (\underbrace{x^3 + 8})(\underbrace{x^3 - 8})$$
$$= (\underbrace{x + 2)(x^2 - 2x + 4})(\underbrace{x - 2)(x^2 + 2x + 4})$$

Verify this factorization by multiplication.   ■

EXERCISE 4.6

In Exercises 1–20, factor each expression.

1. $y^3 + 1$

2. $x^3 + 8$

3. $q^3 - 27$

4. $b^3 + 125$

5. $8 + x^3$

6. $27 - y^3$

7. $s^3 - t^3$

8. $8u^3 + w^3$

9. $27x^3 + y^3$

10. $x^3 - 27y^3$

11. $a^3 + 8b^3$

12. $27a^3 - b^3$

13. $64x^3 - y^3$

14. $27x^3 + 125y^3$

15. $27x^3 - 125y^3$

16. $64x^3 - 27y^3$

17. $a^6 - b^3$

18. $a^3 + b^6$

19. $x^6 + y^6$

20. $x^3 - y^9$

In Exercises 21–36, factor each expression. Factor out any greatest common factors first.

21. $2x^3 + 54$

22. $2x^3 - 2$

23. $-x^3 + 216$

24. $-x^3 - 125$

25. $64m^3x - 8n^3x$

26. $16r^4 + 128rs^3$

27. $x^4y + 216xy^4$

28. $16a^5 - 54a^2b^3$

29. $81r^4s^2 - 24rs^5$

30. $4m^5n + 500m^2n^4$

31. $125a^6b^2 + 64a^3b^5$

32. $216a^4b^4 - 1000ab^7$

33. $y^7z - yz^4$

34. $x^{10}y^2 - xy^5$

35. $2mp^4 + 16mpq^3$

36. $24m^5n - 3m^2n^4$

In Exercises 37–40, factor each expression completely. Factor a difference of two squares first.

37. $x^6 - 1$

38. $x^6 - y^6$

39. $x^{12} - y^6$

40. $a^{12} - 64$

In Exercises 41–54, factor each expression completely. Some exercises do not involve the sum or difference of two cubes.

41. $3(x^3 + y^3) - z(x^3 + y^3)$

42. $x(8a^3 - b^3) + 4(8a^3 - b^3)$

43. $(m^3 + 8n^3) + (m^3x + 8n^3x)$

44. $(a^3x + b^3x) - (a^3y + b^3y)$

45. $(a^4 + 27a) - (a^3b + 27b)$

46. $(x^4 + xy^3) - (x^3y + y^4)$

47. $x^2(y + z) - 4(y + z)$

48. $z^2(x + 1) - 9(x + 1)$

49. $r^2(x - a) - s^2(x - a)$

50. $pq^2(r + s) - p(r + s)$

51. $(x - 1)^2 + 2(x - 1)$

52. $(z + 3)^2 - 5(z + 3)$

53. $y^3(y^2 - 1) - 27(y^2 - 1)$

54. $z^3(y^2 - 4) + 8(y^2 - 4)$

REVIEW EXERCISES

In Review Exercises 1–6, solve each equation.

1. $2x - 6 = 2$

2. $\dfrac{x + 11}{3} = 2$

3. $2(x + 5) = 4$

4. $\dfrac{3(x - 1)}{2} = 3$

5. $\dfrac{2(3a + 4)}{3} = 4$

6. $\dfrac{-3(y - 6)}{2} = -9$

In Review Exercises 7–10, simplify each expression. Write all answers without using negative exponents.

7. $\dfrac{x^2x^3}{x^5}$

8. $\dfrac{y^3y^{-4}}{y^3}$

9. $\left(\dfrac{2x^2y^3}{x^4y^2}\right)^0$

10. $\left(\dfrac{3x^2}{6x^3}\right)^{-4}$

11. A length of one Fermi is 1×10^{-13} centimeter, approximately the radius of a proton. Express this number in standard notation.

12. In the fourteenth century, the Black Plague killed about 25,000,000 people, which was one-quarter of the population of Europe. Express this number in scientific notation.

4.7 SUMMARY OF FACTORING TECHNIQUES

In this brief section we shall discuss ways to approach a randomly chosen factoring problem. For example, suppose we wish to factor the trinomial

$$x^4y + 7x^3y - 18x^2y$$

We begin by attempting to identify the problem type. The first type to look for is **factoring out a common factor.** Because the trinomial has a common factor of x^2y, we factor it out:

$$x^4y + 7x^3y - 18x^2y = x^2y(x^2 + 7x - 18)$$

We can then factor the trinomial $x^2 + 7x - 18$ as $(x + 9)(x - 2)$. Thus,

$$x^4y + 7x^3y - 18x^2y = x^2y(x^2 + 7x - 18)$$
$$= x^2y(x + 9)(x - 2)$$

To identify the type of factoring problem, follow these steps:

1. Factor out all common factors.
2. If an expression has two terms, check to see if the problem type is
   a. the **difference of two squares**: $a^2 - b^2 = (a + b)(a - b)$,
   b. the **sum of two cubes**: $a^3 + b^3 = (a + b)(a^2 - ab + b^2)$, or
   c. the **difference of two cubes**: $a^3 - b^3 = (a - b)(a^2 + ab + b^2)$.
3. If an expression has three terms, check to see if the problem type is
   a **perfect trinomial square**:
$$a^2 + 2ab + b^2 = (a + b)(a + b)$$
$$a^2 - 2ab + b^2 = (a - b)(a - b)$$
   If the trinomial is not a trinomial square, attempt to factor the trinomial as a **general trinomial.**
4. If an expression has four or more terms, try to factor the expression by **grouping.**
5. Continue factoring until each individual factor is prime.
6. Check the results by multiplying.

EXAMPLE 1   Factor $x^5y^2 - xy^6$.

Solution   Begin by factoring out the common factor of xy^2:

$$x^5y^2 - xy^6 = xy^2(x^4 - y^4)$$

Because the expression $x^4 - y^4$ has two terms, check to see if it is the difference of two squares, which it is. As the difference of two squares, it factors as $(x^2 + y^2)(x^2 - y^2)$. Thus,

$$x^5y^2 - xy^6 = xy^2(x^4 - y^4) = xy^2(x^2 + y^2)(x^2 - y^2)$$

The binomial $x^2 + y^2$ is the sum of two squares and cannot be factored. However, the binomial $x^2 - y^2$ is the difference of two squares and factors as $(x + y)(x - y)$. Thus,

$$x^5y^2 - xy^6 = xy^2(x^4 - y^4)$$
$$= xy^2(x^2 + y^2)(x^2 - y^2)$$
$$= xy^2(x^2 + y^2)(x + y)(x - y)$$

Because each of the individual factors is prime, the given expression is in completely factored form.   ∎

EXAMPLE 2   Factor $x^6 - x^4y^2 - x^3y^3 + xy^5$.

Solution   Begin by factoring out the common factor of x.

$$x^6 - x^4y^2 - x^3y^3 + xy^5 = x(x^5 - x^3y^2 - x^2y^3 + y^5)$$

Because the expression $x^5 - x^3y^2 - x^2y^3 + xy^5$ has four terms, try factoring it by grouping:

$$x^6 - x^4y^2 - x^3y^3 + xy^5 = x(x^5 - x^3y^2 - x^2y^3 + y^5)$$
$$= x[x^3(x^2 - y^2) - y^3(x^2 - y^2)]$$
$$= x(x^2 - y^2)(x^3 - y^3) \qquad \text{Factor out } x^2 - y^2.$$

Finally, factor the difference of two squares and the difference of two cubes:

$$x^6 - x^4y^2 - x^3y^3 + xy^5 = x(x + y)(x - y)(x - y)(x^2 + xy + y^2)$$

Because each of the factors is prime, the given expression is in prime-factored form. ∎

EXERCISE 4.7

In Exercises 1–50, factor each expression completely. If the expression is prime, so indicate.

1. $6x + 3$

2. $x^2 - 9$

3. $x^2 - 6x - 7$

4. $a^3 + b^3$

5. $6t^2 + 7t - 3$

6. $3rs^2 - 6r^2 st$

7. $4x^2 - 25$

8. $ac + ad + bc + bd$

9. $t^2 - 2t + 1$

10. $6p^2 - 3p - 2$

11. $a^3 - 8$

12. $2x^2 - 32$

13. $x^2y^2 - 2x^2 - y^2 + 2$

14. $a^2c + a^2d^2 + bc + bd^2$

15. $70p^4q^3 - 35p^4q^2 + 49p^5q^2$

16. $a^2 + 2ab + b^2 - x^2 - 2xy - y^2$

17. $2ab^2 + 8ab - 24a$

18. $t^4 - 16$

19. $-8p^3q^7 - 4p^2q^3$

20. $8m^2n^3 - 24mn^4$

21. $4a^2 - 4ab + b^2 - 9$

22. $3rs + 6r^2 - 18s^2$

23. $x^2 + 7x + 1$

24. $3a^3 + 24b^3$

25. $-2x^5 + 128x^2$

26. $16 - 40z + 25z^2$

27. $14t^3 - 40t^2 + 6t^4$

28. $6x^2 + 7x - 20$

29. $a^2(x - a) - b^2(x - a)$

30. $5x^3y^3z^4 + 25x^2y^3z^2 - 35x^3y^2z^5$

31. $8p^6 - 27q^6$

32. $2c^2 - 5cd - 3d^2$

33. $125p^3 - 64y^3$

34. $8a^2x^3y - 2b^2xy$

35. $-16x^4y^2z + 24x^5y^3z^4 - 15x^2y^3z^7$

36. $2ac + 4ad + bc + 2bd$

37. $81p^4 - 16q^4$

38. $6x^2 - x - 16$

39. $4x^2 + 9y^2$

40. $30a^4 + 5a^3 - 200a^2$

41. $54x^3 + 250y^6$

42. $6a^3 + 35a^2 - 6a$

43. $10r^2 - 13r - 4$

44. $4x^2 + 4x + 1 - y^2$

45. $21t^3 - 10t^2 + t$

46. $16x^2 - 40x^3 + 25x^4$

47. $x^5 - x^3y^2 + x^2y^3 - y^5$

48. $a^3x^3 - a^3y^3 + b^3x^3 - b^3y^3$

49. $2a^2c - 2b^2c + 4a^2d - 4b^2d$

50. $3a^2x^2 + 6a^2x + 3a^2 - 6b^2x^2 - 12b^2x - 6b^2$

REVIEW EXERCISES

In Review Exercises 1–8, write each expression without using parentheses.

1. $2x(x + 2)$

2. $(x - y)(x - y)$

3. $(2a)^3$

4. $3(a + b)(a - b)$

5. $(2x - 3)^2$

6. $(3a + 2b)^2$

7. $\left(\dfrac{3a^3b^{-2}}{6a^{-3}b^2}\right)^2$

8. $\left(\dfrac{8x^{-2}}{4x^3y^{-2}}\right)^{-3}$

In Review Exercises 9–10, solve each equation.

9. $2(t - 5) - t = 3(2 - t)$

10. $5 - 3(2x - 1) = 2(4 + 3x) - 24$

11. The sum of three consecutive even integers is 54. Find the smallest integer.

12. Solve $y = mx + b$ for m.

4.8   SOLVING EQUATIONS BY FACTORING

Equations such as

$$3x + 2 = 0 \qquad \text{and} \qquad -2x + 7 = 0$$

that contain first-degree polynomials are called **linear equations.**
Equations such as

$$3x^2 + 4x - 7 = 0 \qquad \text{and} \qquad -4x^2 + x - 7 = 0$$

that contain second-degree polynomials are called **quadratic equations.**

> **DEFINITION.**   A **quadratic equation** is an equation of the form
> $$ax^2 + bx + c = 0$$
> where a, b, and c are real numbers, and $a \neq 0$.

Many quadratic equations can be solved by factoring. For example, to solve the quadratic equation

$$x^2 + 5x - 6 = 0$$

we begin by factoring the quadratic trinomial and rewriting the equation as

$$(x + 6)(x - 1) = 0$$

This equation indicates that the product of two quantities is 0. However, if the product of two quantities is 0, then at least one of those quantities must be 0. This fact is called the **zero factor theorem.**

The Zero Factor Theorem.   If a and b represent two real numbers and $ab = 0$, then

$$a = 0 \qquad \text{or} \qquad b = 0$$

By applying the zero factor theorem to the equation $(x + 6)(x - 1) = 0$, we have

$$x + 6 = 0 \qquad \text{or} \qquad x - 1 = 0$$

We can solve each of these linear equations to get

$$x = -6 \qquad \text{or} \qquad x = 1$$

To check these answers, we substitute -6 for x, and then 1 for x in the original equation and simplify.

For $x = -6$	For $x = 1$
$x^2 + 5x - 6 = 0$	$x^2 + 5x - 6 = 0$
$(-6)^2 + 5(-6) - 6 \stackrel{?}{=} 0$	$(1)^2 + 5(1) - 6 \stackrel{?}{=} 0$
$36 - 30 - 6 \stackrel{?}{=} 0$	$1 + 5 - 6 \stackrel{?}{=} 0$
$36 - 36 \stackrel{?}{=} 0$	$0 = 0$
$0 = 0$	

Both solutions check.

EXAMPLE 1   Solve the equation $2x^2 + 3x = 2$.

Solution   Write the equation in the form $ax^2 + bx + c = 0$. Then solve for x as follows.

$$2x^2 + 3x = 2$$
$$2x^2 + 3x - 2 = 0 \qquad\qquad \text{Add } -2 \text{ to both sides.}$$
$$(2x - 1)(x + 2) = 0 \qquad\qquad \text{Factor } 2x^2 + 3x - 2.$$
$$2x - 1 = 0 \quad \text{or} \quad x + 2 = 0 \qquad \text{Set each factor equal to 0.}$$
$$2x = 1 \qquad\qquad x = -2 \qquad \text{Solve each linear equation.}$$
$$x = \frac{1}{2}$$

Check each possible solution.

$$\text{For } x = \frac{1}{2}$$

$$2x^2 + 3x = 2$$

$$2\left(\frac{1}{2}\right)^2 + 3\left(\frac{1}{2}\right) \overset{?}{=} 2$$

$$2\left(\frac{1}{4}\right) + \frac{3}{2} \overset{?}{=} 2$$

$$\frac{1}{2} + \frac{3}{2} \overset{?}{=} 2$$

$$2 = 2$$

$$\text{For } x = -2$$

$$2x^2 + 3x = 2$$

$$2(-2)^2 + 3(-2) \overset{?}{=} 2$$

$$2(4) - 6 \overset{?}{=} 2$$

$$8 - 6 \overset{?}{=} 2$$

$$2 = 2$$

Both solutions check.   ■

Quadratic equations such as $3x^2 + 6x = 0$ and $4x^2 - 36 = 0$ are called **incomplete quadratic equations** because they are missing a term. Many incomplete quadratic equations can be solved by factoring.

EXAMPLE 2   Solve the equation $3x^2 + 6x = 0$.

Solution

$$3x^2 + 6x = 0$$

$$3x(x + 2) = 0 \qquad \text{Factor out } 3x.$$

$$3x = 0 \quad \text{or} \quad x + 2 = 0 \qquad \text{Set each factor equal to 0.}$$

$$x = 0 \quad | \quad x = -2 \qquad \text{Solve each linear equation.}$$

Check each possible solution.

$$\text{For } x = 0$$

$$3x^2 + 6x = 0$$

$$3(0)^2 + 6(0) \overset{?}{=} 0$$

$$0 + 0 \overset{?}{=} 0$$

$$0 = 0$$

$$\text{For } x = -2$$

$$3x^2 + 6x = 0$$

$$3(-2)^2 + 6(-2) \overset{?}{=} 0$$

$$3(4) + 6(-2) \overset{?}{=} 0$$

$$12 - 12 \overset{?}{=} 0$$

$$0 = 0$$

Both solutions check.   ■

EXAMPLE 3   Solve the equation $4x^2 - 36 = 0$.

Solution

$$4x^2 - 36 = 0$$

$$x^2 - 9 = 0 \qquad \text{Divide both sides by 4.}$$

$$(x + 3)(x - 3) = 0 \qquad \text{Factor } x^2 - 9.$$

$$x + 3 = 0 \quad \text{or} \quad x - 3 = 0 \qquad \text{Set each factor equal to 0.}$$

$$x = -3 \quad | \quad x = 3 \qquad \text{Solve each linear equation.}$$

Check each possible solution.

$$\begin{array}{c|c}
\textbf{For } x = -3 & \textbf{For } x = 3 \\
4x^2 - 36 = 0 & 4x^2 - 36 = 0 \\
4(-3)^2 - 36 \overset{?}{=} 0 & 4(3)^2 - 36 \overset{?}{=} 0 \\
4(9) - 36 \overset{?}{=} 0 & 4(9) - 36 \overset{?}{=} 0 \\
0 = 0 & 0 = 0
\end{array}$$

Both solutions check.   ■

EXAMPLE 4 Solve the equation $x(2x - 13) = -15$.

Solution

$$x(2x - 13) = -15$$
$$2x^2 - 13x = -15 \qquad \text{Remove parentheses.}$$
$$2x^2 - 13x + 15 = 0 \qquad \text{Add 15 to both sides to get 0 on the right-hand side of the equation.}$$

$$(2x - 3)(x - 5) = 0 \qquad \text{Factor } 2x^2 - 13x + 15.$$

$$\begin{array}{ll}
2x - 3 = 0 \quad \text{or} \quad & x - 5 = 0 \qquad \text{Set each factor equal to 0.} \\
2x = 3 & \qquad\; x = 5 \qquad \text{Solve each linear equation.} \\
x = \dfrac{3}{2} &
\end{array}$$

Check each possible solution.

$$\begin{array}{c|c}
\textbf{For } x = \dfrac{3}{2} & \textbf{For } x = 5 \\
x(2x - 13) = -15 & x(2x - 13) = -15 \\
\dfrac{3}{2}\left(2 \cdot \dfrac{3}{2} - 13\right) \overset{?}{=} -15 & 5(2 \cdot 5 - 13) \overset{?}{=} -15 \\
\dfrac{3}{2}(3 - 13) \overset{?}{=} -15 & 5(10 - 13) \overset{?}{=} -15 \\
\dfrac{3}{2}(-10) \overset{?}{=} -15 & 5(-3) \overset{?}{=} -15 \\
-15 = -15 & -15 = -15
\end{array}$$

Both solutions check.   ■

EXAMPLE 5 Solve the equation $(x - 2)(x^2 - 7x + 6) = 0$.

Solution

$$(x - 2)(x^2 - 7x + 6) = 0$$
$$(x - 2)(x - 6)(x - 1) = 0 \qquad \text{Factor } x^2 - 7x + 6.$$

If the product of these three quantities is 0, then at least one of the quantities must be 0. Hence,

$$\begin{array}{lll}
x - 2 = 0 \quad \text{or} \quad x - 6 = 0 \quad \text{or} \quad x - 1 = 0 & \text{Set each factor equal to 0.} \\
x = 2 \quad | \qquad\quad x = 6 \quad | \qquad\quad x = 1 & \text{Solve each linear equation.}
\end{array}$$

Verify that all three solutions check.   ■

EXAMPLE 6   Solve the equation $6x^3 + 12x = 17x^2$.

Solution

$$6x^3 + 12x = 17x^2$$
$$6x^3 - 17x^2 + 12x = 0 \qquad \text{Add } -17x^2 \text{ to both sides.}$$
$$x(6x^2 - 17x + 12) = 0 \qquad \text{Factor out } x.$$
$$x(2x - 3)(3x - 4) = 0 \qquad \text{Factor } 6x^2 - 17x + 12.$$
$$x = 0 \quad \text{or} \quad 2x - 3 = 0 \quad \text{or} \quad 3x - 4 = 0 \qquad \text{Set each factor equal to 0.}$$
$$x = 0 \quad \bigg| \quad 2x = 3 \quad \bigg| \quad 3x = 4 \qquad \text{Solve the linear equations.}$$
$$x = \frac{3}{2} \quad \bigg| \quad x = \frac{4}{3}$$

Verify that all three solutions check.   ∎

EXERCISE 4.8

In Exercises 1–12, solve each equation by setting each factor equal to 0 and solving the resulting linear equations.

1. $(x - 2)(x + 3) = 0$ **2.** $(x - 3)(x - 2) = 0$ **3.** $(x - 4)(x + 1) = 0$ **4.** $(x + 5)(x + 2) = 0$

5. $(2x - 5)(3x + 6) = 0$ **6.** $(3x - 4)(x + 1) = 0$

7. $(x - 1)(x + 2)(x - 3) = 0$ **8.** $(x + 2)(x + 3)(x - 4) = 0$

9. $(2x + 4)(3x - 12)(x + 7) = 0$ **10.** $(3x - 5)(x + 6)(2x - 1) = 0$

11. $(x - 4)(x + 6)(2x - 3)(3x - 2) = 0$ **12.** $(2x - 4)(3x + 5)(4x - 6)(5x + 10) = 0$

In Exercises 13–36, solve each equation. You may have to rearrange some terms.

13. $x^2 - 13x + 12 = 0$ **14.** $x^2 + 7x + 6 = 0$ **15.** $x^2 - 2x - 15 = 0$ **16.** $x^2 - x - 20 = 0$

17. $6x^2 - x - 2 = 0$ **18.** $2x^2 - 5x - 12 = 0$ **19.** $12x^2 + 5x = 2$ **20.** $4x^2 + 9x = 9$

21. $2x^2 + 6x = 0$ **22.** $3x^2 - 9x = 0$ **23.** $4x^2 - 32x = 0$ **24.** $5x^2 + 125x = 0$

25. $4x^2 - 5 = x$ **26.** $5x^2 - 3 = 14x$ **27.** $x^2 - 16 = 0$ **28.** $x^2 - 25 = 0$

29. $x^2 - 49 = 0$ **30.** $x^2 - 64 = 0$ **31.** $6x^2 - 36x = 0$ **32.** $7x^2 - 63x = 0$

33. $5x^2 - 23x - 10 = 0$ **34.** $6x^2 - 11x + 5 = 0$ **35.** $7x^2 + 19x = 6$ **36.** $16x^2 - 3 = -2x$

In Exercises 37–48, solve each equation.

37. $a(a + 8) = -15$

38. $a(1 + 2a) = 6$

39. $2(y - 4) = -y^2$

40. $-3(y - 6) = y^2$

41. $2x(3x + 10) = -6$

42. $2x^2 = 2(x + 2)$

43. $x^2 + 7x = x - 9$

44. $x(x + 10) = 2(x - 8)$

45. $2y(y + 2) = 3(y + 1)$

46. $2z(z - 3) = 3 - z$

47. $(b - 4)^2 = 1$

48. $(x + 3)^2 = 9$

In Exercises 49–60, solve each equation.

49. $(x - 1)(x^2 + 5x + 6) = 0$

50. $(x - 2)(x^2 - 8x + 7) = 0$

51. $2x^3 - 4x^2 - 6x = 0$

52. $6x^3 + 22x^2 + 12x = 0$

53. $y^3 - 16y = 0$

54. $(y^2 + 6y)(y - 2) = 0$

55. $(3a^2 - 9a)(2a + 1) = 0$

56. $5b^3 - 125b = 0$

57. $21z^3 + z = 10z^2$

58. $6t^3 + 35t^2 = 6t$

59. $(x^2 - 9)(9x^2 - 4) = 0$

60. $(x^2 + 4x + 4)(4x^2 - 25) = 0$

REVIEW EXERCISES

In Review Exercises 1–6, factor each expression completely.

1. $x^2 + 3x$

2. $x^3 - 4x^2$

3. $y^2 - 9$

4. $t^2 - 64$

5. $x^2 + 13x + 12$

6. $x^2 - 11x - 12$

7. One side of a square is s inches long. Find an expression that represents its perimeter.

8. One side of a square is s inches long. Find an expression that represents its area.

9. One side of a rectangle with a perimeter of 24 centimeters is 4 centimeters longer than the other. Find its dimensions.

10. Find an expression that represents the area of a rectangle if one side has a length of $(x + 2)$ inches and another side has a length of $(x + 3)$ inches.

11. The annual interest Jill earns on her investment of $15,000 is $540 less than the annual interest Carol earns on her investment of $21,000. Both are receiving interest at the same annual rate. What is that rate?

12. David bought 35 stamps. He bought three times as many 22-cent stamps as 17-cent stamps, and as many 14-cent stamps as 22-cent stamps. How many of each stamp did he buy?

4.9 APPLICATIONS

The solutions of many word problems involve the use of quadratic equations.

EXAMPLE 1 One negative number is 5 less than another. If their product is 84, find the numbers.

Solution   Let x represent the larger number. Then $x - 5$ represents the smaller number. Because their product is 84, form the equation $x(x - 5) = 84$ and solve it.

$$x(x - 5) = 84$$
$$x^2 - 5x = 84 \qquad \text{Remove parentheses.}$$
$$x^2 - 5x - 84 = 0 \qquad \text{Add } -84 \text{ to both sides.}$$
$$(x - 12)(x + 7) = 0 \qquad \text{Factor.}$$
$$x - 12 = 0 \qquad \text{or} \qquad x + 7 = 0 \qquad \text{Set each factor equal to 0.}$$
$$x = 12 \qquad | \qquad x = -7 \qquad \text{Solve each linear equation.}$$

Because we need two negative numbers, we discard the result $x = 12$. The two required numbers are

$$x = -7 \qquad \text{and} \qquad x - 5 = -7 - 5 = -12$$

Check: The number -12 is five less than -7, and the product of -12 and -7 is 84.   ∎

EXAMPLE 2   If an object is thrown straight up into the air with an initial velocity of 112 feet per second, its height after t seconds is given by the formula

$$h = 112t - 16t^2$$

where h represents the height of the object in feet. After this object has been thrown, in how many seconds will it hit the ground?

Solution   When the object hits the ground, its height will be 0. Hence, set h equal to 0 and solve for t.

$$h = 112t - 16t^2$$
$$0 = 112t - 16t^2$$
$$0 = 16t(7 - t) \qquad \text{Factor out } 16t.$$
$$16t = 0 \qquad \text{or} \qquad 7 - t = 0 \qquad \text{Set each factor equal to 0.}$$
$$t = 0 \qquad | \qquad t = 7 \qquad \text{Solve each linear equation.}$$

When $t = 0$, the object's height above the ground is 0 feet because it has just been released. When $t = 7$, the height is again 0 feet. The object has hit the ground. The solution is 7 seconds.   ∎

Recall that the area of a rectangle is given by the formula

$$A = lw$$

where A represents the area, l the length, and w the width of the rectangle. The perimeter of a rectangle is given by the formula

$$P = 2l + 2w$$

where P represents the perimeter of the rectangle, l the length, and w the width of the rectangle.

EXAMPLE 3 Assume that the rectangle in Figure 4-1 has an area of 52 square centimeters and that its length is 1 centimeter more than three times its width. Find the perimeter of the rectangle.

Solution Let w represent the width of the rectangle. Then $3w + 1$ represents its length. Because the area is 52 square centimeters, substitute 52 for A and $3w + 1$ for l in the formula $A = lw$ and solve for w.

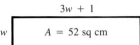

$3w + 1$

w | $A = 52$ sq cm |

FIGURE 4-1

$$A = lw$$
$$52 = (3w + 1)w$$
$$52 = 3w^2 + w \qquad \text{Remove parentheses.}$$
$$0 = 3w^2 + w - 52 \qquad \text{Add } -52 \text{ to both sides.}$$
$$0 = (3w + 13)(w - 4) \qquad \text{Factor.}$$

$3w + 13 = 0$ $\qquad$ or $\qquad$ $w - 4 = 0$ $\qquad$ Set each factor equal to 0.

$3w = -13$ $\qquad\qquad\qquad$ $w = 4$ $\qquad$ Solve each linear equation.

$w = -\dfrac{13}{3}$

Because the length of a rectangle cannot be negative, discard the result $w = -\frac{13}{3}$. Hence, the width of the rectangle is 4, and the length is given by

$$3w + 1 = 3(4) + 1 = 12 + 1 = 13$$

The dimensions of the rectangle are 4 centimeters by 13 centimeters. The perimeter is found by substituting 13 for l and 4 for w in the formula for the perimeter.

$$P = 2l + 2w = 2(13) + 2(4) = 26 + 8 = 34$$

The perimeter of the rectangle is 34 centimeters.

Check: A rectangle with dimensions of 13 centimeters by 4 centimeters does have an area of 52 square centimeters, and the length is 1 centimeter more than three times the width. A rectangle with these dimensions has a perimeter of 34 centimeters. ∎

Recall that the area of a triangle is given by the formula

$$A = \frac{1}{2} bh$$

where A represents the area, b the length of the base, and h the height of the triangle.

EXAMPLE 4 Assume that the triangle in Figure 4-2 has an area of 10 square centimeters and that its height is 3 centimeters less than twice the length of its base. Find the length of the base and the height of the triangle.

Solution   Let b represent the length of the base of the triangle. Then $2b - 3$ represents the height. Because the area is 10 square centimeters, substitute 10 for A along with $2b - 3$ for h in the formula $A = \frac{1}{2}bh$ and solve for b.

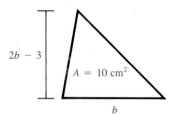

$2b - 3$

$A = 10 \text{ cm}^2$

b

FIGURE 4-2

$$A = \frac{1}{2}bh$$

$$10 = \frac{1}{2}b(2b - 3)$$

$20 = b(2b - 3)$	Multiply both sides by 2.
$20 = 2b^2 - 3b$	Remove parentheses.
$0 = 2b^2 - 3b - 20$	Add -20 to both sides.
$0 = (2b + 5)(b - 4)$	Factor.

$2b + 5 = 0$	or	$b - 4 = 0$	Set both factors equal to 0.
$2b = -5$		$b = 4$	Solve each linear equation.
$b = -\dfrac{5}{2}$			

Because a triangle cannot have a negative number for the length of its base, we must discard the result $b = -\frac{5}{2}$. Hence, the length of the base of the triangle is 4 centimeters. Its height is $2(4) - 3$, or 5 centimeters.

Check: If the base of a triangle has a length of 4 centimeters and the height of the triangle is 5 centimeters, its height is 3 centimeters less than twice the length of its base. Its area is 10 centimeters.

$$A = \frac{1}{2}bh = \frac{1}{2}(4)(5) = 2(5) = 10$$ ∎

EXERCISE 4.9

1. One positive number is 2 more than another. Their product is 35. Find the numbers.

2. One positive number is 5 less than four times another. Their product is 21. Find the numbers.

3. If 4 is added to the square of a composite number, the result is 5 less than ten times that number. Find the number.

4. If three times the square of a certain natural number is added to the number itself, the result is 14. Find the number.

In Exercises 5–8, assume that an object has been thrown straight up into the air. The height of the object above the ground is given by the formula

$$h = vt - 16t^2$$

where h is the height above the ground after t seconds, and v is the velocity with which the object was thrown.

5. In how many seconds will an object hit the ground if it was thrown with a velocity of 144 feet per second?

6. In how many seconds will an object hit the ground if it was thrown with a velocity of 160 feet per second?

7. If an object was thrown with a velocity of 220 feet per second, in how many seconds will the object be at a height of 600 feet?

8. If an object was thrown with a velocity of 128 feet per second, how many seconds will it take for the object to reach a height of 192 feet?

9. The length of a rectangle is 1 meter more than twice its width. Its area is 36 square meters. Find the dimensions of the rectangle.

10. The length of a rectangle is 2 inches less than three times its width. Its area is 21 square inches. Find the dimensions of the rectangle.

11. A room containing 143 square feet is 2 feet longer than it is wide. Find its perimeter.

12. The length of a rectangle is 2 centimeters longer than its width. If the length remains the same, but the width is doubled, the area is 48 square centimeters. Find the perimeter of the rectangle.

13. The length of the base of a triangle is 2 meters more than twice its height. The area is 30 square meters. Find the height and the length of the base of the triangle.

14. The height of a triangle is 2 inches less than five times the length of its base. The area is 36 square inches. Find the length of the base and the height of the triangle.

15. The base of a triangle is numerically 3 less than its area, and the height is numerically 6 less than its area. Find the area of the triangle.

16. The length of the base and the height of a triangle are numerically equal. Their sum is 6 less than the number of square units in the area of the triangle. Find the area of the triangle.

17. The formula for the area of a parallelogram is $A = bh$. The area of the parallelogram in Illustration 1 is 200 square centimeters. If its base is twice its height, how long is the base?

18. The formula for the area of a trapezoid is $A = \dfrac{h(B + b)}{2}$. The area of the trapezoid in Illustration 2 is 24 square centimeters. Find the height of the trapezoid if one base is 8 centimeters and the other has the same length as the height.

ILLUSTRATION 1   **ILLUSTRATION 2**

19. The volume of a rectangular solid is given by the formula $V = lwh$, where l is the length, w is the width, and h is the height. The volume of the rectangular solid in Illustration 3 is 210 cubic centimeters. Find the width of the rectangular solid if its length is 10 centimeters and its height is 1 centimeter longer than twice its width.

20. The volume of a pyramid is given by the formula $V = \dfrac{Bh}{3}$, where B is the area of its base and h is its height. The volume of the pyramid in Illustration 4 is 192 cubic centimeters. Find the dimensions of its rectangular base if one edge of the base is 2 centimeters longer than the other, and the height of the pyramid is 12 centimeters.

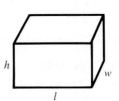
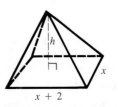

ILLUSTRATION 3   **ILLUSTRATION 4**

21. The volume of a pyramid is 84 cubic centimeters. Its height is 9 centimeters, and one side of its rectangular base is 3 centimeters shorter than the other. Find the dimensions of its base. (See Exercise 20.)

22. The volume of a rectangular solid is 72 cubic centimeters. Its height is 4 centimeters, and its width is 3 centimeters shorter than its length. Find the sum of its length and width. (See Exercise 19.)

REVIEW EXERCISES

In Review Exercises 1–10, solve each inequality and graph the solution on a number line.

1. $x - 3 > 5$

2. $x + 4 \leq 3$

3. $-3x - 5 \geq 4$

4. $2x - 3 < 7$

5. $\dfrac{3(x - 1)}{4} < 12$

6. $\dfrac{-2(x + 3)}{3} \geq 9$

7. $-2 < x \leq 4$

8. $-5 \leq x + 1 < 0$

9. $x + 1 > 2x + 3 > x - 4$

10. $1 < -7x + 8 < 15$

11. The efficiency, E, of a Carnot engine is given by the equation

$$E = 1 - \frac{T_2}{T_1}$$

Solve the equation for T_2.

12. Radioactive tracers are used for diagnostic purposes in nuclear medicine. The *effective half-life, H,* of a radioactive material in a biological organism is given by the formula

$$H = \frac{RB}{R + B}$$

where R is the radioactive half-life, and B is the biological half-life of the tracer. Evaluate H when $R = 12$ hours and $B = 9$ hours.

CHAPTER SUMMARY

Key Words

difference of two cubes (4.6)
difference of two squares (4.3)
factor (4.1)
factoring by grouping (4.2)
factoring out the greatest common factor (4.1)
greatest common factor (or divisor) (4.1)

incomplete quadratic equation (4.8)
irreducible polynomial (4.3)
key number (4.5)
linear equations (4.8)
perfect square trinomials (4.4)
prime factor (4.1)
prime-factored form (4.1)

prime number (4.1)
prime polynomial (4.3)
quadratic equations (4.8)
sum of two cubes (4.6)
sum of two squares (4.3)
zero factor theorem (4.8)

Key Ideas

(4.1) A natural number is in prime-factored form if it is written as the product of prime number factors.

The greatest common factor of several monomials is found by taking each common prime factor and vari-

able factor the fewest number of times that it appears in any one monomial.

To factor a polynomial, first factor out all common factors.

(4.2) If a polynomial has four or more terms, consider factoring it by grouping.

(4.3) To factor the difference of two squares, use the pattern

$$x^2 - y^2 = (x + y)(x - y)$$

(4.4, 4.5) Factor trinomials by trying these steps:

1. Write the trinomial with the exponents of one variable in descending order.
2. Factor out any greatest common factor (including -1 if that is necessary to make the coefficient of the first term positive).
3. If the sign of the third term of the trinomial is $+$, the signs between the terms of each binomial factor are the same as the sign of the trinomial's second term. If the sign of the third term is $-$, the signs between the terms of the binomials are opposite.
4. Mentally try various combinations of first terms and last terms until you find the one that works or you exhaust all the possibilities. In that case, the trinomial does not factor using only integer coefficients.
5. Check the factorization by multiplication.

(4.6) The sum and the difference of two cubes factor according to the patterns

$$x^3 + y^3 = (x + y)(x^2 - xy + y^2)$$
$$x^3 - y^3 = (x - y)(x^2 + xy + y^2)$$

(4.8) **Zero factor theorem.**   If a and b represent two real numbers and if $ab = 0$, then

$$a = 0 \quad \text{or} \quad b = 0$$

CHAPTER 4 REVIEW EXERCISES

(4.1) *In Review Exercises 1–8, find the prime factorization of each number.*

1. 35   **2.** 45   **3.** 96   **4.** 102
5. 87   **6.** 99   **7.** 2050   **8.** 4096

(4.1–4.7) *In Review Exercises 9–52, factor each expression completely.*

9. $3x + 9y$ **10.** $5ax^2 + 15a$ **11.** $7x^2 + 14x$ **12.** $3x^2 - 3x$

13. $2x^3 + 4x^2 - 8x$ **14.** $ax + ay - az$ **15.** $ax + ay - a$ **16.** $x^2yz + xy^2z$

17. $5a^2 + 5ab^2 + 10acd - 15a$ **18.** $7axy + 21x^2y - 35x^3y + 7xy^2$

19. $(x + y)a + (x + y)b$ **20.** $(x + y)^2 + (x + y)$
21. $2x^2(x + 2) + 6x(x + 2)$ **22.** $3x(y + z) - 9x(y + z)^2$
23. $3p + 9q + ap + 3aq$ **24.** $ar - 2as + 7r - 14s$
25. $x^2 + ax + bx + ab$ **26.** $xy + 2x - 2y - 4$
27. $3x^2y - xy^2 - 6xy + 2y^2$ **28.** $5x^2 + 10x - 15xy - 30y$
29. $x^2 - 9$ **30.** $x^2y^2 - 16$
31. $(x + 2)^2 - y^2$ **32.** $z^2 - (x + y)^2$
33. $6x^2y - 24y^3$ **34.** $(x + y)^2 - z^2$
35. $x^2 + 10x + 21$ **36.** $x^2 + 4x - 21$

37. $x^2 + 2x - 24$ **38.** $x^2 - 4x - 12$ **39.** $2x^2 - 5x - 3$ **40.** $3x^2 - 14x - 5$

41. $6x^2 + 7x - 3$ **42.** $6x^2 + 3x - 3$ **43.** $6x^3 + 17x^2 - 3x$ **44.** $4x^3 - 5x^2 - 6x$

45. $x^2 + 2ax + a^2 - y^2$ **46.** $ax^2 + 4ax + 3a - bx - b$

47. $xa + yb + ya + xb$ **48.** $2a^2x + 2abx + a^3 + a^2b$

49. $c^3 - 27$ **50.** $d^3 + 8$

51. $2x^3 + 54$ **52.** $2ab^4 - 2ab$

(4.8) *In Review Exercises 53–68, solve each equation.*

53. $x^2 + 2x = 0$ **54.** $2x^2 - 6x = 0$ **55.** $x^2 - 9 = 0$ **56.** $x^2 - 25 = 0$

57. $a^2 - 7a + 12 = 0$ **58.** $x^2 - 2x - 15 = 0$ **59.** $2x - x^2 + 24 = 0$ **60.** $16 + x^2 - 10x = 0$

61. $2x^2 - 5x - 3 = 0$ **62.** $2x^2 + x - 3 = 0$ **63.** $4x^2 = 1$ **64.** $9x^2 = 4$

65. $x^3 - 7x^2 + 12x = 0$ **66.** $x^3 + 5x^2 + 6x = 0$ **67.** $2x^3 + 5x^2 = 3x$ **68.** $3x^3 - 2x = x^2$

(4.9) *In Review Exercises 69–76, solve each word problem.*

69. The sum of two numbers is 12, and their product is 35. Find the numbers.

70. Two positive numbers differ by 12, and their product is 45. Find the numbers.

71. A rectangle is 2 feet longer than it is wide, and its area is 48 square feet. Find its dimensions.

72. If three times the square of a positive number is added to five times the number, the result is 2. Find the number.

73. A rectangle is 3 feet longer than twice its width, and its area is 27 square feet. Find its dimensions.

74. The base of a triangle is 3 centimeters longer than twice its height. Its area is 45 square centimeters. How long is the base?

75. The area of a square is numerically equal to its perimeter. How long is a side?

76. A rectangle is 3 feet longer than it is wide. Its area is numerically equal to its perimeter. Find its dimensions.

CHAPTER 4 TEST

_____   **1.** Find the prime factorization of 196.

In Problems 2–4, factor out the greatest common factor.

_____   **2.** $5x^3y^2z^3 + 10x^2y^3z^4$

_____   **3.** $60ab^2c^3 + 30a^3b^2c - 25a$

_____   **4.** $3x^2(a + b) - 6xy(a + b)$

In Problems 5–21, factor each expression completely.

_____   **5.** $ax + ay + bx + by$

_____   **6.** $x^2 - 25$

_____   **7.** $3a^2 - 27b^2$

_____   **8.** $16x^4 - 81y^4$

_____   **9.** $x^2 + 4x + 3$

_____   **10.** $x^2 - 9x - 22$

_____   **11.** $x^2 + 10xy + 9y^2$

_____   **12.** $6x^2 - 30xy + 24y^2$

_____   **13.** $3x^2 + 13x + 4$

_____   **14.** $2a^2 + 5a - 12$

_____   **15.** $2x^2 + 3xy - 2y^2$

_____   **16.** $12 - 25x + 12x^2$

_____   **17.** $12a^2 + 6ab - 36b^2$

_____   **18.** $x^3 - 64$

_____   **19.** $216 + 8a^3$

_____   **20.** $x^9z^3 - y^3z^6$

_____   **21.** $16r^3 + 128s^3$

In Problems 22–25, solve each equation.

_____   **22.** $x^2 + 3x = 0$

_____   **23.** $2x^2 + 5x + 3 = 0$

_____   **24.** $9y^2 - 81 = 0$

_____   **25.** $-3(y - 6) + 2 = y^2 + 2$

_____   **26.** One positive number is 10 more than another. Their product is 39. Find their sum.

_____   **27.** An object is fired straight up into the air with a velocity of 192 feet per second. In how many seconds will it hit the ground if its height above the ground is given by the formula $h = vt - 16t^2$, where v is the velocity and t is time?

_____   **28.** The base of a triangle with an area of 40 square meters is 2 meters longer than its height. Find the base of the triangle.

5

Rational Expressions

We have seen that expressions such as $\frac{1}{2}$ and $\frac{-3}{4}$ that indicate the quotient of two integers are called **arithmetic fractions** or **rational numbers.** Expressions such as

$$\frac{x}{x+2} \qquad \text{and} \qquad \frac{5a^2 + b^2}{3a - b}$$

are called **rational expressions.** We begin the study of these algebraic fractions by reviewing the basic properties of fractions.

5.1   THE BASIC PROPERTIES OF FRACTIONS

We have seen that any number that can be written in the form $\frac{a}{b}$, where a and b are integers and $b \neq 0$, is a rational number. The number 0.5, for example, is a rational number because it can be written in the form $\frac{5}{10}$ or $\frac{1}{2}$.

The fraction bar in the symbol $\frac{a}{b}$ indicates that a is to be divided by b. The number above the fraction bar is called the **numerator,** and the number below is called the **denominator.** Remember: *the denominator of a fraction can never be 0.*

There are three signs associated with every fraction: the sign of the fraction, the sign of the numerator, and the sign of the denominator.

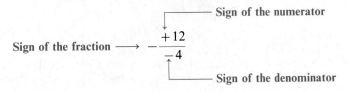

Any two of these signs can be changed without altering the value of the fraction. (If no sign is indicated, a $+$ sign is understood.) For example,

$$-\frac{+12}{-4} = -\frac{-12}{+4} = +\frac{-12}{-4} = +\frac{+12}{+4} = +3$$

In general, we have

$$\frac{a}{b} = \frac{-a}{-b} = -\frac{a}{-b} = -\frac{-a}{b}$$

and

$$-\frac{a}{b} = \frac{-a}{b} = \frac{a}{-b} = -\frac{-a}{-b}$$

We have seen that a fraction can be simplified by dividing out common factors shared by its numerator and denominator. For example,

$$\frac{18}{30} = \frac{3 \cdot 6}{5 \cdot 6} = \frac{3 \cdot \overset{1}{\cancel{6}}}{5 \cdot \underset{1}{\cancel{6}}} = \frac{3}{5}$$

and

$$-\frac{6}{15} = -\frac{3 \cdot 2}{3 \cdot 5} = -\frac{\overset{1}{\cancel{3}} \cdot 2}{\cancel{3} \cdot 5} = -\frac{2}{5}$$

To simplify the fraction $\dfrac{ac}{bc}$, we can divide out the common factor of c to obtain

$$\frac{ac}{bc} = \frac{a\overset{1}{\cancel{c}}}{b\underset{1}{\cancel{c}}} = \frac{a}{b}$$

This fact illustrates the fundamental property of fractions, previously stated in Section 1.2.

> **The Fundamental Property of Fractions.**   If a is a real number and b and c are nonzero real numbers, then
>
> $$\frac{ac}{bc} = \frac{a}{b}$$

The fundamental property of fractions implies that factors common to both the numerator and denominator of a fraction can be divided out. When all common factors have been divided out, we say that the fraction has been **expressed in lowest terms.** To **simplify a fraction** means to write it in lowest terms.

EXAMPLE 1   Simplify $\dfrac{21x^2y}{14xy^2}$.

Solution   To simplify the fraction means to write it in lowest terms.

$$\frac{21x^2y}{14xy^2} = \frac{3 \cdot 7 \cdot x \cdot x \cdot y}{2 \cdot 7 \cdot x \cdot y \cdot y} \qquad \text{Factor the numerator and the denominator.}$$

$$= \frac{3 \cdot \overset{1}{\cancel{7}} \cdot \overset{1}{\cancel{x}} \cdot x \cdot \overset{1}{\cancel{y}}}{2 \cdot \underset{1}{\cancel{7}} \cdot \underset{1}{\cancel{x}} \cdot y \cdot \underset{1}{\cancel{y}}} \qquad \text{Divide out the common factors of 7, } x \text{, and } y.$$

$$= \frac{3x}{2y}$$

Note that this fraction can be simplified by using the rules of exponents:

$$\frac{21x^2y}{14xy^2} = \frac{3 \cdot 7}{2 \cdot 7} x^{2-1} y^{1-2} = \frac{3}{2} xy^{-1} = \frac{3}{2} \cdot \frac{x}{y} = \frac{3x}{2y} \qquad \blacksquare$$

EXAMPLE 2   Write the fraction $\dfrac{x^2 + 3x}{3x + 9}$ in lowest terms.

Solution $$\frac{x^2 + 3x}{3x + 9} = \frac{x(x + 3)}{3(x + 3)}$$ Factor the numerator and the denominator.

$$= \frac{x\overset{1}{\cancel{(x + 3)}}}{3\underset{1}{\cancel{(x + 3)}}}$$ Divide out the common factor of $x + 3$.

$$= \frac{x}{3}$$ ∎

Any number divided by the number 1 remains unchanged. For example,

$$\frac{37}{1} = 37 \quad \text{and} \quad \frac{5x}{1} = 5x$$

In general, for any real number a, we have

$$\frac{a}{1} = a$$

EXAMPLE 3   Simplify $\dfrac{x^3 + x^2}{x + 1}$.

Solution $$\frac{x^3 + x^2}{x + 1} = \frac{x^2(x + 1)}{x + 1}$$ Factor the numerator.

$$= \frac{x^2\overset{1}{\cancel{(x + 1)}}}{\underset{1}{\cancel{x + 1}}}$$ Divide out the common factor of $x + 1$.

$$= \frac{x^2}{1}$$

$$= x^2$$ Denominators of 1 need not be written   ∎

EXAMPLE 4   Simplify **a.** $\dfrac{x - y}{y - x}$ and   **b.** $\dfrac{2a - 1}{1 - 2a}$.

Solution   Rearrange the terms in each numerator, factor out -1, and proceed as follows:

a. $\dfrac{x - y}{y - x} = \dfrac{-y + x}{y - x}$ **b.** $\dfrac{2a - 1}{1 - 2a} = \dfrac{-1 + 2a}{1 - 2a}$

$$= \frac{-(y - x)}{y - x} \qquad\qquad = \frac{-(1 - 2a)}{1 - 2a}$$

$$= \frac{-\overset{1}{\cancel{(y - x)}}}{\underset{1}{\cancel{y - x}}} \qquad\qquad = \frac{-\overset{1}{\cancel{(1 - 2a)}}}{\underset{1}{\cancel{1 - 2a}}}$$

$$= -1 \qquad\qquad\qquad = -1 \qquad\qquad ∎$$

The results of Example 4 illustrate this important fact.

The quotient of any nonzero expression and its negative is -1.

EXAMPLE 5   Simplify $\dfrac{x^2 + 13x + 12}{x^2 - 144}$.

Solution

$$\frac{x^2 + 13x + 12}{x^2 - 144} = \frac{(x + 1)(x + 12)}{(x + 12)(x - 12)}$$

Factor the numerator and the denominator.

$$= \frac{(x + 1)\overset{1}{\cancel{(x + 12)}}}{\underset{1}{\cancel{(x + 12)}}(x - 12)}$$

Divide out the common factor of $x + 12$.

$$= \frac{x + 1}{x - 12}$$ ∎

It is important to remember than only *factors* that are common to the *entire numerator* and the *entire denominator* can be divided out. *Common terms cannot be divided out.* For example, consider the correct simplification

$$\frac{5 + 8}{5} = \frac{13}{5}$$

It would be incorrect to divide out the common *term* of 5 in the above simplification. Note that doing so gives an incorrect answer.

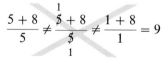

$$\frac{5 + 8}{5} \neq \frac{\overset{1}{\cancel{5}} + 8}{\underset{1}{\cancel{5}}} \neq \frac{1 + 8}{1} = 9$$

EXAMPLE 6   Express the fraction $\dfrac{5(x + 3) - 5}{7(x + 3) - 7}$ in lowest terms.

Solution   Do not divide out the binomials $x + 3$ because $x + 3$ is not a *factor* of the entire numerator, nor is it a *factor* of the entire denominator. Instead, simplify, factor the numerator and the denominator separately, and then divide out any common factors.

$$\frac{5(x + 3) - 5}{7(x + 3) - 7} = \frac{5x + 15 - 5}{7x + 21 - 7}$$

Remove parentheses.

$$= \frac{5x + 10}{7x + 14} \qquad \text{Combine terms.}$$

$$= \frac{5(x + 2)}{7(x + 2)} \qquad \text{Factor the numerator and the denominator.}$$

$$= \frac{5(\cancel{x + 2})^{1}}{7(\cancel{x + 2})_{1}} \qquad \text{Divide out the common factor of } x + 2.$$

$$= \frac{5}{7} \qquad \blacksquare$$

EXAMPLE 7 Simplify $\dfrac{x(x + 3) - 3(x - 1)}{x^2 + 3}$.

Solution
$$\frac{x(x + 3) - 3(x - 1)}{x^2 + 3} = \frac{x^2 + 3x - 3x + 3}{x^2 + 3} \qquad \begin{array}{l}\text{Remove parentheses in the} \\ \text{numerator.}\end{array}$$

$$= \frac{x^2 + 3}{x^2 + 3} \qquad \begin{array}{l}\text{Combine terms in the} \\ \text{numerator.}\end{array}$$

$$= \frac{\cancel{x^2 + 3}^{1}}{\cancel{x^2 + 3}_{1}} \qquad \begin{array}{l}\text{Divide out the common} \\ \text{factor of } x^2 + 3.\end{array}$$

$$= 1 \qquad \blacksquare$$

EXAMPLE 8 Simplify $\dfrac{xy + 2x + 3y + 6}{x^2 + x - 6}$.

Solution Factor the numerator by grouping and the denominator as a general trinomial. Then proceed as follows.

$$\frac{xy + 2x + 3y + 6}{x^2 + x - 6} = \frac{x(y + 2) + 3(y + 2)}{(x - 2)(x + 3)} \qquad \begin{array}{l}\text{Begin to factor the numerator} \\ \text{and factor the denominator.}\end{array}$$

$$= \frac{(y + 2)(x + 3)}{(x - 2)(x + 3)} \qquad \begin{array}{l}\text{Factor out } y + 2 \text{ in the} \\ \text{numerator.}\end{array}$$

$$= \frac{(y + 2)(\cancel{x + 3})^{1}}{(x - 2)(\cancel{x + 3})_{1}} \qquad \begin{array}{l}\text{Divide out the common} \\ \text{factor of } x + 3.\end{array}$$

$$= \frac{y + 2}{x - 2} \qquad \blacksquare$$

Sometimes a fraction does not simplify. Such a fraction is already in lowest terms.

EXAMPLE 9 Simplify $\dfrac{x^2 + x - 2}{x^2 + x}$.

Solution $\dfrac{x^2 + x - 2}{x^2 + x} = \dfrac{(x + 2)(x - 1)}{x(x + 1)}$ Factor the numerator and the denominator.

Because there are no factors common to the numerator and the denominator, this fraction is already in lowest terms. ∎

EXERCISE 5.1

In Exercises 1–68, express each fraction in lowest terms. If a fraction is already in lowest terms, so indicate. Assume that no variable has a value that would make a denominator equal to zero.

1. $\dfrac{8}{12}$

2. $\dfrac{16}{20}$

3. $\dfrac{28}{35}$

4. $\dfrac{14}{20}$

5. $\dfrac{8}{52}$

6. $\dfrac{15}{21}$

7. $\dfrac{10}{45}$

8. $\dfrac{21}{35}$

9. $\dfrac{-18}{54}$

10. $\dfrac{16}{40}$

11. $\dfrac{4x}{2}$

12. $\dfrac{2x}{4}$

13. $\dfrac{-6x}{18}$

14. $\dfrac{-25y}{5}$

15. $\dfrac{45}{9a}$

16. $\dfrac{48}{16y}$

17. $\dfrac{7 + 3}{5z}$

18. $\dfrac{(3 - 18)k}{25}$

19. $\dfrac{(3 + 4)a}{24 - 3}$

20. $\dfrac{x + x}{2}$

21. $\dfrac{2x}{3x}$

22. $\dfrac{5y}{7y}$

23. $\dfrac{6x^2}{4x^2}$

24. $\dfrac{9xy}{6xy}$

25. $\dfrac{2x^2}{3y}$

26. $\dfrac{5y^2}{2y^2}$

27. $\dfrac{15x^2y}{5xy^2}$

28. $\dfrac{12xz}{4xz^2}$

29. $\dfrac{28x}{32y}$

30. $\dfrac{14xz^2}{7x^2z^2}$

31. $\dfrac{x + 3}{3x + 9}$

32. $\dfrac{2x + 14}{x + 7}$

33. $\dfrac{5x + 35}{x + 7}$

34. $\dfrac{x - 9}{3x - 27}$

35. $\dfrac{x^2 + 3x}{2x + 6}$

36. $\dfrac{xz - 2x}{yz - 2y}$

37. $\dfrac{15x - 3x^2}{25y - 5xy}$

38. $\dfrac{3y + xy}{3x + xy}$

39. $\dfrac{6a - 6b + 6c}{9a - 9b + 9c}$

40. $\dfrac{3a - 3b - 6}{2a - 2b - 4}$

41. $\dfrac{x - 7}{7 - x}$

42. $\dfrac{d - c}{c - d}$

43. $\dfrac{6x - 3y}{3y - 6x}$

44. $\dfrac{2c - 4d}{4d - 2c}$

45. $\dfrac{a + b - c}{c - a - b}$

46. $\dfrac{x - y - z}{z + y - x}$

47. $\dfrac{x^2 + 3x + 2}{x^2 + x - 2}$

48. $\dfrac{x^2 + x - 6}{x^2 - x - 2}$

49. $\dfrac{x^2 - 8x + 15}{x^2 - x - 6}$

50. $\dfrac{x^2 - 6x - 7}{x^2 + 8x + 7}$

51. $\dfrac{2x^2 - 8x}{x^2 - 6x + 8}$

52. $\dfrac{3y^2 - 15y}{y^2 - 3y - 10}$

53. $\dfrac{xy + 2x^2}{2xy + y^2}$

54. $\dfrac{3x + 3y}{x^2 + xy}$

55. $\dfrac{x^2 + 3x + 2}{x^3 + x^2}$

56. $\dfrac{6x^2 - 13x + 6}{3x^2 + x - 2}$

57. $\dfrac{x^2 - 8x + 16}{x^2 - 16}$

58. $\dfrac{3x + 15}{x^2 - 25}$

59. $\dfrac{2x^2 - 8}{x^2 - 3x + 2}$

60. $\dfrac{3x^2 - 27}{x^2 + 3x - 18}$

61. $\dfrac{x^2 - 2x - 15}{x^2 + 2x - 15}$

62. $\dfrac{x^2 + 4x - 77}{x^2 - 4x - 21}$

63. $\dfrac{x^2 - 3(2x - 3)}{9 - x^2}$

64. $\dfrac{x(x - 8) + 16}{16 - x^2}$

65. $\dfrac{4(x + 3) + 4}{3(x + 2) + 6}$

66. $\dfrac{4 + 2(x - 5)}{3x - 5(x - 2)}$

67. $\dfrac{(2x + 3) - (x + 6)}{x^2 - 9}$

68. $\dfrac{2(x + 3) - (x + 2)}{x^2 + 5x + 4}$

In Exercises 69–76, simplify each fraction. Assume that no variable has a value that would make a denominator equal to 0. In each exercise, you will have to factor a sum or difference of two cubes or factor by grouping.

69. $\dfrac{x^3 + 1}{x^2 - x + 1}$

70. $\dfrac{x^3 - 1}{x^2 + x + 1}$

71. $\dfrac{2a^3 - 16}{2a^2 + 4a + 8}$

72. $\dfrac{3y^3 + 81}{y^2 - 3y + 9}$

73. $\dfrac{ab + b + 2a + 2}{ab + a + b + 1}$

74. $\dfrac{xy + 2y + 3x + 6}{x^2 + 5x + 6}$

75. $\dfrac{xy + 3y + 3x + 9}{x^2 - 9}$

76. $\dfrac{ab + b^2 + 2a + 2b}{a^2 + 2a + ab + 2b}$

In Review Exercises 1–6, state the following properties of real numbers.

1. the closure properties

2. the commutative properties

3. the associative properties

4. the distributive property

5. the identity properties

6. the inverse properties

7. What is the additive identity?

8. What is the multiplicative identity?

9. Find the additive inverse of -10.

10. Find the multiplicative inverse of -10.

11. What number does not have a multiplicative inverse?

12. Find the additive inverse of 0.

5.2 MULTIPLYING FRACTIONS

Recall that to multiply fractions, we multiply their numerators and multiply their denominators. For example, to find the product of $\frac{4}{7}$ and $\frac{3}{5}$, we proceed as follows:

$$\frac{4}{7} \cdot \frac{3}{5} = \frac{4 \cdot 3}{7 \cdot 5} = \frac{12}{35}$$

In general, we have

> **The Rule for Multiplying Fractions.**   If a and c are real numbers and b and d are nonzero real numbers, then
>
> $$\frac{a}{b} \cdot \frac{c}{d} = \frac{ac}{bd}$$

EXAMPLE 1   Multiply: **a.** $\dfrac{1}{3} \cdot \dfrac{2}{5}$, **b.** $\dfrac{7}{9} \cdot \dfrac{-5}{3x}$, **c.** $\dfrac{x^2}{2} \cdot \dfrac{3}{y^2}$, and **d.** $\dfrac{t+1}{t} \cdot \dfrac{t-1}{t-2}$.

Solution   **a.** $\dfrac{1}{3} \cdot \dfrac{2}{5} = \dfrac{1 \cdot 2}{3 \cdot 5} = \dfrac{2}{15}$ **b.** $\dfrac{7}{9} \cdot \dfrac{-5}{3x} = \dfrac{7(-5)}{9 \cdot 3x} = \dfrac{-35}{27x}$

c. $\dfrac{x^2}{2} \cdot \dfrac{3}{y^2} = \dfrac{x^2 \cdot 3}{2 \cdot y^2} = \dfrac{3x^2}{2y^2}$ **d.** $\dfrac{t+1}{t} \cdot \dfrac{t-1}{t-2} = \dfrac{(+1)(t-1)}{t(t-2)}$ ∎

EXAMPLE 2   Multiply: $\dfrac{35x^2 y}{7y^2 z} \cdot \dfrac{z}{5xy}$.

Solution $\dfrac{35x^2 y}{7y^2 z} \cdot \dfrac{z}{5xy} = \dfrac{5 \cdot 7 \cdot x \cdot x \cdot y \cdot z}{7 \cdot y \cdot y \cdot z \cdot 5 \cdot x \cdot y}$ Multiply the fractions and factor where possible.

$$= \dfrac{\overset{1}{\cancel{5}} \cdot \overset{1}{\cancel{7}} \cdot \overset{1}{\cancel{x}} \cdot x \cdot \overset{1}{\cancel{y}} \cdot \overset{1}{\cancel{z}}}{\underset{1}{\cancel{7}} \cdot y \cdot \underset{1}{\cancel{y}} \cdot \underset{1}{\cancel{z}} \cdot \underset{1}{\cancel{5}} \cdot \underset{1}{\cancel{x}} \cdot y}$$ Divide out all common factors.

$$= \dfrac{x}{y^2}$$ ∎

EXAMPLE 3   Find the product of $\dfrac{x^2 - x}{2x + 4}$ and $\dfrac{x+2}{x}$.

Solution $\dfrac{x^2 - x}{2x + 4} \cdot \dfrac{x+2}{x} = \dfrac{x(x-1)(x+2)}{2(x+2)x}$ Multiply the fractions and factor where possible.

$$= \dfrac{\overset{1}{\cancel{x}}(x-1)\overset{1}{\cancel{(x+2)}}}{2\underset{1}{\cancel{(x+2)}}\underset{1}{\cancel{x}}}$$ Divide out all common factors.

$$= \dfrac{x-1}{2}$$ ∎

EXAMPLE 4 Find the product of $\dfrac{x^2 - 3x}{x^2 - x - 6}$ and $\dfrac{x^2 + x - 2}{x^2 - x}$.

Solution $\dfrac{x^2 - 3x}{x^2 - x - 6} \cdot \dfrac{x^2 + x - 2}{x^2 - x}$

$= \dfrac{x(x - 3)(x + 2)(x - 1)}{(x + 2)(x - 3)x(x - 1)}$ Multiply the fractions and factor where possible.

$= \dfrac{\overset{1}{\cancel{x}}(\overset{1}{\cancel{x - 3}})(\overset{1}{\cancel{x + 2}})(\overset{1}{\cancel{x - 1}})}{(\underset{1}{\cancel{x + 2}})(\underset{1}{\cancel{x - 3}})\underset{1}{\cancel{x}}(\underset{1}{\cancel{x - 1}})}$ Divide out all common factors.

$= 1$ ■

EXAMPLE 5 Multiply the fraction $\dfrac{x^2 + x}{x^2 + 8x + 7}$ by $x + 7$.

Solution $\dfrac{x^2 + x}{x^2 + 8x + 7} \cdot (x + 7) = \dfrac{x^2 + x}{x^2 + 8x + 7} \cdot \dfrac{x + 7}{1}$ Write $x + 7$ as a fraction with a denominator of 1.

$= \dfrac{x(x + 1)(x + 7)}{(x + 1)(x + 7)1}$ Multiply the fractions and factor where possible.

$= \dfrac{x(\overset{1}{\cancel{x + 1}})(\overset{1}{\cancel{x + 7}})}{1(\underset{1}{\cancel{x + 1}})(\underset{1}{\cancel{x + 7}})}$ Divide out all common factors.

$= x$ ■

EXAMPLE 6 Multiply: $\dfrac{x^2 + 2x}{xy - 2y} \cdot \dfrac{x + 1}{x^2 - 4} \cdot \dfrac{x - 2}{x^2 + x}$.

Solution $\dfrac{x^2 + 2x}{xy - 2y} \cdot \dfrac{x + 1}{x^2 - 4} \cdot \dfrac{x - 2}{x^2 + x}$

$= \dfrac{x(x + 2)(x + 1)(x - 2)}{y(x - 2)(x + 2)(x - 2)x(x + 1)}$ Multiply the fractions and factor where possible.

$= \dfrac{\overset{1}{\cancel{x}}(\overset{1}{\cancel{x + 2}})(\overset{1}{\cancel{x + 1}})(\overset{1}{\cancel{x - 2}})}{y(\underset{1}{\cancel{x - 2}})(\underset{1}{\cancel{x + 2}})(x - 2)\underset{1}{\cancel{x}}(\underset{1}{\cancel{x + 1}})}$ Divide out all common factors.

$= \dfrac{1}{y(x - 2)}$ ■

EXERCISE 5.2

In Exercises 1–68, perform the indicated multiplications. Simplify answers if possible.

1. $\dfrac{2}{3} \cdot \dfrac{4}{5}$ **2.** $\dfrac{1}{2} \cdot \dfrac{3}{5}$ **3.** $\dfrac{5}{7} \cdot \dfrac{9}{13}$ **4.** $\dfrac{2}{7} \cdot \dfrac{5}{11}$

5. $\dfrac{2}{3} \cdot \dfrac{3}{5}$

6. $\dfrac{3}{7} \cdot \dfrac{7}{5}$

7. $\dfrac{-3}{7} \cdot \dfrac{14}{9}$

8. $\dfrac{-6}{9} \cdot \dfrac{15}{35}$

9. $\dfrac{25}{35} \cdot \dfrac{21}{55}$

10. $\dfrac{27}{24} \cdot \dfrac{56}{35}$

11. $\dfrac{-21}{18} \cdot \dfrac{-45}{14}$

12. $\dfrac{-33}{7} \cdot \dfrac{-5}{55}$

13. $\dfrac{2}{3} \cdot \dfrac{15}{2} \cdot \dfrac{1}{7}$

14. $\dfrac{2}{5} \cdot \dfrac{10}{9} \cdot \dfrac{3}{2}$

15. $\dfrac{3x}{y} \cdot \dfrac{y}{2}$

16. $\dfrac{2y}{z} \cdot \dfrac{z}{3}$

17. $\dfrac{5y}{7} \cdot \dfrac{7x}{5z}$

18. $\dfrac{4x}{3y} \cdot \dfrac{3y}{7x}$

19. $\dfrac{3y}{4x} \cdot \dfrac{2x}{5}$

20. $\dfrac{5x}{14y} \cdot \dfrac{7y}{x}$

21. $\dfrac{7z}{9z} \cdot \dfrac{4z}{2z}$

22. $\dfrac{8z}{2x} \cdot \dfrac{16x}{3x}$

23. $\dfrac{13x^2}{7x} \cdot \dfrac{28}{2x}$

24. $\dfrac{z^2}{z} \cdot \dfrac{5x}{5}$

25. $\dfrac{2x^2y}{3xy} \cdot \dfrac{3xy^2}{2}$

26. $\dfrac{2x^2z}{z} \cdot \dfrac{5x}{z}$

27. $\dfrac{8x^2y^2}{4x^2} \cdot \dfrac{2xy}{2y}$

28. $\dfrac{9x^2y}{3x} \cdot \dfrac{3xy}{3y}$

29. $\dfrac{-2xy}{x^2} \cdot \dfrac{3xy}{2}$

30. $\dfrac{-3x}{x^2} \cdot \dfrac{2xz}{3}$

31. $\dfrac{ab^2}{a^2b} \cdot \dfrac{b^2c^2}{abc} \cdot \dfrac{abc^2}{a^3c^2}$

32. $\dfrac{x^3y}{z} \cdot \dfrac{xz^3}{x^2y^2} \cdot \dfrac{yz}{xyz}$

33. $\dfrac{10r^2st^3}{6rs^2} \cdot \dfrac{3r^3t}{2rst} \cdot \dfrac{2s^3t^4}{5s^2t^3}$

34. $\dfrac{3a^3b}{25cd^3} \cdot \dfrac{-5cd^2}{6ab} \cdot \dfrac{10abc^2}{2bc^2d}$

35. $\dfrac{z+7}{7} \cdot \dfrac{z+2}{z}$

36. $\dfrac{a-3}{a} \cdot \dfrac{a+3}{5}$

37. $\dfrac{x-2}{2} \cdot \dfrac{2x}{x-2}$

38. $\dfrac{y+3}{y} \cdot \dfrac{3y}{y+3}$

39. $\dfrac{x+5}{5} \cdot \dfrac{x}{x+5}$

40. $\dfrac{y-9}{y+9} \cdot \dfrac{y}{9}$

41. $\dfrac{(x+1)^2}{x+1} \cdot \dfrac{x+2}{x+1}$

42. $\dfrac{(y-3)^2}{y-3} \cdot \dfrac{y-3}{y-3}$

43. $\dfrac{2x+6}{x+3} \cdot \dfrac{3}{4x}$

44. $\dfrac{3y-9}{y-3} \cdot \dfrac{y}{3y^2}$

45. $\dfrac{x^2-x}{x} \cdot \dfrac{3x-6}{3x-3}$

46. $\dfrac{5z-10}{z+2} \cdot \dfrac{3}{3z-6}$

47. $\dfrac{7y-14}{y-2} \cdot \dfrac{x^2}{7x}$

48. $\dfrac{y^2+3y}{9} \cdot \dfrac{3x}{y+3}$

49. $\dfrac{x^2+x-6}{5x} \cdot \dfrac{5x-10}{x+3}$

50. $\dfrac{z^2+4z-5}{5z-5} \cdot \dfrac{5z}{z+5}$

51. $\dfrac{m^2-2m-3}{2m+4} \cdot \dfrac{m^2-4}{m^2+3m+2}$

52. $\dfrac{p^2-p-6}{3p-9} \cdot \dfrac{p^2-9}{p^2+6p+9}$

53. $\dfrac{x^2+7xy+12y^2}{x^2+2xy-8y^2} \cdot \dfrac{x^2-xy-2y^2}{x^2+4xy+3y^2}$

54. $\dfrac{m^2+9mn+20n^2}{m^2-25n^2} \cdot \dfrac{m^2-9mn+20n^2}{m^2-16n^2}$

55. $\dfrac{3r^2+15rs+18s^2}{6r^2-24s^2} \cdot \dfrac{2r-4s}{3r+9s}$

56. $\dfrac{2u^2+8u}{2u+8} \cdot \dfrac{4u^2+8uv+4v^2}{u^2+5uv+4v^2}$

57. $\dfrac{abc^2}{a+1} \cdot \dfrac{c}{a^2b^2} \cdot \dfrac{a^2+a}{ac}$

58. $\dfrac{x^3yz^2}{4x+8} \cdot \dfrac{x^2-4}{2x^2y^2z^2} \cdot \dfrac{8yz}{x-2}$

59. $\dfrac{3x^2 + 5x + 2}{x^2 - 9} \cdot \dfrac{x - 3}{x^2 - 4} \cdot \dfrac{x^2 + 5x + 6}{6x + 4}$

60. $\dfrac{x^2 - 25}{3x + 6} \cdot \dfrac{x^2 + x - 2}{2x + 10} \cdot \dfrac{6x}{3x^2 - 18x + 15}$

61. $\dfrac{x^2 + 5x + 6}{x^2} \cdot \dfrac{x^2 - 2x}{x^2 - 9} \cdot \dfrac{x^2 - 3x}{x^2 - 4}$

62. $\dfrac{x^2 - 1}{1 - x} \cdot \dfrac{4x}{x + x^2} \cdot \dfrac{x^2 + 2x + 1}{2x + 2}$

63. $\dfrac{x^2 + 4x}{xz} \cdot \dfrac{z^2 + z}{x^2 - 16} \cdot \dfrac{z + 3}{z^2 + 4z + 3}$

64. $(x + 1) \cdot \dfrac{x^3 - 1}{x^3 + 1} \cdot \dfrac{x^2 - x + 1}{x^2 - 1}$

65. $\dfrac{x^3 + 8}{x^3 - 8} \cdot \dfrac{x - 2}{x^2 - 4} \cdot (x^2 + 2x + 4)$

66. $(4x^3 - 16x) \cdot \dfrac{1}{12x^2 + 24x} \cdot (3x + 12)$

67. $\dfrac{x^2 + x - 6}{5x^2 + 7x + 2} \cdot \dfrac{5x + 2}{-x^2 + x + 6} \cdot \dfrac{x + 1}{x^2 + 4x + 3}$

68. $\dfrac{x^2 - 3x - 4}{3x^2 - 2x - 1} \cdot \dfrac{3x + 1}{x^2 - 6x + 8} \cdot \dfrac{x^2 - 3x + 2}{x^2 + x}$

In Exercises 69–72, perform the indicated multiplications. You will need to factor by grouping and factor a sum or difference of two cubes to simplify the answers.

69. $\dfrac{ax + bx + ay + by}{x^3 - y^3} \cdot \dfrac{x^2 + xy + y^2}{ax + bx}$

70. $\dfrac{a^2 - ab + b^2}{a^3 + b^3} \cdot \dfrac{ac + ad + bc + bd}{c^2 - d^2}$

71. $\dfrac{x^2 - y^2}{y^2 - xy} \cdot \dfrac{yx^3 - y^4}{ax + ay + bx + by}$

72. $\dfrac{xw - xz + wy - yz}{x^2 + 2xy + y^2} \cdot \dfrac{x^3 - y^3}{z^2 - w^2}$

REVIEW EXERCISES

In Review Exercises 1–6, simplify each expression. Write all answers without using negative exponents.

1. $2x^3y^2(-3x^2y^4z)$

2. $\dfrac{8x^4y^5}{-2x^3y^2}$

3. $(3y)^{-4}$

4. $(a^{-2}a)^{-3}$

5. $\dfrac{x^{3m}}{x^{4m}}$

6. $(3x^2y^3)^0$

7. Write the number 93,000,000 in scientific notation.

8. Write the number 0.00567 in scientific notation.

9. In a mathematics class, there were 12 more women than men. If the total number of people in the class was 34, how many were women?

10. The area of a triangle is 48 square meters. If its height is 8 meters, find the length of its base.

11. Chuck invests $35,000 in each of two accounts, with one paying annual interest at a rate 1% greater than the other. His annual income from the two accounts is $7000. What are the annual rates of these accounts?

12. Diane has 3 more quarters than dimes, and twice as many nickels as quarters. She has 41 coins in all. How many of each does she have?

5.3 DIVIDING FRACTIONS

Recall that division by a nonzero number is equivalent to multiplying by the reciprocal of that number. Thus, to divide two fractions, we must invert the **divisor** (the fraction following the $\div$ sign) and multiply. For example, to divide $\frac{4}{7}$ by $\frac{3}{5}$, we proceed as follows:

$$\frac{4}{7} \div \frac{3}{5} = \frac{4}{7} \cdot \frac{5}{3} = \frac{20}{21}$$

In general, we have

> **The Rule for Dividing Fractions.**   If a is a real number and b, c, and d are nonzero real numbers, then
>
> $$\frac{a}{b} \div \frac{c}{d} = \frac{a}{b} \cdot \frac{d}{c}$$

EXAMPLE 1   Perform the divisions: **a.** $\dfrac{7}{13} \div \dfrac{21}{26}$ and   **b.** $\dfrac{-9x}{35y} \div \dfrac{15x^2}{14}$.

Solution   **a.** $\dfrac{7}{13} \div \dfrac{21}{26} = \dfrac{7}{13} \cdot \dfrac{26}{21}$ Invert the divisor and multiply.

$$= \frac{7 \cdot 2 \cdot 13}{13 \cdot 3 \cdot 7}$$ Multiply the fractions and factor where possible.

$$= \frac{\overset{1}{\cancel{7}} \cdot 2 \cdot \overset{1}{\cancel{13}}}{\underset{1}{\cancel{13}} \cdot 3 \cdot \underset{1}{\cancel{7}}}$$ Divide out common factors.

$$= \frac{2}{3}$$

b. $\dfrac{-9x}{35y} \div \dfrac{15x^2}{14} = \dfrac{-9x}{35y} \cdot \dfrac{14}{15x^2}$ Invert the divisor and multiply.

$$= \frac{-3 \cdot 3 \cdot x \cdot 2 \cdot 7}{5 \cdot 7 \cdot y \cdot 3 \cdot 5 \cdot x \cdot x}$$ Multiply the fractions and factor where possible.

$$= \frac{-3 \cdot \overset{1}{\cancel{3}} \cdot \overset{1}{\cancel{x}} \cdot 2 \cdot \overset{1}{\cancel{7}}}{5 \cdot \underset{1}{\cancel{7}} \cdot y \cdot \underset{1}{\cancel{3}} \cdot 5 \cdot \underset{1}{\cancel{x}} \cdot x}$$ Divide out all common factors.

$$= \frac{-6}{25xy}$$ Multiply the remaining factors.   ∎

EXAMPLE 2   Perform the division: $\dfrac{x^2 + x}{3x - 15} \div \dfrac{x^2 + 2x + 1}{6x - 30}$.

Solution

$$\dfrac{x^2 + x}{3x - 15} \div \dfrac{x^2 + 2x + 1}{6x - 30} = \dfrac{x^2 + x}{3x - 15} \cdot \dfrac{6x - 30}{x^2 + 2x + 1}$$ Invert the divisor and multiply.

$$= \dfrac{x(x + 1) \cdot 2 \cdot 3(x - 5)}{3(x - 5)(x + 1)(x + 1)}$$ Multiply the fractions and factor.

$$= \dfrac{x(\cancel{x + 1}) \cdot 2 \cdot \cancel{3}(\cancel{x - 5})}{\cancel{3}(\cancel{x - 5})(\cancel{x + 1})(x + 1)}$$ Divide out all common factors.

$$= \dfrac{2x}{x + 1}$$ ∎

EXAMPLE 3   Perform the division: $\dfrac{2x^2 - 3x - 2}{2x + 1} \div (4 - x^2)$.

Solution

$$\dfrac{2x^2 - 3x - 2}{2x + 1} \div (4 - x^2) = \dfrac{2x^2 - 3x - 2}{2x + 1} \div \dfrac{4 - x^2}{1}$$ Write $4 - x^2$ as a fraction with a denominator of 1.

$$= \dfrac{2x^2 - 3x - 2}{2x + 1} \cdot \dfrac{1}{4 - x^2}$$ Invert the divisor and multiply.

$$= \dfrac{(2x + 1)(x - 2) \cdot 1}{(2x + 1)(2 + x)(2 - x)}$$ Multiply the fractions and factor where possible.

$$= \dfrac{\cancel{(2x + 1)}\overset{-1}{\cancel{(x - 2)}} \cdot 1}{\cancel{(2x + 1)}(2 + x)\cancel{(2 - x)}}$$ Divide out all common factors.

$$= \dfrac{-1}{2 + x}$$

$$= -\dfrac{1}{x + 2}$$ ∎

Unless parentheses indicate otherwise, multiplications and divisions are to be performed in order from left to right.

EXAMPLE 4   Simplify the expression: $\dfrac{x^2 - x - 6}{x - 2} \div \dfrac{x^2 - 4x}{x^2 - x - 2} \cdot \dfrac{x - 4}{x^2 + x}$.

Solution   There are no parentheses to indicate otherwise, so the operation of division is performed first.

$$\frac{x^2 - x - 6}{x - 2} \div \frac{x^2 - 4x}{x^2 - x - 2} \cdot \frac{x - 4}{x^2 + x}$$

$$= \frac{x^2 - x - 6}{x - 2} \cdot \frac{x^2 - x - 2}{x^2 - 4x} \cdot \frac{x - 4}{x^2 + x}$$ Invert the divisor and multiply.

$$= \frac{(x + 2)(x - 3)(x + 1)(x - 2)(x - 4)}{(x - 2)x(x - 4)x(x + 1)}$$ Multiply the fractions and factor.

$$= \frac{(x + 2)(x - 3)\overset{1}{\cancel{(x + 1)}}\overset{1}{\cancel{(x - 2)}}\overset{1}{\cancel{(x - 4)}}}{\underset{1}{\cancel{(x - 2)}}x\underset{1}{\cancel{(x - 4)}}x\underset{1}{\cancel{(x + 1)}}}$$ Divide out all common factors.

$$= \frac{(x + 2)(x - 3)}{x^2}$$ ∎

EXAMPLE 5   Simplify the expression: $\dfrac{x^2 + 6x + 9}{x^2 - 2x}\left(\dfrac{x^2 - 4}{x^2 + 3x} \div \dfrac{x + 2}{x}\right)$.

Solution   Do the division within the parentheses first.

$$\frac{x^2 + 6x + 9}{x^2 - 2x}\left(\frac{x^2 - 4}{x^2 + 3x} \div \frac{x + 2}{x}\right)$$

$$= \frac{x^2 + 6x + 9}{x^2 - 2x}\left(\frac{x^2 - 4}{x^2 + 3x} \cdot \frac{x}{x + 2}\right)$$ Invert the divisor and multiply.

$$= \frac{(x + 3)(x + 3)(x + 2)(x - 2)x}{x(x - 2)x(x + 3)(x + 2)}$$ Multiply the fractions and factor where possible.

$$= \frac{\overset{1}{\cancel{(x + 3)}}(x + 3)\overset{1}{\cancel{(x + 2)}}\overset{1}{\cancel{(x - 2)}}\overset{1}{\cancel{x}}}{\underset{1}{\cancel{x}}\underset{1}{\cancel{(x - 2)}}x\underset{1}{\cancel{(x + 3)}}\underset{1}{\cancel{(x + 2)}}}$$ Divide out all common factors.

$$= \frac{x + 3}{x}$$ ∎

EXERCISE 5.3

In Exercises 1–44, perform each division. Simplify answers when possible.

1. $\dfrac{1}{3} \div \dfrac{1}{2}$

2. $\dfrac{3}{4} \div \dfrac{1}{3}$

3. $\dfrac{1}{5} \div \dfrac{2}{3}$

4. $\dfrac{1}{7} \div \dfrac{2}{5}$

5. $\dfrac{2}{5} \div \dfrac{1}{3}$

6. $\dfrac{3}{7} \div \dfrac{8}{11}$

7. $\dfrac{8}{5} \div \dfrac{7}{2}$

8. $\dfrac{9}{19} \div \dfrac{4}{7}$

9. $\dfrac{21}{14} \div \dfrac{5}{2}$

10. $\dfrac{14}{3} \div \dfrac{10}{3}$

11. $\dfrac{6}{5} \div \dfrac{6}{7}$

12. $\dfrac{6}{5} \div \dfrac{14}{5}$

13. $\dfrac{35}{2} \div \dfrac{15}{2}$

14. $\dfrac{6}{14} \div \dfrac{10}{35}$

15. $\dfrac{x}{2} \div \dfrac{1}{3}$

16. $\dfrac{y}{3} \div \dfrac{1}{2}$

17. $\dfrac{2}{y} \div \dfrac{4}{3}$

18. $\dfrac{3}{a} \div \dfrac{a}{9}$

19. $\dfrac{3x}{2} \div \dfrac{x}{2}$

20. $\dfrac{y}{6} \div \dfrac{2}{3y}$

21. $\dfrac{3x}{y} \div \dfrac{2x}{4}$

22. $\dfrac{3y}{8} \div \dfrac{2y}{4y}$

23. $\dfrac{4x}{3x} \div \dfrac{2y}{9y}$

24. $\dfrac{14}{7y} \div \dfrac{10}{5z}$

25. $\dfrac{x^2}{3} \div \dfrac{2x}{4}$

26. $\dfrac{z^2}{z} \div \dfrac{z}{3z}$

27. $\dfrac{y^2}{5z} \div \dfrac{3z}{2z}$

28. $\dfrac{xy}{x^2} \div \dfrac{y^2}{5}$

29. $\dfrac{x^2y}{3xy} \div \dfrac{xy^2}{6y}$

30. $\dfrac{2xz}{z} \div \dfrac{4x^2}{z^2}$

31. $\dfrac{x+2}{3x} \div \dfrac{x+2}{2}$

32. $\dfrac{z-3}{3z} \div \dfrac{z+3}{z}$

33. $\dfrac{(z-2)^2}{3z^2} \div \dfrac{z-2}{6z}$

34. $\dfrac{(x+7)^2}{x+7} \div \dfrac{(x-3)^2}{x+7}$

35. $\dfrac{(z-7)^2}{z+2} \div \dfrac{z(z-7)}{5z^2}$

36. $\dfrac{y(y+2)}{y^2(y-3)} \div \dfrac{y^2(y+2)}{(y-3)^2}$

37. $\dfrac{x^2-4}{3x+6} \div \dfrac{x-2}{x+2}$

38. $\dfrac{x^2-9}{5x+15} \div \dfrac{x-3}{x+3}$

39. $\dfrac{x^2-1}{3x-3} \div \dfrac{x+1}{3}$

40. $\dfrac{x^2-16}{x-4} \div \dfrac{3x+12}{x}$

41. $\dfrac{5x^2+13x-6}{x+3} \div \dfrac{5x^2-17x+6}{x-2}$

42. $\dfrac{x^2-x-6}{2x^2+9x+10} \div \dfrac{x^2-25}{2x^2+15x+25}$

43. $\dfrac{2x^2+8x-42}{x-3} \div \dfrac{2x^2+14x}{x^2+5x}$

44. $\dfrac{x^2-2x-35}{3x^2+27x} \div \dfrac{x^2+7x+10}{6x^2+12x}$

In Exercises 45–68, perform the indicated operations. In the absence of grouping symbols, multiplications and divisions are performed as they are encountered from left to right.

45. $\dfrac{2}{3} \cdot \dfrac{15}{5} \div \dfrac{10}{5}$

46. $\dfrac{6}{5} \div \dfrac{3}{5} \cdot \dfrac{5}{15}$

47. $\dfrac{6}{7} \div \dfrac{5}{2} \cdot \dfrac{5}{4}$

48. $\dfrac{15}{7} \div \dfrac{5}{2} \div \dfrac{4}{2}$

49. $\dfrac{x}{3} \cdot \dfrac{9}{4} \div \dfrac{x^2}{6}$

50. $\dfrac{y^2}{2} \div \dfrac{4}{y} \cdot \dfrac{y^2}{8}$

51. $\dfrac{x^2}{18} \div \dfrac{x^3}{6} \div \dfrac{12}{x^2}$

52. $\dfrac{y^3}{3y} \cdot \dfrac{3y^2}{4} \div \dfrac{15}{20}$

53. $\dfrac{x^2-1}{x^2-9} \cdot \dfrac{x+3}{x+2} \div \dfrac{5}{x+2}$

54. $\dfrac{2}{3x-3} \div \dfrac{2x+2}{x-1} \cdot \dfrac{5}{x+1}$

55. $\dfrac{x^2-4}{2x+6} \div \dfrac{x+2}{4} \cdot \dfrac{x+3}{x-2}$

56. $\dfrac{x^2-5x}{x+1} \cdot \dfrac{x+1}{x^2+3x} \div \dfrac{x-5}{x-3}$

57. $\dfrac{x-x^2}{x^2-4}\left(\dfrac{2x+4}{x+2} \div \dfrac{5}{x+2}\right)$

58. $\dfrac{2}{3x-3} \div \left(\dfrac{2x+2}{x-1} \cdot \dfrac{5}{x+1}\right)$

59. $\dfrac{y^2}{x+1} \cdot \dfrac{x^2+2x+1}{x^2-1} \div \dfrac{3y}{xy-y}$

60. $\dfrac{x^2-y^2}{x^4-x^3} \div \dfrac{x-y}{x^2} \div \dfrac{x^2+2xy+y^2}{x+y}$

61. $\dfrac{x^2+x-6}{x^2-4} \cdot \dfrac{x^2+2x}{x-2} \div \dfrac{x^2+3x}{x+2}$

62. $\dfrac{x^2-x-6}{x^2+6x-7} \cdot \dfrac{x^2+x-2}{x^2+2x} \div \dfrac{x^2+7x}{x^2-3x}$

63. $(a+2b) \div \left(\dfrac{a^2+4ab+4b^2}{a+b} \div \dfrac{a^2+7ab+10b^2}{a^2+6ab+5b^2}\right)$

64. $(ab-2b^2) \div \left(\dfrac{a^2-ab}{b-a} \cdot \dfrac{a^2-b^2}{a^3-3a^2b+2ab^2}\right)$

65. $\dfrac{x^2 + 2x - 3}{x^2 + x} \cdot \dfrac{x^2}{x^2 - 1} \div (x^2 + 3x)$

66. $\dfrac{x^2 - 6x + 5}{x + 2} \div (x^2 + 3x - 4) \cdot \dfrac{x^2 + 5x + 6}{x^2 - 2x - 15}$

67. $\dfrac{x^2 + 4x + 3}{x^2 - y^2} \div \dfrac{xy + y}{xy - x^2} \cdot \dfrac{x^2 y + 2xy^2 + y^3}{x^2 + 3x}$

68. $\dfrac{a^2 - b^2}{a^2 - a - 2} \cdot \dfrac{a^2 - 2a - 3}{b - a} \div \dfrac{a^2 + ab}{a^2 - 2a}$

In Exercises 69–72, perform the indicated divisions. You will need to factor by grouping and factor a sum or difference of two cubes to simplify the answers.

69. $\dfrac{ab + 4a + 2b + 8}{b^2 + 4b + 16} \div \dfrac{b^2 - 16}{b^3 - 64}$

70. $\dfrac{r^3 - s^3}{r^2 - s^2} \div \dfrac{r^2 + rs + s^2}{mr + ms + nr + ns}$

71. $\dfrac{p^3 - p^2 q + pq^2}{mp - mq + np - nq} \div \dfrac{q^3 + p^3}{q^2 - p^2}$

72. $\dfrac{s^3 - r^3}{r^2 + rs + s^2} \div \dfrac{pr - ps - qr + qs}{q^2 - p^2}$

REVIEW EXERCISES

In Review Exercises 1–8, perform the indicated operations and simplify.

1. $-4(y^3 - 4y^2 + 3y - 2) + 6(-2y^2 + 4) - 4(-2y^3 - y)$

2. $6(3a^3 + 2a^2 + 3) - (-2a^2 + 4a - 2) + 5(-2a^3 - a^2 + 2a - 3)$

3. $(2r - 3)(r^2 + 5)$

4. $(3x + 4y)^2$

5. $(3m + 2)(-2m + 1)(m - 1)$

6. $(2p - q)(p + q)^2$

7. $y - 5 \overline{)\, 5y^3 - 3y^2 + 4y - 1}$

8. $x + 4 \overline{)\, 6x^3 + 5 - 4x}$

9. On a test the lowest score was half the highest score. Find the highest score if the sum of the two scores was 126.

10. A desk is twice as long as it is wide. To protect the finish, a rectangular piece of glass is used to cover its top. Find the dimensions of the desk if the area of the glass is 18 square feet.

11. Receipts from the sale of 35,750 football tickets totaled $109,500. Regular admission tickets sold for $3.50 each and each student admission ticket cost $1.00. How many of each were sold?

12. One type of candy is worth $4.50 per pound, and another is worth $3.00 per pound. How many pounds of each type are needed to make 150 pounds of a mixture worth $3.90 per pound?

5.4   ADDING AND SUBTRACTING FRACTIONS WITH LIKE DENOMINATORS

Recall that to add fractions with the same denominator, we add their numerators and keep the same denominator. For example, to add $\frac{2}{7}$ and $\frac{3}{7}$, we would proceed as follows:

$$\frac{2}{7} + \frac{3}{7} = \frac{2 + 3}{7} = \frac{5}{7}$$

In general, we have

> **Adding Fractions with Like Denominators.**   If a, b, and c represent real numbers, and $d \neq 0$, then
> $$\frac{a}{d} + \frac{b}{d} = \frac{a + b}{d}$$

EXAMPLE 1   Perform the following additions. Simplify each result when possible.

a. $\dfrac{5}{9} + \dfrac{2}{9} = \dfrac{5 + 2}{9} = \dfrac{7}{9}$

b. $\dfrac{8}{41} + \dfrac{21}{41} = \dfrac{8 + 21}{41} = \dfrac{29}{41}$

c. $\dfrac{x}{7} + \dfrac{y}{7} = \dfrac{x + y}{7}$

d. $\dfrac{x}{7} + \dfrac{3x}{7} = \dfrac{x + 3x}{7} = \dfrac{4x}{7}$

e. $\dfrac{3x + y}{5x} + \dfrac{x + y}{5x} = \dfrac{3x + y + x + y}{5x}$

$$= \dfrac{4x + 2y}{5x}$$

f. $\dfrac{3x}{7y} + \dfrac{4x}{7y} = \dfrac{3x + 4x}{7y}$

$$= \dfrac{7x}{7y}$$

$$= \dfrac{x}{y} \qquad \blacksquare$$

EXAMPLE 2   Add the fractions $\dfrac{3x + 21}{5x + 10}$ and $\dfrac{8x + 1}{5x + 10}$.

Solution   Because the fractions have the same denominator, add their numerators and keep the common denominator.

$$\frac{3x + 21}{5x + 10} + \frac{8x + 1}{5x + 10} = \frac{3x + 21 + 8x + 1}{5x + 10} \qquad \text{Add the fractions.}$$

$$= \frac{11x + 22}{5x + 10} \qquad \text{Combine like terms.}$$

$$= \frac{11\overset{1}{\cancel{(x + 2)}}}{5\underset{1}{\cancel{(x + 2)}}} \qquad \begin{array}{l}\text{Factor and divide out the} \\ \text{common factor of } x + 2.\end{array}$$

$$= \frac{11}{5} \qquad \blacksquare$$

Subtracting Fractions with Like Denominators

Recall that to subtract fractions with the same denominator, we subtract their numerators and keep the same denominator.

> **Subtracting Fractions with Like Denominators.**   If a, b, and d represent real numbers, and $d \neq 0$, then
>
> $$\frac{a}{d} - \frac{b}{d} = \frac{a - b}{d}$$

EXAMPLE 3   Perform each subtraction and simplify: **a.** $\dfrac{5x}{3} - \dfrac{2x}{3}$ and **b.** $\dfrac{5x + 1}{x - 3} - \dfrac{4x - 2}{x - 3}$.

Solution   In each part, both fractions have the same denominator. Thus, subtract the fractions by subtracting their numerators and keeping the common denominator.

a. $\dfrac{5x}{3} - \dfrac{2x}{3} = \dfrac{5x - 2x}{3}$ Subtract the fractions.

$\qquad\qquad = \dfrac{3x}{3}$ Combine like terms.

$\qquad\qquad = \dfrac{x}{1}$ Simplify the fraction.

$\qquad\qquad = x$ Denominators of 1 need not be written.

b. $\dfrac{5x + 1}{x - 3} - \dfrac{4x - 2}{x - 3} = \dfrac{(5x + 1) - (4x - 2)}{x - 3}$ Subtract the fractions.

$\qquad\qquad = \dfrac{5x + 1 - 4x + 2}{x - 3}$ Remove parentheses.

$\qquad\qquad = \dfrac{x + 3}{x - 3}$ Combine like terms.   ■

The denominator of a fraction cannot be 0. However, if the numerator of a fraction is 0, then the value of the fraction is 0. This fact is used in the next example.

EXAMPLE 4   Perform the indicated operations: $\dfrac{3x + 1}{x - 7} - \dfrac{5x + 2}{x - 7} + \dfrac{2x + 1}{x - 7}$.

Solution   This example combines both addition and subtraction of fractions with like denominators. Unless parentheses indicate otherwise, additions and subtractions are accomplished in order from left to right.

$$\dfrac{3x + 1}{x - 7} - \dfrac{5x + 2}{x - 7} + \dfrac{2x + 1}{x - 7}$$

$\qquad = \dfrac{(3x + 1) - (5x + 2) + (2x + 1)}{x - 7}$ Combine the numerators and keep the common denominator.

$\qquad = \dfrac{3x + 1 - 5x - 2 + 2x + 1}{x - 7}$ Remove parentheses.

$\qquad = \dfrac{0}{x - 7}$ Combine like terms.

$\qquad = 0$ Simplify.   ■

EXERCISE 5.4

In Exercises 1–30, perform each addition. Write all answers in lowest terms.

1. $\dfrac{1}{3} + \dfrac{1}{3}$

2. $\dfrac{3}{4} + \dfrac{3}{4}$

3. $\dfrac{1}{5} + \dfrac{2}{5}$

4. $\dfrac{3}{7} + \dfrac{2}{7}$

5. $\dfrac{2}{9} + \dfrac{1}{9}$

6. $\dfrac{5}{7} + \dfrac{9}{7}$

7. $\dfrac{8}{7} + \dfrac{6}{7}$

8. $\dfrac{9}{11} + \dfrac{2}{11}$

9. $\dfrac{21}{14} + \dfrac{7}{14}$

10. $\dfrac{14}{3} + \dfrac{10}{3}$

11. $\dfrac{6}{7} + \dfrac{6}{7}$

12. $\dfrac{6}{5} + \dfrac{14}{5}$

13. $\dfrac{35}{8} + \dfrac{15}{8}$

14. $\dfrac{6}{14} + \dfrac{10}{14}$

15. $\dfrac{14x}{11} + \dfrac{30x}{11}$

16. $\dfrac{6a}{10} + \dfrac{28a}{10}$

17. $\dfrac{-77y}{126} + \dfrac{-7y}{126}$

18. $\dfrac{-39a}{15} + \dfrac{-21a}{15}$

19. $\dfrac{15z}{22} + \dfrac{-15z}{22}$

20. $\dfrac{-30rs}{21} + \dfrac{30rs}{21}$

21. $\dfrac{2x}{y} + \dfrac{2x}{y}$

22. $\dfrac{3y}{5} + \dfrac{2y}{5}$

23. $\dfrac{4y}{3x} + \dfrac{2y}{3x}$

24. $\dfrac{4}{7y} + \dfrac{10}{7y}$

25. $\dfrac{x^2}{4y} + \dfrac{x^2}{4y}$

26. $\dfrac{r^2}{r} + \dfrac{r^2}{r}$

27. $\dfrac{y+2}{5z} + \dfrac{y+4}{5z}$

28. $\dfrac{x+3}{x^2} + \dfrac{x+5}{x^2}$

29. $\dfrac{3x-5}{x-2} + \dfrac{6x-13}{x-2}$

30. $\dfrac{8x-7}{x+3} + \dfrac{2x+37}{x+3}$

In Exercises 31–60, perform each subtraction. Simplify answers when possible.

31. $\dfrac{5}{7} - \dfrac{4}{7}$

32. $\dfrac{5}{9} - \dfrac{3}{9}$

33. $\dfrac{4}{3} - \dfrac{8}{3}$

34. $\dfrac{7}{11} - \dfrac{4}{11}$

35. $\dfrac{17}{13} - \dfrac{15}{13}$

36. $\dfrac{18}{31} - \dfrac{18}{31}$

37. $\dfrac{21}{23} - \dfrac{45}{23}$

38. $\dfrac{35}{72} - \dfrac{44}{72}$

39. $\dfrac{39}{37} - \dfrac{2}{37}$

40. $\dfrac{35}{99} - \dfrac{13}{99}$

41. $\dfrac{-47}{123} - \dfrac{4}{123}$

42. $\dfrac{-23}{17} - \dfrac{11}{17}$

43. $\dfrac{15}{21} - \left(\dfrac{-15}{21}\right)$

44. $\dfrac{-37}{25} - \left(\dfrac{-22}{25}\right)$

45. $\dfrac{2x}{y} - \dfrac{x}{y}$

46. $\dfrac{7y}{5} - \dfrac{4y}{5}$

47. $\dfrac{9y}{3x} - \dfrac{6y}{3x}$

48. $\dfrac{24}{7y} - \dfrac{10}{7y}$

49. $\dfrac{3x^2}{4x} - \dfrac{x^2}{4x}$

50. $\dfrac{5r^2}{2r} - \dfrac{r^2}{2r}$

51. $\dfrac{y+2}{5z} - \dfrac{y+4}{5z}$

52. $\dfrac{x+3}{x^2} - \dfrac{x+5}{x^2}$

53. $\dfrac{6x-5}{3xy} - \dfrac{3x-5}{3xy}$

54. $\dfrac{7x+7}{5y} - \dfrac{2x+7}{5y}$

55. $\dfrac{y+2}{2z} - \dfrac{y+4}{2z}$

56. $\dfrac{2x-3}{x^2} - \dfrac{x-3}{x^2}$

57. $\dfrac{5x+5}{3xy} - \dfrac{2x-4}{3xy}$

58. $\dfrac{8x-7}{2y} - \dfrac{2x+7}{2y}$

59. $\dfrac{3y-2}{y+3} - \dfrac{2y-5}{y+3}$

60. $\dfrac{5x+8}{x+5} - \dfrac{3x-2}{x+5}$

In Exercises 61–76, perform the indicated operations. Simplify answers when possible.

61. $\dfrac{3}{7} - \dfrac{5}{7} + \dfrac{2}{7}$

62. $\dfrac{3}{4} - \dfrac{5}{4} + \dfrac{8}{4}$

63. $\dfrac{3}{5} - \dfrac{2}{5} + \dfrac{7}{5}$

64. $\dfrac{5}{11} - \dfrac{8}{11} + \dfrac{14}{11}$

65. $\dfrac{13x}{15} + \dfrac{12x}{15} - \dfrac{5x}{15}$

66. $\dfrac{13y}{32} + \dfrac{13y}{32} - \dfrac{10y}{32}$

67. $\dfrac{x}{3y} + \dfrac{2x}{3y} - \dfrac{x}{3y}$

68. $\dfrac{5y}{8x} + \dfrac{4y}{8x} - \dfrac{y}{8x}$

69. $\dfrac{3x}{y+2} - \dfrac{3y}{y+2} + \dfrac{x+y}{y+2}$

70. $\dfrac{3y}{x-5} + \dfrac{x}{x-5} - \dfrac{y-x}{x-5}$

71. $\dfrac{x+1}{x-2} - \dfrac{2(x-3)}{x-2} + \dfrac{3(x+1)}{x-2}$

72. $\dfrac{x^2-4}{x+2} + \dfrac{2(x^2-9)}{x+2} - \dfrac{3(x^2-5)}{x+2}$

73. $\dfrac{3xy}{x-y} - \dfrac{x(3y-x)}{x-y} - \dfrac{x(x-y)}{x-y}$

74. $\dfrac{x^2+4x+1}{(x-1)^2} - \dfrac{x(x+1)}{(x-1)^2} - \dfrac{x}{(x-1)^2}$

75. $\dfrac{2(2a+b)}{(a-b)^2} - \dfrac{2(2b+a)}{(a-b)^2} + \dfrac{3(b-a)}{(a-b)^2}$

76. $\dfrac{2(x-2)}{2x-3} + \dfrac{2(2x+1)}{2x-3} - \dfrac{2(5x-4)}{2x-3}$

REVIEW EXERCISES

In Review Exercises 1–6, write each number in prime-factored form.

1. 49

2. 64

3. 136

4. 242

5. 102

6. 315

In Review Exercises 7–12, factor each expression.

7. $x^2 - 2x - 15$

8. $3x^2 - 5xy - 2y^2$

9. $2x^2 - 8$

10. $3y^2 - 27$

11. $ax + ay - 5x - 5y$

12. $xy - x^2 + y - x$

5.5 ADDING AND SUBTRACTING FRACTIONS WITH UNLIKE DENOMINATORS

Recall that to add fractions with unlike denominators, we must first convert them to fractions with the same denominator. To do so, we can use the fundamental property of fractions to multiply both the numerator and denominator of each fraction by some appropriate nonzero number. The fractions can then be added. For example, to add the fractions $\frac{4}{7}$ and $\frac{3}{5}$, we proceed as follows:

$$\frac{4}{7} + \frac{3}{5} = \frac{4 \cdot 5}{7 \cdot 5} + \frac{3 \cdot 7}{5 \cdot 7}$$

Multiply both the numerator and the denominator of the first fraction by 5 and those of the second fraction by 7.

$$= \frac{20}{35} + \frac{21}{35}$$

Do the indicated multiplications.

$$= \frac{41}{35}$$

Add the fractions.

The process of multiplying both the numerator and denominator of a fraction by the same nonzero number is often called **building a fraction.**

EXAMPLE 1 Change **a.** $\dfrac{1}{2}$, **b.** $\dfrac{3}{5}$, and **c.** $\dfrac{7}{10}$ into fractions that have a common denominator of 30.

Solution Build each fraction as follows:

a. $\dfrac{1}{2} = \dfrac{1 \cdot 15}{2 \cdot 15} = \dfrac{15}{30}$ **b.** $\dfrac{3}{5} = \dfrac{3 \cdot 6}{5 \cdot 6} = \dfrac{18}{30}$ **c.** $\dfrac{7}{10} = \dfrac{7 \cdot 3}{10 \cdot 3} = \dfrac{21}{30}$ ∎

To add fractions with unlike denominators, we must express those fractions in equivalent forms with the same denominator.

EXAMPLE 2 Add the fractions $\dfrac{5}{14}$ and $\dfrac{2}{21}$.

Solution Build the fractions so that each has a denominator of 42.

$$\dfrac{5}{14} + \dfrac{2}{21} = \dfrac{5 \cdot 3}{14 \cdot 3} + \dfrac{2 \cdot 2}{21 \cdot 2}$$

Multiply both the numerator and the denominator of the first fraction by 3 and those of the second fraction by 2.

$$= \dfrac{15}{42} + \dfrac{4}{42}$$

Do the indicated multiplications.

$$= \dfrac{19}{42}$$

Add the fractions. ∎

In Example 2, the number 42 was used as the common denominator because it is divisible by both 14 and 21. The number 42 is also the *smallest* number that is divisible by both 14 and 21. The smallest number that can be divided by each of the denominators of a set of fractions is called the **least** (or **lowest**) **common denominator (LCD)** of those fractions.

We have seen that there is a process to find the least common denominator of a set of fractions:

Finding the Least Common Denominator (LCD).

1. Write down each of the different denominators that appear in the given fractions.

2. Factor each of these denominators completely.

3. Form a product using each of the different factors obtained in step 2. Use each different factor the *greatest* number of times it appears in any *single* denominator. The product formed by multiplying the different factors is the least common denominator.

EXAMPLE 3   Several fractions have denominators of 24, 18, and 36. Find the least common denominator.

Solution   First, write down and factor each denominator into products of prime numbers.

$$24 = 2 \cdot 2 \cdot 2 \cdot 3$$
$$18 = 2 \cdot 3 \cdot 3$$
$$36 = 2 \cdot 2 \cdot 3 \cdot 3$$

Then form a product with the factors of 2 and 3. Use each of these factors the greatest number of times it appears in any single denominator. That is, use the factor 2 three times because 2 appears three times as a factor of 24. Use the factor of 3 twice because it occurs twice as a factor of 18 and 36. Thus, the least common denominator of the fraction is

$$\text{LCD} = 2 \cdot 2 \cdot 2 \cdot 3 \cdot 3 = 8 \cdot 9 = 72$$ ■

EXAMPLE 4   Add the fractions $\dfrac{1}{24}$, $\dfrac{5}{18}$, and $\dfrac{7}{36}$.

Solution   The fractions $\frac{1}{24}$, $\frac{5}{18}$, and $\frac{7}{36}$ have different denominators, but each can be written as a fraction with the least common denominator of $2 \cdot 2 \cdot 2 \cdot 3 \cdot 3$, or 72. Proceed as follows:

$$\frac{1}{24} + \frac{5}{18} + \frac{7}{36} = \frac{1}{2 \cdot 2 \cdot 2 \cdot 3} + \frac{5}{2 \cdot 3 \cdot 3} + \frac{7}{2 \cdot 2 \cdot 3 \cdot 3} \qquad \text{Factor each denominator.}$$

In each of these fractions, multiply both the numerator and the denominator by whatever is necessary to build the denominator to the required $2 \cdot 2 \cdot 2 \cdot 3 \cdot 3$.

$$= \frac{1 \cdot 3}{2 \cdot 2 \cdot 2 \cdot 3 \cdot 3} + \frac{5 \cdot 2 \cdot 2}{2 \cdot 3 \cdot 3 \cdot 2 \cdot 2} + \frac{7 \cdot 2}{2 \cdot 2 \cdot 3 \cdot 3 \cdot 2} \qquad \text{Build each fraction.}$$

$$= \frac{3 + 20 + 14}{72} \qquad \qquad \begin{array}{l} \text{Do the indicated} \\ \text{multiplications and} \\ \text{add the fractions.} \end{array}$$

$$= \frac{37}{72} \qquad \qquad \text{Combine like terms.}$$ ■

EXAMPLE 5   Add: $\dfrac{1}{x} + \dfrac{x}{y}$.

Solution   The least common denominator is xy.

$$\frac{1}{x} + \frac{x}{y} = \frac{(1)y}{(x)y} + \frac{x(x)}{x(y)} \qquad \begin{array}{l} \text{Build the fractions to get the common} \\ \text{denominator of } xy. \end{array}$$

$$= \frac{y}{xy} + \frac{x^2}{xy} \qquad \text{Do the indicated multiplications.}$$

$$= \frac{y + x^2}{xy} \qquad \text{Add the fractions.}$$ ■

EXAMPLE 6   Perform the indicated operations: $\dfrac{3}{x^2y} + \dfrac{2}{xy} - \dfrac{1}{xy^2}$.

Solution   Find the least common denominator.

$$\left.\begin{array}{l} x^2y = x \cdot x \cdot y \\ xy = x \cdot y \\ xy^2 = x \cdot y \cdot y \end{array}\right\} \qquad \text{Factor each denominator.}$$

In any one of these denominators, the factor x occurs at most twice, and the factor y occurs at most twice. Thus, the LCD is

$$\text{LCD} = x \cdot x \cdot y \cdot y = x^2y^2$$

Build each given fraction into a new fraction with a denominator of x^2y^2.

$$\frac{3}{x^2y} + \frac{2}{xy} - \frac{1}{xy^2}$$

$$= \frac{3 \cdot y}{x \cdot x \cdot y \cdot y} + \frac{2 \cdot x \cdot y}{x \cdot y \cdot x \cdot y} - \frac{1 \cdot x}{x \cdot y \cdot y \cdot x} \qquad \begin{array}{l}\text{Factor each} \\ \text{denominator and build} \\ \text{each fraction.}\end{array}$$

$$= \frac{3y + 2xy - x}{x^2y^2} \qquad \begin{array}{l}\text{Perform the} \\ \text{multiplications and} \\ \text{add the fractions.} \quad \blacksquare\end{array}$$

EXAMPLE 7   Perform the subtraction: $\dfrac{x}{x+1} - \dfrac{3}{x}$.

Solution   The least common denominator is $(x+1)x$.

$$\frac{x}{x+1} - \frac{3}{x} = \frac{(x)x}{(x+1)x} - \frac{3(x+1)}{x(x+1)} \qquad \begin{array}{l}\text{Build the fractions to get the} \\ \text{common denominator.}\end{array}$$

$$= \frac{x^2}{(x+1)x} - \frac{3x+3}{x(x+1)} \qquad \begin{array}{l}\text{Perform the multiplications in} \\ \text{the numerator.}\end{array}$$

$$= \frac{x^2 - (3x+3)}{x(x+1)} \qquad \begin{array}{l}\text{Subtract the numerators and keep} \\ \text{the common denominator.}\end{array}$$

$$= \frac{x^2 - 3x - 3}{x(x+1)} \qquad \text{Remove parentheses.} \quad \blacksquare$$

EXAMPLE 8   Perform the indicated operations: $\dfrac{3}{x^2 - y^2} + \dfrac{2}{x - y} - \dfrac{1}{x + y}$.

Solution   Find the least common denominator.

$$\begin{array}{l} x^2 - y^2 = (x-y)(x+y) \\ x - y = x - y \\ x + y = x + y \end{array} \qquad \text{Factor each denominator, where possible.}$$

The least common denominator is $(x - y)(x + y)$. Build each given fraction into a new fraction with that common denominator.

$$\frac{3}{x^2 - y^2} + \frac{2}{x - y} - \frac{1}{x + y}$$

$$= \frac{3}{(x + y)(x - y)} + \frac{2}{x - y} - \frac{1}{x + y} \qquad \text{Factor.}$$

$$= \frac{3}{(x + y)(x - y)} + \frac{2(x + y)}{(x - y)(x + y)} - \frac{1(x - y)}{(x + y)(x - y)} \qquad \begin{array}{l}\text{Build the fractions} \\ \text{to get a common} \\ \text{denominator.}\end{array}$$

$$= \frac{3 + 2(x + y) - (x - y)}{(x + y)(x - y)} \qquad \text{Add the fractions.}$$

$$= \frac{3 + 2x + 2y - x + y}{(x + y)(x - y)} \qquad \begin{array}{l}\text{Remove the} \\ \text{parentheses in the} \\ \text{numerator.}\end{array}$$

$$= \frac{3 + x + 3y}{(x + y)(x - y)} \qquad \begin{array}{l}\text{Combine like} \\ \text{terms.} \quad \blacksquare\end{array}$$

EXAMPLE 9 Perform the subtraction $\dfrac{a}{a - 1} - \dfrac{2}{a^2 - 1}$ and simplify.

Solution Factor $a^2 - 1$ and write each fraction as a fraction with a denominator of $(a + 1)(a - 1)$.

$$\frac{a}{a - 1} - \frac{2}{a^2 - 1} = \frac{a(a + 1)}{(a - 1)(a + 1)} - \frac{2}{(a + 1)(a - 1)} \qquad \begin{array}{l}\text{Factor } a^2 - 1 \text{ and} \\ \text{build the first} \\ \text{fraction.}\end{array}$$

$$= \frac{a(a + 1) - 2}{(a - 1)(a + 1)} \qquad \begin{array}{l}\text{Subtract the} \\ \text{fractions.}\end{array}$$

$$= \frac{a^2 + a - 2}{(a - 1)(a + 1)} \qquad \begin{array}{l}\text{Remove parentheses} \\ \text{in the numerator.}\end{array}$$

$$= \frac{(a + 2)\overset{1}{\cancel{(a - 1)}}}{\underset{1}{\cancel{(a - 1)}}(a + 1)} \qquad \text{Factor.}$$

$$= \frac{a + 2}{a + 1} \qquad \begin{array}{l}\text{Divide out the} \\ \text{common factor of} \\ a - 1. \quad \blacksquare\end{array}$$

EXAMPLE 10 Perform the addition $\dfrac{m + 1}{2m + 6} + \dfrac{4 - m^2}{2m^2 + 2m - 12}$.

Solution Factor $2m + 6$ and $2m^2 + 2m - 12$ and write each fraction with a denominator of $2(m + 3)(m - 2)$:

$$\frac{m + 1}{2m + 6} + \frac{4 - m^2}{2m^2 + 2m - 12} = \frac{m + 1}{2(m + 3)} + \frac{4 - m^2}{2(m + 3)(m - 2)}$$

$$= \frac{(m + 1)(m - 2)}{2(m + 3)(m - 2)} + \frac{4 - m^2}{2(m + 3)(m - 2)}$$

Now add the fractions by adding the numerators and keeping the common denominator, and simplify.

$$\frac{m + 1}{2m + 6} + \frac{4 - m^2}{2m^2 + 2m - 12} = \frac{(m + 1)(m - 2) + 4 - m^2}{2(m + 3)(m - 2)}$$

$$= \frac{m^2 - m - 2 + 4 - m^2}{2(m + 3)(m - 2)}$$

$$= \frac{-m + 2}{2(m + 3)(m - 2)}$$

$$= \frac{\overset{1}{-\cancel{(m - 2)}}}{2(m + 3)\underset{1}{\cancel{(m - 2)}}}$$

Factor -1 from $-m + 2$ and divide out $m - 2$.

$$= \frac{-1}{2(m + 3)} \qquad \blacksquare$$

EXERCISE 5.5

In Exercises 1–20, build each fraction into an equivalent fraction with the indicated denominator.

1. $\frac{2}{3}$; 6

2. $\frac{3}{4}$; 12

3. $\frac{25}{4}$; 20

4. $\frac{19}{21}$; 42

5. $\frac{2}{x}$; x^2

6. $\frac{3}{y}$; y^2

7. $\frac{5}{y}$; xy

8. $\frac{3}{x}$; xy

9. $\frac{8}{x}$; $x^2 y$

10. $\frac{7}{y}$; xy^2

11. $\frac{3x}{x + 1}$; $(x + 1)^2$

12. $\frac{5y}{y - 2}$; $(y - 2)^2$

13. $\frac{2y}{x}$; $x^2 + x$

14. $\frac{3x}{y}$; $y^2 - y$

15. $\frac{z}{z - 1}$; $z^2 - 1$

16. $\frac{y}{y + 2}$; $y^2 - 4$

17. $\frac{x + 2}{x - 2}$; $x^2 - 4$

18. $\frac{x - 3}{x + 3}$; $x^2 - 9$

19. $\frac{2}{x + 1}$; $x^2 + 3x + 2$

20. $\frac{3}{x - 1}$; $x^2 + x - 2$

In Exercises 21–38, several denominators are given. Find the least common denominator.

21. 15, 12

22. 18, 24

23. 14, 21, 42

24. 12, 15, 10

25. $2x$, $6x$

26. $3y$, $9y$

27. $x^2 y$, $x^2 y^2$, xy^2

28. xy, x^2, y^2

29. $3x$, $6y$, $9xy$

30. $2x^2$, $6y$, $3xy$

31. $x^2 - 1$, $x + 1$

32. $y^2 - 9$, $y - 3$

33. $x^2 + 6x$, $x + 6$, x

34. $xy^2 - xy$, xy, $y - 1$

35. $x^2 - x - 2$, $(x - 2)^2$

36. $x^2 + 2x - 3$, $(x + 3)^2$

37. $x^2 - 4x - 5$, $x^2 - 25$

38. $x^2 - x - 6$, $x^2 - 9$

In Exercises 39–82, perform the indicated operations. Simplify answers when possible.

39. $\dfrac{1}{2} + \dfrac{2}{3}$

40. $\dfrac{3}{4} + \dfrac{1}{2}$

41. $\dfrac{2}{3} - \dfrac{5}{6}$

42. $\dfrac{4}{9} - \dfrac{2}{3}$

43. $\dfrac{2y}{9} + \dfrac{y}{3}$

44. $\dfrac{3x}{8} + \dfrac{3x}{4}$

45. $\dfrac{8a}{15} - \dfrac{5a}{12}$

46. $\dfrac{2b}{15} - \dfrac{2b}{5}$

47. $\dfrac{21x}{14} - \dfrac{5x}{21}$

48. $\dfrac{7y}{6} + \dfrac{10y}{9}$

49. $\dfrac{4x}{3} + \dfrac{2x}{y}$

50. $\dfrac{2y}{5x} - \dfrac{y}{2}$

51. $\dfrac{2}{x} - 3x$

52. $14 + \dfrac{10}{y^2}$

53. $\dfrac{x^2}{2y^2} + \dfrac{x^2}{3xy}$

54. $\dfrac{r^2}{2rs} + \dfrac{r^2}{6s^2}$

55. $\dfrac{y+2}{5y} + \dfrac{y+4}{15y}$

56. $\dfrac{x+3}{x^2} + \dfrac{x+5}{2x}$

57. $\dfrac{x+5}{xy} - \dfrac{x-1}{x^2y}$

58. $\dfrac{y-7}{y^2} - \dfrac{y+7}{2y}$

59. $\dfrac{x}{x+1} + \dfrac{x-1}{x}$

60. $\dfrac{3x}{xy} + \dfrac{x+1}{y-1}$

61. $\dfrac{3}{x-2} - (x-1)$

62. $a + 1 - \dfrac{3}{a+3}$

63. $\dfrac{x-1}{x} + \dfrac{y+1}{y}$

64. $\dfrac{a+2}{b} + \dfrac{b-2}{a}$

65. $\dfrac{x}{x-2} + \dfrac{4+2x}{x^2-4}$

66. $\dfrac{y}{y+3} - \dfrac{2y-6}{y^2-9}$

67. $\dfrac{x+1}{x-1} + \dfrac{x-1}{x+1}$

68. $\dfrac{2x}{x+2} + \dfrac{x+1}{x-3}$

69. $\dfrac{2x+2}{x-2} - \dfrac{2x}{x+2}$

70. $\dfrac{y+3}{y-1} - \dfrac{y+4}{y+1}$

71. $\dfrac{x}{(x-2)^2} + \dfrac{x-4}{(x+2)(x-2)}$

72. $\dfrac{a-2}{(a+3)^2} - \dfrac{a}{a-3}$

73. $\dfrac{2x}{x^2-3x+2} + \dfrac{2x}{x-1} - \dfrac{x}{x-2}$

74. $\dfrac{4a}{a-2} - \dfrac{3a}{a-3} + \dfrac{4a}{a^2-5a+6}$

75. $\dfrac{2x}{x-1} + \dfrac{3x}{x+1} - \dfrac{x+3}{x^2-1}$

76. $\dfrac{a}{a-1} - \dfrac{2}{a+2} + \dfrac{3(a-2)}{a^2+a-2}$

77. $-2 - \dfrac{y+1}{y-3} + \dfrac{3(y-2)}{y}$

78. $\dfrac{3(a-b)}{a+b} + \dfrac{2(a+b)}{a-b} - 1$

79. $\dfrac{x+1}{2x+4} - \dfrac{x^2}{2x^2-8}$

80. $\dfrac{x+1}{x+2} - \dfrac{x^2+1}{x^2-x-6}$

81. $\dfrac{x-1}{x+2} + \dfrac{x}{3-x} + \dfrac{9x+3}{x^2-x-6}$

82. $\dfrac{x+1}{x-3} - \dfrac{x^2+9x}{x^2-2x-3} - \dfrac{5}{3-x}$

In Exercises 83–84, show that each formula is true.

83. $\dfrac{a}{b} + \dfrac{c}{d} = \dfrac{ad + bc}{bd}$

84. $\dfrac{a}{b} - \dfrac{c}{d} = \dfrac{ad - bc}{bd}$

<hr>

REVIEW EXERCISES

In Review Exercises 1–6, let $a = -2$, $b = -3$, and $c = 4$. Simplify each expression.

1. $c + ab$

2. $ac - bc$

3. $a^2 + bc$

4. $ab^2 - ac^2$

5. $\dfrac{ac + bc}{a + b + c}$

6. $\dfrac{-4a + 3b + 6c}{abc - 1}$

7. Define a prime number.

8. Define a natural number.

9. Define a composite number.

10. Define an integer.

In Review Exercises 11–12, write each statement as an algebraic expression.

11. The quotient obtained when x is divided by a number that is 3 greater than the product of x and y.

12. The product of three consecutive integers of which n is the smallest.

5.6 COMPLEX FRACTIONS

Fractions such as

$$\dfrac{\dfrac{1}{3}}{4}, \qquad \dfrac{\dfrac{5}{3}}{\dfrac{2}{9}}, \qquad \dfrac{x + \dfrac{1}{2}}{3 - x}, \qquad \text{and} \qquad \dfrac{\dfrac{x + 1}{2}}{x + \dfrac{1}{x}}$$

which contain fractions in their numerators or denominators, are called **complex fractions.** Fortunately, complex fractions can often be simplified. For example, to simplify the complex fraction

$$\dfrac{\dfrac{1}{3}}{4}$$

we multiply both the numerator and denominator of the fraction by 3 to eliminate the denominator of the fraction in the numerator, and simplify:

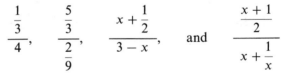

$$\dfrac{\dfrac{1}{3}}{4} = \dfrac{\dfrac{1}{3} \cdot 3}{4 \cdot 3} = \dfrac{\dfrac{1}{3} \cdot \dfrac{3}{1}}{12} = \dfrac{\dfrac{3}{3}}{12} = \dfrac{1}{12}$$

We shall discuss the following two methods for simplifying complex fractions.

> **Methods for Simplifying a Complex Fraction.**
>
> **Method 1**
> Write the numerator and denominator of the complex fraction as single fractions. Then perform the indicated division of the two fractions and simplify.
>
> **Method 2**
> Multiply both the numerator and the denominator of the complex fraction by the LCD of all of the fractions that appear in that numerator and denominator. Then simplify.

To simplify the complex fraction $\dfrac{\frac{3}{5}+1}{2-\frac{1}{5}}$, for example, we can use Method 1 and proceed as follows:

$$\frac{\dfrac{3}{5}+1}{2-\dfrac{1}{5}} = \frac{\dfrac{3}{5}+\dfrac{5}{5}}{\dfrac{10}{5}-\dfrac{1}{5}}$$ Change 1 to $\dfrac{5}{5}$ and 2 to $\dfrac{10}{5}$.

$$= \frac{\dfrac{8}{5}}{\dfrac{9}{5}}$$ Add the fractions in the numerator and subtract the fractions in the denominator.

$$= \frac{8}{5} \div \frac{9}{5}$$ Express the complex fraction as an equivalent division problem.

$$= \frac{8}{5} \cdot \frac{5}{9}$$ Invert the divisor and multiply.

$$= \frac{8 \cdot 5}{5 \cdot 9}$$ Multiply the fractions.

$$= \frac{8}{9}$$ Divide out the common factor of 5.

To use Method 2, we proceed as follows:

$$\frac{\dfrac{3}{5}+1}{2-\dfrac{1}{5}} = \frac{5\left(\dfrac{3}{5}+1\right)}{5\left(2-\dfrac{1}{5}\right)}$$ Multiply both the numerator and the denominator by 5, the LCD of $\dfrac{3}{5}$ and $\dfrac{1}{5}$.

$$= \frac{5 \cdot \dfrac{3}{5}+5 \cdot 1}{5 \cdot 2 - 5 \cdot \dfrac{1}{5}}$$ Remove parentheses.

Hypatia (370 A.D.–415 A.D.)
Hypatia is the earliest known woman in the history of mathematics. She was a professor at the University of Alexandria. Because of her scientific beliefs, she was considered to be a heretic. At the age of 45, she was attacked by a mob and murdered for the beliefs.

$$= \frac{3+5}{10-1} \qquad \text{Simplify.}$$

$$= \frac{8}{9} \qquad \text{Simplify.}$$

In this case, Method 2 is easier than Method 1. Any complex fraction can be simplified by using either method. With practice we will be able to see which method is best to use in any given situation.

EXAMPLE 1   Simplify $\dfrac{\dfrac{x}{3}}{\dfrac{y}{3}}$.

Solution

Method 1

$$\frac{\dfrac{x}{3}}{\dfrac{y}{3}} = \frac{x}{3} \div \frac{y}{3}$$

$$= \frac{x}{3} \cdot \frac{3}{y}$$

$$= \frac{3x}{3y}$$

$$= \frac{x}{y}$$

Method 2

$$\frac{\dfrac{x}{3}}{\dfrac{y}{3}} = \frac{3\left(\dfrac{x}{3}\right)}{3\left(\dfrac{y}{3}\right)}$$

$$= \frac{\dfrac{x}{1}}{\dfrac{y}{1}}$$

$$= \frac{x}{y}$$ ■

EXAMPLE 2   Simplify $\dfrac{\dfrac{x}{x+1}}{\dfrac{y}{x}}$.

Solution

Method 1

$$\frac{\dfrac{x}{x+1}}{\dfrac{y}{x}} = \frac{x}{x+1} \div \frac{y}{x}$$

$$= \frac{x}{x+1} \cdot \frac{x}{y}$$

$$= \frac{x^2}{y(x+1)}$$

Method 2

$$\frac{\dfrac{x}{x+1}}{\dfrac{y}{x}} = \frac{x(x+1)\left(\dfrac{x}{x+1}\right)}{x(x+1)\left(\dfrac{y}{x}\right)}$$

$$= \frac{\dfrac{x^2}{1}}{\dfrac{y(x+1)}{1}}$$

$$= \frac{x^2}{y(x+1)}$$ ■

EXAMPLE 3   Simplify $\dfrac{1 + \dfrac{1}{x}}{1 - \dfrac{1}{x}}$.

Solution

Method 1

$$\frac{1 + \dfrac{1}{x}}{1 - \dfrac{1}{x}} = \frac{\dfrac{x}{x} + \dfrac{1}{x}}{\dfrac{x}{x} - \dfrac{1}{x}}$$

$$= \frac{\dfrac{x + 1}{x}}{\dfrac{x - 1}{x}}$$

$$= \frac{x + 1}{x} \div \frac{x - 1}{x}$$

$$= \frac{x + 1}{x} \cdot \frac{x}{x - 1}$$

$$= \frac{x + 1}{x - 1}$$

Method 2

$$\frac{1 + \dfrac{1}{x}}{1 - \dfrac{1}{x}} = \frac{x\left(1 + \dfrac{1}{x}\right)}{x\left(1 - \dfrac{1}{x}\right)}$$

$$= \frac{x + 1}{x - 1}$$

■

EXAMPLE 4   Simplify the complex fraction $\dfrac{1}{1 + \dfrac{1}{x + 1}}$.

Solution   Use Method 2.

$$\frac{1}{1 + \dfrac{1}{x + 1}} = \frac{(x + 1)1}{(x + 1)\left(1 + \dfrac{1}{x + 1}\right)}$$

Multiply numerator and denominator by $x + 1$.

$$= \frac{x + 1}{1(x + 1) + 1}$$

Simplify.

$$= \frac{x + 1}{x + 2}$$

Simplify.

■

EXAMPLE 5   Simplify the fraction $\dfrac{x^{-1} + y^{-2}}{x^{-2} - y^{-1}}$.

Solution   Write the fraction in complex fraction form and simplify:

$$\frac{x^{-1} + y^{-2}}{x^{-2} - y^{-1}} = \frac{\dfrac{1}{x} + \dfrac{1}{y^2}}{\dfrac{1}{x^2} - \dfrac{1}{y}}$$

$$= \frac{x^2y^2\left(\dfrac{1}{x} + \dfrac{1}{y^2}\right)}{x^2y^2\left(\dfrac{1}{x^2} - \dfrac{1}{y}\right)}$$

Multiply numerator and denominator by x^2y^2.

$$= \frac{xy^2 + x^2}{y^2 - x^2y}$$

Remove parentheses.

$$= \frac{x(y^2 + x)}{y(y - x^2)}$$

Attempt to simplify the fraction by factoring the numerator and the denominator.

The result cannot be simplified.   ■

EXERCISE 5.6

In Exercises 1–34, simplify each complex fraction.

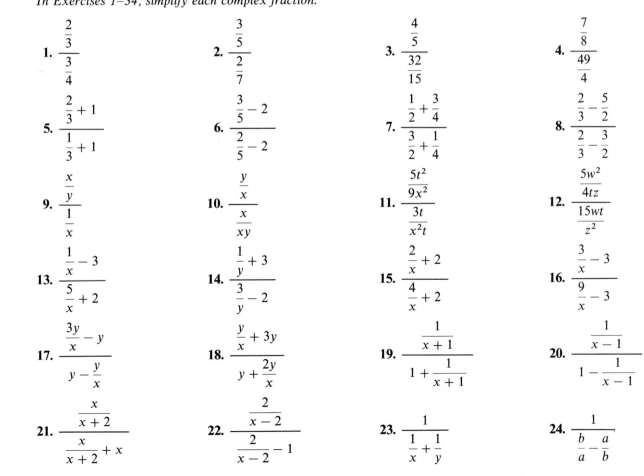

1. $\dfrac{\dfrac{2}{3}}{\dfrac{3}{4}}$

2. $\dfrac{\dfrac{3}{5}}{\dfrac{2}{7}}$

3. $\dfrac{\dfrac{4}{5}}{\dfrac{32}{15}}$

4. $\dfrac{\dfrac{7}{8}}{\dfrac{49}{4}}$

5. $\dfrac{\dfrac{2}{3} + 1}{\dfrac{1}{3} + 1}$

6. $\dfrac{\dfrac{3}{5} - 2}{\dfrac{2}{5} - 2}$

7. $\dfrac{\dfrac{1}{2} + \dfrac{3}{4}}{\dfrac{3}{2} + \dfrac{1}{4}}$

8. $\dfrac{\dfrac{2}{3} - \dfrac{5}{2}}{\dfrac{2}{3} - \dfrac{3}{2}}$

9. $\dfrac{\dfrac{x}{y}}{\dfrac{1}{x}}$

10. $\dfrac{\dfrac{y}{x}}{\dfrac{x}{xy}}$

11. $\dfrac{\dfrac{5t^2}{9x^2}}{\dfrac{3t}{x^2t}}$

12. $\dfrac{\dfrac{5w^2}{4tz}}{\dfrac{15wt}{z^2}}$

13. $\dfrac{\dfrac{1}{x} - 3}{\dfrac{5}{x} + 2}$

14. $\dfrac{\dfrac{1}{y} + 3}{\dfrac{3}{y} - 2}$

15. $\dfrac{\dfrac{2}{x} + 2}{\dfrac{4}{x} + 2}$

16. $\dfrac{\dfrac{3}{x} - 3}{\dfrac{9}{x} - 3}$

17. $\dfrac{\dfrac{3y}{x} - y}{y - \dfrac{y}{x}}$

18. $\dfrac{\dfrac{y}{x} + 3y}{y + \dfrac{2y}{x}}$

19. $\dfrac{\dfrac{1}{x+1}}{1 + \dfrac{1}{x+1}}$

20. $\dfrac{\dfrac{1}{x-1}}{1 - \dfrac{1}{x-1}}$

21. $\dfrac{\dfrac{x}{x+2}}{\dfrac{x}{x+2} + x}$

22. $\dfrac{\dfrac{2}{x-2}}{\dfrac{2}{x-2} - 1}$

23. $\dfrac{1}{\dfrac{1}{x} + \dfrac{1}{y}}$

24. $\dfrac{1}{\dfrac{b}{a} - \dfrac{a}{b}}$

25. $\dfrac{\dfrac{2}{x}}{\dfrac{2}{y} - \dfrac{4}{x}}$

26. $\dfrac{\dfrac{2y}{3}}{\dfrac{2y}{3} - \dfrac{8}{y}}$

27. $\dfrac{3 + \dfrac{3}{x-1}}{3 - \dfrac{3}{x}}$

28. $\dfrac{2 - \dfrac{2}{x+1}}{2 + \dfrac{2}{x}}$

29. $\dfrac{\dfrac{3}{x} + \dfrac{4}{x+1}}{\dfrac{2}{x+1} - \dfrac{3}{x}}$

30. $\dfrac{\dfrac{5}{y-3} - \dfrac{2}{y}}{\dfrac{1}{y} + \dfrac{2}{y-3}}$

31. $\dfrac{\dfrac{2}{x} - \dfrac{3}{x+1}}{\dfrac{2}{x+1} - \dfrac{3}{x}}$

32. $\dfrac{\dfrac{5}{y} + \dfrac{4}{y+1}}{\dfrac{4}{y} - \dfrac{5}{y+1}}$

33. $\dfrac{\dfrac{1}{y^2+y} - \dfrac{1}{xy+x}}{\dfrac{1}{xy+x} - \dfrac{1}{y^2+y}}$

34. $\dfrac{\dfrac{2}{b^2-1} - \dfrac{3}{ab-a}}{\dfrac{3}{ab-a} - \dfrac{2}{b^2-1}}$

In Exercises 35–44, write each expression without using negative exponents. Then simplify the resulting complex fraction.

35. $\dfrac{x^{-2}}{y^{-1}}$

36. $\dfrac{a^{-4}}{b^{-2}}$

37. $\dfrac{1 + x^{-1}}{x^{-1} - 1}$

38. $\dfrac{y^{-2} + 1}{y^{-2} - 1}$

39. $\dfrac{a^{-2} + a}{a + 1}$

40. $\dfrac{t - t^{-2}}{1 - t^{-1}}$

41. $\dfrac{2x^{-1} + 4x^{-2}}{2x^{-2} + x^{-1}}$

42. $\dfrac{x^{-2} - 3x^{-3}}{3x^{-2} - 9x^{-3}}$

43. $\dfrac{1 - 25y^{-2}}{1 + 10y^{-1} + 25y^{-2}}$

44. $\dfrac{1 - 9x^{-2}}{1 - 6x^{-1} + 9x^{-2}}$

REVIEW EXERCISES

In Review Exercises 1–2, identify the base and the exponent in each expression.

1. -3^4

2. $(2x)^5$

In Review Exercises 3–4, write each expression without using exponents.

3. ab^4

4. $(-3y)^5$

In Review Exercises 5–8, write each expression as an expression involving only one exponent.

5. $t^3 t^4 t^2$

6. $(a^0 a^2)^3$

7. $-2r(r^3)^2$

8. $(s^3)^2 (s^4)^0$

In Review Exercises 9–12, write each expression without using parentheses or negative exponents.

9. $\left(\dfrac{3r}{4r^3}\right)^4$

10. $\left(\dfrac{12y^{-3}}{3y^2}\right)^{-2}$

11. $\left(\dfrac{6r^{-2}}{2r^3}\right)^{-2}$

12. $\left(\dfrac{4x^3}{5x^{-3}}\right)^{-2}$

5.7 SOLVING EQUATIONS THAT CONTAIN FRACTIONS

To solve equations containing fractions, it is usually best to eliminate those fractions. To do so, we multiply both sides of the equation by the least common denominator of the fractions that appear in the equation. For example, to solve the equation $\frac{x}{3} + 1 = \frac{x}{6}$, we multiply both sides of the equation by 6:

$$\frac{x}{3} + 1 = \frac{x}{6}$$

$$6\left(\frac{x}{3} + 1\right) = 6\left(\frac{x}{6}\right)$$

We then use the distributive law to remove parentheses on the left-hand side, simplify, and solve the resulting equation for x.

$$6 \cdot \frac{x}{3} + 6 \cdot 1 = 6 \cdot \frac{x}{6}$$

$$2x + 6 = x$$

$$x + 6 = 0 \qquad \text{Add } -x \text{ to both sides.}$$

$$x = -6 \qquad \text{Add } -6 \text{ to both sides.}$$

Check: $\quad \dfrac{x}{3} + 1 = \dfrac{x}{6}$

$$\frac{-6}{3} + 1 \stackrel{?}{=} \frac{-6}{6} \qquad \text{Replace } x \text{ with } -6.$$

$$-2 + 1 \stackrel{?}{=} -1 \qquad \text{Simplify.}$$

$$-1 = -1$$

EXAMPLE 1 Solve the equation $\dfrac{4}{x} + 1 = \dfrac{6}{x}$.

Solution

$$\frac{4}{x} + 1 = \frac{6}{x}$$

$$x\left(\frac{4}{x} + 1\right) = x\left(\frac{6}{x}\right) \qquad \text{Multiply both sides by } x.$$

$$x \cdot \frac{4}{x} + x \cdot 1 = x \cdot \frac{6}{x} \qquad \text{Remove parentheses.}$$

$$4 + x = 6 \qquad \text{Simplify.}$$

$$x = 2 \qquad \text{Add } -4 \text{ to both sides.}$$

Check: $\quad \dfrac{4}{x} + 1 = \dfrac{6}{x}$

$$\frac{4}{2} + 1 \stackrel{?}{=} \frac{6}{2} \qquad \text{Replace } x \text{ with 2.}$$

$$2 + 1 \stackrel{?}{=} 3 \qquad \text{Simplify.}$$

$$3 = 3 \qquad \blacksquare$$

If we multiply both sides of an equation by an expression that involves a variable as we did in Example 1, we *must* check the apparent solutions. The next example shows why.

EXAMPLE 2   Solve the equation $\dfrac{x+3}{x-1} = \dfrac{4}{x-1}$.

Solution   Multiply both sides of the equation by $x-1$, the least common denominator of the fractions contained in the equation.

$$\frac{x+3}{x-1} = \frac{4}{x-1}$$

$$(x-1)\frac{x+3}{x-1} = (x-1)\frac{4}{x-1} \qquad \text{Multiply both sides by } x-1.$$

$$x+3 = 4 \qquad \text{Simplify.}$$

$$x = 1 \qquad \text{Add } -3 \text{ to both sides.}$$

Because both sides of the equation were multiplied by an expression containing a variable, we must check the apparent solution.

$$\frac{x+3}{x-1} = \frac{4}{x-1}$$

$$\frac{1+3}{1-1} \overset{?}{=} \frac{4}{1-1} \qquad \text{Replace } x \text{ with 1.}$$

$$\frac{4}{0} \neq \frac{4}{0} \qquad \text{Simplify.}$$

Because zeros appear in the denominators of fractions, the fractions are undefined. Thus, 1 is a false solution, and the given equation has no solutions. Such false solutions are called **extraneous solutions.** ∎

EXAMPLE 3   Solve the equation $\dfrac{3x+1}{x+1} - 2 = \dfrac{3(x-3)}{x+1}$.

Solution   Multiply both sides of the equation by $x+1$, the least common denominator of the fractions contained in the equation.

$$\frac{3x+1}{x+1} - 2 = \frac{3(x-3)}{x+1}$$

$$(x+1)\left(\frac{3x+1}{x+1} - 2\right) = (x+1)\left[\frac{3(x-3)}{x+1}\right]$$

$$3x + 1 - 2(x+1) = 3(x-3) \qquad \begin{array}{l}\text{Use the distributive property} \\ \text{to remove parentheses.}\end{array}$$

$$3x + 1 - 2x - 2 = 3x - 9 \qquad \text{Remove parentheses.}$$

$$x - 1 = 3x - 9 \qquad \text{Combine like terms.}$$

$$-2x = -8 \qquad \text{Add } -3x \text{ and 1 to both sides.}$$

$$x = 4 \qquad \text{Divide both sides by } -2.$$

Check: $\dfrac{3x + 1}{x + 1} - 2 = \dfrac{3(x - 3)}{x + 1}$

$$\dfrac{3(4) + 1}{4 + 1} - 2 \overset{?}{=} \dfrac{3(4 - 3)}{4 + 1}$$

$$\dfrac{13}{5} - \dfrac{10}{5} \overset{?}{=} \dfrac{3(1)}{5}$$

$$\dfrac{3}{5} = \dfrac{3}{5}$$

∎

EXAMPLE 4 Solve the equation $\dfrac{x + 2}{x + 3} + \dfrac{1}{x^2 + 2x - 3} = 1.$

Solution

$$\dfrac{x + 2}{x + 3} + \dfrac{1}{x^2 + 2x - 3} = 1$$

$$\dfrac{x + 2}{x + 3} + \dfrac{1}{(x + 3)(x - 1)} = 1 \qquad \text{Factor } x^2 + 2x - 3.$$

$$(x + 3)(x - 1)\left[\dfrac{x + 2}{x + 3} + \dfrac{1}{(x + 3)(x - 1)}\right] = (x + 3)(x - 1)1 \qquad \text{Multiply both sides by } (x + 3)(x - 1).$$

$$(x + 3)(x - 1)\left(\dfrac{x + 2}{x + 3}\right) + (x + 3)(x - 1)\left[\dfrac{1}{(x + 3)(x - 1)}\right] = (x + 3)(x - 1)1 \qquad \text{Remove brackets.}$$

$$(x - 1)(x + 2) + 1 = (x + 3)(x - 1) \qquad \text{Simplify.}$$

$$x^2 + x - 2 + 1 = x^2 + 2x - 3 \qquad \text{Remove parentheses.}$$

$$x - 2 + 1 = 2x - 3 \qquad \text{Subtract } x^2 \text{ from both sides.}$$

$$x - 1 = 2x - 3 \qquad \text{Combine like terms.}$$

$$-x - 1 = -3 \qquad \text{Add } -2x \text{ to both sides.}$$

$$-x = -2 \qquad \text{Add 1 to both sides.}$$

$$x = 2 \qquad \text{Divide both sides by } -1.$$

Verify that 2 is a solution of the given equation. ∎

EXAMPLE 5 Solve the equation $\dfrac{4}{5} + y = \dfrac{4y - 50}{5y - 25}.$

Solution

$$\dfrac{4}{5} + y = \dfrac{4y - 50}{5y - 25}$$

$$\dfrac{4}{5} + y = \dfrac{4y - 50}{5(y - 5)} \qquad \text{Factor } 5y - 25.$$

$$5(y - 5)\left[\dfrac{4}{5} + y\right] = 5(y - 5)\left[\dfrac{4y - 50}{5(y - 5)}\right] \qquad \text{Multiply both sides by } 5(y - 5).$$

$$4(y - 5) + 5y(y - 5) = 4y - 50 \qquad \text{Remove brackets.}$$

$$4y - 20 + 5y^2 - 25y = 4y - 50 \qquad \text{Remove parentheses.}$$

$$5y^2 - 25y - 20 = -50 \qquad \text{Add } -4y \text{ to both sides and rearrange terms.}$$

$$5y^2 - 25y + 30 = 0 \qquad \text{Add 50 to both sides.}$$
$$y^2 - 5y + 6 = 0 \qquad \text{Divide both sides by 5.}$$
$$(y - 3)(y - 2) = 0 \qquad \text{Factor } y^2 - 5y + 6.$$
$$y - 3 = 0 \quad \text{or} \quad y - 2 = 0 \qquad \text{Set each factor equal to 0.}$$
$$y = 3 \quad | \quad y = 2$$

Verify that 3 and 2 both satisfy the original equation.   ■

EXERCISE 5.7

In Exercises 1–58, solve each equation and check the solution. If an equation has no solution, so indicate.

1. $\dfrac{x}{2} + 4 = \dfrac{3x}{2}$

2. $\dfrac{y}{3} + 6 = \dfrac{4y}{3}$

3. $\dfrac{2y}{5} - 8 = \dfrac{4y}{5}$

4. $\dfrac{3x}{4} - 6 = \dfrac{x}{4}$

5. $\dfrac{x}{3} + 1 = \dfrac{x}{2}$

6. $\dfrac{x}{2} - 3 = \dfrac{x}{5}$

7. $\dfrac{x}{5} - \dfrac{x}{3} = -8$

8. $\dfrac{2}{3} + \dfrac{x}{4} = 7$

9. $\dfrac{3a}{2} + \dfrac{a}{3} = -22$

10. $\dfrac{x}{2} + x = \dfrac{9}{2}$

11. $\dfrac{x - 3}{3} + 2x = -1$

12. $\dfrac{x + 2}{2} - 3x = x + 8$

13. $\dfrac{z - 3}{2} = z + 2$

14. $\dfrac{b + 2}{3} = b - 2$

15. $\dfrac{5(x + 1)}{8} = x + 1$

16. $\dfrac{3(x - 1)}{2} + 2 = x$

17. $\dfrac{c - 4}{4} = \dfrac{c + 4}{8}$

18. $\dfrac{t + 3}{2} = \dfrac{t - 3}{3}$

19. $\dfrac{x + 1}{3} + \dfrac{x - 1}{5} = \dfrac{2}{15}$

20. $\dfrac{y - 5}{7} + \dfrac{y - 7}{5} = \dfrac{-2}{5}$

21. $\dfrac{3x - 1}{6} - \dfrac{x + 3}{2} = \dfrac{3x + 4}{3}$

22. $\dfrac{2x + 3}{3} + \dfrac{3x - 4}{6} = \dfrac{x - 2}{2}$

23. $\dfrac{3}{x} + 2 = 3$

24. $\dfrac{2}{x} + 9 = 11$

25. $\dfrac{5}{a} - \dfrac{4}{a} = 8 + \dfrac{1}{a}$

26. $\dfrac{11}{b} + \dfrac{13}{b} = 12$

27. $\dfrac{2}{y + 1} + 7 = \dfrac{12}{y + 1}$

28. $\dfrac{1}{t - 3} = \dfrac{-2}{t - 3} + 1$

29. $\dfrac{1}{x - 1} + \dfrac{3}{x - 1} = 1$

30. $\dfrac{3}{p + 6} - 2 = \dfrac{7}{p + 6}$

31. $\dfrac{a^2}{a + 2} - \dfrac{4}{a + 2} = a$

32. $\dfrac{z^2}{z + 1} + 2 = \dfrac{1}{z + 1}$

33. $\dfrac{x}{x - 5} - \dfrac{5}{x - 5} = 3$

34. $\dfrac{3}{y - 2} + 1 = \dfrac{3}{y - 2}$

35. $\dfrac{3r}{2} - \dfrac{3}{r} = \dfrac{3r}{2} + 3$

36. $\dfrac{2p}{3} - \dfrac{1}{p} = \dfrac{2p - 1}{3}$

37. $\dfrac{1}{3} + \dfrac{2}{x - 3} = 1$

38. $\dfrac{3}{5} + \dfrac{7}{x + 2} = 2$

39. $\dfrac{u}{u - 1} + \dfrac{1}{u} = \dfrac{u^2 + 1}{u^2 - u}$

40. $\dfrac{v}{v + 2} + \dfrac{1}{v - 1} = 1$

41. $\dfrac{3}{x - 2} + \dfrac{1}{x} = \dfrac{2(3x + 2)}{x^2 - 2x}$

42. $\dfrac{5}{x} + \dfrac{3}{x + 2} = \dfrac{-6}{x(x + 2)}$

43. $\dfrac{7}{q^2 - q - 2} + \dfrac{1}{q + 1} = \dfrac{3}{q - 2}$

44. $\dfrac{-5}{s^2 + s - 2} + \dfrac{3}{s + 2} = \dfrac{1}{s - 1}$

45. $\dfrac{3y}{3y - 6} + \dfrac{8}{y^2 - 4} = \dfrac{2y}{2y + 4}$

46. $\dfrac{x - 3}{4x - 4} + \dfrac{1}{9} = \dfrac{x - 5}{6x - 6}$

47. $y + \dfrac{2}{3} = \dfrac{2y - 12}{3y - 9}$

48. $y + \dfrac{3}{4} = \dfrac{3y - 50}{4y - 24}$

49. $\dfrac{5}{4y + 12} - \dfrac{3}{4} = \dfrac{5 - y^2 - 3y}{4y + 12}$

50. $\dfrac{3}{5x - 20} + \dfrac{4}{5} = \dfrac{3 + 4x - x^2}{5x - 20}$

51. $\dfrac{x}{x - 1} - \dfrac{12}{x^2 - x} = \dfrac{-1}{x - 1}$

52. $1 - \dfrac{3}{b} = \dfrac{-8b}{b^2 + 3b}$

53. $\dfrac{z - 4}{z - 3} = \dfrac{z + 2}{z + 1}$

54. $\dfrac{a + 2}{a + 8} = \dfrac{a - 3}{a - 2}$

55. $\dfrac{n}{n^2 - 9} + \dfrac{n + 8}{n + 3} = \dfrac{n - 8}{n - 3}$

56. $\dfrac{x - 3}{x - 2} - \dfrac{1}{x} = \dfrac{x - 3}{x}$

57. $\dfrac{b + 2}{b + 3} + 1 = \dfrac{-7}{b - 5}$

58. $\dfrac{x - 4}{x - 3} + \dfrac{x - 2}{x - 3} = x - 3$

REVIEW EXERCISES

In Review Exercises 1–4, factor each sum or difference of two cubes.

1. $a^3 + 125$

2. $y^3 - 27$

3. $8x^3 - 125y^6$

4. $27a^6 + 1000b^3$

In Review Exercises 5–10, factor each expression by grouping.

5. $ab + 3a + 2b + 6$

6. $yz + z^2 + y + z$

7. $mr + ms + nr + ns$

8. $ac + bc - ad - bd$

9. $2a + 2b - a^2 - ab$

10. $xa - x + ya - y$

11. Multiply: $(3rs^2 - 2)(2rs^2 + 5)$

12. Divide: $y + 3 \,\overline{)\, 2y - 4y^2 + 3y^3}$

5.8 APPLICATIONS OF EQUATIONS THAT CONTAIN FRACTIONS

Many applications involve equations that contain fractions. In this section we shall consider several of these applications.

EXAMPLE 1   In electronics the formula $\dfrac{1}{r} = \dfrac{1}{r_1} + \dfrac{1}{r_2}$ is used to calculate the combined resistance r of two resistors wired in parallel. If the combined resistance r is 4 ohms and the resistance of one of the resistors r_1 is 12 ohms, find the resistance r_2 of the other resistor.

Solution   Substitute 4 for r and 12 for r_1 in the formula and solve for r_2.

$$\frac{1}{r} = \frac{1}{r_1} + \frac{1}{r_2}$$

$$\frac{1}{4} = \frac{1}{12} + \frac{1}{r_2}$$

$$12r_2\left(\frac{1}{4}\right) = 12r_2\left(\frac{1}{12} + \frac{1}{r_2}\right) \qquad \text{Multiply both sides by } 12r_2.$$

$$3r_2 = r_2 + 12 \qquad \text{Remove parentheses and simplify.}$$

$$2r_2 = 12 \qquad \text{Add } -r_2 \text{ to both sides.}$$

$$r_2 = 6 \qquad \text{Divide both sides by 2.}$$

The resistance of the second resistor is 6 ohms.   ∎

EXAMPLE 2   Solve the formula $\dfrac{1}{r} = \dfrac{1}{r_1} + \dfrac{1}{r_2}$ for r.

Solution   Proceed as follows:

$$\frac{1}{r} = \frac{1}{r_1} + \frac{1}{r_2}$$

$$rr_1r_2\left(\frac{1}{r}\right) = rr_1r_2\left(\frac{1}{r_1} + \frac{1}{r_2}\right) \qquad \text{Multiply both sides by } rr_1r_2.$$

$$\frac{rr_1r_2}{r} = \frac{rr_1r_2}{r_1} + \frac{rr_1r_2}{r_2} \qquad \text{Remove parentheses.}$$

$$r_1r_2 = rr_2 + rr_1 \qquad \text{Simplify.}$$

$$r_1r_2 = r(r_2 + r_1) \qquad \text{Factor out an } r.$$

$$\frac{r_1r_2}{r_2 + r_1} = r \qquad \text{Divide both sides by } r_2 + r_1.$$

or

$$r = \frac{r_1r_2}{r_2 + r_1}$$ ∎

EXAMPLE 3   If the same number is added to both the numerator and denominator of the fraction $\frac{3}{5}$, the result is $\frac{4}{5}$. Find the number.

Solution   Let n represent the number, add n to both the numerator and denominator of $\frac{3}{5}$, and set the result equal to $\frac{4}{5}$. Then solve the equation for n.

$$\frac{3+n}{5+n} = \frac{4}{5}$$

$$5(5+n)\left(\frac{3+n}{5+n}\right) = 5(5+n)\left(\frac{4}{5}\right) \qquad \text{Multiply both sides by } 5(5+n).$$

$$5(3+n) = (5+n)4 \qquad\qquad \text{Simplify.}$$

$$15 + 5n = 20 + 4n \qquad\qquad \text{Remove parentheses.}$$

$$5n = 5 + 4n \qquad\qquad\quad \text{Add } -15 \text{ to both sides.}$$

$$n = 5 \qquad\qquad\qquad\;\; \text{Add } -4n \text{ to both sides.}$$

The number is 5.

Check: Add 5 to both the numerator and denominator of $\frac{3}{5}$ and get

$$\frac{3+5}{5+5} = \frac{8}{10} = \frac{4}{5} \qquad\qquad\qquad\qquad\qquad\qquad\qquad \blacksquare$$

EXAMPLE 4   An inlet pipe can fill an oil tank in 7 days, and a second inlet pipe can fill the same tank in 9 days. If both pipes are used, how long will it take to fill the tank?

Analysis   The key in this shared-work problem is to note what each inlet pipe can do in 1 day. If you add what the first inlet pipe can do in 1 day to what the second inlet pipe can do in 1 day, the sum is what they can do together in 1 day. Since the first inlet pipe can fill the tank in 7 days, it can do $\frac{1}{7}$ of the job in 1 day. Since it takes the second inlet pipe 9 days, it can do $\frac{1}{9}$ of the job in 1 day. Since it takes x days for both inlet pipes to fill the tank, together they can do $\frac{1}{x}$ of the job in 1 day.

Solution   Let x represent the number of days it will take to fill the tank if both inlet pipes are used. Then form the equation

What the first inlet pipe can do in 1 day	$+$	**what the second inlet pipe can do in 1 day**	$=$	**what they can do together in 1 day.**
$\dfrac{1}{7}$	$+$	$\dfrac{1}{9}$	$=$	$\dfrac{1}{x}$

$$63x\left(\frac{1}{7} + \frac{1}{9}\right) = 63x\left(\frac{1}{x}\right) \qquad \text{Multiply both sides by } 63x.$$

$$9x + 7x = 63 \qquad\qquad\qquad \text{Remove parentheses and simplify.}$$

$$16x = 63 \qquad\qquad\qquad\quad \text{Combine terms.}$$

$$x = \frac{63}{16} \qquad\qquad\qquad\quad\; \text{Divide both sides by 16.}$$

It will take $\frac{63}{16}$ or $3\frac{15}{16}$ days for both inlet pipes to fill the tank.

Check: In $\frac{63}{16}$ days, the first inlet pipe does $\frac{1}{7}(\frac{63}{16})$ of the total job and the second inlet pipe does $\frac{1}{9}(\frac{63}{16})$ of the total job. The sum of these efforts, $\frac{9}{16} + \frac{7}{16}$, is equal to one complete job.   ■

EXAMPLE 5   Tom can jog 10 miles in the same amount of time that his wife Gail can jog 12 miles. If Gail can run 1 mile per hour faster than Tom, how fast can Gail run?

Analysis   This is a uniform motion problem that is based on the formula $d = rt$, where d is the distance traveled, r is the rate, and t is the time. If we solve the formula for t, we obtain

$$t = \frac{d}{r}$$

Since Tom can run 10 miles at some unknown rate of r, it will take him $\frac{10}{r}$ hours. Since Gail can run 12 miles at a rate of $r + 1$ miles per hour, it will take her $\frac{12}{r+1}$ hours. As before, organize the information of the problem in chart form as in Figure 5-1.

	d	$=$	r	$\cdot$	t
Gail	12		$r + 1$		$\dfrac{12}{r+1}$
Tom	10		r		$\dfrac{10}{r}$

FIGURE 5-1

Because the times are given to be equal, we know that $\frac{12}{r+1} = \frac{10}{r}$.

Solution   Let r be the rate that Tom can run.
Then $r + 1$ is the rate that Gail can run.

We can form the equation

The time it takes Gail to run 12 miles	$=$	the time it takes Tom to run 10 miles.

$$\frac{12}{r+1} = \frac{10}{r}$$

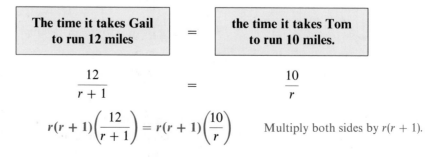

$$r(r+1)\left(\frac{12}{r+1}\right) = r(r+1)\left(\frac{10}{r}\right) \qquad \text{Multiply both sides by } r(r+1).$$

$$12r = 10(r + 1) \qquad \text{Simplify.}$$
$$12r = 10r + 10 \qquad \text{Remove parentheses.}$$
$$2r = 10 \qquad \text{Add } -10r \text{ to both sides.}$$
$$r = 5 \qquad \text{Divide both sides by 2.}$$

Thus, Tom can run 5 miles per hour. Since Gail can run 1 mile per hour faster, she can run 6 miles per hour. Verify that these results check.   ■

EXAMPLE 6 At a bank, a sum of money invested for one year will earn $96 interest. If invested in bonds, that same money would earn $108. If the interest rate paid by the bonds is 1% greater than that paid by the bank, find the bank's rate.

Analysis The interest paid by either investment is the product of the principal (the amount invested) and the interest rate. If you let r represent the bank's rate of interest, then $r + 0.01$ represents the rate paid by the bonds. For each investment, the principal is the interest paid divided by the interest rate. See Figure 5-2.

	Interest	=	Principal	·	Rate
Bank	96		$\dfrac{96}{r}$		r
Bonds	108		$\dfrac{108}{r + 0.01}$		$r + 0.01$

FIGURE 5-2

Because the principal that would be invested in either account is the same, we can set up and solve the following equation:

$$\frac{96}{r} = \frac{108}{r + 0.01}$$

Solution Solve the equation as follows:

$$\frac{96}{r} = \frac{108}{r + 0.01}$$

$$r(r + 0.01)\left(\frac{96}{r}\right) = r(r + 0.01)\left(\frac{108}{r + 0.01}\right) \qquad \begin{array}{l}\text{Multiply both sides by} \\ r(r + 0.01).\end{array}$$

$$96(r + 0.01) = 108r$$

$$96r + 0.96 = 108r \qquad \text{Remove parentheses.}$$

$$0.96 = 12r \qquad \text{Subtract } 96r \text{ from both sides.}$$

$$0.08 = r \qquad \text{Divide both sides by 12.}$$

The bank's interest rate is 0.08, or 8%. The bonds pay 9% interest, a rate greater than that paid by the bank. Verify that these rates check.

EXERCISE 5.8

1. In the equation

$$\frac{1}{a} + \frac{1}{b} = 1$$

$a = 10$. Find b.

2. In the equation

$$\frac{1}{a} + \frac{1}{b} = 1$$

$b = 7$. Find a.

3. Solve the formula in Exercise 1 for a.

4. Solve the formula in Exercise 1 for b.

5. In optics, the focal length f of a lens is given by the formula

$$\frac{1}{f} = \frac{1}{d_1} + \frac{1}{d_2}$$

where d_1 is the distance from the object to the lens and d_2 is the distance from the lens to the image. Find d_2 if $f = 8$ meters and $d_1 = 12$ meters.

6. Solve the formula in Exercise 5 for f.

7. Solve the formula in Exercise 5 for d_1.

8. Solve the formula in Exercise 5 for d_2.

9. If the denominator of the fraction $\frac{3}{4}$ is increased by a number and the numerator of the fraction is doubled, the result is 1. Find the number.

10. If a number is added to the numerator of the fraction $\frac{7}{8}$ and the same number is subtracted from the denominator, the result is 2. Find the number.

11. If a number is added to the numerator of the fraction $\frac{3}{4}$ and twice as much is added to the denominator, the result is $\frac{4}{7}$. Find the number.

12. If a number is added to the numerator of the fraction $\frac{5}{7}$ and twice as much is subtracted from the denominator, the result is 8. Find the number.

13. The sum of a number and its reciprocal is $\frac{13}{6}$. Find the numbers.

14. The sum of the reciprocals of two consecutive even integers is $\frac{7}{24}$. Find the integers.

15. An inlet pipe can fill an empty swimming pool in 5 hours, and another inlet pipe can fill the pool in 4 hours. How long will it take both pipes to fill the pool?

16. One inlet pipe can fill an empty pool in 4 hours, and a drain can empty the pool in 8 hours. How long will it take the pipe to fill the pool if the drain is left open?

17. A homeowner estimates it will take 7 days to roof his house. A professional roofer estimates that he could roof the house in 4 days. How long will it take if the homeowner helps the roofer?

18. A pond is filled by two inlet pipes. One pipe can fill the pond in 15 days, and the other pipe can fill the pond in 21 days. However, evaporation can empty the pond in 36 days. How long will it take the two inlet pipes to fill an empty pond?

Juan can bicycle 28 miles in the same time he can walk 8 miles. If he can ride 10 miles per hour faster than he an walk, how fast can he walk?

plane can fly 300 miles in the same time it takes a car to go 120 miles. If the car travels 90 miles per hour er than the plane, how fast is the plane?

 t that can travel 18 miles per hour in still water can travel 33 miles downstream in the same amount of t it can travel 21 miles upstream. Find the speed of the current in the stream.

22. A plane can fly 300 miles downwind in the same amount of time it can travel 210 miles upwind. Find the velocity of the wind if the plane can fly at 255 miles per hour in still air.

23. Two certificates of deposit pay interest at rates that differ by 1%. Money invested for one year in the first CD earns $175 interest. The same principal invested in the other CD earns $200. Find the two rates of interest.

24. Two bond funds pay interest at rates that differ by 2%. Money invested for one year in the first fund earns $315 interest. The same amount invested in the other fund earns $385. Find the lower rate of interest.

25. The office workers bought a $35 gift for their boss. If there had been two more employees to contribute, everyone's cost would have been $2 less. How many workers contributed to the gift?

26. A dealer bought some radios for a total of $1200. She gave away 6 radios as gifts, sold each of the rest for $10 more than she paid for each radio, and broke even. How many radios did she buy?

27. A college bookstore can purchase several calculators for a total cost of $120. If each calculator cost $1 less, the bookstore could purchase 10 additional calculators at the same total cost. How many calculators can be purchased at the regular price?

28. A repairman purchased several furnace-blower motors for a total cost of $210. If his cost per motor had been $5 less, he could have purchased 1 additional motor. How many motors did he buy at the regular rate?

REVIEW EXERCISES

In Review Exercises 1–8, solve each equation.

1. $x^2 + 5x + 6 = 0$

2. $x^2 - 25 = 0$

3. $(t + 2)(t^2 + 7t + 12) = 0$

4. $2(y - 4) = -y^2$

5. $y^3 - y^2 = 0$

6. $5a^3 - 125a = 0$

7. $(x^2 - 1)(x^2 - 4) = 0$

8. $6t^3 + 35t^2 = 6t$

9. A room containing 168 square feet is 2 feet longer than it is wide. Find its perimeter.

10. The base of a triangle is 4 meters longer than twice its height. Its area is 48 square meters. Find the height and the length of the base of the triangle.

11. On a trip from Edens to Grandville, Tina averaged 30 miles per hour. She returned at 50 miles per hour, and the round trip took her 8 hours. How far is Grandville from Edens?

12. On the expressway Rick could average 60 miles per hour, but on the back roads he could average only 45 miles per hour. If his trip of 180 miles took $3\frac{1}{2}$ hours, how far did he travel on the expressway?

5.9 RATIO AND PROPORTION

An indicated quotient of two numbers is often called a **ratio.**

> **DEFINITION.** A **ratio** is the comparison of two numbers by cated quotient.

The previous definition implies that a ratio is a fraction. Some examples of ratios are

$$\frac{7}{8}, \qquad \frac{21}{24}, \qquad \text{and} \qquad \frac{117}{223}$$

The fraction $\frac{7}{8}$ can be read as "the ratio of 7 to 8," the fraction $\frac{21}{24}$ can be read as "the ratio of 21 to 24," and the ratio $\frac{117}{223}$ can be read as "the ratio of 117 to 223." Because the fractions $\frac{7}{8}$ and $\frac{21}{24}$ represent equal numbers, they are called **equal ratios.**

EXAMPLE 1   Express each phrase as a ratio in lowest terms:
   a. the ratio of 15 to 12   **b.** the ratio of 3 inches to 7 inches
   c. the ratio of 2 feet to 1 yard   **d.** the ratio of 6 ounces to 1 pound

Solution   **a.** The ratio of 15 to 12 can be written as the fraction $\frac{15}{12}$. Expressed in lowest terms, it is $\frac{5}{4}$.

   b. The ratio of 3 inches to 7 inches can be written as the fraction $\frac{3\text{ inches}}{7\text{ inches}}$, or just $\frac{3}{7}$.

   c. The ratio of 2 feet to 1 yard should not be written as the fraction $\frac{2}{1}$. To express the ratio as a pure number, you must use the same units. Because there are 3 feet in 1 yard, the proper ratio is $\frac{2\text{ feet}}{3\text{ feet}}$, or just $\frac{2}{3}$.

   d. The ratio 6 ounces to 1 pound should not be written as the fraction $\frac{6}{1}$. To express the ratio as a pure number, you must use the same units. Because there are 16 ounces in 1 pound, the proper ratio is $\frac{6\text{ ounces}}{16\text{ ounces}}$, which simplifies to $\frac{3}{8}$.   ∎

DEFINITION.   A **proportion** is a statement indicating that two ratios are equal.

Some examples of proportions are

$$\frac{1}{2} = \frac{3}{6}, \qquad \frac{3}{7} = \frac{9}{21}, \qquad \text{and} \qquad \frac{8}{1} = \frac{40}{5}$$

The proportion $\frac{1}{2} = \frac{3}{6}$ can be read as "1 is to 2 as 3 is to 6," the proportion $\frac{3}{7} = \frac{9}{21}$ can be read as "3 is to 7 as 9 is to 21," and the proportion $\frac{8}{1} = \frac{40}{5}$ can be read as "8 is to 1 as 40 is to 5."

In the proportion $\frac{1}{2} = \frac{3}{6}$, the numbers 1 and 6 are called the **extremes** of the proportion, and the numbers 2 and 3 are called the **means.** If we find the product of the extremes and the product of the means in this proportion, we see that the products are equal:

$$1 \cdot 6 = 6 \qquad \text{and} \qquad 3 \cdot 2 = 6$$

This is not a coincidence. To show that this is always true, we consider the proportion

$$\frac{a}{b} = \frac{c}{d} \qquad (b \neq 0, d \neq 0)$$

where a and d are the extremes and b and c are the means. Then

$$\frac{a}{b} = \frac{c}{d}$$

$$bd\,\frac{a}{b} = bd\,\frac{c}{d} \qquad \text{Multiply both sides by } bd.$$

$$ad = bc \qquad \text{Simplify.}$$

and the product of the extremes, ad, is equal to the product of the means, bc. Thus, we have this important theorem.

> **Theorem.** In any proportion, the product of the extremes is equal to the product of the means.

EXAMPLE 2 Determine if the equation $\dfrac{x}{3y} = \dfrac{xy + 3x}{3y^2 + 9y}$ is a proportion.

Solution Check to see if the product of the extremes is equal to the product of the means.

$$x(3y^2 + 9y) = 3xy^2 + 9xy$$
$$3y(xy + 3x) = 3xy^2 + 9xy$$

Because the products are equal, the equation is a proportion. ∎

Solving Proportions

EXAMPLE 3 Solve the proportion $\dfrac{12}{18} = \dfrac{3}{x}$ for x.

Solution Proceed as follows:

$$\frac{12}{18} = \frac{3}{x}$$

$$12 \cdot x = 3 \cdot 18 \qquad \text{The product of the extremes equals the product of the means.}$$

$$12x = 54 \qquad \text{Simplify.}$$

$$x = \frac{54}{12} \qquad \text{Divide both sides by 12.}$$

$$x = \frac{9}{2} \qquad \text{Simplify.}$$

Thus, x represents the fraction $\frac{9}{2}$. ∎

EXAMPLE 4   Solve the proportion $\dfrac{y + 1}{y + 7} = \dfrac{2}{y + 2}$ for y.

Solution

$$\frac{y + 1}{y + 7} = \frac{2}{y + 2}$$

$(y + 1)(y + 2) = 2(y + 7)$ The product of the extremes is equal to the product of the means.

$y^2 + 3y + 2 = 2y + 14$ Remove parentheses.

$y^2 + y - 12 = 0$ Add $-2y$ and -14 to both sides.

$(y - 3)(y + 4) = 0$ Factor.

$y - 3 = 0$ or $y + 4 = 0$ Set each factor equal to 0.

$y = 3$ | $y = -4$

The proportion has two solutions: $y = 3$ and $y = -4$.   ■

EXAMPLE 5   If 5 tomatoes cost \$1.15, how much will 16 tomatoes cost?

Solution   Let c represent the cost of 16 tomatoes. The ratio of the numbers of tomatoes is the same as the ratio of their costs. Express this relationship as a proportion and find c.

$$\frac{5}{16} = \frac{1.15}{c}$$

$5c = 1.15(16)$ The product of the extremes is equal to the product of the means.

$5c = 18.4$ Do the multiplication.

$c = \dfrac{18.4}{5}$ Divide both sides by 5.

$c = 3.68$ Simplify.

Sixteen tomatoes will cost \$3.68.   ■

Similar Triangles

If two triangles have the same shape, they are said to be **similar triangles.** The following theorem points out an important fact about similar triangles.

> **Theorem.**   If two triangles are similar, then all pairs of corresponding sides are in proportion.

This theorem often enables us to measure sides of triangles indirectly. For example, on a sunny day we can find the height of a tree and stay safely on the ground.

EXAMPLE 6 On a sunny day, a large tree casts a shadow of 24 feet at the same time a vertical yardstick casts a shadow of 2 feet. Find the height of the tree.

Solution Refer to Figure 5-3, which shows the triangles determined by the tree and its shadow, and the yardstick and its shadow. Because the triangles have the same shape, they are similar, and the measures of their corresponding sides are in proportion. If we let h represent the height of the tree, we can find h by solving the following proportion: the height of the tree is to the height of the yardstick as the length of the shadow of the tree is to the length of the shadow of the yardstick.

$$\frac{h}{3} = \frac{24}{2}$$

$2h = 3(24)$ The product of the extremes is equal to the product of the means.

$2h = 72$ Simplify.

$h = 36$ Divide both sides by 2.

The tree is 36 feet tall.

 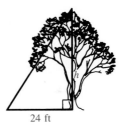

3 ft

2 ft

24 ft

FIGURE 5-3 ∎

EXERCISE 5.9

In Exercises 1–20, express each phrase as a ratio in lowest terms.

1. 5 to 7

2. 3 to 5

3. 17 to 34

4. 19 to 38

5. 22 to 33

6. 14 to 21

7. 4 ounces to 12 ounces

8. 3 inches to 15 inches

9. 12 minutes to 1 hour

10. 8 ounces to 1 pound

11. 3 days to 1 week

12. 2 quarts to 1 gallon

13. 4 inches to 2 yards

14. 1 mile to 5280 feet

15. 3 pints to 2 quarts

16. 4 dimes to 8 pennies

17. 6 nickels to 1 quarter

18. 3 people to 12 people

19. 3 meters to 12 centimeters

20. 3 dollars to 3 quarters

In Exercises 21–34, tell whether each statement is a proportion.

21. $\dfrac{9}{7} = \dfrac{81}{70}$

22. $\dfrac{5}{2} = \dfrac{20}{8}$

23. $\dfrac{-7}{3} = \dfrac{14}{-6}$

24. $\dfrac{13}{-19} = \dfrac{-65}{95}$

25. $\dfrac{9}{19} = \dfrac{38}{80}$

26. $\dfrac{40}{29} = \dfrac{29}{22}$

27. $\dfrac{x^2}{y} = \dfrac{x}{y^2}$

28. $\dfrac{x^2 y}{x^2 z} = \dfrac{y}{z}$

29. $\dfrac{3x^2 y}{3xy^2} = \dfrac{x}{y}$

30. $\dfrac{5y}{25x} = \dfrac{5x}{y}$

31. $\dfrac{x+2}{x(x+2)} = \dfrac{1}{x}$

32. $\dfrac{x+y}{x-2} = \dfrac{y}{2}$

33. $\dfrac{xy+x}{xy} = \dfrac{y+1}{y}$

34. $\dfrac{x^2+x}{x+1} = \dfrac{x}{1}$

In Exercises 35–60, solve for the variable in each proportion.

35. $\dfrac{2}{3} = \dfrac{x}{6}$

36. $\dfrac{3}{6} = \dfrac{x}{8}$

37. $\dfrac{5}{10} = \dfrac{3}{c}$

38. $\dfrac{7}{14} = \dfrac{2}{b}$

39. $\dfrac{-6}{x} = \dfrac{8}{4}$

40. $\dfrac{4}{x} = \dfrac{2}{8}$

41. $\dfrac{x}{3} = \dfrac{9}{3}$

42. $\dfrac{x}{2} = \dfrac{-18}{6}$

43. $\dfrac{x+1}{5} = \dfrac{3}{15}$

44. $\dfrac{x-1}{7} = \dfrac{2}{21}$

45. $\dfrac{x+3}{12} = \dfrac{-7}{6}$

46. $\dfrac{x+7}{-4} = \dfrac{3}{12}$

47. $\dfrac{4-x}{13} = \dfrac{11}{26}$

48. $\dfrac{5-x}{17} = \dfrac{13}{34}$

49. $\dfrac{2x+1}{9} = \dfrac{x}{27}$

50. $\dfrac{3x-2}{7} = \dfrac{x}{28}$

51. $\dfrac{3(x+5)}{2} = \dfrac{5(x-2)}{3}$

52. $\dfrac{7(x+6)}{6} = \dfrac{6(x+3)}{5}$

53. $\dfrac{2(x+3)}{3} = \dfrac{4(x-4)}{5}$

54. $\dfrac{x+4}{5} = \dfrac{3(x-2)}{3}$

55. $\dfrac{1}{x+3} = \dfrac{-2x}{x+5}$

56. $\dfrac{x-1}{x+1} = \dfrac{2}{3x}$

57. $\dfrac{2}{x+6} = \dfrac{-2x}{5}$

58. $\dfrac{x-3}{x-2} = \dfrac{x+3}{2x}$

59. $\dfrac{x+1}{x} = \dfrac{10}{2x}$

60. $\dfrac{2x}{x+4} = \dfrac{5x+2}{18}$

In Exercises 61–80, set up and solve the required proportion.

61. Three pints of yogurt cost \$1. How much will 51 pints cost?

62. Sport shirts are on sale at two for \$25. How much will 5 shirts cost?

63. Garden seeds are on sale at three packets for 50 cents. How much will 39 packets cost?

64. A recipe for spaghetti sauce requires four 16-ounce bottles of ketchup to make two gallons of sauce. How many bottles of ketchup are needed to make 10 gallons of sauce?

65. A car gets 42 miles per gallon of gas. How much gas is needed to drive 315 miles?

66. A truck gets 12 miles per gallon of gas. How far can the truck go on 17 gallons of gas?

67. Bill earns \$412 for a 40-hour week. Last week he missed 10 hours of work. How much did he get paid?

68. An HO-scale model railroad engine is 9 inches long. The HO scale is 87 feet to 1 foot. How long is a real engine?

69. An N-scale model railroad caboose is 3.5 inches long. The N scale is 169 feet to 1 foot. How long is a real caboose?

70. Standard doll house scale is 1 inch to 1 foot. Heidi's doll house is 32 inches wide. How wide would it be if it were a real house?

71. A school board has determined that there should be 3 teachers for every 50 students. How many teachers are needed for an enrollment of 2700 students?

72. In a scale drawing, a 280-foot antenna tower is drawn 7 inches high. The building next to it is drawn 2 inches high. How tall is the actual building?

73. The instructions on a can of oil intended to be added to lawn mower gasoline read:

Recommended	Gasoline	Oil
50 to 1	6 gal	16 oz

Are these instructions correct? (*Hint:* There are 128 ounces in 1 gallon.)

74. A tree casts a shadow of 26 feet at the same time a 6-foot man casts a shadow of 4 feet. Find the height of the tree.

75. A man places a mirror on the ground and sees the reflection of the top of a flagpole, as in Illustration 1. The two triangles in the illustration are similar. Find the height, *h*, of the flagpole.

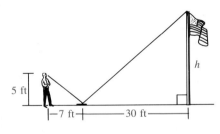

ILLUSTRATION 1

76. Use the dimensions in Illustration 2 to find *w*, the width of the river. The two triangles in the illustration are similar.

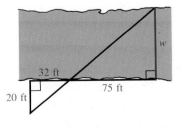

ILLUSTRATION 2

77. An airplane ascends 100 feet as it flies a horizontal distance of 1000 feet. How much altitude will it gain as it flies a horizontal distance of 1 mile? (*Hint:* 5280 feet = 1 mile.)

78. An airplane descends 1350 feet as it flies a horizontal distance of 1 mile. How much altitude is lost as it flies a horizontal distance of 5 miles?

79. A $\frac{1}{2}$-mile-long ski course falls 100 feet in every 300 feet of horizontal run. Find the height of the hill.

80. A mountain road ascends 750 feet in every 2500 feet of travel. By how much will the road rise in a trip of 10 miles?

REVIEW EXERCISES

In Review Exercises 1–4, find the perimeter or circumference of each figure.

1. A square with sides measuring 6 inches.

2. A rectangle with sides measuring 2 by 7 centimeters.

3. A trapezoid with sides measuring 6, 5, 4, and 7 centimeters.

4. A circle with a diameter of 8 feet.

In Review Exercises 5–8, find the area of each figure.

5. A square with sides measuring 6 inches.

6. A rectangle with sides measuring 2 by 7 centimeters.

7. A trapezoid with a height of 5 centimeters and bases measuring 6 and 7 centimeters.

8. A circle with a diameter of 8 feet.

In Review Exercises 9–12, find the volume of each solid.

9. A cylinder with a height of 5 meters and a circular base with a diameter of 10 meters.

10. A sphere with a diameter of 28 meters.

11. A cone with a height of 64 centimeters and a base 42 centimeters in diameter.

12. A pyramid 12 feet high and with a square base with a perimeter of 24 feet.

CHAPTER SUMMARY

Key Words

arithmetic fraction (5.1)
common denominator (5.4)
complex fractions (5.6)
denominator (5.1)
divisor (5.3)
extraneous solutions (5.7)
extremes of a proportion (5.9)

fraction (5.1)
least common denominator
  (LCD) (5.5)
lowest terms (5.1)
means of a proportion (5.9)
numerator (5.1)

proportion (5.9)
ratio (5.9)
rational expression (5.1)
rational number (5.1)
similar triangles (5.9)
simplifying a fraction (5.1)

Key Ideas

(5.1) If b and c are not equal to zero, then $\dfrac{a}{b} = \dfrac{a \cdot c}{b \cdot c}$.

$\dfrac{a}{1} = a, \qquad \dfrac{a}{0}$ is not defined.

(5.2) $\dfrac{a}{b} \cdot \dfrac{c}{d} = \dfrac{a \cdot c}{b \cdot d}$

(5.3) $\dfrac{a}{b} \div \dfrac{c}{d} = \dfrac{a}{b} \cdot \dfrac{d}{c}$

(5.4) $\dfrac{a}{d} + \dfrac{b}{d} = \dfrac{a+b}{d}$ $\qquad$ $\dfrac{a}{d} - \dfrac{b}{d} = \dfrac{a-b}{d}$

If $d \neq 0$, then $\dfrac{0}{d} = 0$.

(5.5) To add or subtract fractions with unlike denominators, first find the least common denominator of those fractions. Then express each of the fractions in equivalent form with the same common denominator. Finally, add or subtract the fractions.

(5.6) To simplify a complex fraction, use either of these methods:

1. Write the numerator and the denominator of the complex fraction as single fractions; then simplify.

2. Multiply both the numerator and the denominator of the complex fraction by the LCD of the fractions that appear in the numerator and the denominator; then simplify.

(5.7) To solve an equation that contains fractions, transform it into another equation without fractions. Do so by multiplying both sides by the LCD of the fractions. Check all solutions.

(5.9) In any proportion, the product of the extremes is equal to the product of the means.

The measures of corresponding sides of similar triangles are in proportion.

CHAPTER 5 REVIEW EXERCISES

(5.1) *In Review Exercises 1–14, write each fraction in lowest terms. If a fraction is already in lowest terms, so indicate.*

1. $\dfrac{10}{25}$
2. $\dfrac{-12}{18}$
3. $\dfrac{-51}{153}$
4. $\dfrac{105}{45}$

5. $\dfrac{3x^2}{6x^3}$
6. $\dfrac{5xy^2}{2x^2y^2}$
7. $\dfrac{x^2}{x^2+x}$
8. $\dfrac{x+2}{x^2+2x}$

9. $\dfrac{6xy}{3xy}$
10. $\dfrac{8x^2y}{2x(4xy)}$
11. $\dfrac{x^2+4x+3}{x^2-4x-5}$
12. $\dfrac{x^2-x-56}{x^2-5x-24}$

13. $\dfrac{2x^2-16x}{2x^2-18x+16}$
14. $\dfrac{x^2+x-2}{x^2-x-2}$

(5.2) *In Review Exercises 15–18, perform each multiplication and simplify.*

15. $\dfrac{3xy}{2x} \cdot \dfrac{4x}{2y^2}$
16. $\dfrac{3x}{x^2-x} \cdot \dfrac{2x-2}{x^2}$

17. $\dfrac{x^2+3x+2}{x^2+2x} \cdot \dfrac{x}{x+1}$
18. $\dfrac{x^3-y^3}{x^2+xy+y^2} \cdot \dfrac{x}{x^2-y^2}$

(5.3) *In Review Exercises 19–22, perform the indicated division and simplify.*

19. $\dfrac{3x^2}{5x^2y} \div \dfrac{6x}{15xy^2}$
20. $\dfrac{x^2+5x}{x^2+4x-5} \div \dfrac{x^2}{x-1}$

21. $\dfrac{x^2-x-6}{2x-1} \div \dfrac{x^2-2x-3}{2x^2+x-1}$
22. $\dfrac{x^3+125}{x+5} \div \dfrac{x^2-5x+25}{x^2+10x+25}$

(5.4–5.5) *In Review Exercises 23–28, several denominators are given. Find the least common denominator.*

23. $4, 8$
24. $35, 14$
25. $3x^2y, xy^2$
26. $2x+1, 2x^2+x$

27. $x+2, x-3$
28. x^2+4x+3, x^2+x

In Review Exercises 29–36, perform the indicated operation. Simplify all answers.

29. $\dfrac{x}{x+y} + \dfrac{y}{x+y}$

30. $\dfrac{3x}{x-7} - \dfrac{x-2}{x-7}$

31. $\dfrac{x}{x-1} + \dfrac{1}{x}$

32. $\dfrac{1}{7} - \dfrac{1}{x}$

33. $\dfrac{3}{x+1} - \dfrac{2}{x}$

34. $\dfrac{x+2}{2x} - \dfrac{2-x}{x^2}$

35. $\dfrac{x}{x+2} + \dfrac{3}{x} - \dfrac{4}{x^2+2x}$

36. $\dfrac{2}{x-1} - \dfrac{3}{x+1} + \dfrac{x-5}{x^2-1}$

(5.6) *In Review Exercises 37–42, simplify each complex fraction.*

37. $\dfrac{\dfrac{3}{2}}{\dfrac{2}{3}}$

38. $\dfrac{\dfrac{3}{2}+1}{\dfrac{2}{3}+1}$

39. $\dfrac{\dfrac{1}{x}+1}{\dfrac{1}{x}-1}$

40. $\dfrac{1+\dfrac{3}{x}}{2-\dfrac{1}{x^2}}$

41. $\dfrac{\dfrac{2}{x-1}+\dfrac{x-1}{x+1}}{\dfrac{1}{x^2-1}}$

42. $\dfrac{\dfrac{a}{b}+c}{\dfrac{b}{a}+c}$

(5.7) *In Review Exercises 43–48, solve each equation. Check all answers.*

43. $\dfrac{3}{x} = \dfrac{2}{x-1}$

44. $\dfrac{5}{x+4} = \dfrac{3}{x+2}$

45. $\dfrac{2}{3x} + \dfrac{1}{x} = \dfrac{5}{9}$

46. $\dfrac{2x}{x+4} = \dfrac{3}{x-1}$

47. $\dfrac{2}{x-1} + \dfrac{3}{x+4} = \dfrac{-5}{x^2+3x-4}$

48. $\dfrac{4}{x+2} - \dfrac{3}{x+3} = \dfrac{6}{x^2+5x+6}$

(5.8) *In Review Exercises 49–52, solve each problem.*

49. Solve the equation $\dfrac{1}{x} - \dfrac{1}{y} = 1$ for x.

50. If Luiz can paint a house in 14 days and Desi can paint the house in 10 days, how long will it take if they work together?

51. Tony can bicycle 30 miles in the same time he can jog 10 miles. If he can ride 10 miles per hour faster than he can jog, how fast can he jog?

52. A plane can fly 400 miles downwind in the same amount of time it can travel 320 miles upwind. Find the velocity of the wind if the plane can fly at 360 miles per hour in still air.

(5.9) *In Review Exercises 53–56, write each ratio as a fraction in lowest terms.*

53. 3 to 6

54. $12x$ to $15x$

55. 2 feet to 1 yard

56. 5 pints to 3 quarts

In Review Exercises 57–60, solve each proportion.

57. $\dfrac{3}{x} = \dfrac{6}{9}$

58. $\dfrac{x}{3} = \dfrac{x}{5}$

59. $\dfrac{x-2}{5} = \dfrac{x}{7}$

60. $\dfrac{x+1}{4} = \dfrac{3}{x}$

In Review Exercises 61–66, set up and solve a proportion.

61. If 5 tons of iron ore yields 3 tons of pig iron, how much iron ore is needed to make 18 tons of pig iron?

62. On a certain map, 1 inch represents 60 miles. If two cities are 3.5 inches apart on the map, what is the distance between the cities?

63. A pharmacist mixes 3 grams of medicine with 300 milliliters of sugar syrup. Find how much medicine is in a single dose of 5 milliliters.

64. Ten feet of copper wire weighs 1.2 pounds. How long is a 564-pound roll of this wire?

65. How tall is a building that casts a shadow of 53 feet at the same time that a 5-foot woman casts a shadow of 2 feet?

66. A plane gains 1350 feet in altitude as it travels a horizontal distance of 10,000 feet. How much will it gain in altitude as it flies a horizontal distance of 5 miles?

CHAPTER 5 TEST

1. Simplify: $\dfrac{48x^2y}{54xy^2}$

2. Simplify: $\dfrac{2x^2 - x - 3}{4x^2 - 9}$

3. Simplify: $\dfrac{3(x + 2) - 3}{2x - 4 - (x - 5)}$

4. Multiply and simplify: $\dfrac{12x^2y}{15xyz} \cdot \dfrac{25y^2z}{16xt}$

5. Multiply and simplify: $\dfrac{x^2 + 3x + 2}{3x + 9} \cdot \dfrac{x + 3}{x^2 - 4}$

6. Divide and simplify: $\dfrac{8x^2y}{25xt} \div \dfrac{16x^2y^3}{30xyt^3}$

7. Divide and simplify: $\dfrac{x^2 - x}{3x^2 + 6x} \div \dfrac{3x - 3}{3x^3 + 6x^2}$

8. Simplify: $\dfrac{x^2 + xy}{x - y} \cdot \dfrac{x^2 - y^2}{x^2 - 2x} \div \dfrac{x^2 + 2xy + y^2}{x^2 - 4}$

9. Add: $\dfrac{5x - 4}{x - 1} + \dfrac{5x + 3}{x - 1}$

_____   **10.** Subtract: $\dfrac{3y + 7}{2y + 3} - \dfrac{3(y - 2)}{2y + 3}$

_____   **11.** Add: $\dfrac{x + 1}{x} + \dfrac{x - 1}{x + 1}$

_____   **12.** Subtract: $\dfrac{5x}{x - 2} - 3$

_____   **13.** Simplify: $\dfrac{\dfrac{8x^2}{xy^3}}{\dfrac{4y^3}{x^2y^3}}$

_____   **14.** Simplify: $\dfrac{1 + \dfrac{y}{x}}{\dfrac{y}{x} - 1}$

_____   **15.** Solve for x: $\dfrac{x}{10} - \dfrac{1}{2} = \dfrac{x}{5}$

_____   **16.** Solve for x: $3x - \dfrac{2(x + 3)}{3} = 16 - \dfrac{x + 2}{2}$

_____   **17.** Solve for x: $\dfrac{7}{x + 4} - \dfrac{1}{2} = \dfrac{3}{x + 4}$

_____   **18.** Express as a ratio in lowest terms: 6 feet to 3 yards.

_____   **19.** Is the equation $\dfrac{3xy}{5xy} = \dfrac{3xt}{5xt}$ a proportion?

_____   **20.** Solve the proportion for y: $\dfrac{y}{y - 1} = \dfrac{y - 2}{y}$

_____   **21.** If George could pick up all the trash on a strip of highway in 9 hours and Maria could pick up the trash in only 7 hours, how long will it take them if they work together?

_____   **22.** A boat can motor 28 miles downstream in the same amount of time it can motor 18 miles upstream. Find the speed of the current if the boat can motor at 23 miles per hour in still water.

_____   **23.** A plane drops 575 feet as it flies a horizontal distance of $\frac{1}{2}$ mile. How much altitude will it lose as it flies a horizontal distance of 7 miles?

6

Graphing
Linear Equations
and Inequalities

Most equations encountered thus far have involved only one variable. However, many applications of mathematics involve equations that contain two or more variables. In this chapter we shall discuss equations and inequalities that contain two variables.

6.1   GRAPHING LINEAR EQUATIONS

The equation $3x + 2 = 5$ contains the single variable x, and its only solution is 1. This solution can be graphed (or plotted) on the number line as in Figure 6-1.

An equation such as $x + 2y = 5$, however, contains the two variables x and y. The solutions of such equations are pairs of numbers. For example, the pair of numbers $x = 1$ and $y = 2$ is a solution, because the equation is satisfied if we substitute 1 for x and 2 for y:

$$x + 2y = 5$$
$$\mathbf{1} + 1(\mathbf{2}) = 5 \qquad \text{Substitute 1 for } x \text{ and 2 for } y.$$
$$1 + 4 = 5$$
$$5 = 5$$

The pair of numbers $x = 5$ and $y = 0$ is also a solution, because this pair also satisfies the equation:

$$x + 2y = 5$$
$$\mathbf{5} + 2(\mathbf{0}) = 5 \qquad \text{Substitute 5 for } x \text{ and 0 for } y.$$
$$5 + 0 = 5$$
$$5 = 5$$

Solutions of equations that contain two variables can be plotted on a **rectangular coordinate system,** sometimes called a **Cartesian coordinate system** after the seventeenth-century French mathematician René Descartes. The rectangular coordinate system consists of two number lines, called the **x-axis** and the **y-axis,** drawn at right angles to each other as in Figure 6-2**a**. The two axes intersect at a point called the **origin,** which is the 0 point on each axis. The positive direction

René Descartes (1596–1650) Descartes is famous for his work in philosophy as well as for his work in mathematics. His philosophy is expressed in the words "I think, therefore I am." He is best known in mathematics for his invention of a coordinate system and his work with conic sections.

FIGURE 6-1

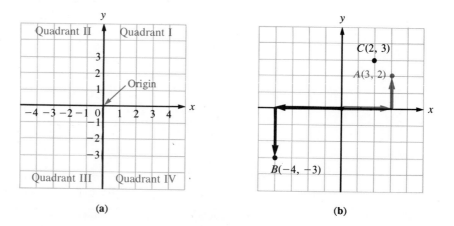

(a)   (b)

FIGURE 6-2

on the *x*-axis is to the right, and the positive direction on the *y*-axis is upward. The two axes divide the plane into four regions, called **quadrants,** which are numbered in a counterclockwise direction as shown in Figure 6-2**a**.

To plot the pair of numbers *x* = 3 and *y* = 2, we start at the origin. Because 3 is the value of *x*, we move 3 units to the right along the *x*-axis as in Figure 6-2**b**. Because 2 is the corresponding value of *y*, we then move 2 units upward in the positive *y* direction. This locates point *A* in the figure. Point *A* has an **x-coordinate** or **abscissa** of 3, and a **y-coordinate** or **ordinate** of 2. This information is denoted concisely by the pair of numbers (3, 2), called the **coordinates of point *A*.** Point *A* is the **graph of the pair** (3, 2).

To plot the pair of numbers (−4, −3), we start at the origin, move 4 units to the left along the *x*-axis, and then move 3 units down. This locates point *B* in the figure. Point *C* has coordinates of (2, 3).

The order of the coordinates of a point is important. Point *A* in Figure 6-2**b** with coordinates of (3, 2) is not the same as point *C* with coordinates (2, 3). For this reason, the pair of coordinates (*x*, *y*) of a point is often called an **ordered pair.** The *x*-coordinate is the first number in the ordered pair, and the *y*-coordinate is the second number.

EXAMPLE 1   On a rectangular coordinate system, plot the points **a.** *A*(−1, 2), **b.** *B*(0, 0), **c.** *C*(5, 0), **d.** *D*(−$\frac{5}{2}$, −3), and **e.** *E*(3, −2).

Solution   **a.** The point *A*(−1, 2) has an *x*-coordinate of −1 and a *y*-coordinate of 2. To plot point *A*, start at the origin and move 1 unit to the *left* and then 2 units *up*. (See Figure 6-3.) Point *A* lies in quadrant II.

b. To plot point *B*(0, 0), start at the origin and move 0 units to the *right* and 0 units *up*. Because there is no movement, point *B* is the origin.

c. To plot point *C*(5, 0), start at the origin and move 5 units to the *right* and 0 units *up*. Point *C* does not lie in any quadrant. It lies on the *x*-axis, 5 units to the right of the origin.

d. To plot point *D*(−$\frac{5}{2}$, −3), start at the origin and move $\frac{5}{2}$ units ($\frac{5}{2}$ units is $2\frac{1}{2}$ units) to the *left* and then 3 units *down*. Point *D* lies in quadrant III.

e. To plot point *E*(3, −2), start at the origin and move 3 units to the *right* and then 2 units *down*. Point *E* lies in quadrant IV.

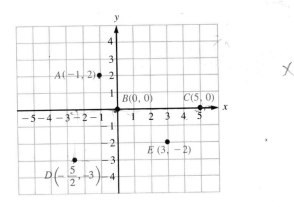

FIGURE 6-3

Graphing Linear Equations

We have seen that the solutions of an equation in the two variables x and y are pairs of numbers. These solutions can be written as ordered pairs. For example, to find some ordered pairs (x, y) that satisfy the equation $y = 5 - x$, we can pick numbers at random, substitute these numbers for x, and calculate the corresponding values of y. If we pick $x = 1$, we have

$$y = 5 - x$$
$$y = 5 - 1 \qquad \text{Substitute 1 for } x.$$
$$y = 4$$

The ordered pair (1, 4) satisfies the equation.
   If we let $x = 2$, we have

$$y = 5 - x$$
$$y = 5 - 2 \qquad \text{Substitute 2 for } x.$$
$$y = 3$$

A second solution of the equation is (2, 3).
   If we let $x = 5$, we have

$$y = 5 - x$$
$$y = 5 - 5 \qquad \text{Substitute 5 for } x.$$
$$y = 0$$

A third solution of the equation is (5, 0).
   If we let $x = -1$, we have

$$y = 5 - x$$
$$y = 5 - (-1) \qquad \text{Substitute } -1 \text{ for } x.$$
$$y = 6$$

A fourth solution is $(-1, 6)$.
   As a final example, if we let $x = 6$, we have

$$y = 5 - x$$
$$y = 5 - 6 \qquad \text{Substitute 6 for } x.$$
$$y = -1$$

A fifth solution is $(6, -1)$. The graphs of the ordered pairs (1, 4), (2, 3), (5, 0), $(-1, 6)$, and $(6, -1)$ appear in Figure 6-4.

The five points lie on the line that also appears in Figure 6-4. This line, called the **graph** of the equation $y = 5 - x$, passes through many other points than the five that we have plotted. The coordinates of *every* point on this line determine a solution of the equation $y = 5 - x$. For example, the line passes through the point (4, 1), and the pair $x = 4$ and $y = 1$ is a solution of the equation because these numbers satisfy the equation:

$$y = 5 - x$$
$$1 = 5 - 4 \qquad \text{Substitute 4 for } x \text{ and 1 for } y.$$
$$1 = 1$$

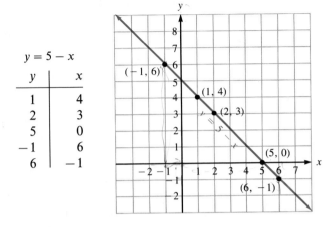

$y = 5 - x$

y	x
1	4
2	3
5	0
−1	6
6	−1

FIGURE 6-4

An equation, such as $y = 5 - x$, whose graph is a line is called a **linear equation in two variables.** Any point on that line has coordinates that satisfy the equation of the line, and the graph of any pair (x, y) that satisfies the equation is a point on the line.

Although only two points are needed to determine the graph of a linear equation, it is wise to plot a third point as a check. If the three points do not lie on a line, then at least one of the points is in error. Follow these steps to graph a linear equation.

> **Procedure for Graphing Linear Equations in the Variables x and y.**
>
> **1.** Determine two pairs (x, y) that satisfy the equation. To do so, pick arbitrary numbers for x and then solve the equation for the corresponding values of y. A third point provides a check.
> **2.** Plot each resulting pair (x, y) on a rectangular coordinate system. If they do not appear to be on a line, check your calculations.
> **3.** Draw the line that passes through the points.

EXAMPLE 2   Graph the equation $y = 3x - 4$.

Solution   Substitute numbers for x and find the corresponding values of y. For example, let $x = 1$ and determine y.

$$y = 3x - 4$$
$$y = 3(1) - 4 \qquad \text{Substitute 1 for } x.$$
$$y = 3 - 4$$
$$y = -1$$

Thus, the pair $(1, -1)$ is a solution. To find another solution, let $x = 2$ and determine y.

$$y = 3x - 4$$
$$y = 3(2) - 4 \qquad \text{Substitute 2 for } x.$$
$$y = 6 - 4$$
$$y = 2$$

A second solution is $(2, 2)$. To find a third solution, let $x = 3$ and determine y.

$$y = 3x - 4$$
$$y = 3(3) - 4 \qquad \text{Substitute 3 for } x.$$
$$y = 9 - 4$$
$$y = 5$$

A third solution is $(3,5)$. Plot these points as in Figure 6-5 and join them with a line. This line is the graph of the equation $y = 3x - 4$.

$y = 3x - 4$

x	y
1	-1
2	2
3	5

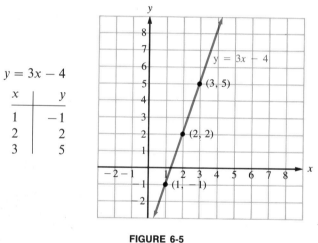

FIGURE 6-5   ■

★**EXAMPLE 3**   Graph the equation $y - 4 = \dfrac{1}{2}(x - 8)$.

Solution   It is easier to find pairs (x, y) that satisfy this equation if we remove parentheses and solve the equation for y.

$$y - 4 = \frac{1}{2}(x - 8)$$

$$y - 4 = \frac{1}{2}x - 4 \qquad \text{Use the distributive property to remove parentheses.}$$

$$y = \frac{1}{2}x \qquad \text{Add 4 to both sides.}$$

If $x = 0$, then $y = 0$. Thus, the pair $(0, 0)$ is a solution of the equation. If $x = 2$, then $y = 1$. Thus, the pair $(2, 1)$ is a second solution. Finally, if $x = -4$, then $y = -2$, and $(-4, -2)$ is a third solution. The graph of the equation appears in Figure 6-6.

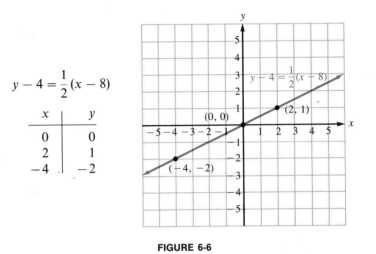

$$y - 4 = \frac{1}{2}(x - 8)$$

x	y
0	0
2	1
-4	-2

FIGURE 6-6   ■

The Intercept Method of Graphing a Line

The y-coordinate of the point at which a line crosses the y-axis is called the **y-intercept** of the line. If we substitute the number 0 for x, the value determined for y is the y-intercept. Similarly, the **x-intercept** of a line is the x-coordinate of the point at which the line crosses the x-axis. If we substitute the number 0 for y, the number determined for x is the x-intercept. Plotting these two points and drawing a line through them is called the **intercept method of graphing a line.**

This method is useful in graphing linear equations that are written in **general form.**

> **General Form of the Equation of a Line.**   If A, B, and C are real numbers and A and B are not both 0, then the equation
>
> $$Ax + By = C$$
>
> is called the **general form** of the equation of a line.

Whenever possible, we will write the general form $Ax + By = C$ so that A, B, and C are integers and $A \geq 0$.

EXAMPLE 4   Use the intercept method to graph the equation $3x + 2y = 6$.

Solution   To find the y-intercept, let $x = 0$ and solve for y.

$$3x + 2y = 6$$
$$3(0) + 2y = 6 \qquad \text{Substitute 0 for } x.$$
$$2y = 6 \qquad \text{Simplify.}$$
$$y = 3 \qquad \text{Divide both sides by 2.}$$

The y-intercept is 3, and the pair $(0, 3)$ is a solution of the equation.
To find the x-intercept, let $y = 0$ and solve for x.

$$3x + 2y = 6$$
$$3x + 2(0) = 6 \qquad \text{Substitute 0 for } y.$$
$$3x = 6 \qquad \text{Simplify.}$$
$$x = 2 \qquad \text{Divide both sides by 3.}$$

The x-intercept is 2, and the pair $(2, 0)$ is a solution of the equation.
As a check, plot one other point. Choose some value of x, like $x = 4$, and find the corresponding value of y:

$$3x + 2y = 6$$
$$3(4) + 2y = 6 \qquad \text{Substitute 4 for } x.$$
$$12 + 2y = 6 \qquad \text{Simplify.}$$
$$2y = -6 \qquad \text{Add } -12 \text{ to both sides.}$$
$$y = -3 \qquad \text{Divide both sides by 2.}$$

Thus, the point $(4, -3)$ also lies on the graph of the line. Plot these three points and join them with a line. The graph of the equation $3x + 2y = 6$ appears in Figure 6-7.

$3x + 2y = 6$

x	y
0	3
2	0
4	−3

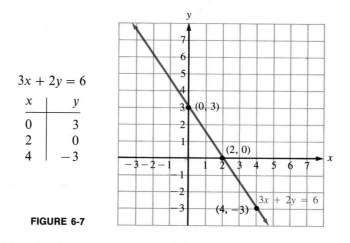

FIGURE 6-7

Graphing Lines Parallel to the *x*- and *y*-Axes

EXAMPLE 5   Graph the equations   **a.** $y = 3$ and   **b.** $x = -2$.

Solution   **a.** Write the equation $y = 3$ in general form as $0x + y = 3$. Because the coefficient of x is 0, the numbers assigned to x have no effect on y. The value of y is always 3. For example, if $x = -3$, we replace x in the equation $0x + y = 3$ with -3 and determine y:

$$0x + y = 3$$
$$0(-3) + y = 3$$
$$0 + y = 3$$
$$y = 3$$

A table of values in Figure 6-8**a** gives several possible ordered pairs that satisfy the equation $y = 3$. After plotting these pairs (x, y) and joining them with a line, note that the graph of the equation $y = 3$ is a horizontal line, parallel to the x-axis and intersecting the y-axis at 3. Note also that the y-intercept of the line is 3 and that the line has no x-intercept.

b. Write the equation $x = -2$ in general form as $x + 0y = -2$. Because the coefficient of y is 0, y can take on any value; that is, as long as $x = -2$, y can be any number. A table of values and the graph appear in Figure 6-8**b**.

The graph of $x = -2$ is a vertical line, parallel to the y-axis, which intersects the x-axis at -2. Note that the x-intercept of the line is -2, and that the line has no y-intercept.

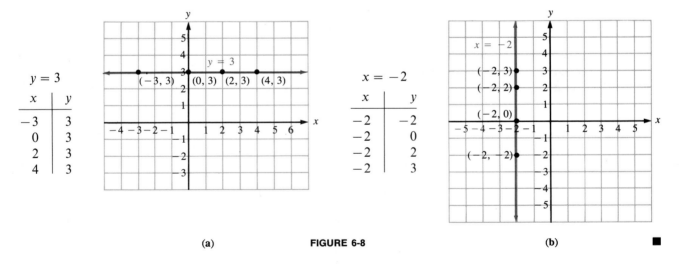

(a)   **FIGURE 6-8**   (b)   ∎

From the results of Example 5, we can conclude the following facts.

Equations of Lines Parallel to the Coordinate Axes.

The equation $x = a$ represents a vertical line that intersects the x-axis at a. If $a = 0$, the line is the y-axis.

The equation $y = b$ represents a horizontal line that intersects the y-axis at b. If $b = 0$, the line is the x-axis.

EXERCISE 6.1

In Exercises 1–8, plot each point on a rectangular coordinate system. Indicate in which quadrant each point lies.

1. $A(2, 5)$

2. $B(5, 2)$

3. $C(-3, 1)$

4. $D(1, -3)$

5. $E(-2, -3)$

6. $F(-3, -2)$

7. $G(3, -2)$

8. $H(-4, 5)$

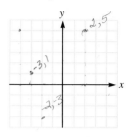

In Exercises 9–16, plot each point on a rectangular coordinate system. Indicate in which quadrant each point lies.

9. $A(-3, 5)$

10. $B(-5, 3)$

11. $C(3, -5)$

12. $D(5, -3)$

13. $E\left(-\dfrac{3}{2}, -4\right)$

14. $F\left(-5, \dfrac{9}{2}\right)$

15. $G\left(\dfrac{5}{2}, \dfrac{7}{2}\right)$

16. $H\left(\dfrac{7}{2}, -\dfrac{7}{2}\right)$

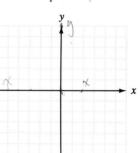

In Exercises 17–24, plot each point on a rectangular coordinate system. Indicate on which axis each point lies.

17. $A(0, 6)$

18. $B(0, -2)$

19. $C(2, 0)$

20. $D(-3, 0)$

21. $E(-5, 0)$

22. $F(0, -5)$

23. $G(0, 0)$

24. $H(-6, 0)$

In Exercises 25–32, refer to Illustration 1 and determine the coordinates of each point.

25. A 2,3

26. B

27. C -2, -3

28. D

29. E 0,0

30. F

31. G -5, -5

32. H

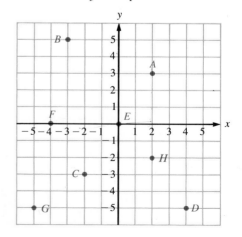

ILLUSTRATION 1

In Exercises 33–44, complete the table of solutions for each equation, and then graph the equation.

33. $y = x + 2$

x	y
3	5
1	3
−2	0

34. $y = x - 3$

x	y
3	0
1	−2
−3	−6

35. $y = x - 4$

x	y
5	1
4	0
−1	−5

36. $y = x + 1$

x	y
5	6
1	2
−1	0

37. $y = -2x$

x	y
2	−4
1	−2
−3	+6

38. $y = 3x$

x	y
2	6
0	0
−2	−6

39. $y = \dfrac{x}{2}$

x	y
1	$\frac{1}{2}$
−1	$-\frac{1}{2}$
−4	−2

40. $y = -\dfrac{x}{3}$

x	y
1	$-\frac{1}{3}$
−1	$\frac{1}{3}$
−3	1

41. $y = 2x - 1$

x	y
3	5
−1	−3
−2	−5

42. $y = 3x + 1$

x	y
−2	−5
0	1
1	4

43. $y = \dfrac{x}{2} - 2$

x	y
8	2
0	−2
−2	−3

44. $y = \dfrac{x}{3} - 3$

x	y
6	−1
0	−3
−3	−4

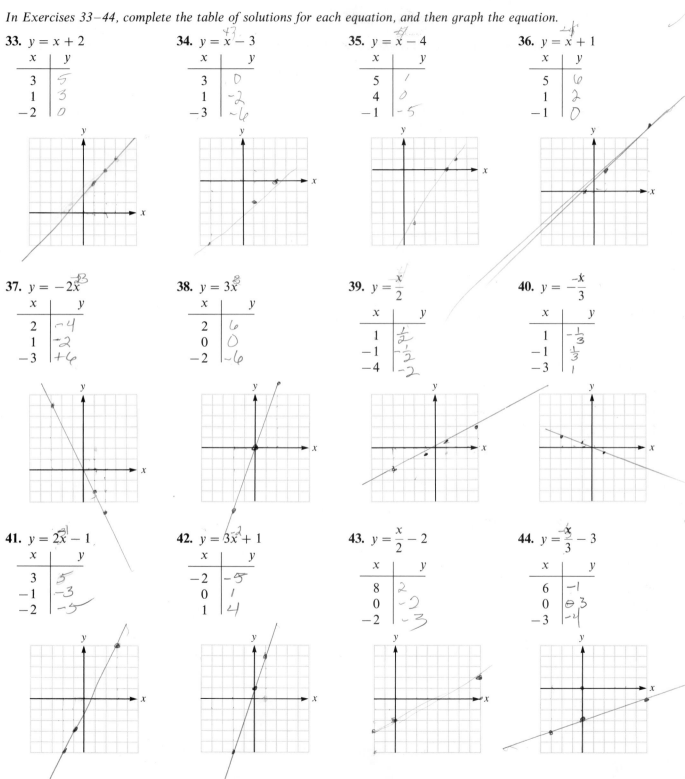

In Exercises 45–60, write each equation in general form if necessary. Then graph it by using the intercept method.

45. $x + y = 7$

46. $x + y = -2$

47. $x - y = 7$

48. $x - y = -2$

49. $2x + y = 5$

50. $3x + y = -1$

51. $2x + 3y = 12$

52. $3x - 2y = 6$

53. $3x + 12 = 4y$

54. $2x + 12 = 9y$

55. $5x + 10 = -2y$

56. $8 - 3y = 2x$

57. $2(x + 2) - y = 4$

58. $3(y + 1) - x = 4$

59. $3(2 - y) + 2(x + 1) = 4$

60. $2(x + 2) - (y - 2) = 4$

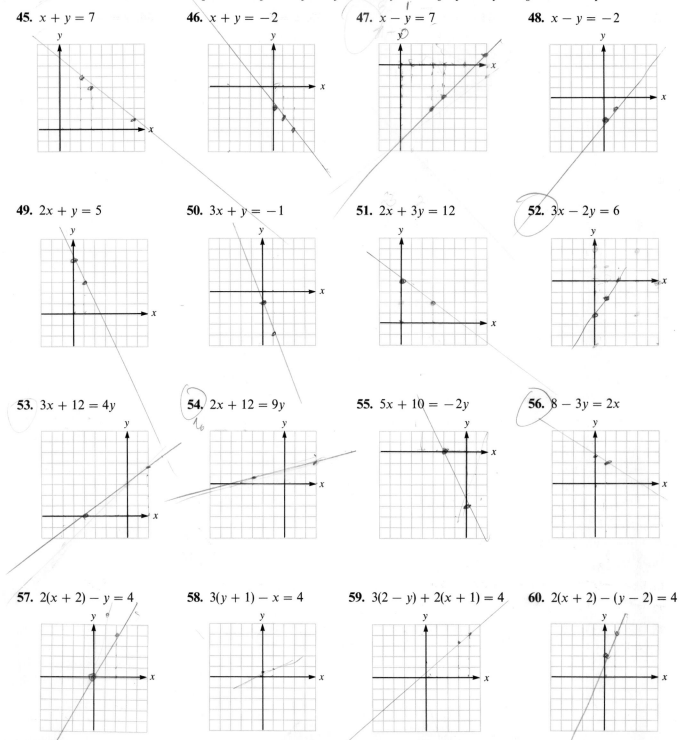

In Exercises 61–72, graph each equation. You may have to simplify the equation first.

61. $y = -5$ **62.** $x = 4$ **63.** $x = 5$ **64.** $y = -5$

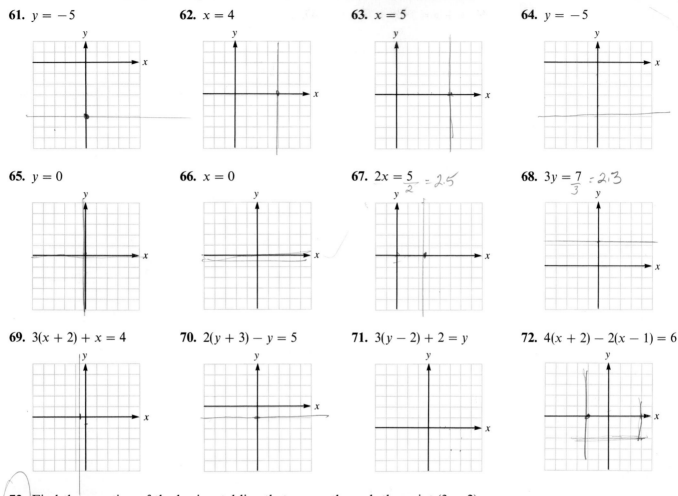

65. $y = 0$ **66.** $x = 0$ **67.** $2x = \frac{5}{2} = 2.5$ **68.** $3y = \frac{7}{3} = 2.3$

69. $3(x + 2) + x = 4$ **70.** $2(y + 3) - y = 5$ **71.** $3(y - 2) + 2 = y$ **72.** $4(x + 2) - 2(x - 1) = 6$

73. Find the equation of the horizontal line that passes through the point $(3, -2)$.

74. Find the equation of the vertical line that passes through the point $(3, -2)$.

75. Find the equation of the x-axis.

76. Find the equation of the y-axis.

In Exercises 77–78, assume that Carlos charges $12 per hour in his part-time job tutoring mathematics.

77. Find Carlos's fee for working 1 hour, 2 hours, 3 hours, and 5 hours.

78. Let y represent Carlos's fee for working x hours. Plot the pairs (x, y) calculated in Exercise 77. Do the points lie on a line?

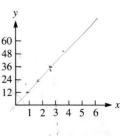

In Exercises 79–80, assume that Wendy charges $10 for materials and $40 per hour for installing computer systems.

79. Find Wendy's fee for working 1 hour, 2 hours, 3 hours, and $5\frac{1}{2}$ hours.

80. Let y represent Wendy's total fee for working x hours. Plot the pairs (x, y) that were calculated in Exercise 79. Do the points lie on a line?

In Exercises 81–82, assume that a person's maximum heart rate for safe aerobic exercise is approximately 220 minus that person's age.

81. Find the maximum heart rate for persons 20 years old, 40 years old, and 60 years old.

82. Let y represent the maximum heart rate for a person x years old. Write an equation that relates x and y, and graph it.

If points $P(a, b)$ and $Q(c, d)$ are two points on a rectangular coordinate system and point M is midway between them, then point M is called the **midpoint** of the line segment joining P and Q. (See Illustration 2.) To find the coordinates of the midpoint M of a line segment PQ, find the average of the x-coordinates and the average of the y-coordinates of P and Q:

$$x\text{-coordinate of } M = \frac{a + c}{2} \quad \text{and} \quad y\text{-coordinate of } M = \frac{b + d}{2}$$

In Exercises 83–88, find the coordinates of the midpoint of the line segment with the given endpoints.

83. $P(5, 3)$ and $Q(7, 9)$

84. $R(5, 6)$ and $S(7, 10)$

85. $R(2, -7)$ and $S(-3, 12)$

86. $P(-8, 12)$ and $Q(3, -9)$

87. $A(4, 6)$ and $B(10, 6)$

88. $A(8, -6)$ and the origin

ILLUSTRATION 2

REVIEW EXERCISES

In Review Exercises 1–2, write each expression without using absolute value symbols.

1. $|-8 - (-3)|$

2. $-|-(-3)|$

In Review Exercises 3–4, let $x = -6$, $y = -3$, and $z = 4$. Evaluate each expression.

3. $\dfrac{2z^2 + 2x}{x + z}$

4. $\dfrac{z^2 - 2yz + y^2}{z - x + y}$

In Review Exercises 5–6, solve each equation.

5. $\dfrac{5(y - 2)}{3} = -(6 + y)$

6. $7(y + 14) = 35(y - 2)$

7. Solve the equation $A = \frac{1}{2}bh$ for h.

8. Solve the equation $S = 180(n - 2)$ for n.

9. The sum of three consecutive even integers is 72. Find the integers.

10. The product of two consecutive integers is 7 greater than the square of the smaller. Find the integers.

In Review Exercises 11–12, perform each division.

11. $\dfrac{8x^2y^3 + 12x^3y^2}{6x^2y^2}$

12. $2x + 4 \overline{)\, 6x^2 + 8x - 8}$

6.2   THE SLOPE OF A LINE

We have seen that two points can be used to graph a line. We can also graph a line if we know the coordinates of only one point and also know the slant or the steepness of the line. A measure of this slant is called the **slope** of the line. The slope indicates how rapidly the line rises or falls.

In Figure 6-9, a line passes through the points $P(1, 2)$ and $Q(3, 6)$. Moving along the line from P to Q causes the value of y to change from $y = 2$ to $y = 6$, an increase of $6 - 2$, or 4 units. In that same move, the value of x increases $3 - 1$, or 2 units. The slope of the line is the ratio of the change in y to the change in x. Thus,

$$\text{The slope of line } PQ = \frac{\text{change in the } y \text{ values}}{\text{change in the } x \text{ values}}$$

$$= \frac{6 - 2}{3 - 1}$$

$$= \frac{4}{2}$$

$$= 2$$

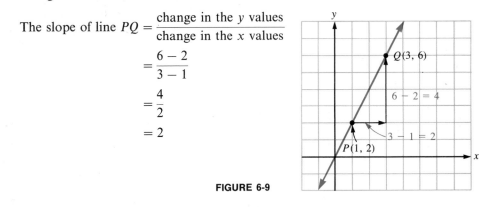

FIGURE 6-9

In general, if we know the coordinates of any two points on a nonvertical line, we can calculate the slope of the line. In Figure 6-10, a line passes through points P and Q. To distinguish between the coordinates of these two points, we use **subscript notation.** In the figure, point P is denoted as $P(x_1, y_1)$, read as "point P with coordinates of x sub 1 and y sub 1." The second point Q is denoted as $Q(x_2, y_2)$, read as "point Q with coordinates of x sub 2 and y sub 2."

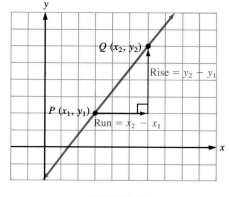

FIGURE 6-10

As a point on the line in Figure 6-10 moves from P to Q, its y-coordinate changes by the amount $y_2 - y_1$, while its x-coordinate changes by $x_2 - x_1$. The slope of the line is the change in y divided by the change in x.

Thus, we have the following definition.

DEFINITION. If $P(x_1, y_1)$ and $Q(x_2, y_2)$ are two points on a nonvertical line, the slope m of line PQ is given by the formula

$$m = \frac{y_2 - y_1}{x_2 - x_1}$$

In Figure 6-10, the difference in the y-coordinates of points P and Q is the vertical distance the line *rises* between points P and Q. The difference of the x-coordinates is the *run*—the horizontal distance between points P and Q. Thus, the slope of a nonvertical line is the ratio of its *rise* to its *run*.

$$m = \frac{y_2 - y_1}{x_2 - x_1} = \frac{\text{rise}}{\text{run}}$$

EXAMPLE 1 Find the slope of the line that passes through the points $(-3, 2)$ and $(2, -5)$, and draw its graph.

Solution Pick one point to be (x_1, y_1) and the other to be (x_2, y_2). For the sake of argument, let (x_1, y_1) be $(-3, 2)$ and let (x_2, y_2) be $(2, -5)$. Then

$$\begin{array}{ccc} x_1 = -3 & & x_2 = 2 \\ & \text{and} & \\ y_1 = 2 & & y_2 = -5 \end{array}$$

To find the slope of the line, substitute these values into the formula for slope and simplify.

$$\text{Slope} = \frac{y_2 - y_1}{x_2 - x_1}$$

$$= \frac{-5 - 2}{2 - (-3)} \qquad \text{Substitute } -5 \text{ for } y_2, \text{ 2 for } y_1, \text{ 2 for } x_2, \\ \text{and } -3 \text{ for } x_1.$$

$$= -\frac{7}{5} \qquad \text{Simplify.}$$

The slope of the line passing through the points $(-3, 2)$ and $(2, -5)$ is $-\frac{7}{5}$. We would obtain the same result if we had let $(x_1, y_1) = (2, -5)$ and $(x_2, y_2) = (-3, 2)$. The graph of this line appears in Figure 6-11.

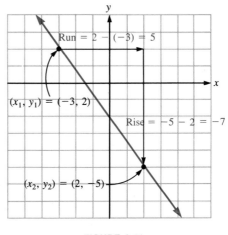

FIGURE 6-11   ■

In the previous example the slope of the line was negative because as the value of x increased 5 units, the y-value *decreased* 7 units. Whenever increasing values of x produce decreasing values of y, the slope of the line will be negative, and the graph of the line will *drop* as we move to the right. If the line neither rises nor falls (if the line is horizontal), its slope is 0. See Figure 6-12.

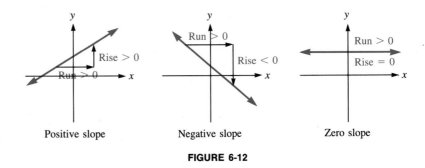

Positive slope   Negative slope   Zero slope

FIGURE 6-12

We have graphed a line after finding the coordinates of two points on that line. We can also graph lines if we know the coordinates of one point on the

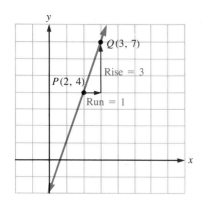

FIGURE 6-13

line and the slope of the line. As an example, we shall graph a line that passes through the point $P(2, 4)$ and has a slope of 3.

We begin by plotting the point $P(2, 4)$ as in Figure 6-13. Because the slope of the line is 3, the line rises 3 units for every 1 unit it moves to the right. Thus, we can find a second point on the line by starting at $P(2, 4)$ and moving **1** unit to the right and **3** units up. This brings us to point Q with coordinates $(2 + \mathbf{1}, 4 + \mathbf{3})$, or $(3, 7)$. The required line passes through points P and Q.

Slope-Intercept Form of the Equation of a Line

We can determine the slope of a line from its equation by first finding the coordinates of two points on the line, and using the definition of slope. For example, to find the slope of the line with equation $y = 2x + 7$, we first find the coordinates of two points that are on the graph. One of these is the y-intercept of the line, found by substituting 0 for x in the equation and solving for y.

$$y = 2x + 7$$
$$= 2(\mathbf{0}) + 7 \qquad \text{Substitute 0 for } x.$$
$$= 7 \qquad\qquad \text{Simplify.}$$

Thus, the y-intercept of the line is 7, and the line passes through the point $(0, 7)$.

To determine the coordinates of a second point on the line, replace x in the equation with some number other than 0, and determine y. We will let $x = 1$.

$$y = 2x + 7$$
$$= 2(\mathbf{1}) + 7 \qquad \text{Substitute 1 for } x.$$
$$= 9 \qquad\qquad \text{Simplify.}$$

Thus, the line passes through the point $(1, 9)$.

To find the slope of the line, we use the definition of slope. Letting (x_1, y_1) be the point $(0, 7)$ and (x_2, y_2) be the point $(1, 9)$, we have

$$\text{Slope of the line} = \frac{y_2 - y_1}{x_2 - x_1} = \frac{9 - 7}{1 - 0} = \frac{2}{1} = 2$$

Thus, the slope of the line is 2.

We see that the slope of the line is **2**, the same as the coefficient of the variable x in the equation $y = 2x + 7$. The y-intercept of the line is **7**, the same as the constant in the equation $y = 2x + 7$. This is more than coincidence. Mathematicians have established the following fact:

> **The Slope-Intercept Form of the Equation of a Line.**   If a linear equation is written in the form
>
> $$y = mx + b$$
>
> where m and b are constants, then the graph of that equation is a line with slope m and with y-intercept b.

EXAMPLE 2   **a.** Find the slope and the y-intercept of the line determined by $3x + 5y - 15 = 0$, and   **b.** graph the line.

Solution   **a.** Write the equation in slope-intercept form by solving it for y:

$$3x + 5y - 15 = 0$$

$$5y = -3x + 15 \qquad \text{Add } -3x \text{ and } 15 \text{ to both sides.}$$

$$y = -\frac{3}{5}x + \frac{15}{5} \qquad \text{Divide both sides by 5.}$$

$$y = -\frac{3}{5}x + 3$$

This equation is in the form $y = mx + b$ with $m = -\frac{3}{5}$ and $b = 3$. The slope of the line is $-\frac{3}{5}$, and its y-intercept is 3.

b. To graph the equation, begin by graphing its y-intercept, the point $(0, 3)$, as in Figure 6-14. Then use the slope to find a second point on the line. Because the slope is $\frac{\text{rise}}{\text{run}} = \frac{-3}{5}$, we can find a second point on the line by starting at the y-intercept, moving 5 units to the right, and 3 units *down*. This brings us to the point

$$(0 + 5, 3 + (-3)) \qquad \text{or} \qquad (5, 0)$$

Plot the point $(5, 0)$ and draw the line passing through both points. The complete graph appears in Figure 6-14.

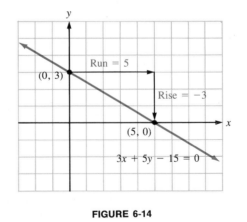

FIGURE 6-14

We can determine the slope and the y-intercept of a line such as $y = 3$ by writing the equation in the equivalent form

$$y = 0x + 3$$

We see that the slope of the line is 0, and its y-intercept is 3. Because its slope is 0, the line is horizontal. Its graph appears in Figure 6-15**a**.

The graph of an equation such as $x = 3$ is also a line. Because the equation $x = 3$ does not contain the variable y, we cannot find the slope or y-intercept by writing the equation in the form $y = mx + b$. Instead, we will find two points on the line and try to determine the slope by using the definition for slope. The points $(3, 1)$ and $(3, 2)$ lie on the given line because *any* point with an x-coordinate of 3 lies on the line. Thus,

$$m = \frac{y_2 - y_1}{x_2 - x_1}$$

$$= \frac{2 - 1}{3 - 3} \qquad \text{Substitute 2 for } y_2, \text{ 1 for } y_1, \text{ 3 for } x_2, \text{ and 3 for } x_1.$$

$$= \frac{1}{0}$$

Because division by zero is not defined, the symbol $\frac{1}{0}$ has no meaning. The slope of the line $x = 3$ is undefined. The graph is the vertical line in Figure 6-15**b**. It has no y-intercept, and its x-intercept is 3.

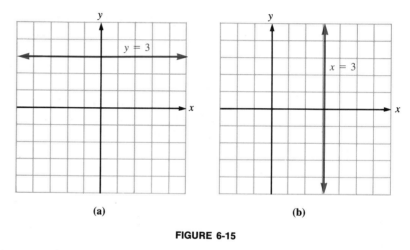

(a)   (b)

FIGURE 6-15

The previous discussion suggests the following facts:

> **All horizontal lines (lines with equations of the form $y = b$) have a slope of zero.**
>
> **All vertical lines (lines with equations of the form $x = a$) have undefined slope.**

Applications

Economists have learned that increasing the price of merchandise often causes sales to decrease. The equation of a line can describe the relationship between the price of a product and the number of units that can be sold. For example, the owners of a bicycle store have determined that for the stock they have on hand, the number of bicycles they will sell is related to the price they charge according to the equation of a line. If x is the price, in dollars, of one bicycle, and y is the number of bicycles that will be sold, the equation is

$$y = 500 - \frac{3}{5}x$$

To determine the number of bicycles they will sell at a unit price of $100, for example, let $x = 100$ in the equation $y = 500 - \frac{3}{5}x$ and determine y.

$$y = 500 - \frac{3}{5}x$$

$$y = 500 - \frac{3}{5}(100) \qquad \text{Substitute 100 for } x.$$

$$= 500 - 60$$

$$= 440$$

At a price of $100, they would expect to sell 440 bicycles.

Increasing the price will decrease sales. For example, if we substituted $x = 150$ into the equation, we would find that $y = 410$. Thus, at a price of $150, the owners would expect to sell 410 bicycles. Similarly, at a price of $200, they would expect to sell 380 bicycles.

To graph this equation, we plot several points (x, y) that satisfy the equation, and join them with a line. We have seen that the pairs (100, 440), (150, 410), and (200, 380) satisfy the equation. We plot these, and graph the equation as in Figure 6-16.

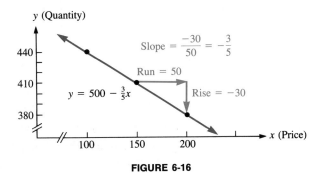

FIGURE 6-16

The slope of this line is $-\frac{3}{5}$, which is the coefficient of x in the equation $y = 500 - \frac{3}{5}x$. This slope represents the ratio of a change in the number of bicycles sold to the change in the price of a bicycle. The fraction $-\frac{3}{5}$ is equal to $-\frac{30}{50}$. Refer to Figure 6-16 to see that for each increase of $50 in the cost of a bicycle, the store expects to sell 30 fewer bicycles.

In economics, the graph of this equation is called a **demand curve,** and its slope is called the **marginal demand.**

The next example illustrates another fact of economics: As machinery wears out, it is worth less. For tax purposes, accountants estimate the decreasing value of aging equipment using **linear depreciation,** a method based on the equation of a line.

EXAMPLE 3   An insurance company buys a $12,500 computer system with an estimated useful life of 6 years. After x years of use, the value y of the computer is linearly depreciated according to the equation

$$y = 12,500 - 2000x$$

a. Determine the value of the computer after $3\frac{1}{2}$ years.
b. Determine the economic meaning of the y-intercept of the graph of the equation.
c. Determine the economic meaning of the slope of the graph of the equation.
d. Determine the value of the computer at the end of its useful life.

Solution   **a.** To determine the computer's value after $3\frac{1}{2}$ years, substitute 3.5 for x in the equation, and calculate y.

$$y = 12,500 - 2000x$$
$$y = 12,500 - 2000(\mathbf{3.5})$$
$$= 12,500 - 7000$$
$$= 5500$$

When the computer is $3\frac{1}{2}$ years old, its value will be $5500.

b. The y-intercept of a graph is the value of y found by letting $x = 0$. Thus, in this example the y-intercept is the value of a zero-year-old computer—that is, the computer's original cost. Because the y-intercept is 12,500, the computer cost $12,500 when it was new.

c. Each year, the value of the computer decreases by $2000, because the slope of the line is -2000. The slope of the depreciation line is called the **annual depreciation rate.**

d. To determine the computer's value at the end of its 6-year useful life, substitute 6 for x in the equation and calculate y.

$$y = 12,500 - 2000x$$
$$y = 12,500 - 2000(\mathbf{6})$$
$$= 12,500 - 12,000$$
$$= 500$$

At the end of its useful life, the computer is worth $500. This is known as the **salvage value** of the equipment.   ■

EXERCISE 6.2

In Exercises 1–16, find the slope of the line passing through the two given points. If the slope is not defined, write "undefined slope."

1. (1, 3)   (2, 4)   **2.** (1, 4)   (2, 3)   **3.** (2, 5)   (3, 4)   **4.** (3, 6)   (5, 2)

5. (2, 6)   (3, 7)   **6.** (5, 8)   (2, 9)   **7.** (0, −5)   (4, 3)   **8.** (3, −2)   (3, 5)

9. (2, 3)   (−3, 2)   **10.** (−7, 3)   (−3, 7)

11. (−5, −7)   (−4, −7)   **12.** (−6, −8)   (−5, −8)

13. $\left(-2, \frac{1}{2}\right)$ $\left(0, -\frac{3}{2}\right)$ 14. $\left(\frac{2}{3}, -9\right)$ $\left(\frac{5}{3}, 0\right)$ 15. $\left(\frac{5}{7}, \frac{1}{2}\right)$ $\left(-\frac{2}{7}, 0\right)$ 16. $\left(0, \frac{7}{2}\right)$ $\left(-\frac{5}{2}, 0\right)$

In Exercises 17–28, graph the line that passes through the given point and has the given slope.

17. $(0, 3)$, slope 1 18. $(3, 2)$, slope 3 19. $(-3, 2)$, slope 4 20. $(-1, 0)$, slope 2

21. $(1, -3)$, slope -1 22. $(1, -3)$, slope -2 23. $(3, 5)$, slope -4 24. $(0, 0)$, slope -5

25. $(0, 0)$, slope $\frac{1}{2}$ 26. $(2, 3)$, slope $-\frac{1}{2}$ 27. $(-1, 3)$, slope $-\frac{5}{3}$ 28. $(0, 0)$, slope $\frac{3}{5}$

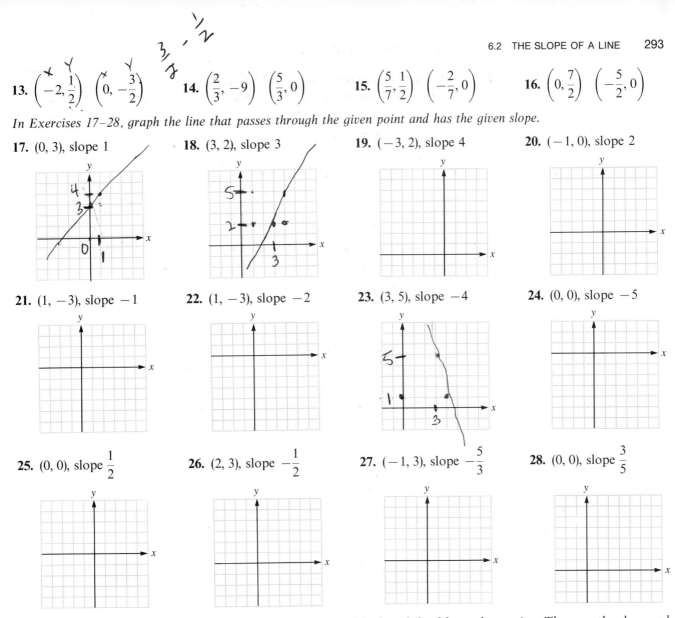

In Exercises 29–40, indicate the slope and the y-intercept of the line defined by each equation. Then use the slope and y-intercept to graph the line.

29. $y = 3x + 3$ 30. $y = 4x - 5$ 31. $y = 5x + 1$ 32. $y = -3x + 2$

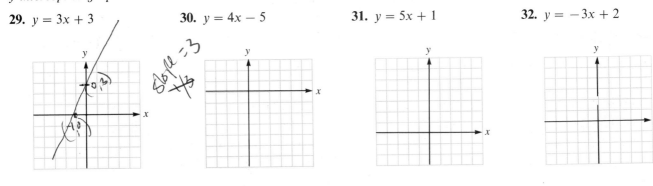

33. $y = -3x$ **34.** $y = -3x + 5$ **35.** $y = 3x - 2$ **36.** $y = -5x + 1$

37. $y = \dfrac{x}{3}$ **38.** $y = -\dfrac{x}{2} + 2$ **39.** $y = \dfrac{5}{3}x + \dfrac{1}{2}$ **40.** $y = \dfrac{3}{5}x - \dfrac{1}{2}$

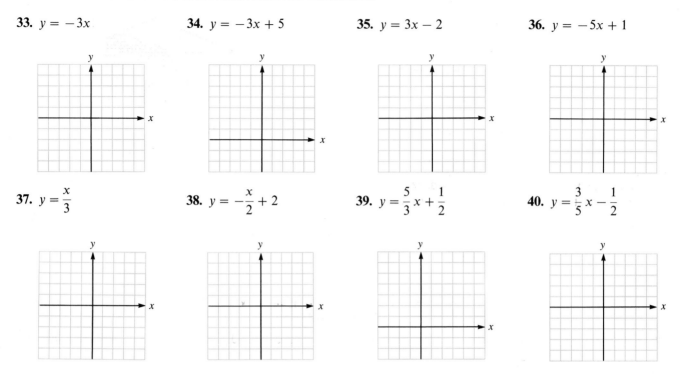

In Exercises 41–52, write each equation in slope-intercept form, and then indicate the slope and the y-intercept of the line.

41. $3(x - 2) + y = 1$ **42.** $2(x + 2) - 3y = 2$

43. $5(x - 5) = 2y + 1$ **44.** $2(y + 7) - x = 3$

45. $2(y - 1) = y$ **46.** $y - 7(y + 5) = 6$

47. $\dfrac{2y + 7}{2} = x$ **48.** $\dfrac{2 - x}{4} = y$

49. $\dfrac{3(y - 5)}{2} = x + 3$ **50.** $\dfrac{5(3 + x)}{2} = y - 5$

51. $x = 3y + 5$ **52.** $x = -\dfrac{1}{5}y - \dfrac{1}{2}$

53. The slope of the road in Illustration 1 is the vertical rise divided by the horizontal run. If the vertical rise is 24 feet for a horizontal run of 1 mile, determine the slope of the road. (*Hint:* 1 mile = 5280 feet.)

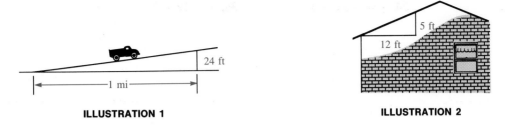

ILLUSTRATION 1   **ILLUSTRATION 2**

54. The demand, y, for television sets depends on the sales price, x, according to the equation $y = 5500 - 6x$. Determine the impact on sales for each $1 increase in price.

55. The pitch of a roof is defined to be the vertical rise divided by the horizontal run. If the rise of the roof in Illustration 2 is 5 feet for a run of 12 feet, determine the pitch of the roof.

56. The total daily cost C to an electronics company for manufacturing x television sets is given by the equation

$$C = 1200 + 130x$$

The y-intercept of the graph of this equation is called the company's **fixed cost,** and the slope of the line is called its **marginal cost.** Determine the company's fixed and marginal costs.

57. A large truck is depreciated linearly by the formula $y = 57,000 - 5500x$, where x represents its age in years. Find the value of the truck after $5\frac{1}{2}$ years.

58. Find the annual rate of depreciation of the truck in Exercise 57.

59. An office copy machine is valued annually by the linear depreciation formula $y = 6000 - 1200x$. Find the copier's original cost.

60. Find the salvage value of the copier in Exercise 59 if its useful life is $4\frac{1}{2}$ years.

61. A $450-drill press will have no salvage value after 3 years. Find the annual rate of depreciation.

62. A $750 video camera has a useful life of 5 years and a salvage value of $0. Find its annual rate of depreciation.

REVIEW EXERCISES

In Review Exercises 1–6, factor each expression.

1. $3x^2 - 6x$ **2.** $y^2 - 25$ **3.** $2z^2 - 5z - 3$ **4.** $9t^2 - 15t + 6$

5. $9u^2 + 24u + 16$ **6.** $y^4 - 16x^4$

In Review Exercises 7–10, solve each equation.

7. $4y^2 + 8y = 0$ **8.** $r^2 - 36 = 0$ **9.** $x^2 - 7x + 6 = 0$ **10.** $12s^2 + 13s = 4$

In Review Exercises 11–12, write each statement as an algebraic expression.

11. The product obtained when the sum of a and b is multiplied by a number that is 3 greater than z.

12. The product of three consecutive even integers, where x represents the largest integer.

6.3 **WRITING EQUATIONS OF LINES**

We have seen how to find the graph of a linear equation. We now consider the reverse problem of determining a linear equation from its graph.

To begin, we must recall that a linear equation written in the form $y = mx + b$ is written in slope-intercept form, that m is the slope of its straight-line graph, and that b is its y-intercept. If we know both the slope and the y-intercept of a line, we can always write its equation. For example, to write the equation of the line with slope 7 and y-intercept -2, we simply substitute 7 for m and -2 for b in the slope-intercept form of the equation of a line and simplify.

$$y = mx + b$$
$$y = 7x + (-2) \qquad \text{Substitute 7 for } m \text{ and } -2 \text{ for } b.$$
$$y = 7x - 2$$

Thus, the equation of the line with slope 7 and y-intercept -2 is $y = 7x - 2$.

EXAMPLE 1 Find the equation of the line with slope $-\frac{3}{2}$ and y-intercept 3. Express the equation in general form.

Solution Proceed as follows:

$$y = mx + b \qquad \text{Use the slope-intercept form.}$$
$$y = -\frac{3}{2}x + 3 \qquad \text{Substitute } -\frac{3}{2} \text{ for } m \text{ and 3 for } b.$$
$$2y = -3x + 6 \qquad \text{Multiply both sides by 2.}$$
$$3x + 2y = 6 \qquad \text{Add } 3x \text{ to both sides.}$$

The equation $3x + 2y = 6$ is written in the form $Ax + By = C$, which is the general form of the equation of a line. ■

Suppose we know the slope of a line, but instead of its y-intercept we know the coordinates of a second point on the line. It is still possible to find the equation of the line. For example, we will find the equation of the line that has a slope of 3 and passes through the point $P(2, 1)$. See Figure 6-17. We will then use the same method to develop the point-slope form of the equation of a line.

Point-Slope Form of the Equation of a Line

We know that $P(2, 1)$ are the coordinates of one point on the line. We can then let $Q(x, y)$ be the coordinates of some other point on the line. To find the slope of the line that passes through the points $P(2, 1)$ and $Q(x, y)$, we use the definition of slope.

$$m = \frac{y_2 - y_1}{x_2 - x_1}$$

1. $\quad m = \dfrac{y - 1}{x - 2} \qquad$ Substitute y for y_2, 1 for y_1, x for x_2, and 2 for x_1.

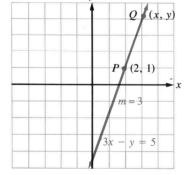

FIGURE 6-17

However, we are given that the slope m of the line is 3. Thus, we have

$$3 = \frac{y-1}{x-2}$$ Substitute 3 for m in Equation 1.

$$3(x-2) = y-1$$ Multiply both sides by $x-2$.

$$3x-6 = y-1$$ Use the distributive property to remove parentheses.

$$3x-y = 5$$ Add $-y$ and 6 to both sides.

The equation of the required line, written in general form, is $3x - y = 5$.

We will now use this method to find the equation of any nonvertical line with known slope passing through a known point. To do so, we refer to Figure 6-18 and suppose that the slope of the line is m and that the known point has coordinates of (x_1, y_1). If (x, y) are the coordinates of any point on the line, then the slope m of the line is

$$m = \frac{y - y_1}{x - x_1}$$

If we multiply both sides of this equation by $x_2 - x_1$, we have

$$m(x - x_1) = y - y_1$$

or

$$y - y_1 = m(x - x_1)$$

This equation is called the **point-slope form** of the equation of a line.

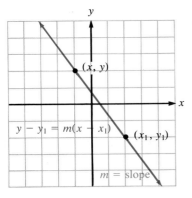

FIGURE 6-18

The Point-Slope Form of the Equation of a Line.   If a line with slope m passes through the point (x_1, y_1), then the equation of the line is

$$y - y_1 = m(x - x_1)$$

EXAMPLE 2   A line has a slope of $\frac{3}{4}$ and passes through the point $(-1, \frac{1}{2})$. Write the equation of the line in general form.

Solution   Substitute $\frac{3}{4}$ for m, -1 for x_1, and $\frac{1}{2}$ for y_1 in the point-slope form of a linear equation and simplify.

$$y - y_1 = m(x - x_1)$$

$$y - \frac{1}{2} = \frac{3}{4}[x - (-1)]$$

$$4y - 2 = 3(x + 1)$$ Multiply both sides by 4 and simplify.

$$4y - 2 = 3x + 3$$ Remove parentheses.

$$-5 = 3x - 4y$$ Add -3 and $-4y$ to both sides.

In general form, the equation of the required line is $3x - 4y = -5$.   ∎

Parallel and Perpendicular Lines

Because all horizontal lines are parallel, all distinct lines with a slope of 0 are parallel. Likewise, because all vertical lines are parallel, all distinct lines with undefined slopes are parallel. Because the slope of a line is a measure of how rapidly the line rises, it is reasonable to assume that lines with equal slopes are parallel.

> **Two lines with the same slope are parallel.**

Thus, the equations $y = 5x + 7$ and $y = 5x - 8$ represent parallel lines, because the slope of each line is **5**.

EXAMPLE 3   Graph the equation $y = -3x + 4$ and then find the equation of the line parallel to it that passes through the point $(5, -2)$. Write the result in general form and graph it.

Solution   Because the given equation is written in slope-intercept form, we can easily find that its slope is -3 and its y-intercept is 4. To graph the given line, first plot the y-intercept $(0, 4)$. Because the slope of the line is -3, the line must *drop* 3 units for every 1 unit it moves to the right. Begin at the y-intercept, move 1 unit to the right, and 3 units *down*. This locates another point on the line, the point $(1, 1)$. The line through these points is the graph of the given equation. It appears in Figure 6-19.

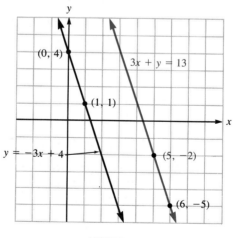

FIGURE 6-19

Because the required line is parallel to the given line, it will have the same slope as the given line. Thus, the slope of the required line is also -3, and it passes through the point $(5, -2)$. Substitute these values into the point-slope form of the equation of a line, and simplify.

$$y - y_1 = m(x - x_1)$$
$$y - (-2) = -3(x - 5) \qquad \text{Substitute } -3 \text{ for } m, 5 \text{ for } x_1, \text{ and } -2 \text{ for } y_1.$$
$$y + 2 = -3x + 15 \qquad \text{Remove parentheses.}$$
$$3x + y = 13 \qquad \text{Add } 3x \text{ and } -2 \text{ to both sides.}$$

In general form, the equation of the required line is $3x + y = 13$. To graph this line, begin by plotting the given point $(5, -2)$. Because the slope of the line is -3, move 1 unit to the right and 3 units down, to the point $(6, -5)$. The graph of the equation is the line through these two points. It appears in Figure 6-19. ■

Lines that meet at right angles are called **perpendicular lines.** For example, a vertical line is perpendicular to a horizontal line. Mathematicians have shown that the slopes of other perpendicular lines are related by the following fact.

The product of the slopes of perpendicular lines is -1.

Two numbers that have a product of -1 are called **negative reciprocals** of each other. The numbers 3 and $-\frac{1}{3}$, for example, are negative reciprocals, because their product is -1:

$$3\left(-\frac{1}{3}\right) = -1$$

EXAMPLE 4   One line has a slope of 5 and another has a slope of $-\frac{1}{5}$. Determine whether the lines are parallel, perpendicular, or neither.

Solution   Because the slopes of the two lines are not equal ($5 \neq -\frac{1}{5}$), the lines are not parallel. To determine whether the lines are perpendicular, find the product of their slopes. Because the product of their slopes is $5(-\frac{1}{5})$ or -1, the lines are perpendicular. ■

EXAMPLE 5   Determine whether the graphs of the following equations are parallel, perpendicular, or neither:

   1. $y = 2x - 5$ **2.** $y = 2(3 - x)$

Solution   The slope of the line $y = 2x - 5$ is 2, the coefficient of x. To find the slope of the second line, solve its equation for y.

$$y = 2(3 - x)$$
$$y = 6 - 2x \qquad \text{Remove parentheses.}$$
$$y = -2x + 6$$

The slope of the second line is -2, which is the coefficient of x in the equation $y = -2x + 6$. Because the slopes (2 and -2) of the two lines are not equal, the lines are not parallel. Because the product of the slopes is not -1, the lines are not perpendicular either. Thus, the lines are neither parallel nor perpendicular. ■

EXAMPLE 6   Find the equation of the line that is perpendicular to the line $y = \frac{1}{3}x + 5$ and passes through the point $(2, -1)$. Write the result in general form and graph both equations.

Solution   The slope of the line with equation $y = \frac{1}{3}x + 5$ is $\frac{1}{3}$. Because the required line is to be perpendicular to the given line, its slope will be the negative reciprocal of $\frac{1}{3}$.

$$\text{Slope of the required line} = -\frac{1}{\frac{1}{3}} = -(1)\left(\frac{3}{1}\right) = -3$$

Use the point-slope form to find the equation of the line with a slope of -3 and passing through the point $(2, -1)$, as follows:

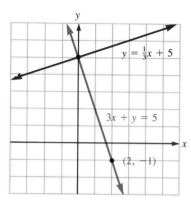

$$y - y_1 = m(x - x_1)$$
$$y - (-1) = -3(x - 2) \qquad \text{Substitute } -3 \text{ for } m, 2 \text{ for } x_1, \text{ and } -1 \text{ for } y_1.$$
$$y + 1 = -3x + 6 \qquad \text{Remove parentheses.}$$
$$3x + y = 5 \qquad \text{Add } 3x \text{ and } -1 \text{ to both sides.}$$

Written in general form, the equation of the required line is $3x + y = 5$. The graphs of both lines appear in Figure 6-20. ∎

FIGURE 6-20

The various forms of the equation of a line are summarized as follows:

Forms of a Line.	
$Ax + By = C$	**General form** of a linear equation. A and B cannot both be zero.
$y = mx + b$	**Slope-intercept form** of a linear equation. The slope is m, and the y-intercept is b.
$y - y_1 = m(x - x_1)$	**Point-slope form** of a linear equation. The slope is m, and the line passes through (x_1, y_1).
$y = b$	**Horizontal line.** The slope is 0, and the y-intercept is b.
$x = a$	**Vertical line.** The slope is undefined, and the x-intercept is a.

EXAMPLE 7   George's monthly water bill is related to the number of gallons used by a linear equation. If George is billed $12 for using 1000 gallons and $16 for using 1800 gallons, find his bill for 2000 gallons.

Solution   When George uses 1000 gallons, his bill is $12 and when he uses 1800 gallons, his bill is $16. If y is the cost of using x gallons, then x and y are related by a linear equation. The points $P(1000, 12)$ and $Q(1800, 16)$ lie on the graph of this equation, as shown in Figure 6-21. Thus, we must write the equation of the line passing through $P(1000, 12)$ and $Q(1800, 16)$. To do so, we first compute the slope of the line passing through those points:

$$m = \frac{y_2 - y_1}{x_2 - x_1}$$

$$= \frac{16 - 12}{1800 - 1000}$$

$$= \frac{4}{800}$$

$$= \frac{1}{200}$$

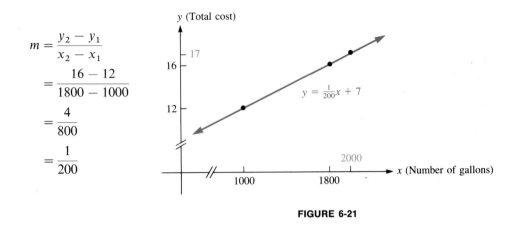

FIGURE 6-21

We then substitute $\frac{1}{200}$ for m and the coordinates of one of the known points (say $P(1000, 12)$) into the point-slope form of the equation of a line and proceed as follows:

$$y - y_1 = m(x - x_1)$$

$$y - 12 = \frac{1}{200}(x - 1000)$$

$$y - 12 = \frac{1}{200}x - \frac{1}{200}(1000)$$

$$y - 12 = \frac{1}{200}x - 5$$

$$y = \frac{1}{200}x + 7$$

To find George's water bill for 2000 gallons, substitute 2000 for x in the billing formula and determine y:

$$y = \frac{1}{200}(2000) + 7$$

$$y = 10 + 7$$

$$y = 17$$

The bill for use of 2000 gallons of water is \$17.   ■

EXERCISE 6.3

In Exercises 1–12, use the slope-intercept form to find the equation of each line with the given properties. Write the equation in general form.

1. Slope 4, y-intercept 5

2. Slope 2, y-intercept -3

3. Slope -7, y-intercept -2

4. Slope -3, y-intercept 10

5. Slope $-\dfrac{1}{2}$, y-intercept $-\dfrac{1}{2}$

6. Slope $\dfrac{2}{3}$, y-intercept -5

7. Slope $-\dfrac{3}{5}$, y-intercept $-\dfrac{2}{5}$

8. Slope $-\dfrac{5}{7}$, y-intercept $-\dfrac{3}{7}$

9. Slope 0, y-intercept 3

10. Slope 3, y-intercept 0

11. Slope 0, y-intercept 0

12. Slope $-\dfrac{4}{5}$, y-intercept 0

In Exercises 13–24, use the point-slope form to find the equation of each line with the given properties. Write the equation in general form.

13. Slope 2, passing through $(1, 1)$

14. Slope 3, passing through $(2, 3)$

15. Slope 5, passing through $(1, -1)$

16. Slope -3, passing through $(0, -3)$

17. Slope -2, passing through $(0, 0)$

18. Slope $\dfrac{1}{2}$, passing through $(2, 3)$

19. Slope $-\dfrac{1}{2}$, passing through $(-2, 4)$

20. Slope $-\dfrac{2}{3}$, passing through $(0, 3)$

21. Slope $\dfrac{7}{5}$, passing through $\left(3, \dfrac{2}{5}\right)$

22. Slope $-\dfrac{3}{7}$, passing through $\left(1, \dfrac{5}{7}\right)$

23. Slope 0, passing through $(739, 3)$

24. Slope 0, passing through $(0, 0)$

In Exercises 25–32, determine whether the lines with the given slopes are parallel, perpendicular, or neither.

25. Slopes of 3 and -3

26. Slopes of 2 and $-\dfrac{1}{2}$

27. Slopes of 4 and $\dfrac{8}{2}$

28. Slopes of $-\dfrac{3}{2}$ and $\dfrac{2}{3}$

29. Slopes of 3 and $-\dfrac{1}{3}$

30. Slopes of 0.5 and $\dfrac{1}{2}$

31. Slopes of 1.5 and $\dfrac{3}{2}$

32. Slopes of 1.5 and -1.5

In Exercises 33–42, indicate whether the given pairs of lines are parallel, perpendicular, or neither.

33. $y = 3x + 2$ and $y = 3x - \dfrac{1}{2}$

34. $y = 2x + 5$ and $y = 2x - 7$

35. $y = \dfrac{1}{3}x + 1$ and $y = 3x - 1$

36. $y = 5x + 5$ and $y = -\dfrac{1}{5}x + 5$

37. $y = x + \dfrac{7}{2}$ and $y = -x' + \dfrac{2}{7}$

38. $y = 5$ and $x = 5$

39. $y = -\dfrac{3}{4}x$ and $y = \dfrac{4}{3}x - \dfrac{1}{2}$

40. $y = 7x + 1$ and $y = -\dfrac{1}{7}x - 7$

41. $2x + 3y = 5$ and $y = -\dfrac{2}{3}x - 5$

42. $3x - y = 8$ and $x + 3y = 8$

In Exercises 43–62, write the equation of each line with the given properties in general form.

43. Slope $\dfrac{2}{15}$, passing through $(15, 10)$

44. Slope $\dfrac{3}{11}$, y-intercept $\dfrac{1}{11}$

45. Slope $-\dfrac{4}{9}$, y-intercept $-\dfrac{2}{9}$

46. Slope $-\dfrac{13}{5}$, passing through $\left(3, -\dfrac{2}{5}\right)$

47. Parallel to the line $y = 5x - 8$, passing through $(0, 0)$

48. Parallel to the line $y = -2(7 - x)$, passing through $(0, 5)$

49. Parallel to the line $y = 5(x + 3)$, with y-intercept -4

50. Parallel to the line $y = -6(x - 2)$, with y-intercept 5

51. Parallel to the line $5x - 2y = 3$, passing through $(2, 1)$

52. Parallel to the line $3x + 4y = -9$, with y-intercept -3

53. Passing through $(2, 5)$ and $(3, 7)$ (*Hint:* First find the slope.)

54. Passing through $(-3, 0)$ and $(3, -1)$ (*Hint:* First find the slope.)

55. Passing through $(-5, 2)$ and $(7, 2)$ (*Hint:* First find the slope.)

56. Passing through $(-5, 2)$ and $(-5, 7)$ (*Hint:* First find the slope.)

57. Passing through $(2, 5)$ and perpendicular to a line with slope $\frac{1}{5}$

58. Passing through $(2, 1)$ and perpendicular to a line with slope -3

59. Passing through the origin and perpendicular to the line $y = 5x + 11$

60. Passing through the point $(3, -5)$ and perpendicular to the line $y = -\frac{1}{2}x - 7$

61. Having a y-intercept of 3 and perpendicular to the line $y = \frac{1}{5}x + 12$

62. Having a y-intercept of $-\frac{1}{2}$ and perpendicular to the line $y = -\frac{1}{3}x$

63. Seamless aluminum rain gutters can be installed for a fixed charge, plus an additional per-foot cost. If an installation of 200 feet costs $350 and 250 feet costs $425, find the cost to install 500 feet of gutter.

64. Fahrenheit temperature, F, is related to the Celsius temperature, C, by an equation of the form

$$F = mC + b$$

If water freezes at $0°$ Celsius and $32°$ Fahrenheit, and water boils at $100°$ Celsius and $212°$ Fahrenheit, determine m and b.

65. The accounting department uses the linear method to depreciate a word-processing system. After 3 years, the system is worth $2000. After 4 years, it is worth nothing. What was the purchase price of the system?

66. A company estimates the useful life of a lathe to be 10 years. After 3 years, it is worth $330. After 9 years, it is worth $90. What is the salvage value of the lathe?

In Review Exercises 1–6, perform each operation.

1. $\dfrac{x^2 - 25}{x^2 + 10x + 25} \cdot \dfrac{x^2 + 6x + 5}{x^2 - 5x}$

2. $\dfrac{x^2 - x - 2}{x^2 + 4x + 3} \div \dfrac{x^2 - 2x}{x + 3}$

3. $\dfrac{3 + x}{x + 1} + \dfrac{3 - x}{x + 1}$

4. $\dfrac{2}{x} + \dfrac{x}{3}$

5. $\dfrac{x + 1}{x - 1} - \dfrac{x - 1}{x + 1}$

6. $\dfrac{x + 3}{x - 1} - \dfrac{x + 4}{x + 1}$

In Review Exercises 7–10, write each number in scientific notation.

7. 73,000,000,000

8. 0.0000000245

9. 37.2×10^{-2}

10. 0.0043×10^5

11. Write the expression $(x - 4)^3$ without using parentheses.

12. Solve the equation $\dfrac{5(x - 4)}{2} = \dfrac{4(x - 5)}{3}$.

6.4 GRAPHING INEQUALITIES

The graph of an equation such as $y = x - 5$ consists of only those points on a line whose coordinates satisfy the equation. It is possible to graph an *inequality* such as $y \geq x - 5$. This graph, too, will contain only those points whose coordinates satisfy the inequality. The graph of an inequality will not be a line, but an area bounded by a line. Such areas are called **half-planes.**

EXAMPLE 1 Graph the inequality $y \geq x - 5$.

Solution Because the symbol $\geq$ allows the possibility that the two sides of the inequality are actually equal, begin by graphing the equation $y = x - 5$. The graph appears in Figure 6-22**a**.

Because the symbol $\geq$ also allows the possibility that the left-hand side of the inequality $y \geq x - 5$ is greater than the right-hand side, the coordinates of points other than those indicated by the graph in Figure 6-22**a** satisfy the inequality $y \geq x - 5$. For example, the coordinates of the origin satisfy the inequality. Verify this by letting x and y be 0 in the given inequality:

$$y \geq x - 5$$
$$0 \geq 0 - 5$$
$$0 \geq -5$$

Because 0 is greater than or equal to -5, the coordinates of the origin satisfy the original inequality. As a matter of fact, the coordinates of *every* point on the same side of the line as the origin satisfy the inequality. The graph of the inequality $y \geq x - 5$ is the half-plane that is shaded in Figure 6-22**b**. Because the bondary line $y = x - 5$ is included also, it is drawn with a solid line.

$y = x - 5$

x	y
0	-5
5	0

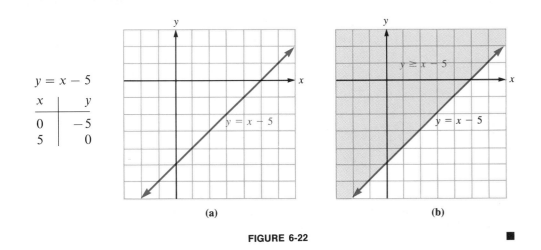

(a)   (b)

FIGURE 6-22

EXAMPLE 2   Graph the inequality $2(x - 3) - (x - y) \leq -1$.

Solution   Begin by simplifying the given inequality as follows:

$$2(x - 3) - (x - y) \leq -1$$
$$2x - 6 - x + y \leq -1 \qquad \text{Remove parentheses.}$$
$$x - 6 + y \leq -1 \qquad \text{Combine terms.}$$
$$x + y \leq \ \ \ 5 \qquad \text{Add 6 to both sides.}$$

Graph the inequality $x + y \leq 5$ by first graphing the equation $x + y = 5$, as in Figure 6-23**a**.

The symbol $\leq$ allows the possibility that the left-hand side of the inequality $x + y \leq 5$ is less than the right-hand side. Again, the coordinates of the origin satisfy the inequality. Verify this by letting x and y be 0 in the given inequality.

$$x + y \leq 5$$
$$0 + 0 \leq 5 \qquad \text{Substitute 0 for } x \text{ and for } y.$$
$$0 \leq 5$$

Thus, the coordinates of the origin satisfy the original inequality. As a matter of fact, the coordinates of *every* point on the same side of the line as the origin satisfy the inequality. The graph of the inequality $x + y \leq 5$ is the half-plane that appears in color in Figure 6-23**b**. Because the boundary line $x + y = 5$ is included also, it is drawn as a solid line.

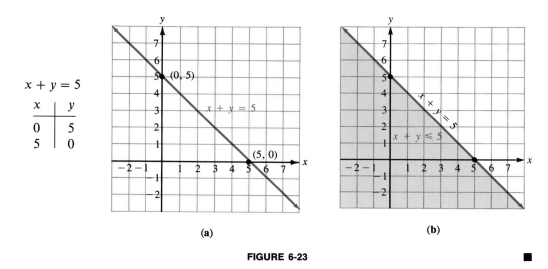

$x + y = 5$

x	y
0	5
5	0

(a)   **(b)**

FIGURE 6-23

■

EXAMPLE 3   Graph the inequality $y > 2x$.

Solution   Although the symbol $>$ does not allow the possibility that y and $2x$ are equal, begin by graphing the line $y = 2x$ anyway. Draw the graph as a broken line to indicate that the points on the line are not part of the solution, as in Figure 6-24**a**.

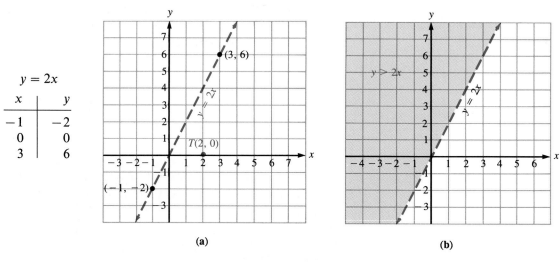

$y = 2x$

x	y
-1	-2
0	0
3	6

(a)   **(b)**

FIGURE 6-24

Substitute into the inequality $y > 2x$ the coordinates of some convenient point on one side of the line. Point $T(2, 0)$, for example, is obviously *below* the line $y = 2x$ (see Figure 6-24**a**). Check to see if point $T(2, 0)$ satisfies the original inequality by substituting 2 for x and 0 for y in the inequality.

$$y > 2x$$

$$\mathbf{0} > 2(\mathbf{2}) \qquad \text{Substitute 2 for } x \text{ and 0 for } y.$$

$$0 > 4$$

The inequality $0 > 4$ is a false statement, so the coordinates of point T do not satisfy the inequality. Thus, point T is not on the side of the line we wish to shade. The coordinates of points on the other side of the line are the ones that do satisfy the inequality. Therefore, we must shade the other side of the line. The graph of the solution set of the inequality $y > 2x$ appears in Figure 6-24**b**. ∎

EXAMPLE 4 Graph the inequalities **a.** $x + 2y < 6$ and **b.** $y \geq 0$.

Solution   **a.** Find the boundary by graphing the equation $x + 2y = 6$. Graph this boundary as a broken line to show that the line itself is not part of the solution. Then choose any point that is not on the boundary line and see if its coordinates satisfy the original inequality. The origin is a convenient choice.

$$x + 2y < 6$$

$$\mathbf{0} + 2(\mathbf{0}) < 6 \qquad \text{Substitute 0 for } x \text{ and for } y.$$

$$0 < 6$$

Because $0 < 6$ is a true statement, shade the side of the line that includes the origin. The graph of $x + 2y < 6$ appears in Figure 6-25**a**.

b. First find the boundary by graphing the equation $y = 0$. The equation $y = 0$ represents the x-axis. Graph this boundary as a solid line to show that the line is part of the solution. Then choose any point that is not on the boundary line and see if its coordinates satisfy the original inequality. The point $T(0, 1)$ is above the x-axis and is a convenient choice.

$$y \geq 0$$

$$1 \geq 0 \qquad \text{Substitute 1 for } y.$$

Because $1 \geq 0$ is a true statement, shade the side of the line that includes the point T. This is the half-plane above the x-axis. The graph of $y \geq 0$ appears in Figure 6-25**b**.

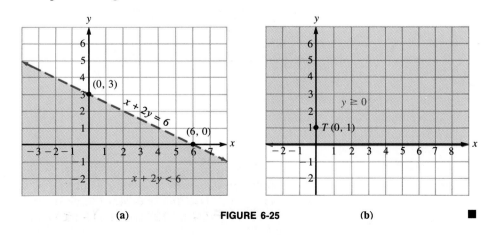

(a)   **FIGURE 6-25**   (b)   ∎

The inequalities of Examples 1–4 are called **linear inequalities.** The following is a summary of the procedure for graphing linear inequalities.

> 1. Graph the boundary line of the region. If the inequality allows the possibility of equality (the symbol is either $\leq$ or $\geq$), then draw the boundary line as a solid line. If equality is not allowed ($<$ or $>$), draw the boundary line as a broken line.
> 2. Pick a point that is obviously on one side of the boundary line. (Use the origin if possible.) Replace x and y with the coordinates of that point. If the inequality is satisfied, shade the side that contains that point. If the inequality is *not* satisfied, shade the other side.

EXERCISE 6.4

In Exercises 1–8, the boundary of a graph has been drawn. Complete the graph by shading the correct side of the boundary.

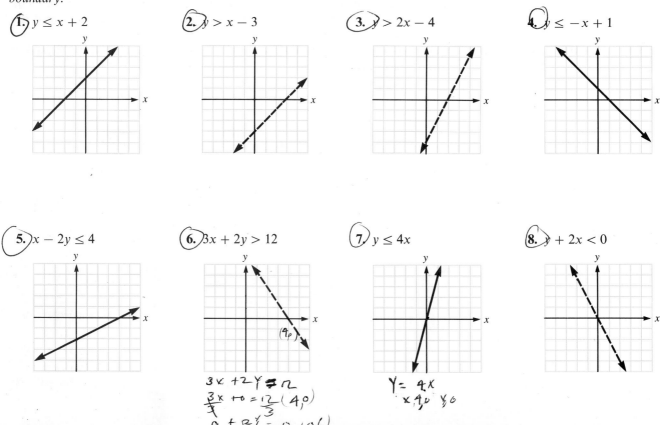

1. $y \leq x + 2$

2. $y > x - 3$

3. $y > 2x - 4$

4. $y \leq -x + 1$

5. $x - 2y \leq 4$

6. $3x + 2y > 12$

$(4, 0)$

7. $y \leq 4x$

8. $y + 2x < 0$

$3x + 2y \neq 12$

$\dfrac{3x + 0}{7} = \dfrac{12}{3} (4, 0)$

$0 + \dfrac{2y}{2} = \dfrac{12}{2} (0, 6)$

$Y = 4x$

$x \ 1, 0 \ y \ 0$

In Exercises 9–24, graph each linear inequality.

9. $y \geq 3 - x$

10. $y < 2 - x$

11. $y < 2 - 3x$

12. $y \geq 5 - 2x$

13. $y \geq 2x$

14. $y < 3x$

15. $2y - x < 8$

16. $y + 9x \geq 3$

17. $y - x \geq 0$

18. $y + x < 0$

19. $2x + y > 2$

20. $3x - 2y > 6$

21. $3x - 4y > 12$

22. $4x + 3y \leq 12$

23. $5x + 4y \geq 20$

24. $7x - 2y < 21$

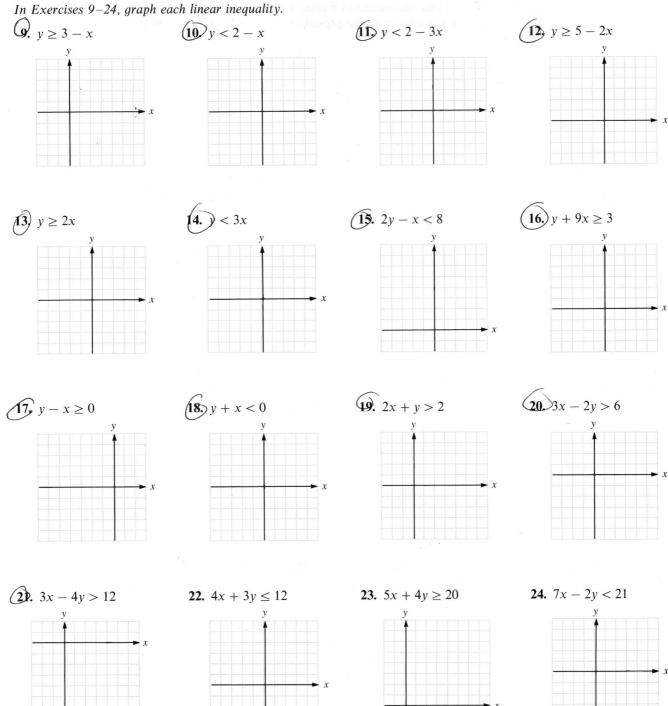

In Exercises 25–40, simplify each inequality and construct its graph.

25. $3(x + y) + x < 6$ **26.** $2(x - y) - y \geq 4$ **27.** $4x - 3(x + 2y) \geq -6y$ **28.** $3y + 2(x + y) < 5y$

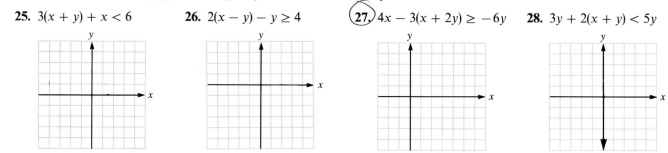

29. $7(x + 2y) - 2(x - 3y) < 50$ **30.** $3(6x + 5y) - 5(3x + 4y) \geq 15$

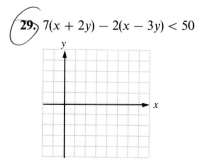

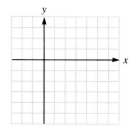

31. $5(x - 2y) > 5x$ **32.** $3(y - x) \leq x - 3$ **33.** $x(x + 2) \leq x^2 + 3x + 1$ **34.** $y(y - 5) > y^2 - 7$

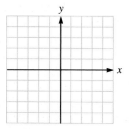

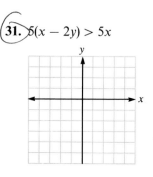

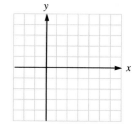

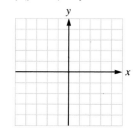

35. $x^2 + 3y \leq x(x + 2) - 1$ **36.** $x + y(y - 3) < (y + 1)(y - 2)$

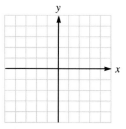

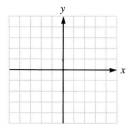

37. $3x + 7 \leq 5y + 7$

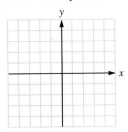

38. $5x \leq x + 5(y + x)$

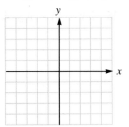

39. $(x + 1)(x - 2) + y^2 < x^2 + y(y - 2)$

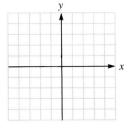

40. $(x + 2)(x - 2) + y^2 \geq (y - 2)(y + 1) + x^2$

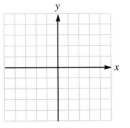

REVIEW EXERCISES

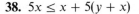

1. List the prime numbers between 40 and 50.
2. State the associative property of addition.
3. State the commutative property of multiplication.
4. What is the additive identity element?
5. What is the multiplicative identity element?
6. What is the multiplicative inverse of $\frac{5}{3}$?

In Review Exercises 7–12, let $P(x) = 3x^2 - 4x + 3$. *Find each quantity.*

7. $P(2)$

8. $P(-3)$

9. $P(-x)$

10. $P(x + 1)$

11. $P(x^2)$

12. $P(y^2)$

6.5 RELATIONS AND FUNCTIONS

We have seen that equations and inequalities determine ordered pairs of numbers. Each ordered pair indicates a correspondence between two numbers. The equation $y = -x + 2$, for example, determines ordered pairs (x, y) in which each number y corresponds to a value x. If $x = 3$ in this equation, the corresponding value of y is $-3 + 2$, or -1, and the ordered pair $(3, -1)$ indicates this correspondence. The graphs of *all* such ordered pairs is the line in Figure 6-26a.

The inequality $y \leq -x + 2$ also determines a set of ordered pairs. For this inequality, however, to each value of x there correspond *many* values of y. Because

Sonya Kovalevskaya (1850–1891)
This talented young Russian
woman hoped to study mathematics
at the University of Berlin, but
strict rules prohibited women from
attending lectures. Undaunted, she
studied privately with the great
mathematician Karl Weierstrauss
and published several important
papers.

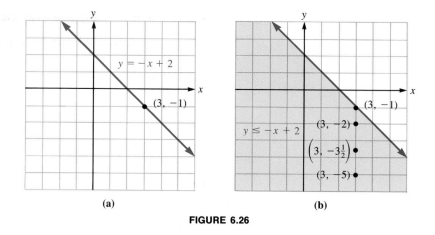

(a)   **(b)**

FIGURE 6.26

the boundary of the graph of $y \leq -x + 2$ is the line $y = -x + 2$, the inequality determines the ordered pair $(3, -1)$. The inequality also determines many other pairs with an x-coordinate of 3: $(3, -2)$, $(3, -3\frac{1}{2})$, and $(3, -5)$ are examples. Thus, to the number $x = 3$ there correspond *several* values of y. The graph of the inequality appears in Figure 6-26**b**.

Because any ordered pair of numbers indicates that certain values of x and y are related, any set of ordered pairs is called a **relation**. Thus, the set of ordered pairs graphed in either part of Figure 6-26 is a relation. The set of all possible values of x is called the **domain** of the relation, and the set of all possible values of y is called the **range**. In the relations graphed in Figure 6-26, both x and y could be any real number. The domain and the range of both relations are the set of real numbers.

The two relations in Figure 6-26 differ in an important way. Several ordered pairs in part **b** of the figure have the same value of x but different values of y: The relation contains, for example, the pairs $(3, -1)$ and $(3, -2)$. In part **a**, however, *exactly one* value of y corresponds to each value of x. The relation in part **a** is called a **function**.

> **DEFINITION.**   A **function** is a correspondence that assigns to each number x exactly one value y. The variable x is called the **independent variable,** and the variable y is called the **dependent variable.**

EXAMPLE 1   Does the equation $y = 3x + 1$ define a function?

Solution   For a function to exist, *each* number x must determine a *single* value y. To determine y in this example, a number x is multiplied by 3 and then added to 1. Because for each x the multiplication and addition will give a single result, a single value y is produced. Thus, the equation $y = 3x + 1$ does define a function.   ∎

EXAMPLE 2   Does the equation $y^2 = x$ define a function?

Solution   For a function to exist, *each* number x must determine a *single* value y. Let x be the number 9, for example. The variable y could be 3 or -3, because $3^2 = 9$ and $(-3)^2 = 9$. Thus, the relation contains the two ordered pairs $(9, 3)$ and $(9, -3)$, which have the same value of x but different values of y. Because there are two values of y determined when $x = 9$, the equation does not represent a function.   ∎

Function Notation

The concept of function is so important that special notation has been developed for its use.

> **Function Notation.**   The notation
>
> $$y = f(x)$$
>
> denotes that the variable y is a function of x.

The notation $f(x)$ is read as "f of x." It does not read as, nor does it mean, "f times x."

The notation $y = f(x)$ is similar to the notation $y = P(x)$ we have used for polynomials. It provides a way of denoting the values of y that correspond to individual numbers x. If $y = f(x)$, then the value of y that is determined by $x = 3$ is denoted by $f(3)$. Similarly, $f(-1)$ represents the value of y that corresponds to $x = 1$.

EXAMPLE 3   Let $f(x) = 3x + 1$. Find   **a.** $f(3)$,   **b.** $f(-1)$,   **c.** $f(0)$,   and   **d.** $f(r)$.

Solution   **a.** Replace x with 3:

$$f(x) = 3x + 1$$
$$f(3) = 3(3) + 1$$
$$= 9 + 1$$
$$= 10$$

b. Replace x with -1:

$$f(x) = 3x + 1$$
$$f(-1) = 3(-1) + 1$$
$$= -3 + 1$$
$$= -2$$

c. Replace x with 0:

$$f(x) = 3x + 1$$
$$f(0) = 3(0) + 1$$
$$= 0 + 1$$
$$= 1$$

d. Replace x with r:

$$f(x) = 3x + 1$$
$$f(r) = 3r + 1$$

∎

The letter f is used in the notation $y = f(x)$ to represent the word *function,* but other letters can be used. The notations $y = g(x)$ and $y = h(x)$ also denote functions involving the variable x.

EXAMPLE 4   Let $g(x) = x^2 + 2x$. Calculate **a.** $g(\frac{2}{5})$, **b.** $g(s)$, **c.** $g(s - 1)$, and **d.** $g(-t)$.

Solution   **a.** Replace x with $\frac{2}{5}$:

$$g(x) = x^2 + 2x$$

$$g\left(\frac{2}{5}\right) = \left(\frac{2}{5}\right)^2 + 2\left(\frac{2}{5}\right)$$

$$= \frac{4}{25} + \frac{4}{5}$$

$$= \frac{24}{25}$$

b. Replace x with s:

$$g(x) = x^2 + 2x$$

$$g(s) = s^2 + 2s$$

c. Replace x with $s - 1$:

$$g(x) = x^2 + 2x$$

$$g(s - 1) = (s - 1)^2 + 2(s - 1)$$

$$= (s^2 - 2s + 1) + 2s - 2$$

$$= s^2 - 1$$

d. Replace x with $-t$:

$$g(x) = x^2 + 2x$$

$$g(-t) = (-t)^2 + 2(-t)$$

$$= t^2 - 2t$$ ■

EXAMPLE 5   Let $f(x) = 3x + 2$. Calculate **a.** $f(3) + f(2)$ and **b.** $f(a) - f(b)$.

Solution   **a.** First calculate $f(3)$.

$$f(x) = 3x + 2$$

$$f(3) = 3(3) + 2$$

$$= 9 + 2$$

$$= 11$$

Then calculate $f(2)$.

$$f(x) = 3x + 2$$

$$f(2) = 3(2) + 2$$

$$= 6 + 2$$

$$= 8$$

Finally, add the results to obtain

$$f(3) + f(2) = 11 + 8$$

$$= 19$$

b. First calculate $f(a)$.

$$f(x) = 3x + 2$$

$$f(a) = 3a + 2$$

Then calculate $f(b)$.

$$f(x) = 3x + 2$$

$$f(b) = 3b + 2$$

Finally, subtract the results to obtain

$$f(a) - f(b) = (3a + 2) - (3b + 2)$$

$$= 3a + 2 - 3b - 2$$

$$= 3a - 3b$$ ■

EXERCISE 6.5

In Exercises 1–14, indicate whether the equation or inequality determines y to be a function of x. If it does not, indicate some numbers x for which there is more than one corresponding value of y.

1. $y = x$

2. $y = 2x$

3. $y = x + 3$

4. $y = 2x - 1$

5. $y = 3x^2$

6. $y^2 = x + 1$

7. $y = 3 + 7x$

8. $y = 3 - 2x$

9. $y \leq x$

10. $y > x$

11. $x + y = 2$

12. $x - y = 3$

13. $x^2 = y^2$

14. $x^2 = 4y$

In Exercises 15–22, find a. $f(3)$, b. $f(0)$, and c. $f(-1)$.

15. $f(x) = 3x$

16. $f(x) = -4x$

17. $f(x) = 2x - 3$

18. $f(x) = 3x - 5$

19. $f(x) = 7 + 5x$

20. $f(x) = 3 + 3x$

21. $f(x) = 9 - 2x$

22. $f(x) = 12 + 3x$

In Exercises 23–30, find a. $f(1)$, b. $f(-2)$, and c. $f(3)$.

23. $f(x) = x^2$

24. $f(x) = x^2 - 2$

25. $f(x) = x^3 - 1$

26. $f(x) = x^3$

27. $f(x) = (x + 1)^2$

28. $f(x) = (x - 3)^2$

29. $f(x) = 2x^2 - x$

30. $f(x) = 5x^2 + 2x - 1$

In Exercises 31–38, find a. $f(2)$, b. $f(1)$, and c. $f(-2)$.

31. $f(x) = |x| + 2$

32. $f(x) = |x| - 5$

33. $f(x) = x^2 - 2$

34. $f(x) = x^2 + 3$

35. $f(x) = \dfrac{1}{x + 3}$

36. $f(x) = \dfrac{3}{x - 4}$

37. $f(x) = \dfrac{x}{x - 3}$

38. $f(x) = \dfrac{x}{x^2 + 2}$

In Exercises 39–46, find a. $g(w)$ and b. $g(w + 1)$.

39. $g(x) = 2x$

40. $g(x) = -3x$

41. $g(x) = 3x - 5$

42. $g(x) = 2x - 7$

43. $g(x) = x^2 + x$

44. $g(x) = x^2 - 2x$

45. $g(x) = x^2 - 1$

46. $g(x) = |x - 1|$

In Exercises 47–54, let $f(x) = 2x + 1$. Then calculate the value requested.

47. $f(3) + f(2)$

48. $f(1) - f(-1)$

49. $f(b) - f(a)$

50. $f(b) + f(a)$

51. $f(b) - 1$

52. $f(b) - f(1)$

53. $f(0) + f(-\frac{1}{2})$

54. $f(a) + f(2a)$

REVIEW EXERCISES

In Review Exercises 1–6, solve each equation.

1. $\dfrac{y + 2}{2} = 4(y + 2)$

2. $\dfrac{3z - 1}{6} - \dfrac{3z + 4}{3} = \dfrac{z + 3}{2}$

3. $\dfrac{2x + 1}{9} = \dfrac{x}{27}$

4. $\dfrac{y + 4}{5} = \dfrac{3y - 6}{3}$

5. $\dfrac{2}{x - 3} - 1 = -\dfrac{1}{3}$

6. $\dfrac{5}{x} + \dfrac{6}{x^2 + 2x} = \dfrac{-3}{x + 2}$

In Review Exercises 7–12, solve each inequality and graph the solution set on a number line.

7. $3(x - 2) + 3x < 24$

8. $\dfrac{4(x + 3)}{2} - 3(x - 4) \geq 6$

9. $x + 5 \leq 2x - 4 < 10 + x$

10. $\dfrac{-3(x - 1)}{4} \geq -x + 4$

11. $3x^2 + 5x - 1 \leq (3x + 2)(x - 1)$

12. $(x - 5)(x + 1) \geq x^2 - 3x + 2$

6.6 VARIATION

Because functions are useful in the area of science, we now introduce some special terminology that scientists use to describe these functions. The first type is called **direct variation.**

> **DEFINITION.**   The words "y varies directly with x" mean that
>
> $$y = kx$$
>
> for some constant k. The constant k is called the **constant of variation.**

If a force is applied to a spring, the spring will stretch. The greater the force, the more the spring will stretch. A physicist would call this fact Hooke's Law, and say that the *distance a spring will stretch varies directly with the force applied.* If d represents distance and f represents force, this relationship can be expressed mathematically as

1. $d = kf$

where k is the constant of variation. If a spring stretches 5 inches when a weight of 2 pounds is attached, the constant of variation can be computed by substituting 5 for d and 2 for f in Equation 1 and solving for k.

$$d = kf$$
$$5 = k(2)$$
$$\frac{5}{2} = k$$

To find the distance that the spring will stretch when a weight of 6 pounds is attached, we substitute $\frac{5}{2}$ for k and 6 for f in Equation 1 and evaluate d.

$$d = kf$$
$$d = \frac{5}{2}(6)$$
$$d = 15$$

The spring will stretch 15 inches when a weight of 6 pounds is attached.

EXAMPLE 1   At a constant speed, the distance traveled varies directly with the time. If Carlos can drive 105 miles in 3 hours, how far could he drive in 5 hours?

Solution   Let d represent the distance traveled, and let t represent the time. Translate the words "distance varies directly with time" into the equation

2. $d = kt$

To find the constant of variation, k, substitute 105 for d and 3 for t in Equation 2, and solve for k.

$$d = kt$$
$$105 = k(3)$$
$$35 = k \qquad \text{Divide both sides by 3.}$$

Substitute $k = 35$ into Equation 2.

3. $d = 35t$

To find the distance traveled in 5 hours, substitute $t = 5$ into Equation 3.

$$d = 35t$$
$$d = 35(5)$$
$$d = 175$$

In 5 hours, Carlos could travel 175 miles.   ■

Inverse Variation

DEFINITION.   The words "y varies inversely with x" mean that

$$y = \frac{k}{x}$$

for some constant k. The constant k is called the **constant of variation.**

Under a constant temperature, the volume occupied by a gas varies inversely with its pressure. If V represents volume and p represents pressure, this relationship can be expressed mathematically by the formula

4. $V = \dfrac{k}{p}$

EXAMPLE 2   A gas occupies a volume of 15 cubic inches when it is placed under 4 pounds per square inch of pressure. Find how much pressure is needed to compress the gas into a volume of 10 cubic inches.

Solution   To find the constant of variation, substitute 15 for V and 4 for p in Equation 4 and solve for k.

$$V = \frac{k}{p}$$
$$15 = \frac{k}{4}$$
$$60 = k \qquad \text{Multiply both sides by 4.}$$

To find the pressure needed to compress the gas into a volume of 10 cubic inches, substitute 60 for k and 10 for V in Equation 4 and solve for p.

$$V = \frac{k}{p}$$

$$10 = \frac{60}{p}$$

$10p = 60$ Multiply both sides by p.

$p = 6$ Divide both sides by 10.

It will take a pressure of 6 pounds per square inch to compress the gas into a volume of 10 cubic inches.   ■

Joint Variation

> **DEFINITION.** The words "y varies jointly with x and z" mean that
> $$y = kxz$$
> for some constant k. The constant k is called the **constant of variation.**

The area A of a rectangle depends on its length l and its width w by the formula

$$A = lw$$

We could say, therefore, that the area of a rectangle varies jointly with its length and its width. In this example, the constant of variation, k, is 1.

EXAMPLE 3   The area of a triangle varies jointly with the length of its base and its height. If a triangle with an area of 63 square inches has a base of 18 inches and a height of 7 inches, find the area of a triangle with a base of 12 inches and a height of 10 inches.

Solution   Let A represent the area of a triangle, let b represent the length of the base, and let h represent the height. Translate the words "Area varies jointly with the length of the base and the height" into the formula

5. $A = kbh$

We are given that $A = 63$ when $b = 18$ and $h = 7$. To determine the constant of variation, k, substitute these known values into Equation 5, and solve for k:

$$A = kbh$$

$$63 = k(18)(7)$$

$$63 = k(126)$$

$$\frac{63}{126} = k$$

$$\frac{1}{2} = k$$

Thus, $k = \frac{1}{2}$, and the complete formula for finding the area of any triangle is

6. $A = \dfrac{1}{2} bh$

To find the area of a triangle with a base of 12 inches and a height of 10 inches, substitute 12 for b and 10 for h in Equation 6, and calculate A:

$$A = \frac{1}{2} bh$$

$$A = \frac{1}{2} (12)(10)$$

$$A = 60$$

The area of the triangle is 60 square inches.   ■

Combined Variation

Many applied problems involve a combination of direct and inverse variation. Such applications are called **combined variation.**

EXAMPLE 4 The pressure of a fixed amount of gas varies directly with its temperature and inversely with its volume. A sample of gas at a pressure of 1 atmosphere occupies a volume of 3 cubic meters when its temperature is 273 degrees Kelvin (about 0° Celsius). The gas is heated to 364°K and compressed to 1 cubic meter. Find the pressure of the gas.

Solution Let P represent the pressure of the gas, let T be its temperature, and let V be the volume. Then the words "the pressure varies directly with temperature and inversely with volume" translate into the equation

7. $P = \dfrac{kT}{V}$

To determine the constant of variation, k, substitute the given information into Equation 7 and solve for k. Let $P = 1$, let $T = 273$, and let $V = 3$:

$$P = \frac{kT}{V}$$

$$1 = \frac{k \cdot 273}{3}$$

$$1 = 91k$$

$$k = \frac{1}{91}$$

Thus, $k = \frac{1}{91}$, and the complete formula is

$$P = \frac{\frac{1}{91} T}{V}$$

or

8. $P = \dfrac{T}{91V}$

To determine the pressure of the gas under the new conditions, substitute 364 for T and 1 for V into Equation 8, and calculate P:

$$P = \frac{T}{91V}$$

$$P = \frac{364}{91 \cdot 1}$$

$$= 4$$

The pressure of the heated and compressed gas is now 4 atmospheres. ∎

EXERCISE 6.6

In Exercises 1–10, express each sentence as a formula.

1. The distance d a car can travel while moving at a constant speed varies directly with n, the number of gallons of gasoline it consumes.

2. A farmer's harvest h varies directly with a, the number of acres he plants.

3. For a fixed area, the length l of a rectangle varies inversely with its width w.

4. The value v of a car varies inversely with its age a.

5. The area A of a circle varies directly with the square of its radius r.

6. The distance s that a body falls varies directly with the square of the time t.

7. Distance D traveled varies jointly with the speed s and the time t.

8. The interest I on a savings account varies jointly with the interest rate r and the time t that the money is left on deposit.

9. The current I varies directly with the voltage V and inversely with the resistance R.

10. The force of gravity F varies directly with the product of the masses m_1 and m_2, and inversely with the square of the distance d between them.

In Exercises 11–28, assume that all variables represent positive numbers.

11. Assume that y varies directly with x. If $y = 10$ when $x = 2$, find y when $x = 7$.

12. Assume that A varies directly with z. If $A = 30$ when $z = 5$, find A when $z = 9$.

13. Assume that r varies directly with s. If $r = 21$ when $s = 6$, find r when $s = 12$.

14. Assume that d varies directly with t. If $d = 15$ when $t = 3$, find t when $d = 3$.

15. Assume that s varies directly with t^2. If $s = 12$ when $t = 4$, find s when $t = 30$.

16. Assume that y varies directly with x^3. If $y = 16$ when $x = 2$, find y when $x = 3$.

17. Assume that y varies inversely with x. If $y = 8$ when $x = 1$, find y when $x = 8$.

18. Assume that V varies inversely with p. If $V = 30$ when $p = 5$, find V when $p = 6$.

19. Assume that r varies inversely with s. If $r = 40$ when $s = 10$, find r when $s = 15$.

20. Assume that J varies inversely with v. If $J = 90$ when $v = 5$, find J when $v = 45$.

21. Assume that y varies inversely with x^2. If $y = 6$ when $x = 4$, find y when $x = 2$.

22. Assume that i varies inversely with d^2. If $i = 6$ when $d = 3$, find i when $d = 2$.

23. Assume that y varies jointly with r and s. If $y = 4$ when $r = 2$ and $s = 6$, find y when $r = 3$ and $s = 4$.

24. Assume that A varies jointly with x and y. If $A = 18$ when $x = 3$ and $y = 3$, find A when $x = 7$ and $y = 9$.

25. Assume that D varies jointly with p and q. If $D = 20$ when p and q are both 5, find D when p and q are both 10.

26. Assume that z varies jointly with r and the square of s. If $z = 24$ when r and s are both 2, find z when $r = 3$ and $s = 4$.

27. Assume that y varies directly with a and inversely with b. If $y = 1$ when $a = 2$ and $b = 10$, find y when $a = 7$ and $b = 14$.

28. Assume that y varies directly with the square of x and inversely with z. If $y = 1$ when $x = 2$ and $z = 10$, find y when $x = 4$ and $z = 5$.

29. The distance traveled by an object in free fall varies directly with the square of the time that it falls. If the object falls 256 feet in 4 seconds, how far will it fall in 6 seconds?

30. The distance that a car can travel without refueling varies directly with the number of gallons of gasoline in the tank. If a car can go 360 miles on 12 gallons of gas, how far could the car go on 7 gallons?

31. For a fixed rate and principal, the interest earned in a bank account paying simple interest varies directly with the length of time the principal is left on deposit. If an investment of $5000 earns $700 in 2 years, how much will it earn in 7 years?

32. The force of gravity acting on an object varies directly with the mass of the object. The force on a mass of 5 kilograms is 49 newtons. (A newton is a unit of force.) What is the force acting on a mass of 12 kilograms?

33. The time it takes a car to travel a certain distance varies inversely with its rate of speed. If a certain trip takes 3 hours when the driver travels at 50 miles per hour, how long will the trip take when the driver travels at 60 miles per hour?

34. For a fixed area, the length of a rectangle varies inversely with its width. A rectangle has a width of 12 feet and a length of 20 feet. If the length is increased to 24 feet, find the width of the rectangle.

35. If the temperature of a gas is constant, the volume occupied varies inversely with the pressure. If a gas occupies a volume of 40 cubic meters under a pressure of 8 atmospheres, what will the volume be if the pressure is changed to 6 atmospheres?

36. Assume that the value of a machine varies inversely with the machine's age. If a drill press is worth $300 when it is 2 years old, find its worth when it is 6 years old.

37. The interest earned on a fixed amount of money varies jointly with the annual interest rate and the time that the money is left on deposit. If an account earns $120 at 8% annual interest when left on deposit for 2 years, find the interest that would be earned in 3 years if the annual rate were 12%.

38. The total cost of a quantity of identical items varies jointly with the unit cost and the quantity. If the unit cost is tripled and the quantity is doubled, by what factor is the total cost multiplied?

39. The current in a circuit varies directly with the voltage and inversely with the resistance. If a current of 4 amperes flows when 36 volts is applied to a 9-ohm resistance, find the current if the voltage is 42 volts and the resistance is 11 ohms.

40. The deflection of a beam is inversely proportional to its width and the cube of its depth. If the deflection is 2 inches when the width is 4 inches and the depth is 3 inches, find the deflection if the width is 3 inches and the depth is 4 inches.

REVIEW EXERCISES

In Review Exercises 1–4, remove parentheses and simplify each expression.

1. $2(x + 4) + 3(2x - 1)$

2. $-3(3x + 5) - 2(2x + 4)$

3. $3x(x^2 - 2) - 6x^2(x - 1)$

4. $-5a^2(a + 1) - 3a(a^2 + 4a - 3)$

In Review Exercises 5–8, simplify each fraction.

5. $\dfrac{y^2 + 2 + 3y}{y^3 + y^2}$

6. $\dfrac{a^2 - 9}{18 - 3a - a^2}$

7. $\dfrac{\dfrac{1}{t} + 1}{1 - \dfrac{1}{t}}$

8. $\dfrac{\dfrac{1}{r} - \dfrac{1}{s}}{\dfrac{1}{r} + \dfrac{1}{s}}$

In Review Exercises 9–10, perform each operation and simplify, if possible.

9. $\dfrac{x^2 + 5x + 6}{x + 3} \cdot \dfrac{x^2 + 2x - 8}{x^2 - 4}$

10. $\dfrac{3x + 1}{x + 2} + \dfrac{2x}{x + 1}$

11. If one pair of opposite sides of a square are each increased by 8 inches, and the other sides are each decreased by 4 inches, the area remains unchanged. Find the dimensions of the original square.

12. The perimeter of a triangle is 19 feet. The first side is twice as long as the second side, and the third side is 3 feet greater than the second. Find the length of each side.

CHAPTER SUMMARY

Key Words

abscissa (6.1)
cartesian coordinate system   (6.1)
combined variation (6.6)
constant of variation (6.6)
dependent variable (6.5)
direct variation (6.6)
domain (6.5)
function (6.5)
general form of a linear
   equation (6.1)
graph (6.1)
graph of a line (6.1)
half-plane (6.4)
independent variable (6.5)

intercept method of graphing   (6.1)
inverse variation (6.6)
joint variation (6.6)
linear equation (6.1)
linear inequality (6.4)
ordered pair (6.1)
ordinate (6.1)
origin (6.1)
parallel lines (6.3)
perpendicular lines (6.3)
point-slope form of a linear
   equation (6.3)
quadrant (6.1)
range (6.5)

rectangular coordinate
   system (6.1)
relation (6.5)
slope (6.2)
slope-intercept form of a linear
   equation (6.2)
subscript notation (6.2)
x-axis (6.1)
x-coordinate (6.1)
x-intercept (6.1)
y-axis (6.1)
y-coordinate (6.1)
y-intercept (6.1)

Key Ideas

(6.1) Ordered pairs of numbers are associated with points in a rectangular coordinate system.

To graph an equation in the variables x and y, choose several numbers for x (at least three), calculate the corresponding value of y, plot the points (x, y), and draw the line that passes through them.

General form of a linear equation: $Ax + By = C$, where A and B are not both zero.

To graph a linear equation by the intercept method, plot the points corresponding to the x- and y-intercepts and draw the line through them. Plot a third point as a check.

The equation $x = a$ represents the y-axis or a line parallel to the y-axis.

The equation $y = b$ represents the x-axis or a line parallel to the x-axis.

(6.2) The slope of a line passing through (x_1, y_1) and (x_2, y_2) is given by the formula

$$\text{Slope} = \frac{y_2 - y_1}{x_2 - x_1} \qquad (x_2 \neq x_1)$$

The graph of the equation $y = mx + b$ is a line with a slope of m and y-intercept of b.

Slope-intercept form of a linear equation: $y = mx + b$

Horizontal lines have a slope of zero.

The slope of a vertical line is undefined.

(6.3) **Point-slope form of a linear equation:**

$$y - y_1 = m(x - x_1)$$

Nonvertical parallel lines have the same slope.

The product of the slopes of perpendicular lines is -1, provided neither line is vertical.

(6.4) To graph an inequality in the variables x and y, first graph the boundary and then use a convenient point not on that boundary to determine which side of the line to shade.

(6.5) A relation is any set of ordered pairs of numbers.

A function is a correspondence that assigns to each number x exactly one value y. To indicate such a correspondence, the notation $y = f(x)$ is used.

(6.6) A formula of the form $y = kx$ represents direct variation.

A formula of the form $y = \frac{k}{x}$ represents inverse variation.

A formula of the form $y = kxz$ represents joint variation.

Direct and inverse variation are used together in combined variation.

CHAPTER 6 REVIEW EXERCISES

(6.1) *In Review Exercises 1–6, plot each point on a rectangular coordinate system.*

1. $A(1, 3)$

2. $B(1, -3)$

3. $C(-3, 1)$

4. $D(-3, -1)$

5. $E(0, 5)$

6. $F(-5, 0)$

In Review Exercises 7–14, find the coordinates of each indicated point in Illustration 1.

7. A

8. B

9. C

10. D

11. E

12. F

13. G

14. H

ILLUSTRATION 1

In Review Exercises 15–22, graph each equation on a rectangular coordinate system.

15. $y = x - 5$ **16.** $y = 2x + 1$ **17.** $y = \dfrac{x}{2} + 2$ **18.** $y = 3$

19. $x + y = 4$ **20.** $x - y = -3$ **21.** $3x + 5y = 15$ **22.** $7x - 4y = 28$

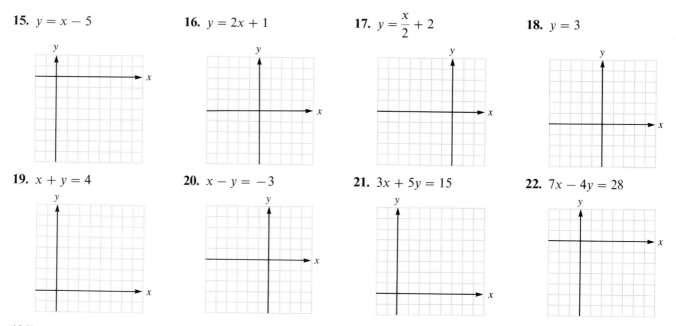

(6.2) *In Review Exercises 23–26, find the slope of the line passing through the two given points. If the slope is undefined, write "undefined slope."*

23. $(1, 4)$ $(2, 3)$ **24.** $(-1, 3)$ $(3, -2)$ **25.** $(-1, -1)$ $(-3, 0)$ **26.** $(-8, 2)$ $(3, 2)$

In Review Exercises 27–30, graph the line that passes through the given point and has the given slope.

27. $(-1, 4)$, slope 2   **28.** $(1, -2)$, slope -2 **29.** $\left(0, \dfrac{1}{2}\right)$, slope $\dfrac{3}{2}$ **30.** $(-3, 0)$, slope $-\dfrac{5}{2}$

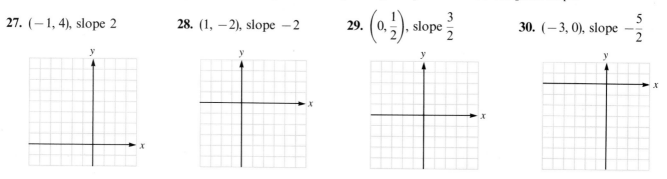

In Review Exercises 31–34, find the slope and the y-intercept of the line defined by the given equation, and graph the equation. If the line has no defined slope, write "undefined slope."

31. $y = 5x + 2$ **32.** $y = -\dfrac{x}{2} + 4$

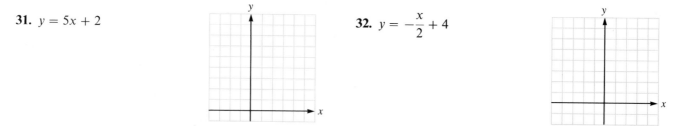

33. $y + 3 = 0$

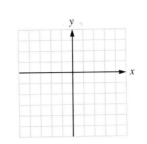

34. $x + 3y = 1$

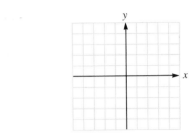

(6.3) *In Review Exercises 35–38, use the slope-intercept form of a linear equation to find the equation of each line with the given properties. Write the equation in general form.*

35. slope -3, y-intercept 2

36. slope 0, y-intercept -7

37. slope 7, y-intercept 0

38. slope $\frac{1}{2}$, y-intercept $-\frac{3}{2}$

In Review Exercises 39–42, use the point-slope form of a linear equation to find the equation of each line with the given properties. Write the equation in general form.

39. slope 3, passing through $(0, 0)$

40. slope $-\frac{1}{3}$, passing through $\left(1, \frac{2}{3}\right)$

41. slope $\frac{1}{9}$, passing through $(-27, -2)$

42. slope $-\frac{3}{5}$, passing through $\left(1, -\frac{1}{5}\right)$

In Review Exercises 43–46, determine if lines with the given slopes are parallel, perpendicular, or neither.

43. 5 and $\frac{1}{5}$

44. $\frac{2}{4}$ and 0.5

45. -5 and $\frac{1}{5}$

46. 0.25 and $\frac{1}{4}$

In Review Exercises 47–50, determine whether the graphs of the given equations are parallel, perpendicular, or neither.

47. $3x = y$
  $x = 3y$

48. $3x = y$
  $x = -3y$

49. $x + 2y = y - x$
  $2x + y = 3$

50. $3x + 2y = 7$
  $2x - 3y = 8$

In Review Exercises 51–54, find the equation of each line with the given properties. Write the equation in general form.

51. parallel to $y = 7x - 18$ and passing through $(2, 5)$

52. parallel to $3x + 2y = 7$ and passing through $(-3, 5)$

53. perpendicular to $2x - 5y = 12$ and passing through the origin

54. perpendicular to $y = \frac{x}{3} + 17$ and having a y-intercept of -4

(6.4) *In Review Exercises 55–58, graph each linear inequality.*

55. $y \le 3x + 1$

56. $y > x - 5$

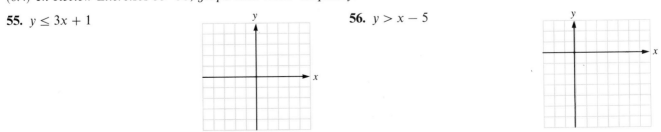

57. $2x - 3y \geq 6$

58. $2y + 3(x - y) < 5y$

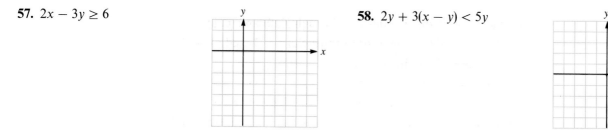

(6.5) *In Review Exercises 59–62, indicate whether the equation determines y as a function of x. If it does not, indicate some number x for which there are two corresponding values of y.*

59. $y = 2x$

60. $y = 5x^2$

61. $|y| = x$

62. $y^2 = 4x^2$

In Review Exercises 63–70, let $f(x) = x^2 - x + 1$. Then find the value indicated.

63. $f(0)$

64. $f(-2)$

65. $f(3)$

66. $f(w)$

67. $f(1) - f(-1)$

68. $f(2) + f(-2)$

69. $f(a) + f(0)$

70. $f(1) - f(a)$

(6.6) *In Review Exercises 71–74, express each variation as an equation. Then find the value requested.*

71. Assume that s varies directly with the square of t. Find s when $t = 10$ if $s = 64$ when $t = 4$.

72. Assume that l varies inversely with w. Find the constant of variation if $l = 30$ when $w = 20$.

73. Assume that R varies jointly with b and c. If $R = 72$ when $b = 4$ and $c = 24$, find R when $b = 6$ and $c = 18$.

74. Assume that s varies directly with w and inversely with the square of m. If $s = \frac{7}{4}$ when w and m are both 4, find s when $w = 5$ and $m = 7$.

CHAPTER 6 TEST

In Problems 1–4, graph each equation.

1. $y = \dfrac{x}{2} + 1$

2. $2(x + 1) - y = 4$

3. $x = 1$

4. $2y = 8$

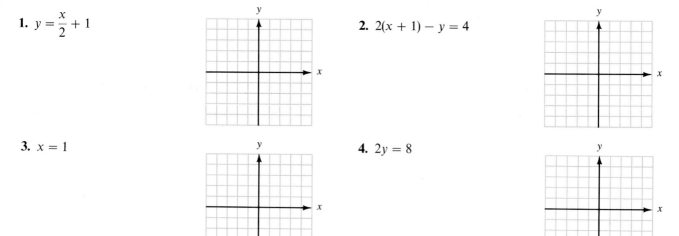

_____ **5.** Find the slope of the line passing through $(0, 0)$ and $(6, 8)$.

_____ **6.** Find the slope of the line passing through $(-1, 3)$ and $(3, -1)$.

_____ **7.** Find the slope of the line determined by $2x + y = 3$.

_____ **8.** Find the y-intercept of the line determined by $2y - 7(x + 5) = 7$.

_____ **9.** Find the slope of a line parallel to the line determined by $y = \frac{3}{5}x + 3$.

_____ **10.** Find the slope of a line perpendicular to a line with a slope of 2.

_____ **11.** In general form, write the equation of a line with a slope of $\frac{1}{2}$ and a y-intercept of 3.

_____ **12.** In general form, write the equation of a line with a slope of 7 that passes through the point $(-2, 5)$.

_____ **13.** Write the equation of a line that is parallel to the y-axis and passes through $(-3, 17)$.

_____ **14.** Write the equation of a line passing through $(3, -5)$ and perpendicular to the line with the equation $y = \frac{1}{3}x + 11$.

In Problems 15–16, graph each inequality.

15. $y \geq x + 2$

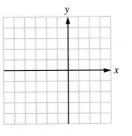

16. $x < 3$

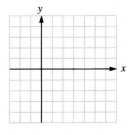

_____ **17.** Does the equation $x = 7y - 8$ determine y to be a function of x?

_____ **18.** Does the equation $y^2 = x$ determine y to be a function of x?

In Problems 19–22, assume that $f(x) = 3x + 2$.

_____ **19.** Find $f(3)$.

_____ **20.** Find $f(-2) + f(0)$.

_____ **21.** Find $f(a) - f(b)$.

_____ **22.** Find $f(t^2)$.

_____ **23.** If y varies directly with x and $y = 32$ when $x = 8$, find x when $y = 4$.

_____ **24.** If i varies inversely with the square of d, find the constant of variation if $i = 100$ when $d = 2$.

7

Solving Systems of Equations and Inequalities

Pierre de Fermat (1601–1665)
Pierre de Fermat shares the honor
with Descartes for discovering
analytic geometry, and with Pascal
for developing the theory of
probability. But to Fermat alone goes
credit for founding number theory.
He is probably most famous for an
unproved theorem called *Fermat's
last theorem*. It states that if n
represents a number greater than 2
there are no whole numbers, a, b,
and c that satisfy the equation
$a^n + b^n = c^n$. Although this theorem
is probably true, no one has been
able to prove it.

We have considered equations such as $x + y = 3$ that contain two variables. Because there are infinitely many pairs of numbers whose sum is 3, there are infinitely many ordered pairs (x, y) that will satisfy this equation. Some of these pairs are

$$x + y = 3$$

x	y
0	3
1	2
2	1
3	0

Likewise, there are infinitely many ordered pairs (x, y) that will satisfy the equation $3x - y = 1$. Some of these pairs are

$$3x - y = 1$$

x	y
0	-1
1	2
2	5
3	8

Although there are infinitely many ordered pairs that satisfy each of these equations, only the pair (1, 2) satisfies both equations at the same time. The pair of equations

$$\begin{cases} x + y = 3 \\ 3x - y = 1 \end{cases}$$

is called a **system of equations.** Because the ordered pair (1, 2) satisfies both equations simultaneously, it is called a **simultaneous solution,** or just a **solution** of the system of equations. We shall discuss three methods for finding the simultaneous solution of a system of two equations, each containing two variables.

7.1 SOLVING SYSTEMS OF EQUATIONS BY GRAPHING

To use the graphing method to solve the system

$$\begin{cases} x + y = 3 \\ 3x - y = 1 \end{cases}$$

we graph both equations on a single set of coordinate axes as in Figure 7-1. Although there are infinitely many pairs of real numbers (x, y) that satisfy the equation $x + y = 3$, and infinitely many pairs of real numbers (x, y) that satisfy the equation $3x - y = 1$, only the coordinates of the point where their graphs intersect satisfy both equations simultaneously. Thus, the solution of this system is $x = 1$ and $y = 2$, or just (1, 2).

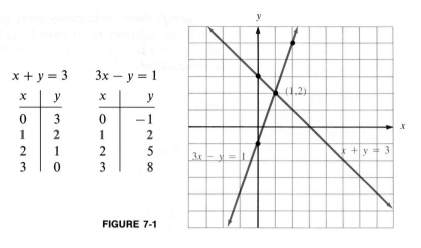

$$x + y = 3 \qquad 3x - y = 1$$

x	y
0	3
1	2
2	1
3	0

x	y
0	-1
1	2
2	5
3	8

FIGURE 7-1

To check this solution, we substitute 1 for x and 2 for y in each equation and verify that the pair $(1, 2)$ satisfies each equation.

$$x + y = 3 \qquad\qquad 3x - y = 1$$
$$1 + 2 \overset{?}{=} 3 \qquad\qquad 3(1) - 2 \overset{?}{=} 1$$
$$3 = 3 \qquad\qquad 3 - 2 \overset{?}{=} 1$$
$$1 = 1$$

The graphs of the two equations in this system are different lines. When this is so, the equations of the system are called **independent equations.** The system of equations in this example has a solution. When a system of equations has at least one solution, the system is called a **consistent system of equations.**

EXAMPLE 1 Use the graphing method to solve the system $\begin{cases} 2x + 3y = 2 \\ 3x = 2y + 16 \end{cases}$.

Solution Begin by graphing both equations on a single set of coordinate axes as in Figure 7-2.

$$2x + 3y = 2 \qquad 3x = 2y + 16$$

x	y
1	0
-2	2
7	-4

x	y
6	1
0	-8
2	-5

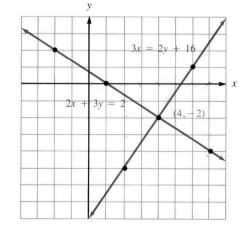

FIGURE 7-2

Although there are infinitely many pairs (x, y) that satisfy the equation $2x + 3y = 2$ and infinitely many pairs (x, y) that satisfy the equation $3x = 2y + 16$, only the coordinates of the point where the graphs intersect satisfy both equations simultaneously. Thus, the solution is $x = 4$ and $y = -2$, or just $(4, -2)$.

To check this solution, substitute 4 for x and -2 for y in each equation and verify that the pair $(4, -2)$ satisfies each equation.

$$2x + 3y = 2 \qquad\qquad 3x = 2y + 16$$
$$2(4) + 3(-2) \overset{?}{=} 2 \qquad\qquad 3(4) \overset{?}{=} 2(-2) + 16$$
$$8 - 6 \overset{?}{=} 2 \qquad\qquad 12 \overset{?}{=} -4 + 16$$
$$2 = 2 \qquad\qquad 12 = 12$$

The equations in this system are *independent equations* (their graphs are different lines), and the system is a *consistent system of equations* (it has a solution). ∎

EXAMPLE 2   Solve the system $\begin{cases} 2x + y = -6 \\ 4x + 2y = 8 \end{cases}$.

Solution   Graph both equations on one set of coordinate axes as in Figure 7-3.

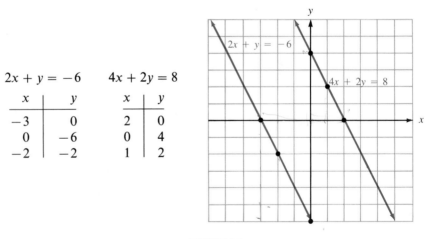

$2x + y = -6$		$4x + 2y = 8$	
x	y	x	y
-3	0	2	0
0	-6	0	4
-2	-2	1	2

FIGURE 7-3

The lines in Figure 7-3 are parallel. We can verify that this is true by writing each equation in slope-intercept form and observing that the coefficients of x are equal.

$$2x + y = -6 \qquad\qquad 4x + 2y = 8$$
$$y = -2x - 6 \qquad\qquad 2y = -4x + 8$$
$$\qquad\qquad\qquad\qquad y = -2x + 4$$

Because parallel lines do not intersect, there is no simultaneous solution to this system. ∎

In Example 2, the graphs of the two equations are different lines. Thus, the equations of the system are **independent equations.** However, the system does not have a solution. Such a system is called an **inconsistent system of equations.**

EXAMPLE 3   Solve the system $\begin{cases} y - 2x = 4 \\ 4x + 8 = 2y \end{cases}$.

Solution   Graph both equations on one set of coordinate axes as in Figure 7-4.

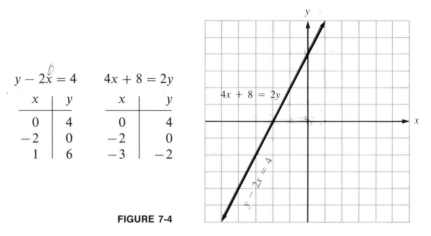

$y - 2x = 4$			$4x + 8 = 2y$	
x	y		x	y
0	4		0	4
-2	0		-2	0
1	6		-3	-2

FIGURE 7-4

The lines in Figure 7-4 coincide (they are the same line). Because these lines intersect at infinitely many points, there is an unlimited number of solutions. Any pair (x, y) that satisfies one of the equations satisfies the other also. Some possible solutions are $(0, 4)$, $(1, 6)$, and $(2, 8)$, because each of these ordered pairs satisfies each equation.   ■

When the graphs of two lines coincide, as in Example 3, the equations of the system are called **dependent equations.** Because the system of Example 3 has at least one solution, the system is consistent.

The possibilities that can occur when two equations, each with two variables, are graphed are summarized as follows:

Possible graph	if the	then
	lines are different and intersect,	the equations are independent, and the system is consistent. One solution exists.
	lines are different and parallel,	the equations are independent, and the system is inconsistent. No solutions exist.
	lines coincide,	the equations are dependent, and the system is consistent. An unlimited number of solutions exist.

EXAMPLE 4 Solve the system $\begin{cases} \dfrac{2}{3}x - \dfrac{1}{2}y = 1 \\ \dfrac{1}{2}x + \dfrac{1}{3}y = 5 \end{cases}$.

Solution Multiply both sides of the first equation by 6 to clear it of fractions.

$$\frac{2}{3}x - \frac{1}{2}y = 1$$

$$6\left(\frac{2}{3}x - \frac{1}{2}y\right) = 6(1)$$

1. $\qquad 4x - 3y = 6$

Then multiply both sides of the second given equation by 6 to clear it of fractions.

$$\frac{1}{2}x + \frac{1}{3}y = 5$$

$$6\left(\frac{1}{2}x + \frac{1}{3}y\right) = 6(5)$$

2. $\qquad 3x + 2y = 30$

Equations 1 and 2 form the following equivalent system of equations, which has the same solutions as the original system.

$$\begin{cases} 4x - 3y = 6 \\ 3x + 2y = 30 \end{cases}$$

Graph each of the equations as in Figure 7-5 and find that their point of intersection is the point with coordinates $(6, 6)$. Thus, the solution of the given system of equations is $x = 6$ and $y = 6$, or just $(6, 6)$.

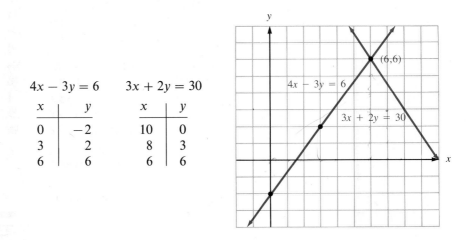

$4x - 3y = 6$			$3x + 2y = 30$	
x	y		x	y
0	-2		10	0
3	2		8	3
6	6		6	6

FIGURE 7-5

To verify that the pair $(6, 6)$ satisfies each equation of the original system, substitute 6 for x and 6 for y in each of the original equations and simplify.

$$\frac{2}{3}x - \frac{1}{2}y = 1 \qquad\qquad \frac{1}{2}x + \frac{1}{3}y = 5$$

$$\frac{2}{3}(6) - \frac{1}{2}(6) \overset{?}{=} 1 \qquad\qquad \frac{1}{2}(6) + \frac{1}{3}(6) \overset{?}{=} 5$$

$$4 - 3 \overset{?}{=} 1 \qquad\qquad 3 + 2 \overset{?}{=} 5$$

$$1 = 1 \qquad\qquad 5 = 5$$

The equations in this system are independent, and the system itself is consistent.

■

EXERCISE 7.1

In Exercises 1–12, determine whether the given ordered pair is a simultaneous solution for the given system of equations.

1. $(1, 1)$; $\begin{cases} x + y = 2 \\ 2x - y = 1 \end{cases}$ yes

2. $(1, 3)$; $\begin{cases} 2x + y = 5 \\ 3x - y = 0 \end{cases}$

3. $(3, -2)$; $\begin{cases} 2x + y = 4 \\ x + y = 1 \end{cases}$ yes

4. $(-2, 4)$; $\begin{cases} 2x + 2y = 4 \\ x + 3y = 10 \end{cases}$

5. $(4, 5)$; $\begin{cases} 2x - 3y = -7 \\ 4x - 5y = 25 \end{cases}$ NO

6. $(2, 3)$; $\begin{cases} 3x - 2y = 0 \\ 5x - 3y = -1 \end{cases}$

7. $(-2, -3)$; $\begin{cases} 4x + 5y = -23 \\ -3x + 2y = 0 \end{cases}$ yes

8. $(-5, 1)$; $\begin{cases} -2x + 7y = 17 \\ 3x - 4y = -19 \end{cases}$

9. $\left(\frac{1}{2}, 3\right)$; $\begin{cases} 2x + y = 4 \\ 4x - 3y = 11 \end{cases}$ No

10. $\left(2, \frac{1}{3}\right)$; $\begin{cases} x - 3y = 1 \\ -2x + 6y = -6 \end{cases}$

11. $\left(-\frac{2}{5}, \frac{1}{4}\right)$; $\begin{cases} 5x - 4y = -6 \\ 8y = 10x + 12 \end{cases}$ No

12. $\left(-\frac{1}{3}, \frac{3}{4}\right)$; $\begin{cases} 3x + 4y = 2 \\ 12y = 3(2 - 3x) \end{cases}$

In Exercises 13–24, use the graphing method to solve each system of equations. Write each answer as an ordered pair where possible. If a system is inconsistent, or if the equations of a system are dependent, so indicate.

13. $\begin{cases} x + y = 2 \\ x - y = 0 \end{cases}$

14. $\begin{cases} x + y = 4 \\ x - y = 0 \end{cases}$

1,1

2,2

15. $\begin{cases} x + y = 2 \\ x - y = 4 \end{cases}$

16. $\begin{cases} x + y = 1 \\ x - y = -5 \end{cases}$

17. $\begin{cases} 3x + 2y = -8 \\ 2x - 3y = -1 \end{cases}$

18. $\begin{cases} x + 4y = -2 \\ x + y = -5 \end{cases}$

19. $\begin{cases} 4x - 2y = 8 \\ y = 2x - 4 \end{cases}$

20. $\begin{cases} 3x - 6y = 18 \\ x = 2y + 3 \end{cases}$

21. $\begin{cases} 2x - 3y = -18 \\ 3x + 2y = -1 \end{cases}$

22. $\begin{cases} -x + 3y = -11 \\ 3x - y = 17 \end{cases}$

23. $\begin{cases} 4x = 3(4 - y) \\ 2y = 4(3 - x) \end{cases}$

24. $\begin{cases} 2x = 3(2 - y) \\ 3y = 2(3 - x) \end{cases}$

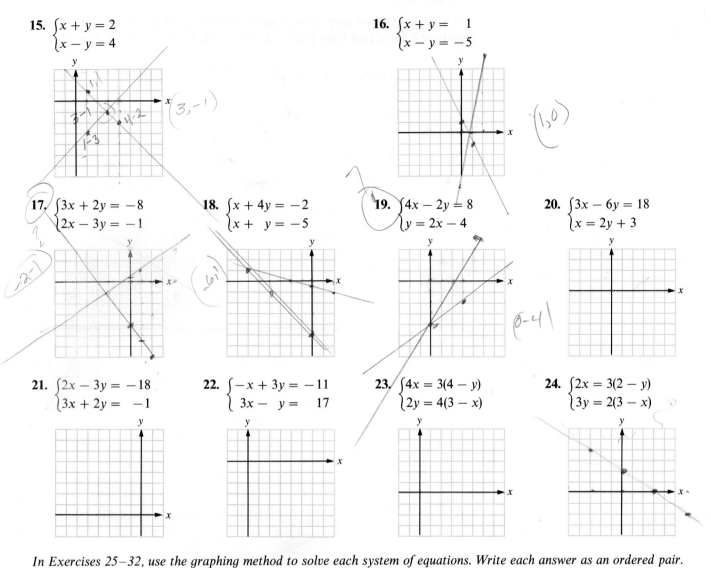

In Exercises 25–32, use the graphing method to solve each system of equations. Write each answer as an ordered pair.

25. $\begin{cases} x + 2y = -4 \\ x - \dfrac{1}{2}y = 6 \end{cases}$

26. $\begin{cases} \dfrac{2}{3}x - y = -3 \\ 3x + y = 3 \end{cases}$

27. $\begin{cases} -\dfrac{3}{4}x + y = 3 \\ \dfrac{1}{4}x + y = -1 \end{cases}$

28. $\begin{cases} \dfrac{1}{3}x + y = 7 \\ \dfrac{2}{3}x - y = -4 \end{cases}$

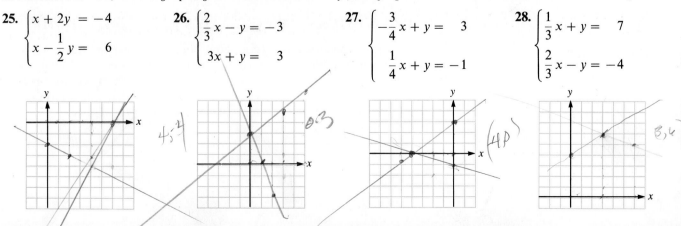

29. $\begin{cases} \dfrac{1}{2}x + \dfrac{1}{4}y = 0 \\ \dfrac{1}{4}x - \dfrac{3}{8}y = -2 \end{cases}$

30. $\begin{cases} \dfrac{1}{2}x + \dfrac{2}{3}y = -5 \\ \dfrac{3}{2}x - y = 3 \end{cases}$

31. $\begin{cases} \dfrac{1}{3}x - \dfrac{1}{2}y = \dfrac{1}{6} \\ \dfrac{2}{5}x + \dfrac{1}{2}y = \dfrac{13}{10} \end{cases}$

32. $\begin{cases} \dfrac{3}{4}x + \dfrac{2}{3}y = -\dfrac{19}{6} \\ y - x = -\dfrac{4x}{3} \end{cases}$

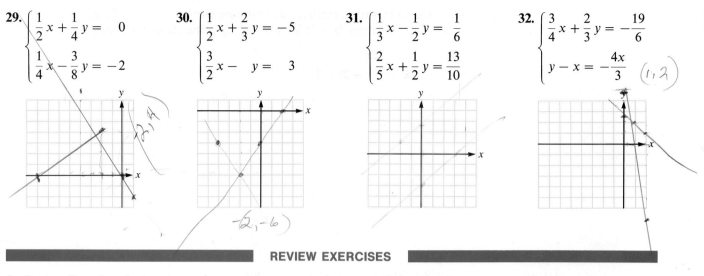

REVIEW EXERCISES

In Review Exercises 1–4, write each expression as a single exponential expression.

1. $x^3x^4x^5$

2. $\dfrac{y^7}{y^2y^3}$

3. $\dfrac{(a^3)^2}{(a^2)^3}$

4. $\dfrac{(t^2t^3)^3}{(t^2t)^4}$

In Review Exercises 5–10, perform the indicated operations and simplify.

5. $2(x^2 + 3) + 3(2x^2 - 3x + 2)$

6. $-4(2y^2 - 3y + 2) - 2(3y^2 + 4y - 3)$

7. $3x^2y(-4xy^2 - 3x^2y^2)$

8. $(2z^2 - 5)(3z^2 + 2)$

9. $\dfrac{6a^2b^2 + 8ab^2 - 10a^2b}{2ab}$

10. $3x - 1 \overline{)6x^2 + 7x - 3}$

11. The volume V of a cylinder varies directly with its height, h. If $V = 27$ when $h = 8$, find V when $h = \frac{4}{3}$.

12. The volume V of a cone varies directly with the square of r, the radius of its base. If $V = 27$ when $r = 3$, find V when $r = 6$.

7.2   SOLVING SYSTEMS OF EQUATIONS BY SUBSTITUTION

The graphing method for solving equations does not always provide exact solutions. For example, if the solution to a system of equations is $x = \frac{11}{97}$ and $y = \frac{13}{97}$, the graphs of the equations do not intersect at a point where we can read the solutions exactly. Fortunately, there are other methods that will determine exact solutions. We now consider one called the **substitution method.**

To solve the system

$$\begin{cases} y = 3x - 2 \\ 2x + y = 8 \end{cases}$$

by the substitution method, we first note that $y = 3x - 2$. Because $y = 3x - 2$, we can substitute $3x - 2$ for y in the equation $2x + y = 8$ to get

$$2x + y = 8$$
$$2x + (3x - 2) = 8$$

The resulting equation is an equation with only one variable that can be solved for x.

$$
\begin{aligned}
2x + (3x - 2) &= 8 \\
2x + 3x - 2 &= 8 \qquad \text{Use the distributive property to remove parentheses.} \\
5x - 2 &= 8 \qquad \text{Combine like terms.} \\
5x &= 10 \qquad \text{Add 2 to both sides.} \\
x &= 2 \qquad \text{Divide both sides by 5.}
\end{aligned}
$$

We can find the value of y by substituting 2 for x in either equation of the given system. Because the equation $y = 3x - 2$ is already solved for y, it is easiest to substitute in this equation.

$$y = 3x - 2 = 3(2) - 2 = 6 - 2 = 4$$

The solution to the given system is $x = 2$ and $y = 4$, or just $(2, 4)$.

$$
\begin{array}{l|l}
Check: \quad y = 3x - 2 & 2x + y = 8 \\
\qquad 4 \overset{?}{=} 3(2) - 2 & 2(2) + 4 \overset{?}{=} 8 \\
\qquad 4 \overset{?}{=} 6 - 2 & 4 + 4 \overset{?}{=} 8 \\
\qquad 4 = 4 & 8 = 8
\end{array}
$$

Because the pair of numbers $x = 2$ and $y = 4$ is a solution, the two lines represented by the equations of the given system would intersect at the point $(2, 4)$. The equations of this system are independent and the system is consistent.

EXAMPLE 1   Use the method of substitution to solve the system $\begin{cases} 2x + y = -5 \\ 3x + 5y = -4 \end{cases}$.

Solution   First solve one of the equations for one of the variables. Because the term y in the first equation has a coefficient of 1, solve the first equation for y. Then substitute this value of y into the second equation and solve for x.

$$\begin{cases} 2x + y = -5 \longrightarrow y = \boxed{-5 - 2x} \\ 3x + 5y = -4 \end{cases}$$

$$
\begin{aligned}
3x + 5(-5 - 2x) &= -4 \\
3x - 25 - 10x &= -4 \qquad \text{Remove parentheses.} \\
-7x - 25 &= -4 \qquad \text{Combine like terms.} \\
-7x &= 21 \qquad \text{Add 25 to both sides.} \\
x &= -3 \qquad \text{Divide both sides by } -7.
\end{aligned}
$$

Find y by substituting -3 for x in the equation $y = -5 - 2x$.

$$y = -5 - 2x = -5 - 2(-3) = -5 + 6 = 1$$

The solution to the given system is $x = -3$ and $y = 1$, or just $(-3, 1)$.

Check:

$2x + y = -5$	$3x + 5y = -4$
$2(-3) + 1 \stackrel{?}{=} -5$	$3(-3) + 5(1) \stackrel{?}{=} -4$
$-6 + 1 \stackrel{?}{=} -5$	$-9 + 5 \stackrel{?}{=} -4$
$-5 = -5$	$-4 = -4$

The equations of this system are independent and the system is consistent.

∎

EXAMPLE 2   Use the substitution method to solve the system $\begin{cases} 2x + 3y = 5 \\ 3x + 2y = 0 \end{cases}$.

Solution   Perhaps it is most convenient to solve the second equation for x and proceed as follows:

$$\begin{cases} 2x + 3y = 5 \\ 3x + 2y = 0 \longrightarrow 3x = -2y \end{cases}$$

$$x = \frac{-2y}{3}$$

$$2\left(\frac{-2y}{3}\right) + 3y = 5$$

$$\frac{-4y}{3} + 3y = 5 \qquad \text{Remove parentheses.}$$

$$3\left(\frac{-4y}{3}\right) + 3(3y) = 3(5) \qquad \text{Multiply both sides by 3.}$$

$$-4y + 9y = 15 \qquad \text{Remove parentheses.}$$

$$5y = 15 \qquad \text{Combine like terms.}$$

$$y = 3 \qquad \text{Divide both sides by 5.}$$

Find x by substituting 3 for y in the equation $x = \dfrac{-2y}{3}$.

$$x = \frac{-2y}{3} = \frac{-2(3)}{3} = -2$$

The solution to this system is $(-2, 3)$. Check this solution in each equation. The equations are independent and the system is consistent.   ∎

EXAMPLE 3   Use the substitution method to solve the system $\begin{cases} x = 4(3 - y) \\ 2x = 4(3 - 2y) \end{cases}$.

Solution   Substitute $4(3 - y)$ for x in the second equation and solve for y.

$$\begin{cases} x = \boxed{4(3 - y)} \\ 2\underset{\leftarrow}{x} = \boxed{4(3 - 2y)} \end{cases}$$

$$2 \cdot 4(3 - y) = 4(3 - 2y)$$
$$8(3 - y) = 4(3 - 2y) \qquad \text{Simplify.}$$
$$24 - 8y = 12 - 8y \qquad \text{Remove parentheses.}$$
$$24 = 12 \qquad \text{Add } 8y \text{ to both sides.}$$

Of course, 24 is not equal to 12. This impossible result indicates that the equations in this system are independent but the system itself is inconsistent. If each equation in this system were graphed, these graphs would be parallel lines. This system has no solution. ∎

EXAMPLE 4 Use the substitution method to solve the system $\begin{cases} 3x = 4(6 - y) \\ 4y + 3x = 24 \end{cases}$.

Solution Substitute $4(6 - y)$ for $3x$ in the second equation and proceed as follows:

$$4y + 3x = 24$$
$$4y + 4(6 - y) = 24$$
$$4y + 24 - 4y = 24 \qquad \text{Remove parentheses.}$$
$$24 = 24 \qquad \text{Combine like terms.}$$

Although the result $24 = 24$ is true, we have failed to determine a value for y. This result indicates that the equations of this system are dependent. If each equation in this system were graphed, the same line would result. Any ordered pair that satisfies one equation of this system satisfies the other also. Hence, this system has an unlimited number of solutions. Some of them are $(8, 0)$, $(0, 6)$, and $(4, 3)$, because each of these ordered pairs satisfies both equations. ∎

EXAMPLE 5 Use the substitution method to solve the system $\begin{cases} 3(x - y) = 5 \\ x + 3 = -\dfrac{5}{2} y \end{cases}$.

Solution Begin by writing each equation in general form:

$$3(x - y) = 5 \qquad \text{First equation.}$$
1. $\qquad 3x - 3y = 5 \qquad \text{Remove parentheses.}$

$$x + 3 = -\frac{5}{2} y \qquad \text{Second equation.}$$
$$2x + 6 = -5y \qquad \text{Multiply both sides by 2.}$$
2. $\qquad 2x + 5y = -6 \qquad \text{Add } 5y \text{ and } -6 \text{ to both sides.}$

Equations 1 and 2 form a new system that can be solved as follows:

1. $3x - 3y = 5 \longrightarrow 3x = 5 + 3y$

$$x = \boxed{\dfrac{5 + 3y}{3}}$$

2. $2x + 5y = -6$

$2\left(\dfrac{5 + 3y}{3}\right) + 5y = -6$

$2(5 + 3y) + 15y = -18$ Multiply both sides by 3.

$10 + 6y + 15y = -18$ Remove parentheses.

$10 + 21y = -18$ Combine like terms.

$21y = -28$ Add -10 to both sides.

$y = \dfrac{-28}{21}$ Divide both sides by 21.

$y = -\dfrac{4}{3}$ Simplify $\dfrac{-28}{21}$.

To find x, substitute $-\dfrac{4}{3}$ for y in the equation $x = \dfrac{5 + 3y}{3}$ and simplify.

$$x = \frac{5 + 3y}{3} = \frac{5 + 3\left(-\dfrac{4}{3}\right)}{3} = \frac{5 - 4}{3} = \frac{1}{3}$$

The solution is $(\tfrac{1}{3}, -\tfrac{4}{3})$. Check this solution in each equation.   ■

EXERCISE 7.2

Use the substitution method to solve each system of equations. Write each answer as an ordered pair, where possible. If a system is inconsistent or if the equations of a system are dependent, so indicate.

1. $\begin{cases} y = 2x \\ x + y = 6 \end{cases}$ **2.** $\begin{cases} y = 3x \\ x + y = 4 \end{cases}$ **3.** $\begin{cases} y = 2x - 6 \\ 2x + y = 6 \end{cases}$ **4.** $\begin{cases} y = 2x - 9 \\ x + 3y = 8 \end{cases}$

5. $\begin{cases} y = 2x + 5 \\ x + 2y = -5 \end{cases}$ **6.** $\begin{cases} y = -2x \\ 3x + 2y = -1 \end{cases}$ **7.** $\begin{cases} 2a + 4b = -24 \\ a = 20 - 2b \end{cases}$ **8.** $\begin{cases} 3a + 6b = -15 \\ a = -2b - 5 \end{cases}$

9. $\begin{cases} 2a = 3b - 13 \\ b = 2a + 7 \end{cases}$ **10.** $\begin{cases} a = 3b - 1 \\ b = 2a + 2 \end{cases}$ **11.** $\begin{cases} r + 3s = 9 \\ 3r + 2s = 13 \end{cases}$ **12.** $\begin{cases} x - 2y = 2 \\ 2x + 3y = 11 \end{cases}$

13. $\begin{cases} 4x + 5y = 2 \\ 3x - y = 11 \end{cases}$ **14.** $\begin{cases} 5u + 3v = 5 \\ 4u - v = 4 \end{cases}$ **15.** $\begin{cases} 2x + y = 0 \\ 3x + 2y = 1 \end{cases}$ **16.** $\begin{cases} 3x - y = 7 \\ 2x + 3y = 1 \end{cases}$

17. $\begin{cases} 3x + 4y = -7 \\ 2y - x = -1 \end{cases}$ **18.** $\begin{cases} 4x + 5y = -2 \\ x + 2y = -2 \end{cases}$ **19.** $\begin{cases} 9x = 3y + 12 \\ 4 = 3x - y \end{cases}$ **20.** $\begin{cases} 8y = 15 - 4x \\ x + 2y = 4 \end{cases}$

21. $\begin{cases} 2x + 3y = 5 \\ 3x + 2y = 5 \end{cases}$

22. $\begin{cases} 3x - 2y = -1 \\ 2x + 3y = -5 \end{cases}$

23. $\begin{cases} 2x + 5y = -2 \\ 4x + 3y = 10 \end{cases}$

24. $\begin{cases} 3x + 4y = -6 \\ 2x - 3y = -4 \end{cases}$

25. $\begin{cases} 2x - 3y = -3 \\ 3x + 5y = -14 \end{cases}$

26. $\begin{cases} 4x - 5y = -12 \\ 5x - 2y = 2 \end{cases}$

27. $\begin{cases} 7x - 2y = -1 \\ -5x + 2y = -1 \end{cases}$

28. $\begin{cases} -8x + 3y = 22 \\ 4x + 3y = -2 \end{cases}$

29. $\begin{cases} 2a + 3b = 2 \\ 8a - 3b = 3 \end{cases}$

30. $\begin{cases} 3a - 2b = 0 \\ 9a + 4b = 5 \end{cases}$

31. $\begin{cases} y - x = 3x \\ 2(x + y) = 14 - y \end{cases}$

32. $\begin{cases} y + x = 2x + 2 \\ 2(3x - 2y) = 21 - y \end{cases}$

33. $\begin{cases} 3(x - 1) + 3 = 8 + 2y \\ 2(x + 1) = 4 + 3y \end{cases}$

34. $\begin{cases} 4(x - 2) = 19 - 5y \\ 3(x + 1) - 2y = 2y \end{cases}$

35. $\begin{cases} 6a = 5(3 + b + a) - a \\ 3(a - b) + 4b = 5(1 + b) \end{cases}$

36. $\begin{cases} 5(x + 1) + 7 = 7(y + 1) \\ 5(y + 1) = 6(1 + x) + 5 \end{cases}$

37. $\begin{cases} \dfrac{1}{2}x + \dfrac{1}{2}y = -1 \\ \dfrac{1}{3}x - \dfrac{1}{2}y = -4 \end{cases}$

38. $\begin{cases} \dfrac{2}{3}y + \dfrac{1}{5}z = 1 \\ \dfrac{1}{3}y - \dfrac{2}{5}z = 3 \end{cases}$

39. $\begin{cases} 5x = \dfrac{1}{2}y - 1 \\ \dfrac{1}{4}y = 10x - 1 \end{cases}$

40. $\begin{cases} \dfrac{2}{3}x = 1 - 2y \\ 2(5y - x) + 11 = 0 \end{cases}$

41. $\begin{cases} \dfrac{6x - 1}{3} - \dfrac{5}{3} = \dfrac{3y + 1}{2} \\ \dfrac{1 + 5y}{4} + \dfrac{x + 3}{4} = \dfrac{17}{2} \end{cases}$

42. $\begin{cases} \dfrac{5x - 2}{4} + \dfrac{1}{2} = \dfrac{3y + 2}{2} \\ \dfrac{7y + 3}{3} = \dfrac{x}{2} + \dfrac{7}{3} \end{cases}$

REVIEW EXERCISES

In Review Exercises 1–6, factor each expression completely.

1. $8x^2y^2 - 32xy^2z + 16xyz^2$

2. $(x - y)a - (x - y)b$

3. $a^6 - 25$

4. $b^4 - 625$

5. $r^2 + 2rs - 15s^2$

6. $4m^2 - 15mn + 9n^2$

In Review Exercises 7–10, simplify each fraction.

7. $\dfrac{21ab^2c^3}{14abc^4}$

8. $\dfrac{x^2 + xy}{x^2 - y^2}$

9. $\dfrac{-2x + 2y + 2z}{4x - 4y - 4z}$

10. $\dfrac{2t^2 + t - 3}{2t^2 + 7t + 6}$

11. If the voltage is constant, then the current I passing through a resistance R varies inversely with R. If $I = 17$ when $R = 2$, find I when $R = 17$.

12. If the power dissipated by a resistance R is constant, then R is inversely proportional to the square of the current I. If $R = 9$ when $I = 2$, find R when $I = 3$.

7.3 SOLVING SYSTEMS OF EQUATIONS BY ADDITION

Another method used to solve systems of equations is called the **addition method.** To use the addition method to solve the system

$$\begin{cases} x + y = 8 \\ x - y = -2 \end{cases}$$

we add the left-hand sides of the equations and the right-hand sides of the equations to eliminate the variable y. The resulting equation can then be solved for x.

$$\begin{array}{r} x + y = 8 \\ x - y = -2 \\ \hline 2x = 6 \end{array}$$

We can now solve the equation $2x = 6$ for x.

$$2x = 6$$
$$x = 3 \qquad \text{Divide both sides by 2.}$$

To find the value of y, we multiply the first equation of the system by -1 to obtain the system

$$\begin{cases} -x - y = -8 \\ x - y = -2 \end{cases}$$

When we add these equations, the terms involving x are eliminated, and we can solve the resulting equation for y.

$$\begin{array}{r} -x - y = -8 \\ x - y = -2 \\ \hline -2y = -10 \\ y = 5 \qquad \text{Divide both sides by } -2. \end{array}$$

Thus, the solution of this system of equations is $(3, 5)$.

We could have found the value of y by substituting 3 for x in either of the equations of the system and solving the resulting equation for y. For example, substituting 3 for x in the equation $x + y = 8$ and solving for y gives

$$x + y = 8$$
$$3 + y = 8$$
$$y = 5 \qquad \text{Add } -3 \text{ to both sides.}$$

Check the solution by verifying that the pair $(3, 5)$ satisfies each equation of the system.

EXAMPLE 1   Use the addition method to solve the system $\begin{cases} 3y = 14 + x \\ x + 22 = 5y \end{cases}$.

Solution   Write the equations in the form $\begin{cases} -x + 3y = 14 \\ x - 5y = -22 \end{cases}$.

When these equations are added, the terms involving x are eliminated. Solve the resulting equation for y.

$$\begin{array}{r} -x + 3y = 14 \\ x - 5y = -22 \\ \hline -2y = -8 \end{array}$$

$$y = 4 \qquad \text{Divide both sides by } -2.$$

To find the value of x, substitute 4 for y in either equation of the system. For example, we could substitute 4 for y in the equation $-x + 3y = 14$ and solve for x.

$$\begin{aligned} -x + 3y &= 14 \\ -x + 3(4) &= 14 \\ -x + 12 &= 14 \qquad \text{Simplify.} \\ -x &= 2 \qquad \text{Add } -12 \text{ to both sides.} \\ x &= -2 \qquad \text{Divide both sides by } -1. \end{aligned}$$

The solution to this system is $(-2, 4)$.

Check this solution by verifying that the pair $(-2, 4)$ satisfies each equation of the system.   ∎

EXAMPLE 2   Use the addition method to solve the system $\begin{cases} 2x - 5y = 10 \\ 3x - 2y = -7 \end{cases}$.

Solution   In this system each equation must be adjusted so that one of the variables will be eliminated when the equations are added. To eliminate the x variable, for example, multiply the first equation by 3 and the second equation by -2. This gives the system

$$\begin{cases} 6x - 15y = 30 \\ -6x + 4y = 14 \end{cases}$$

When these equations are added, the terms involving the variable x are eliminated.

$$\begin{array}{r} 6x - 15y = 30 \\ -6x + 4y = 14 \\ \hline -11y = 44 \end{array}$$

$$y = -4 \qquad \text{Divide both sides by } -11.$$

To find x, substitute -4 for y in the equation $2x - 5y = 10$, for example.

$$2x - 5y = 10$$
$$2x - 5(-4) = 10$$
$$2x + 20 = 10 \qquad \text{Simplify.}$$
$$2x = -10 \qquad \text{Add } -20 \text{ to both sides.}$$
$$x = -5 \qquad \text{Divide both sides by 2.}$$

The solution to this system is $(-5, -4)$. Check this solution.   ∎

EXAMPLE 3   Use the addition method to solve the system $\begin{cases} x - \dfrac{2}{3}y = \dfrac{8}{3} \\ -\dfrac{3}{2}x + y = -6 \end{cases}$.

Solution   Multiply both sides of the first equation by 3 and both sides of the second equation by 2 to clear the equations of fractions. This gives the system

$$\begin{cases} 3x - 2y = 8 \\ -3x + 2y = -12 \end{cases}$$

Add these equations to eliminate the x variable.

$$\begin{array}{r} 3x - 2y = 8 \\ -3x + 2y = -12 \\ \hline 0 = -4 \end{array}$$

In this case *both* variables are eliminated, and the *false* result $0 = -4$ is obtained. This indicates that the equations of the system are independent but the system itself is inconsistent. The given system has no solutions.   ∎

EXAMPLE 4   Use the addition method to solve the system $\begin{cases} x - \dfrac{5}{2}y = \dfrac{19}{2} \\ -\dfrac{2}{5}x + y = -\dfrac{19}{5} \end{cases}$.

Solution   Multiply both sides of the first equation by 2 and both sides of the second equation by 5 to clear the equations of fractions. This gives the system

$$\begin{cases} 2x - 5y = 19 \\ -2x + 5y = -19 \end{cases}$$

Add these equations to eliminate a variable.

$$\begin{array}{r} 2x - 5y = 19 \\ -2x + 5y = -19 \\ \hline 0 = 0 \end{array}$$

As in Example 3, both the x and y variables were eliminated. However, this time the *true* result $0 = 0$ was obtained. This indicates that the equations of this system are dependent, and the system has an unlimited number of solutions. Any ordered pair that satisfies one of the equations satisfies the other also. Some solutions are $(2, -3)$, $(12, 1)$, and $(0, -\frac{19}{5})$.   ∎

EXAMPLE 5   Use the addition method to solve the system $\begin{cases} \dfrac{5}{6}x + \dfrac{2}{3}y = \dfrac{7}{6} \\ \dfrac{10}{7}x - \dfrac{4}{9}y = \dfrac{17}{21} \end{cases}$.

Solution   To clear the equations of fractions, multiply both sides of the first equation by 6 and both sides of the second equation by 63. This gives the system

1. $\begin{cases} 5x + 4y = 7 \end{cases}$
2. $\begin{cases} 90x - 28y = 51 \end{cases}$

Solve for x by eliminating the terms involving y. To do so, multiply Equation 1 by 7 and add the result to Equation 2.

$$\begin{array}{r} 35x + 28y = 49 \\ 90x - 28y = 51 \\ \hline 125x \qquad = 100 \end{array}$$

$$x = \frac{100}{125} \qquad \text{Divide both sides by 125.}$$

$$x = \frac{4}{5} \qquad \text{Simplify.}$$

To solve for y, substitute $\frac{4}{5}$ for x in any convenient equation, such as Equation 1, and simplify.

$$5x + 4y = 7$$

$$5\left(\frac{4}{5}\right) + 4y = 7$$

$$4 + 4y = 7 \qquad \text{Simplify.}$$

$$4y = 3 \qquad \text{Add } -4 \text{ to both sides.}$$

$$y = \frac{3}{4} \qquad \text{Divide both sides by 4.}$$

The solution of this system is $(\frac{4}{5}, \frac{3}{4})$. Check this solution.   ■

EXERCISE 7.3

In Exercises 1–12, use the addition method to solve each system of equations. Write each answer as an ordered pair.

1. $\begin{cases} x + y = 5 \\ x - y = -3 \end{cases}$

2. $\begin{cases} x - y = 1 \\ x + y = 7 \end{cases}$

3. $\begin{cases} x - y = -5 \\ x + y = 1 \end{cases}$

4. $\begin{cases} x + y = 1 \\ x - y = 5 \end{cases}$

5. $\begin{cases} 2x + y = -1 \\ -2x + y = 3 \end{cases}$

6. $\begin{cases} 3x + y = -6 \\ x - y = -2 \end{cases}$

7. $\begin{cases} 2x - 3y = -11 \\ 3x + 3y = 21 \end{cases}$

8. $\begin{cases} 3x - 2y = 16 \\ -3x + 8y = -10 \end{cases}$

9. $\begin{cases} 2x + y = -2 \\ -2x - 3y = -6 \end{cases}$

10. $\begin{cases} 3x + 4y = 8 \\ 5x - 4y = 24 \end{cases}$

11. $\begin{cases} 4x + 3y = 24 \\ 4x - 3y = -24 \end{cases}$

12. $\begin{cases} 5x - 4y = 8 \\ -5x - 4y = 8 \end{cases}$

In Exercises 13–42, use the addition method to solve each system of equations. Write each answer as an ordered pair, where possible. If the equations of a system are dependent or if a system is inconsistent, so indicate.

13. $\begin{cases} x + y = 5 \\ x + 2y = 8 \end{cases}$
14. $\begin{cases} x + 2y = 0 \\ x - y = -3 \end{cases}$
15. $\begin{cases} 2x + y = 4 \\ 2x + 3y = 0 \end{cases}$
16. $\begin{cases} 2x + 5y = -13 \\ 2x - 3y = -5 \end{cases}$

17. $\begin{cases} 3x + 29 = 5y \\ 4y - 34 = -3x \end{cases}$
18. $\begin{cases} 3x - 16 = 5y \\ 33 - 5y = 4x \end{cases}$
19. $\begin{cases} 2x = 3(y - 2) \\ 2(x + 4) = 3y \end{cases}$
20. $\begin{cases} 3(x - 2) = 4y \\ 2(2y + 3) = 3x \end{cases}$

21. $\begin{cases} -2(x + 1) = 3(y - 2) \\ 3(y + 2) = 6 - 2(x - 2) \end{cases}$
22. $\begin{cases} 5(x - 1) = 8 - 3(y + 2) \\ 4(x + 2) - 7 = 3(2 - y) \end{cases}$

23. $\begin{cases} 4(x + 1) = 17 - 3(y - 1) \\ 2(x + 2) + 3(y - 1) = 9 \end{cases}$
24. $\begin{cases} 3(x + 3) + 2(y - 4) = 5 \\ 3(x - 1) = -2(y + 2) \end{cases}$

25. $\begin{cases} 2x + y = 10 \\ x + 2y = 10 \end{cases}$
26. $\begin{cases} 3x + 2y = 0 \\ 2x - 3y = -13 \end{cases}$
27. $\begin{cases} 2x - y = 16 \\ 3x + 2y = 3 \end{cases}$
28. $\begin{cases} 3x + 4y = -17 \\ 4x - 3y = -6 \end{cases}$

29. $\begin{cases} 4x + 5y = -20 \\ 5x - 4y = -25 \end{cases}$
30. $\begin{cases} 3x - 5y = 4 \\ 7x + 3y = 68 \end{cases}$
31. $\begin{cases} 6x = -3y \\ 5y = 2x + 12 \end{cases}$
32. $\begin{cases} 3y = 4x \\ 5x = 4y - 2 \end{cases}$

33. $\begin{cases} 4(2x - y) = 18 \\ 3(x - 3) = 2y - 1 \end{cases}$
34. $\begin{cases} 2(2x + 3y) = 5 \\ 8x = 3(1 + 3y) \end{cases}$
35. $\begin{cases} \dfrac{3}{5}x + \dfrac{4}{5}y = 1 \\ -\dfrac{1}{4}x + \dfrac{3}{8}y = 1 \end{cases}$
36. $\begin{cases} \dfrac{1}{2}x - \dfrac{1}{4}y = 1 \\ \dfrac{1}{3}x + y = 3 \end{cases}$

37. $\begin{cases} \dfrac{3}{5}x + y = 1 \\ \dfrac{4}{5}x - y = -1 \end{cases}$
38. $\begin{cases} \dfrac{1}{2}x + \dfrac{4}{7}y = -1 \\ 5x - \dfrac{4}{5}y = -10 \end{cases}$

39. $\begin{cases} \dfrac{x}{2} - \dfrac{y}{3} = -2 \\ \dfrac{2x - 3}{2} + \dfrac{6y + 1}{3} = \dfrac{17}{6} \end{cases}$
40. $\begin{cases} \dfrac{x + 2}{4} + \dfrac{y - 1}{3} = \dfrac{1}{12} \\ \dfrac{x + 4}{5} - \dfrac{y - 2}{2} = \dfrac{5}{2} \end{cases}$

41. $\begin{cases} \dfrac{x - 3}{2} + \dfrac{y + 5}{3} = \dfrac{11}{6} \\ \dfrac{x + 3}{3} - \dfrac{5}{12} = \dfrac{y + 3}{4} \end{cases}$
42. $\begin{cases} \dfrac{x + 2}{3} = \dfrac{3 - y}{2} \\ \dfrac{x + 3}{2} = \dfrac{2 - y}{3} \end{cases}$

REVIEW EXERCISES

In Review Exercises 1–6, solve each equation.

1. $8(3x - 5) - 12 = 4(2x + 3)$

2. $3y + \dfrac{y + 2}{2} = \dfrac{2(y + 3)}{3} + 16$

3. $3z^2 - 5z + 2 = 0$ **4.** $3t^2 + 4 = 8t$ **5.** $10y^2 + 21y = 10$ **6.** $x(9x - 24) + 12 = 0$

7. Solve the formula $P = 2l + 2w$ for w.   **8.** Solve the formula $A = p + prt$ for p.

In Review Exercises 9–12, solve each word problem.

9. Find three consecutive integers whose sum is 318.

10. The length of a rectangle is 6 feet greater than its width, and its perimeter is 72 feet. Find its dimensions.

11. Don had $14,000 to invest, some at 7% annual interest and the rest at 10%. How much did he invest at the lower rate if his annual income from the two investments is $1280?

12. In a triangle, one angle is twice the second angle, and the third angle is three times the second. How large is each angle? (*Hint:* There are 180° in the sum of the angles of a triangle.)

7.4 APPLICATIONS OF SYSTEMS OF EQUATIONS

Previously we have set up equations involving one variable to solve word problems. In this section we consider ways to solve word problems by using two variables. The following steps are helpful when solving a problem involving two unknown quantities.

---

1. Read the problem several times and analyze the facts. Occasionally, a sketch, chart, or diagram will help you visualize the facts of the problem.
2. Pick different variables to represent two unknown quantities, and write a sentence stating what each variable represents.
3. Find two equations involving each of the two variables. This will give a system of two equations in two variables.
4. Solve the system using the most convenient method.
5. State the solution or solutions.
6. Check the answer in the words of the problem.

EXAMPLE 1   A farmer raises wheat and soybeans on 215 acres. He plants wheat on 31 more acres than he plants soybeans. How many acres of each does he plant?

Solution   Let w represent the number of acres of wheat planted, and let s equal the number of acres of soybeans planted. Because the total land used by both crops together is 215 acres, we have the equation

$$w + s = 215$$

Because the area devoted to wheat exceeds that devoted to soybeans by 31 acres, we also have the equation

$$w - s = 31$$

We can now solve the system

1. $\begin{cases} w + s = 215 \\ w - s = 31 \end{cases}$
2.

by using the addition method.

$$w + s = 215$$
$$\underline{w - s = 31}$$
$$2w = 246$$
$$w = 123 \qquad \text{Divide both sides by 2.}$$

To find the value of s, substitute 123 for w in one of the original equations of the system, such as Equation 1, and solve for s.

$$w + s = 215$$
$$123 + s = 215 \qquad \text{Substitute 123 for } w.$$
$$s = 92 \qquad \text{Add } -123 \text{ to both sides.}$$

Thus, the farmer plants 123 acres of wheat and 92 acres of soybeans.

Check: The total acreage planted is $123 + 92$, or 215 acres. The area devoted to wheat is 31 acres greater than that used for soybeans, because $123 - 92 = 31$.

■

EXAMPLE 2   Gail wants to cut a 19-foot pole into two pieces. The longer piece is to be 1 foot longer than twice the shorter piece. Find the length of each piece.

Solution   Let s represent the length of the shorter piece and l represent the length of the longer piece (see Figure 7-6).
   Because the pole is 19 feet long, we have the equation

$$s + l = 19$$

Because the longer piece is 1 foot longer than twice the shorter piece, we have the equation

$$l = 2s + 1$$

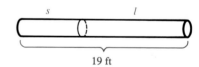

FIGURE 7-6

Use the substitution method to solve this system.

1. $\begin{cases} s + l = 19 \\ l = 2s + 1 \end{cases}$
2.

$$s + 2s + 1 = 19$$
$$3s + 1 = 19 \qquad \text{Combine like terms.}$$
$$3s = 18 \qquad \text{Add } -1 \text{ to both sides.}$$
$$s = 6 \qquad \text{Divide both sides by 3.}$$

The shorter piece should be 6 feet long.

To find the length of the longer piece, substitute 6 for s in Equation 1 and solve for l.

$$s + l = 19$$
$$6 + l = 19$$
$$l = 13 \qquad \text{Add } -6 \text{ to both sides.}$$

Thus, the longer piece should be 13 feet long.

Check: The sum of 13 and 6 is 19. 13 is 1 more than twice 6.   ■

EXAMPLE 3   Tom intends to use 150 feet of fencing to enclose a rectangular garden. The length of the garden is 5 feet less than 3 times its width. Find the area of the garden.

Solution   Let l represent the length of the garden and w represent the width (see Figure 7-7). Because the perimeter (the distance around the garden) is 150 feet, the sum of two lengths and two widths must equal 150. Thus,

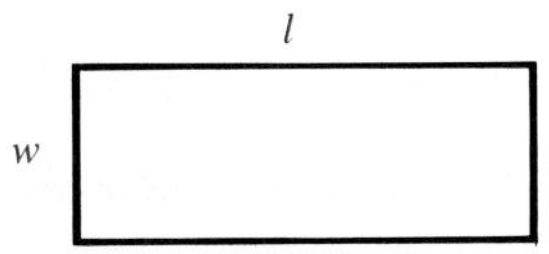

$$2l + 2w = 150$$

Because the length is 5 feet less than 3 times the width,

$$l = 3w - 5$$

Use the substitution method to solve this system.

1. $\begin{cases} 2l + 2w = 150 \\ l = 3w - 5 \end{cases}$
2.

$$2(3w - 5) + 2w = 150$$
$$6w - 10 + 2w = 150 \qquad \text{Remove parentheses.}$$
$$8w - 10 = 150 \qquad \text{Combine like terms.}$$
$$8w = 160 \qquad \text{Add 10 to both sides.}$$
$$w = 20 \qquad \text{Divide both sides by 8.}$$

Thus, the width is 20 feet.

To find the length, substitute 20 for w in Equation 2 and simplify.

$$l = 3w - 5 = 3(20) - 5 = 60 - 5 = 55$$

Because the dimensions of the rectangle are 55 feet by 20 feet, and the area of a rectangle is given by the formula

$$A = l \cdot w \qquad \text{Area = length times width.}$$

we have

$$A = 55 \cdot 20$$
$$= 1100$$

The garden covers an area of 1100 square feet.

Check: Because the dimensions of the garden are 55 feet by 20 feet, the perimeter can be found as follows:

FIGURE 7-7

$$P = 2l + 2w = 2(55) + 2(20) = 110 + 40 = 150$$

It is also true that 55 feet is 5 feet less than 3 times 20 feet. ∎

EXAMPLE 4 Nancy has a shoe box containing 1056 coins, consisting of nickels and dimes. The coins are worth $84.55. How many of each coin are in the shoe box?

Solution Let n represent the number of nickels and d represent the number of dimes. Because there are 1056 coins in all, we can form the equation

$$n + d = 1056$$

Because n nickels are worth $0.05n$ and d dimes are worth $0.10d$, the sum of $0.05n$ and $0.10d$ represents the total value of the coins. This fact enables us to form the equation

$$0.05n + 0.10d = 84.55$$

We now have the system

1.
2. $\begin{cases} n + d = 1056 \\ 0.05n + 0.10d = 84.55 \end{cases}$

Multiply Equation 2 by 100 to eliminate the decimal fractions. Then multiply Equation 1 by -5 to get a system that can be solved by addition.

$$\begin{array}{r} -5n - 5d = -5280 \\ 5n + 10d = 8455 \\ \hline 5d = 3175 \\ d = 635 \end{array}$$ Divide both sides by 5.

To find the number of nickels, substitute 635 for d in Equation 1 and simplify.

$$\begin{array}{l} n + d = 1056 \\ n + 635 = 1056 \\ n = 421 \end{array}$$ Add -635 to both sides.

Nancy has 421 nickels and 635 dimes.

Check: $421 + 635 = 1056$

The value of 421 nickels is $21.05
The value of 635 dimes is $63.50
The total value is $84.55 ∎

EXAMPLE 5 Jeffrey invested some money at 8% annual interest, and Grant invested some at 10%. The first year's income on their combined investment of $15,000 is $1340. How much did each invest?

Solution Let x represent the amount of money invested by Jeffrey and y represent the amount of money invested by Grant. Because their total investment was $15,000, we can form the equation

$$x + y = 15,000$$

The income on the x dollars invested at 8% is $0.08x$. The income on the y

dollars invested at 10% is $0.10y$. The combined income is $1340. Hence, we can form the equation

$$0.08x + 0.10y = 1340$$

to get the system

1. $\begin{cases} x + y = 15{,}000 \\ 0.08x + 0.10y = 1340 \end{cases}$
2.

Use the addition method and proceed as follows:

$$
\begin{array}{ll}
-8x - 8y = -120{,}000 & \text{Multiply both sides of Equation 1 by } -8. \\
\underline{8x + 10y = 134{,}000} & \text{Multiply both sides of Equation 2 by 100.} \\
2y = 14{,}000 & \\
y = 7000 & \text{Divide both sides by 2.}
\end{array}
$$

To find x, substitute 7000 for y in Equation 1 and simplify.

$$
\begin{aligned}
x + y &= 15{,}000 \\
x + 7000 &= 15{,}000 \\
x &= 8000 \qquad \text{Add } -7000 \text{ to both sides.}
\end{aligned}
$$

Jeffrey invested $8000 and Grant invested $7000.

Check: $8000 + 7000 = 15{,}000$

$$
\begin{aligned}
&8\% \text{ of } \$8000 \text{ is } \ \ \$640 \\
&10\% \text{ of } \$7000 \text{ is } \ \ \underline{\$700} \\
&\text{The total interest is } \$1340
\end{aligned}
$$

■

EXAMPLE 6 A boat can travel 30 kilometers downstream in 3 hours and can make the return trip in 5 hours. Find the speed of the boat in still water.

Solution Let s represent the speed of the boat in still water and let c represent the speed of the current. Then the rate of speed of the boat while going downstream is $s + c$. The rate of the boat while going upstream is $s - c$. Organize the information of the problem as in Figure 7-8.

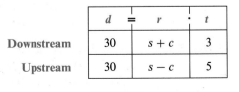

	d =	r ·	t
Downstream	30	$s + c$	3
Upstream	30	$s - c$	5

FIGURE 7-8

Because $d = r \cdot t$, the information in the table gives two equations in two variables.

$$\begin{cases} 30 = 3(s + c) \\ 30 = 5(s - c) \end{cases}$$

After removing parentheses and rearranging terms, we have

1. $\begin{cases} 3s + 3c = 30 \\ 5s - 5c = 30 \end{cases}$
2.

To solve this system by addition, we multiply Equation 1 by 5, Equation 2 by 3, add the equations, and solve for s.

$$\begin{aligned} 15s + 15c &= 150 \\ \underline{15s - 15c} &= \underline{90} \\ 30s &= 240 \end{aligned}$$

$$s = 8 \qquad \text{Divide both sides by 30.}$$

The speed of the boat in still water is 8 kilometers per hour. Check the result. ■

EXAMPLE 7   A laboratory technician has one batch of solution that is 40% alcohol and a second batch that is 60% alcohol. She would like to make 8 liters of solution that is 55% alcohol. How many liters of each batch should she use?

Solution   Let x represent the number of liters to be used from batch 1 and y represent the number of liters to be used from batch 2. Organize the information of the problem as in Figure 7-9.

	Fractional part that is alcohol	Number of liters of solution	Number of liters of alcohol
Batch 1	0.40	x	0.40x
Batch 2	0.60	y	0.60y
Mixture	0.55	8	0.55(8)

FIGURE 7-9

The information in Figure 7-9 provides two equations.

1. $x + y = 8$ The number of liters of batch 1 plus the number of liters of batch 2 equals the total number of liters in the mixture.

2. $0.40x + 0.60y = 0.55(8)$ The amount of alcohol in batch 1 plus the amount of alcohol in batch 2 equals the amount of alcohol in the mixture.

Use addition to solve this system.

$$\begin{aligned} -40x - 40y &= -320 \qquad && \text{Multiply both sides of Equation 1 by } -40. \\ \underline{40x + 60y} &= \underline{440} \qquad && \text{Multiply both sides of Equation 2 by 100.} \\ 20y &= 120 \end{aligned}$$

$$y = 6 \qquad \text{Divide both sides by 20.}$$

To find x, substitute 6 for y in Equation 1 and simplify.

$$x + y = 8$$
$$x + 6 = 8$$
$$x = 2 \qquad \text{Add } -6 \text{ to both sides.}$$

The technician should use 2 liters of the 40% solution and 6 liters of the 60% solution. Check the result.   ■

EXERCISE 7.4

Use two equations in two variables to solve each word problem.

1. One number is twice another. Their sum is 96. Find the numbers.

2. If the sum of two numbers is 38 and the difference is 12, what are the numbers?

3. Three times a certain number plus another number is 29, but the first number plus twice the second number is 18. Find the numbers.

4. Twice a certain number plus another number is 21, but the first number plus 3 times the second number is 33. Find the numbers

5. Eight cans of paint and three paint brushes cost $135. How much does each cost if six cans of paint and two brushes cost $100?

6. One catcher's mitt and 10 outfielder's gloves cost $239.50. One catcher's mitt and 5 outfielder's gloves cost $134.50. How much does each cost?

7. Two bottles of contact lens cleaner and three bottles of soaking solution cost $29.40, and three bottles of cleaner and two bottles of soaking solution cost $28.60. Find the cost of each.

8. Two pairs of shoes and four pairs of socks cost $109, and three pairs of shoes and five pairs of socks cost $160. Find the cost of a pair of socks.

9. Jerry wants to cut a 25-foot pole into two pieces. He wants one piece to be 5 feet longer than the other. How long should each piece be?

10. A carpenter wants to cut a 20-foot pole into two pieces so that one piece is 4 times as long as the other. How long should each piece be?

11. The perimeter of a rectangle is 110 feet. The length is 5 feet longer than its width. Find its dimensions.

12. A rectangle is 3 times as long as it is wide. Its perimeter is 80 centimeters. Find its dimensions.

13. A rectangle has a length that is 2 feet more than twice its width. Its perimeter is 34 feet. Find its area.

14. The perimeter of a rectangle is 50 meters. Its width is two-thirds its length. Find its area.

15. A girl has 80 coins that are quarters and dimes. The coins are worth $14. How many quarters and how many dimes does she have?

16. A girl has some dimes and some nickels. There are 25 coins in all, and they are worth $1.75. How many of each type of coin does she have?

17. David has equal numbers of dimes and quarters worth $3.50. How many of each does he have?

18. A girl has twice as many nickels as dimes. Her coins are worth $12.40. How many coins does she have?

19. Bill invested some money at 5% annual interest, and Janette invested some at 7%. If their combined interest was $310 on a total investment of $5000, how much did each invest?

20. Peter invested some money at 6% annual interest, and Martha invested some at twice that rate. If their combined investment was $6000 and their combined interest was $540, how much money did each invest?

21. Students can buy tickets to a basketball game for $1. However, the admission for nonstudents is $2. If 350 tickets are sold and the total receipts are $450, how many student tickets and how many nonstudent tickets were sold?

22. General admission movie tickets cost $4, but senior citizens are admitted for $3. If the receipts were $720 for a total audience of 190 people, how many senior citizens attended?

23. A boat can travel 24 miles downstream in 2 hours and can make the return trip in 3 hours. Find the speed of the boat in still water.

24. With the wind, a plane can fly 3000 miles in 5 hours. Against the same wind, the trip takes 6 hours. What is the airspeed of the plane (the speed in still air)?

25. An airplane can fly downwind a distance of 600 miles in 2 hours. However, the return trip against the same wind takes 3 hours. What is the speed of the wind?

26. It takes a motorboat 4 hours to travel 56 miles down a river, but it takes 3 hours longer to make the return trip. Find the speed of the current.

27. A chemist has a solution that is 40% alcohol and another solution that is 55% alcohol. How much of each must she use to make 15 liters of a solution that is 50% alcohol?

28. A pharmacist has a solution that is 25% alcohol and another that is 50% alcohol. How much of each must he use to make 20 liters of a solution that is 40% alcohol?

29. A merchant wants to mix peanuts worth $3 per pound with cashews worth $6 per pound to get 48 pounds of mixed nuts to sell at $4 per pound. How many pounds of peanuts and how many pounds of cashews must the merchant use?

30. A merchant wants to mix peanuts worth $3 per pound with jelly beans worth $1.50 per pound to make 30 pounds of a mixture worth $2.10 per pound. How many pounds of each should he use?

31. Lisa wants to buy gifts for seven friends. On some she will only spend $7, but on each of her best friends she will spend $9. Lisa spends $53, total. How many of each gift did she buy?

32. Three soft drinks and a sandwich cost $4.15. One soft drink and three sandwiches cost $6.45. Find the cost of each.

33. An electronics store put two types of car radio on sale. One model costs $87, and the other costs $119. During the sale, the receipts for the 25 radios sold were $2495. How many of each model were sold?

34. At a restaurant, ice cream cones cost $.90 and sundaes cost $1.65. One day the receipts for a total of 148 cones and sundaes were $180.45. Determine how many cones and how many sundaes were sold.

35. An investment of $950 at one rate of interest and $1200 at a higher rate together generate an annual income of $205.50. The investment rates differ by 1%. Find the two rates of interest. (*Hint:* Treat 1% as 0.01.)

36. A man drives for a while at 45 miles per hour. Realizing that he is running late, he increases his speed to 60 miles per hour, and completes his 405-mile trip in 8 hours. How long does he travel at each speed?

REVIEW EXERCISES

In Review Exercises 1–4, graph each equation.

1. $y = 2x - 4$ **2.** $2x + y = 6$ **3.** $2x = 3y + 1$ **4.** $\dfrac{x}{2} - \dfrac{y}{3} = 2$

In Review Exercises 5–8, graph each inequality.

5. $2x - y \le 4$ **6.** $x + 3y > -2$ **7.** $2x + 3y > 0$ **8.** $\dfrac{1}{3}x + 2y \ge 1$

In Review Exercises 9–12, graph each expression on a rectangular coordinate system.

9. $y = 5$ **10.** $x < 2$ **11.** $y \ge -3$ **12.** $x = -1$

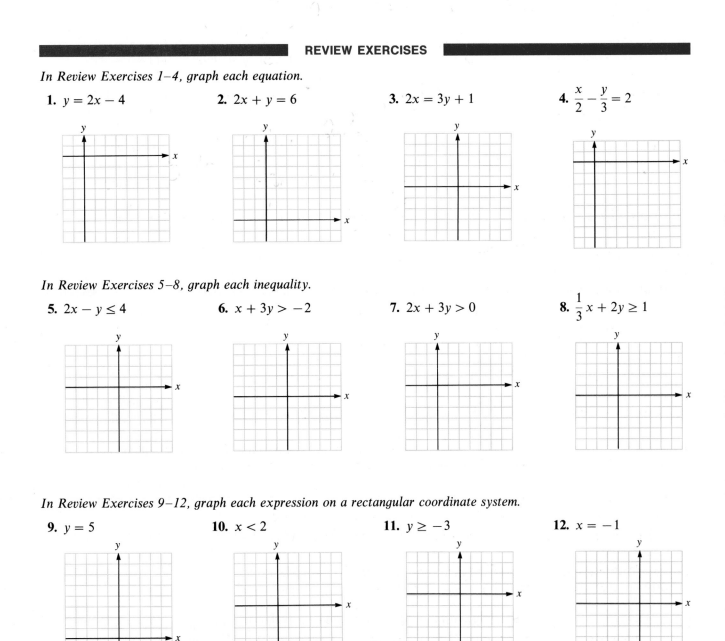

7.5   SOLVING SYSTEMS OF INEQUALITIES

In Section 6.4, we graphed inequalities that contain two variables. We now consider how to solve **systems of inequalities.** To solve the system

$$\begin{cases} x + y \geq 1 \\ x - y \geq 1 \end{cases}$$

for example, we graph each inequality on a set of coordinate axes as in Figure 7-10.

$x + y = 1$		$x - y = 1$	
x	y	x	y
0	1	0	−1
1	0	1	0
2	−1	2	1

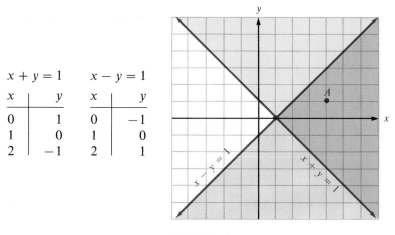

FIGURE 7-10

The graph of $x + y \geq 1$ includes the graph of the equation $x + y = 1$ and all points above it. Because the boundary line is included, we draw it with a solid line. The graph of the inequality $x - y \geq 1$ includes the graph of the equation $x - y = 1$ and all points below it. Because the boundary line is included also, we draw it with a solid line.

The area that is shaded blue represents the set of simultaneous solutions of the given system of inequalities. Any point in the blue-shaded region has coordinates that will satisfy both of the inequalities of the system.

To see that this is true, we can pick points, such as point A, that lie in the blue-shaded region and show that their coordinates satisfy both inequalities. Because point A has coordinates (4, 1), we have

$$\begin{array}{ccc} x + y \geq 1 & \text{and} & x - y \geq 1 \\ 4 + 1 \geq 1 & & 4 - 1 \geq 1 \\ 5 \geq 1 & & 3 \geq 1 \end{array}$$

Since the coordinates of point A satisfy each equation, point A is a solution of the system. If we pick a point that is not in the blue-shaded region, its coordinates will not satisfy one of the equations.

In general, to solve systems of inequalities, we will

> **1.** Graph each inequality in the system on the same coordinate axes.
> **2.** Find the region where the graphs overlap.
> **3.** Pick a test point from the region to verify the solution. Its coordinates must satisfy each inequality in the system.

EXAMPLE 1 Graph the solution set of the system $\begin{cases} 2x + y < 4 \\ -2x + y > 2 \end{cases}$.

Solution Graph each inequality in the system on the same set of coordinate axes as in Figure 7-11.

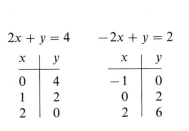

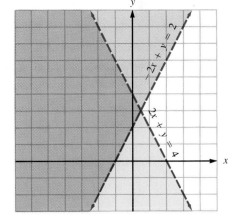

$$2x + y = 4 \qquad -2x + y = 2$$

x	y
0	4
1	2
2	0

x	y
−1	0
0	2
2	6

FIGURE 7-11

The graph of $2x + y < 4$ includes all points below the line $2x + y = 4$. Because the boundary line itself is *not* included, draw it as a broken line. The graph of $-2x + y > 2$ includes all points above the line $-2x + y = 2$. Because the boundary line itself is again *not* included, draw it as a broken line.

The area that is shaded blue represents the set of simultaneous solutions of the given system of inequalities. Any point in the blue-shaded region has coordinates that will satisfy both inequalities of the system. Pick a point in the blue-shaded region and show that it satisfies both inequalitites. ■

EXAMPLE 2 Graph the solution set of the system $\begin{cases} x \le 2 \\ y > 3 \end{cases}$.

Solution Graph each inequality on the same set of coordinate axes as in Figure 7-12.

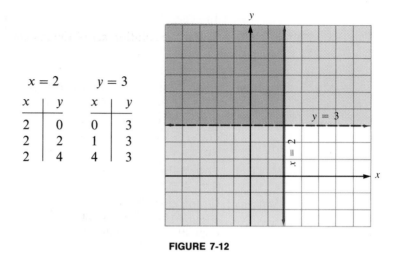

$x = 2$		$y = 3$	
x	y	x	y
2	0	0	3
2	2	1	3
2	4	4	3

FIGURE 7-12

The graph of $x \leq 2$ includes all points to the left of the line $x = 2$. Because the boundary line *is* included, draw it as a solid line. The graph $y > 3$ includes all points above the line $y = 3$. Because the boundary line is *not* included, draw it as a broken line.

The area that is shaded blue represents the set of simultaneous solutions of the given system of inequalities. Any point in the blue-shaded region has coordinates that will satisfy both inequalities of the system. Pick a point in the blue-shaded region and show that this is true. ∎

EXAMPLE 3   Graph the solution set of the system $\begin{cases} y < 3x - 1 \\ y \geq 3x + 1 \end{cases}$.

Solution   Graph each inequality as in Figure 7-13. The graph of $y < 3x - 1$ includes all of the points below the broken line $y = 3x - 1$. The graph of $y \geq 3x + 1$ includes all of the points on and above the solid line $y = 3x + 1$. Because the graphs of these inequalities do not intersect, the solution set is empty. There are no solutions.

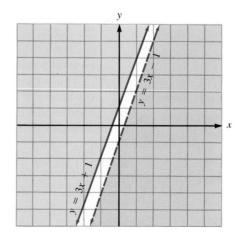

FIGURE 7-13   ∎

EXAMPLE 4   Graph the solution set of the system $\begin{cases} x \geq 0 \\ y \geq 0 \\ x + 2y \leq 6 \end{cases}$

Solution   Graph each inequality as in Figure 7-14. The graph of $x \geq 0$ includes all of the points on the y-axis and to the right. The graph of $y \geq 0$ includes all of the points on the x-axis and above. The graph of $x + 2y \leq 6$ includes all of the points on the line $x + 2y = 6$ and below. The solution is the region where all three of these graphs overlap. This includes triangle OPQ and the triangular region it encloses.

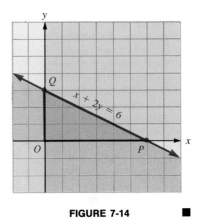

FIGURE 7-14   ∎

EXERCISE 7.5

In Exercises 1–26, use the method of graphing to find the solution set of each system of inequalities. If a system has no solutions, so indicate.

1. $\begin{cases} x + 2y \leq 3 \\ 2x - y \geq 1 \end{cases}$

2. $\begin{cases} 2x + y \geq 3 \\ x - 2y \leq -1 \end{cases}$

3. $\begin{cases} x + y < -1 \\ x - y > -1 \end{cases}$

4. $\begin{cases} x + y > 2 \\ x - y < -2 \end{cases}$

5. $\begin{cases} 2x - y < 4 \\ x + y \geq -1 \end{cases}$

6. $\begin{cases} x - y \geq 5 \\ x + 2y < -4 \end{cases}$

7. $\begin{cases} x > 2 \\ y \leq 3 \end{cases}$

8. $\begin{cases} x \geq -1 \\ y > -2 \end{cases}$

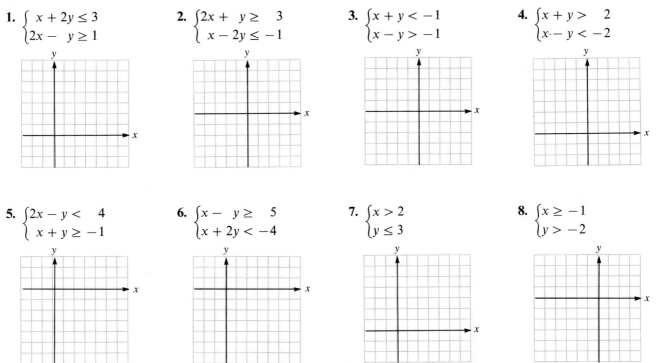

9. $\begin{cases} 2x - 3y < 0 \\ y > x - 1 \end{cases}$

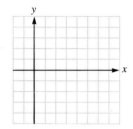

10. $\begin{cases} 3x - y \geq -1 \\ y \geq 3x + 1 \end{cases}$

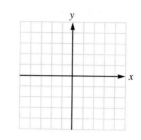

11. $\begin{cases} x + y < 1 \\ x + y > 3 \end{cases}$

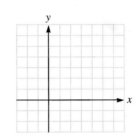

12. $\begin{cases} x + y > 2 \\ x + y < 4 \end{cases}$

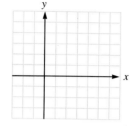

13. $\begin{cases} x > 0 \\ y > 0 \end{cases}$

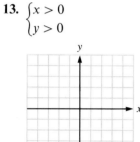

14. $\begin{cases} x \leq 0 \\ y < 0 \end{cases}$

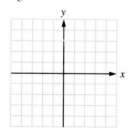

15. $\begin{cases} 3x + 4y > -7 \\ 2x - 3y \geq 1 \end{cases}$

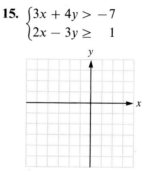

16. $\begin{cases} 3x + y \leq 1 \\ 4x - y > -8 \end{cases}$

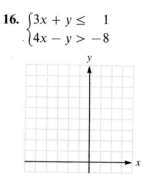

17. $\begin{cases} x < 3y - 1 \\ y \geq 2x - 3 \end{cases}$

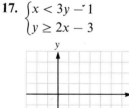

18. $\begin{cases} y \geq x + 2 \\ x \leq y - 2 \end{cases}$

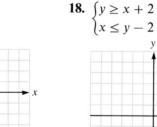

19. $\begin{cases} 2x + y < 7 \\ y > 2(1 - x) \end{cases}$

20. $\begin{cases} 2x + y \geq 6 \\ y \leq 2(2x - 3) \end{cases}$

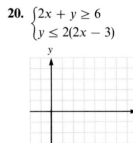

21. $\begin{cases} 2x - 4y > -6 \\ 3x + y \geq 5 \end{cases}$

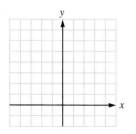

22. $\begin{cases} 2x - 3y < 0 \\ 2x + 3y \geq 12 \end{cases}$

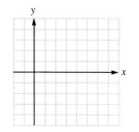

23. $\begin{cases} 3x - y \leq -4 \\ 3y > -2(x + 5) \end{cases}$

24. $\begin{cases} 3x + y < -2 \\ y > 3(1 - x) \end{cases}$

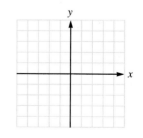

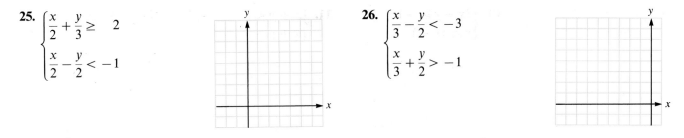

25. $\begin{cases} \dfrac{x}{2} + \dfrac{y}{3} \geq 2 \\[2mm] \dfrac{x}{2} - \dfrac{y}{2} < -1 \end{cases}$

26. $\begin{cases} \dfrac{x}{3} - \dfrac{y}{2} < -3 \\[2mm] \dfrac{x}{3} + \dfrac{y}{2} > -1 \end{cases}$

In Exercises 27–30, use the graphing method to find the region that satisfies all of the inequalities of the system.

27. $\begin{cases} x \geq 0 \\ y \geq 0 \\ x + y \leq 3 \end{cases}$

28. $\begin{cases} x - y \leq 6 \\ x + 2y \leq 6 \\ x \geq 0 \end{cases}$

29. $\begin{cases} x \geq 0 \\ y \geq 0 \\ x \leq 5 \\ y \leq x \end{cases}$

30. $\begin{cases} x \geq 0 \\ y \geq 0 \\ y \leq 2 + x \\ y \geq 4x - 2 \end{cases}$

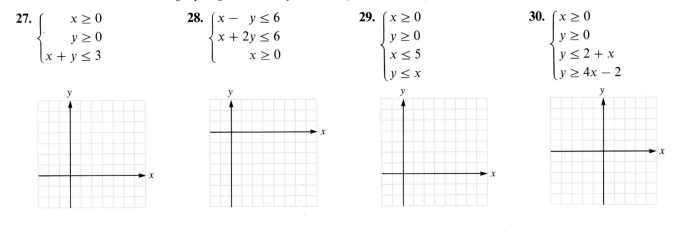

REVIEW EXERCISES

In Review Exercises 1–8, simplify each expression. Write each answer without using negative exponents.

1. $x^5 x^2$

2. $\dfrac{y^6}{y^2}$

3. $(a^2)^5$

4. a^{-3}

5. $(z^3 z^{-2})^{-3}$

6. $\dfrac{t^2 t^{-1}}{t^3}$

7. $\left(\dfrac{3m^2}{n^3}\right)^4$

8. $\left(\dfrac{y^4}{y^{-2}}\right)^3$

In Review Exercises 9–10, simplify each complex fraction.

9. $\dfrac{\dfrac{x^2}{y^3 z}}{\dfrac{x}{yz^2}}$

10. $\dfrac{\dfrac{x}{y} + \dfrac{y}{x}}{\dfrac{y}{x} - \dfrac{x}{y}}$

In Review Exercises 11–12, factor each expression.

11. $15x^2 - 27x$

12. $18x^3 - 50x$

7.6 SOLVING SYSTEMS OF THREE EQUATIONS IN THREE VARIABLES

We have seen that a solution to a system of equations such as

$$\begin{cases} 2x + 3y = 13 \\ 3x + 2y = 12 \end{cases}$$

is an ordered pair of real numbers like $(2, 3)$ that satisfies both of the given equations simultaneously. Likewise, a solution to the system

$$\begin{cases} 2x + 3y + 4z = 9 \\ 3x + 2y + 7z = 12 \\ 5x + 2y + 2z = 9 \end{cases}$$

is an ordered triple of numbers (x, y, z) that satisfies each of the three given equations simultaneously. The solution is $(1, 1, 1)$.

A linear equation in two variables has a graph that is a straight line. A system of two linear equations in two variables is consistent or inconsistent depending on whether a pair of lines intersect or are parallel.

The graph of an equation in three variables of the form $ax + by + cz = d$ is a flat surface called a **plane.** A system of three equations in three variables is consistent or inconsistent depending on how the three planes corresponding to the three equations intersect. The drawings in Figure 7-15 illustrate some of the possibilities.

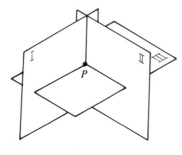

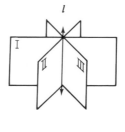

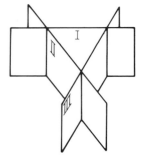

The three planes intersect at a single point P: One solution

The three planes have a line l in common: An infinite number of solutions

The three planes have no point in common: No solutions

(a)

(b)

(c)

FIGURE 7-15

Example 1 discusses a consistent system of three equations in three variables. Example 2 discusses a system that is inconsistent.

EXAMPLE 1 Solve the system $\begin{cases} 2x + y + 4z = 12 \\ x + 2y + 2z = 9 \\ 3x - 3y - 2z = 1 \end{cases}$.

Solution We are given the following system of equations in three variables:

1. $\begin{cases} 2x + y + 4z = 12 \\ \end{cases}$
2. $\begin{cases} x + 2y + 2z = 9 \end{cases}$
3. $\begin{cases} 3x - 3y - 2z = 1 \end{cases}$

Use the addition method to eliminate the variable z and, thereby, obtain a system of two equations in two variables. If Equations 2 and 3 are added, the variable z is eliminated:

2. $\quad x + 2y + 2z = 9$
3. $\quad 3x - 3y - 2z = 1$
4. $\quad \overline{4x - y \qquad = 10}$

Now pick a different pair of equations and eliminate the variable z again. If each side of Equation 3 is multiplied by 2 and the resulting equation is added to equation 1, the variable z is eliminated again:

1. $\quad 2x + y + 4z = 12$
   $\quad 6x - 6y - 4z = 2$
5. $\quad \overline{8x - 5y \qquad = 14}$

Equations 4 and 5 form a system of two equations in two variables:

4. $\begin{cases} 4x - y = 10 \\ \end{cases}$
5. $\begin{cases} 8x - 5y = 14 \end{cases}$

To solve this system, multiply Equation 4 by -5, add the resulting equation to Equation 5 to eliminate the variable y, and solve for x:

$\quad -20x + 5y = -50$
5. $\quad 8x - 5y = 14$
6. $\quad \overline{-12x \qquad = -36}$
$\quad\qquad x = 3 \qquad$ Divide both sides by -12.

To find the variable y, substitute 3 for x in an equation containing the variables x and y, such as Equation 5, and solve for y:

5. $\quad 8x - 5y = 14$
   $\quad 8(3) - 5y = 14$
   $\quad 24 - 5y = 14 \qquad$ Simplify.
   $\quad -5y = -10 \qquad$ Add -24 to both sides.
   $\quad y = 2 \qquad$ Divide both sides by -5.

To find the variable z, substitute 3 for x and 2 for y in an equation that contains the variables x, y, and z, such as Equation 1, and solve for z:

1. $2x + y + 4z = 12$

$2(3) + 2 + 4z = 12$

$8 + 4z = 12$ Simplify.

$4z = 4$ Add -8 to both sides.

$z = 1$ Divide both sides by 4.

The solution of the system is $(x, y, z) = (3, 2, 1)$. Verify that these values satisfy each of the equations in the system.   ∎

EXAMPLE 2 Solve the system $\begin{cases} 2x + y - 3z = -3 \\ 3x - 2y + 4z = 2 \\ 4x + 2y - 6z = -7 \end{cases}$.

Solution We are given the following system of equations in three variables:

1. $\begin{cases} 2x + y - 3z = -3 \\ 3x - 2y + 4z = 2 \\ 4x + 2y - 6z = -7 \end{cases}$
2.
3.

Begin by multiplying Equation 1 by 2 and adding the resulting equation to Equation 2 to eliminate the variable y:

$\begin{array}{rl} & 4x + 2y - 6z = -6 \\ \textbf{2.} & 3x - 2y + 4z = 2 \\ \hline \textbf{4.} & 7x \quad\quad - 2z = -4 \end{array}$

Now add Equations 2 and 3 to eliminate the variable y again:

$\begin{array}{rl} \textbf{2.} & 3x - 2y + 4z = 2 \\ \textbf{3.} & 4x + 2y - 6z = -7 \\ \hline \textbf{5.} & 7x \quad\quad - 2z = -5 \end{array}$

Equations 4 and 5 form the system

4. $7x - 2z = -4$

5. $7x - 2z = -5$

Because no values of x and z can cause $7x - 2z$ to equal both -4 and -5 at the same time, this system must be inconsistent. Thus, the original system has no solutions, either; the original system is inconsistent.   ∎

EXAMPLE 3 The sum of three integers is 2. The third integer is 2 greater than the second and 17 greater than the first. Find the three integers.

Solution Let a, b, and c represent the three integers. Because their sum is 2, we have

1. $a + b + c = 2$

Because the third integer is 2 greater than the second, we know that

$c = b + 2,$ or

2. $-b + c = 2$

Because the third integer is 17 greater than the first, we have $c = a + 17$ or

3. $-a + c = 17$

Put these three equations together to form a system of three equations in three variables:

1.
2. $\begin{cases} a + b + c = 2 \\ \quad\ \ -b + c = 2 \\ -a \quad\ \ + c = 17 \end{cases}$
3.

Add Equations 1 and 2 to get Equation 4:

4. $a + 2c = 4$

Equations 3 and 4 form a system of two equations in two variables:

3.
4. $\begin{cases} -a + \ c = 17 \\ \ \ a + 2c = 4 \end{cases}$

Add Equations 3 and 4 to get the equation

$$3c = 21$$
$$c = 7$$

Substitute 7 for c in Equation 4 to find a:

4. $a + 2c = 4$
$a + 2(7) = 4$
$a + 14 = 4$ Simplify.
$a = -10$ Add -14 to both sides.

Substitute 7 for c in Equation 2 to find b:

2. $-b + c = 2$
$-b + 7 = 2$
$-b = -5$ Add -7 to both sides.
$b = 5$ Divide both sides by -1.

Thus, the three integers are -10, 5, and 7. Note that these three integers have a sum of 2, that 7 is 2 greater than 5, and that 7 is 17 greater than -10. ■

EXERCISE 7.6

In Exercises 1–12, solve each system of equations. If a system of equations is inconsistent, or if the equations are dependent, so indicate.

1. $\begin{cases} x + \ y + z = 4 \\ 2x + \ y - z = 1 \\ 2x - 3y + z = 1 \end{cases}$

2. $\begin{cases} x + y + z = 4 \\ x - y + z = 2 \\ x - y - z = 0 \end{cases}$

3. $\begin{cases} 2x + 2y + 3z = 10 \\ 3x + y - z = 0 \\ x + y + 2z = 6 \end{cases}$

4. $\begin{cases} x - y + z = 4 \\ x + 2y - z = -1 \\ x + y - 3z = -2 \end{cases}$

5. $\begin{cases} x + y + 2z = 7 \\ x + 2y + z = 8 \\ 2x + y + z = 9 \end{cases}$

6. $\begin{cases} x + 2y + 2z = 10 \\ 2x + y + 2z = 9 \\ 2x + 2y + z = 11 \end{cases}$

7. $\begin{cases} 2x + y - z = 1 \\ x + 2y + 2z = 2 \\ 4x + 5y + 3z = 3 \end{cases}$

8. $\begin{cases} 4x + 3z = 4 \\ 2y - 6z = -1 \\ 8x + 4y + 3z = 9 \end{cases}$

9. $\begin{cases} 2x + 3y + 4z = 6 \\ 2x - 3y - 4z = -4 \\ 4x + 6y + 8z = 12 \end{cases}$

10. $\begin{cases} x - 3y + 4z = 2 \\ 2x + y + 2z = 3 \\ 4x - 5y + 10z = 7 \end{cases}$

11. $\begin{cases} x + \dfrac{1}{3}y + z = 13 \\ \dfrac{1}{2}x - y + \dfrac{1}{3}z = -2 \\ x + \dfrac{1}{2}y - \dfrac{1}{3}z = 2 \end{cases}$

12. $\begin{cases} x - \dfrac{1}{5}y - z = 9 \\ \dfrac{1}{4}x + \dfrac{1}{5}y - \dfrac{1}{2}z = 5 \\ 2x + y + \dfrac{1}{6}z = 12 \end{cases}$

In Exercises 13–22, solve each word problem.

13. The sum of three numbers is 18. The third number is four times the second, and the second number is 6 more than the first. Find the numbers.

14. The sum of three numbers is 48. If the first number is doubled, the sum is 60. If the second number is doubled, the sum is 63. Find the numbers.

15. Three numbers have a sum of 30. The third number is 8 less than the sum of the first and second, and the second number is one-half the sum of the first and third. Find the numbers.

16. The sum of the three angles in any triangle is 180°. In triangle *ABC*, angle *A* is 100° less than the sum of angles *B* and *C*, and angle *C* is 40° less than twice angle *B*. Find each angle.

17. A collection of 17 nickels, dimes, and quarters has a value of $1.50. There are twice as many nickels as dimes. How many of each kind are there?

18. A unit of food contains 1 gram of fat, 1 gram of carbohydrate, and 2 grams of protein. A second contains 2 grams of fat, 1 gram of carbohydrate, and 1 gram of protein. A third contains 2 grams of fat, 1 gram of carbohydrate, and 2 grams of protein. How many units of each must be used to provide exactly 11 grams of fat, 6 grams of carbohydrate, and 10 grams of protein?

19. A factory manufactures three types of footballs at a monthly cost of $2425 for 1125 footballs. The manufacturing costs for the three types of footballs are $4, $3, and $2. These footballs sell for $16, $12, and $10, respectively. How many of each type are manufactured if the monthly profit is $9275? (*Hint:* Profit = income − cost.)

20. A retailer purchased 105 radios from sources A, B, and C. Five fewer units were purchased from C than from A and B combined. If twice as many had been purchased from A, the total would have been 130. Find the number purchased from each source.

21. Tickets for a concert cost $5, $3, and $2. Twice as many $5 tickets were sold as $2 tickets. The receipts for 750 tickets were $2625. How many tickets of each price were sold?

22. The owner of a candy store wants to mix some peanuts worth $3 per pound, some cashews worth $9 per pound, and some brazil nuts worth $9 per pound to get 50 pounds of a mixture that will sell for $6 per pound. She used 15 fewer pounds of cashews than peanuts. How many pounds of each did she use?

REVIEW EXERCISES

1. If $3x - 2 = 5x + 8$, find the value of $3x - 1$.

2. If $5(y + 3) = 4y - 7$, find the value of $2y + 5$.

3. If $\dfrac{3 - x}{2} = \dfrac{2 - x}{3}$, find the value of $x + 1$.

4. If $h + 6.7 = 3(h - 2.9)$, find the value of $h + 0.3$.

5. If $2(x - 5y) = 5(x - 3y)$, what is the value of x in terms of y?

6. If $2(x - 5y) = 5(x - 3y)$, what is the value of y in terms of x?

7. A line with slope of 3 and y-intercept of -5 passes through the point $(a, 1)$. Find the value of a.

8. A line passes through the points $(2, 3)$, $(5, y)$, and $(-7, 3)$. Find the value of y.

9. Find the slope of the line that passes through the points $(-3, 5)$ and $(3, -5)$.

10. Find the y-intercept of the line that passes through the points $(-3, 5)$ and $(3, -5)$.

CHAPTER SUMMARY

Key Words

addition method (7.3)
consistent system of equations (7.1)
dependent equations (7.1)
inconsistent system of equations (7.1)

independent equations (7.1)
simultaneous solution (7.1)
solution of a system of equations (7.1)
substitution method (7.2)

system of equations (7.1)
system of inequalities (7.5)

Key Ideas

(7.1) To solve a system of equations graphically, carefully graph each equation of the system. If the lines intersect, the coordinates of the point of intersection give the solution of the system.

(7.2) To solve a system of equations by substitution, solve one of the equations of the system for one of its variables, substitute the resulting expression into the other equation, and solve for the other variable.

(7.3) To solve a system of equations by addition, first multiply one or both of the equations by suitable constants, if necessary, to eliminate one of the variables when the equations are added. The equation that results can be solved for its single variable. Then substitute the

value obtained back into one of the original equations and solve for the other variable.

(7.4) Systems of equations are useful in solving many different types of word problems.

(7.5) To graph a system of inequalities, first graph the individual inequalities of the system. The final solution, if one exists, is that region where all the individual graphs intersect.

(7.6) To solve a system of three equations in three variables, eliminate one variable from two of the equations. From another pair, eliminate the same variable. Solve the resulting system of two equations in two variables using the methods of Sections 7.2 or 7.3.

CHAPTER 7 REVIEW EXERCISES

(7.1) *In Review Exercises 1–4, determine whether the given ordered pair is a solution of the given system of equations.*

1. $(1, 5)$; $\begin{cases} 3x - y = -2 \\ 2x + 3y = 17 \end{cases}$

2. $(-2, 4)$; $\begin{cases} 5x + 3y = 2 \\ -3x + 2y = 16 \end{cases}$

3. $\left(14, \dfrac{1}{2}\right)$; $\begin{cases} 2x + 4y = 30 \\ \dfrac{x}{4} - y = 3 \end{cases}$

4. $\left(\dfrac{7}{2}, -\dfrac{2}{3}\right)$; $\begin{cases} 4x - 6y = 18 \\ \dfrac{x}{3} + \dfrac{y}{2} = \dfrac{5}{6} \end{cases}$

In Review Exercises 5–8, use the graphing method to solve each system. If the system is inconsistent, or if the equations are dependent, so indicate.

5. $\begin{cases} x + y = 7 \\ 2x - y = 5 \end{cases}$

6. $\begin{cases} \dfrac{x}{3} + \dfrac{y}{5} = -1 \\ x - 3y = -3 \end{cases}$

7. $\begin{cases} 3x + 6y = 6 \\ x + 2y = 2 \end{cases}$

8. $\begin{cases} 6x + 3y = 12 \\ 2x + y = 2 \end{cases}$

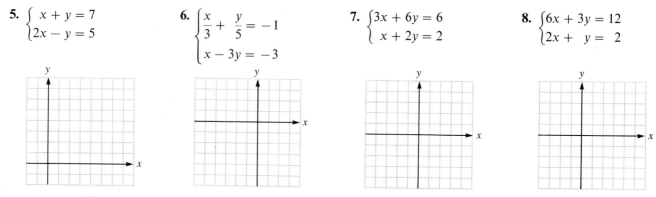

(7.2) *In Review Exercises 9–12, use the substitution method to solve each system of equations.*

9. $\begin{cases} x = 3y + 5 \\ 5x - 4y = 3 \end{cases}$

10. $\begin{cases} 3x - \dfrac{2y}{5} = 2(x - 2) \\ 2x - 3 = 3 - 2y \end{cases}$

11. $\begin{cases} 8x + 5y = 3 \\ 5x - 8y = 13 \end{cases}$

12. $\begin{cases} 6(x + 2) = y - 1 \\ 5(y - 1) = x + 2 \end{cases}$

(7.3) *In Review Exercises 13–20, use the addition method to solve each system of equations. If the equations of a system are dependent, or if the system is inconsistent, so indicate.*

13. $\begin{cases} 2x + y = 1 \\ 5x - y = 20 \end{cases}$

14. $\begin{cases} x + 8y = 7 \\ x - 4y = 1 \end{cases}$

15. $\begin{cases} 5x + y = 2 \\ 3x + 2y = 11 \end{cases}$

16. $\begin{cases} x + y = 3 \\ 3x = 2 - y \end{cases}$

17. $\begin{cases} 11x + 3y = 27 \\ 8x + 4y = 36 \end{cases}$

18. $\begin{cases} 9x + 3y = 5 \\ 3x = 4 - y \end{cases}$

19. $\begin{cases} 9x + 3y = 5 \\ 3x + y = \dfrac{5}{3} \end{cases}$

20. $\begin{cases} \dfrac{x}{3} + \dfrac{y + 2}{2} = 1 \\ \dfrac{x + 8}{8} + \dfrac{y - 3}{3} = 0 \end{cases}$

(7.4) *In Review Exercises 21–26, use a system of equations to solve each word problem.*

21. One number is 5 times another, and their sum is 18. Find the numbers.

22. The length of a rectangle is 3 times its width. The perimeter is 24 feet. Find the dimensions of the rectangle.

23. A grapefruit costs 15 cents more than an orange. Together, they cost 85 cents. Find the cost of each.

24. A man's electric bill for January was $23 less than his gas bill. These two utilities cost him a total of $109. Find the amount of each bill.

25. Two gallons of milk and 3 dozen eggs cost $6.80. Three gallons of milk and 2 dozen eggs cost $7.35. How much does each gallon of milk and each dozen eggs cost?

26. Carlos invested part of $3000 in a 10% certificate account and the rest in a 6% passbook account. The total annual interest from both accounts is $270. How much did he invest in each account?

(7.5) *In Review Exercises 27–30, solve each system of inequalities.*

27. $\begin{cases} 5x + 3y < 15 \\ 3x - y > 3 \end{cases}$

28. $\begin{cases} 5x - 3y \geq 5 \\ 3x + 2y \geq 3 \end{cases}$

29. $\begin{cases} x \geq 3y \\ y < 3x \end{cases}$

30. $\begin{cases} x > 0 \\ x \leq 3 \end{cases}$

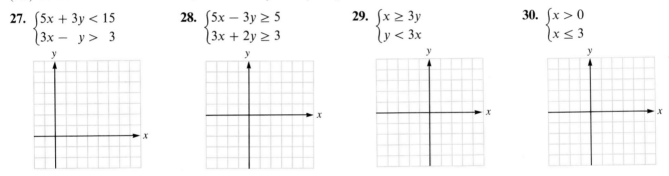

(7.6) *In Review Exercises 31–34, solve each system of three equations in three variables. If the equations are dependent or the system is inconsistent, so indicate.*

31. $\begin{cases} x - 2y - 2z = 7 \\ 2x - y + z = -1 \\ x + 5y + 3z = -8 \end{cases}$

32. $\begin{cases} x + 2z = 4 \\ y - 3z = -4 \\ 2x + 3y + z = 2 \end{cases}$

33. $\begin{cases} x + y - z = 1 \\ 3x - y + z = 11 \\ x - 2y + 2z = 7 \end{cases}$

34. $\begin{cases} 2x - y + z = 1 \\ x + 2y + 2z = 2 \\ x - 3y + 2z = 2 \end{cases}$

CHAPTER 7 TEST

In Problems 1–2, determine whether the given ordered pair is a solution of the given system.

_____ **1.** $(2, -3)$; $\begin{cases} 3x - 2y = 12 \\ 2x + 3y = -5 \end{cases}$

_____ **2.** $(-2, -1)$; $\begin{cases} 4x + y = -9 \\ 2x - 3y = -7 \end{cases}$

In Problems 3–4, solve each system by graphing.

3. $\begin{cases} 3x + y = 7 \\ x - 2y = 0 \end{cases}$

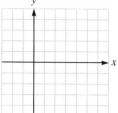

4. $\begin{cases} x + \dfrac{y}{2} = 1 \\ y = 1 - 3x \end{cases}$

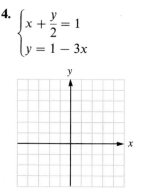

In Problems 5–6, solve each system by substitution.

_____   **5.** $\begin{cases} y = x - 1 \\ 2x + y = -7 \end{cases}$

_____   **6.** $\begin{cases} \dfrac{x}{6} + \dfrac{y}{10} = 3 \\ \dfrac{5x}{16} - \dfrac{3y}{16} = \dfrac{15}{8} \end{cases}$

In Problems 7–8, solve each system by addition.

_____   **7.** $\begin{cases} 3x - y = 2 \\ 2x + y = 8 \end{cases}$

_____   **8.** $\begin{cases} 4x + 3 = -3y \\ \dfrac{-x}{7} + \dfrac{4y}{21} = 1 \end{cases}$

In Problems 9–10, identify each system as consistent or inconsistent.

_____   **9.** $\begin{cases} 2x + 3(y - 2) = 0 \\ -3y = 2(x - 4) \end{cases}$

_____   **10.** $\begin{cases} \dfrac{x}{3} + y - 4 = 0 \\ -3y = x - 12 \end{cases}$

In Problems 11–12, use a system of equations in two variables to solve each word problem.

_____   **11.** The sum of two numbers is -18. One number is 2 greater than 3 times the other. Find the product of the numbers.

_____   **12.** A woman invested some money at 8% and some at 9%. The interest on the combined investment of $10,000 was $840. How much was invested at 9%?

In Problems 13–14, solve each system of inequalities by graphing.

13. $\begin{cases} x + y < 3 \\ x - y > 1 \end{cases}$

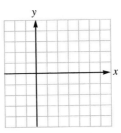

14. $\begin{cases} 2x + 3y \leq 6 \\ x \geq 2 \end{cases}$

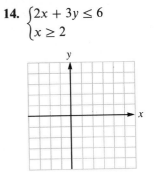

In Problems 15–16, solve each system of equations.

_____ **15.** $\begin{cases} 2x + y - 3z = 5 \\ x + 2y - z = 7 \\ x + y + 5z = 4 \end{cases}$

_____ **16.** $\begin{cases} 3x + 2y - 2z = 1 \\ x + 2y - 3z = 5 \\ -x + y - 3z = 7 \end{cases}$

8

Roots and Radical Expressions

The product $b \cdot b$ is called the **square of b** and is usually denoted by the exponential expression b^2. For example,

The square of 3 is 9 because $3^2 = 9$.
The square of -3 is 9 because $(-3)^2 = 9$.
The square of 12 is 144 because $12^2 = 144$.
The square of -12 is 144 because $(-12)^2 = 144$.
The square of 0 is 0 because $0^2 = 0$.

In this chapter we shall reverse the squaring process and find *square roots* of numbers. We shall also discuss how to find other roots of numbers.

8.1   RADICALS

A number b is a **square root of a** if the square of b is equal to a. For example,

3 is a square root of 9 because $3^2 = 9$.
-3 is a square root of 9 because $(-3)^2 = 9$.
12 is a square root of 144 because $12^2 = 144$.
-12 is a square root of 144 because $(-12)^2 = 144$.
0 is a square root of 0 because $0^2 = 0$.

In general, we have

> **DEFINITION.**   The number b is a **square root of a** if $b^2 = a$.

We have seen that 3 and -3 are both square roots of 9 and that 12 and -12 are both square roots of 144. In fact, all positive numbers have two square roots, one that is positive and one that is negative. The number 0 is the only number that has a single square root, which is 0. The symbol $\sqrt{}$, called a **radical sign,** is used to represent the *positive* square root of a number.

> **DEFINITION.**   If a is a positive number, then the **principal square root of a,** denoted by the symbol $\sqrt{a}$, is the positive square root of a.
> The principal square root of 0 is 0: $\sqrt{0} = 0$.

The expression under a radical sign is called a **radicand.**

The principal square root of a positive number a must be a positive number also. Although 3 and -3 are both square roots of 9, only 3 can be called the principal square root. The symbol $\sqrt{9}$ represents 3. To designate -3, we must place a $-$ sign in front of the radical. Thus,

$$\sqrt{9} = 3 \qquad \text{and} \qquad -\sqrt{9} = -3$$

Likewise,

$$\sqrt{144} = 12 \qquad \text{and} \qquad -\sqrt{144} = -12$$

EXAMPLE 1 Simplify each square root.

a. $\sqrt{0} = 0$ b. $\sqrt{1} = 1$ c. $\sqrt{49} = 7$ d. $\sqrt{121} = 11$

e. $-\sqrt{4} = -2$ f. $-\sqrt{81} = -9$ g. $\sqrt{225} = 15$ h. $\sqrt{169} = 13$

i. $-\sqrt{625} = -25$ j. $-\sqrt{900} = -30$ k. $\sqrt{576} = 24$ l. $\sqrt{1600} = 40$ ■

Square roots of certain numbers, such as 7, are difficult to compute by hand. However, we can find the approximate value of $\sqrt{7}$ by using a calculator or a table of square roots.

Using a Calculator to Find Square Roots

To find the principal square root of 7, we enter 7 into a calculator and press the $\boxed{\sqrt{}}$ key. The approximate value of $\sqrt{7}$ will appear on the calculator's display.

$\sqrt{7} \approx 2.6457513$ Read $\approx$ as "is approximately equal to."

Using a Table to Find Square Roots

To find the principal square root of 7, we refer to the table of square roots in Appendix III. In the left column, headed by n, locate the number 7. The column headed $\sqrt{n}$ contains the approximate value of $\sqrt{7}$.

$\sqrt{7} \approx 2.646$

Any number that is the square of an integer is called a **perfect integer square.** Numbers such as 4, 9, 16, and 49 are perfect integer squares. Because $\sqrt{4} = 2$ and 2 is a rational number, $\sqrt{4}$ must be a rational number. Similarly, $\sqrt{9}$, $\sqrt{16}$, and $\sqrt{49}$ are also rational numbers. In fact, the square root of any perfect integer square is a rational number.

However, some numbers are not rational numbers. One such number is $\sqrt{7}$. Numbers such as $\sqrt{7}$ that are not square roots of perfect integer squares are examples from a set of numbers called the **irrational numbers.** Recall the set of rational numbers together with the set of irrational numbers is the set of real numbers.

Because the square of any real number is nonnegative, there is no real number whose square can be a negative number such as -4. Thus, $\sqrt{-4}$ is not a real number. The number $\sqrt{-4}$ is an example from a set of numbers called the **imaginary numbers.** (This set of numbers is discussed in Chapter 9.) We emphasize this important fact:

> **The square root of a negative number is not a real number.**

In this chapter we shall assume that *all radicands under the square root symbols are either positive numbers or 0.* Thus, all square roots will be real numbers.

A square root of a number a is any number whose square is a. Likewise, the **cube root of a** is any number whose *cube is a.*

> **DEFINITION.**   The **cube root of a** is denoted as $\sqrt[3]{a}$, and
> $$\sqrt[3]{a} = b \qquad \text{if} \qquad b^3 = a$$

EXAMPLE 2   **a.** $\sqrt[3]{8} = 2$ because $2^3 = 8$.

b. $\sqrt[3]{-8} = -2$ because $(-2)^3 = 8$. Note that the cube root of a negative number is another negative number. It is the *square root* of a negative number that is an imaginary number.

c. $\sqrt[3]{27} = 3$ because $3^3 = 27$.   ∎

EXAMPLE 3   **a.** $\sqrt[3]{0} = 0$ **b.** $\sqrt[3]{-1} = -1$ **c.** $\sqrt[3]{64} = 4$

d. $\sqrt[3]{-64} = -4$ **e.** $\sqrt[3]{125} = 5$ **f.** $\sqrt[3]{-125} = -5$ ∎

Just as there are square roots and cube roots, there are also fourth roots, fifth roots, sixth roots, and so on. In general,

> **DEFINITION.**   The ***n*th root of a** is denoted by $\sqrt[n]{a}$, and
> $$\sqrt[n]{a} = b \qquad \text{if} \qquad b^n = a$$
> The number n is called the **index** of the radical. If n is an even natural number, then both a and $\sqrt[n]{a}$ must be either positive numbers or 0.

In the square root symbol $\sqrt{}$, the unwritten index is understood to be 2.

EXAMPLE 4   **a.** $\sqrt[4]{81} = 3$ because $3^4 = 81$.

b. $\sqrt[5]{32} = 2$ because $2^5 = 32$.

c. $\sqrt[5]{-32} = -2$ because $(-2)^5 = -32$.

d. $\sqrt[4]{-81}$ is not a real number because no real number raised to the fourth power can equal -81.   ∎

We can also find the square roots of certain quantities that contain variables, provided we know that these variables represent positive numbers or 0.

EXAMPLE 5   Assume that each variable represents a positive number and find each root.

a. $\sqrt{x^2} = x$ because $x^2 = x^2$.

b. $\sqrt{x^4} = x^2$ because $(x^2)^2 = x^4$.

c. $\sqrt{x^4 y^2} = x^2 y$ because $(x^2 y)^2 = x^4 y^2$.

d. $\sqrt[3]{x^6 y^3} = x^2 y$ because $(x^2 y)^3 = x^6 y^3$.

e. $\sqrt[4]{x^{12} y^8} = x^3 y^2$ because $(x^3 y^2)^4 = x^{12} y^8$.

f. $\sqrt[5]{x^{10} y^{25}} = x^2 y^5$ because $(x^2 y^5)^5 = x^{10} y^{25}$.   ∎

EXAMPLE 6   Assume that each variable represents a positive number and find each root.

a. $\sqrt{(x+1)^2} = x + 1$ because $(x+1)^2 = (x+1)^2$.

b. $\sqrt{x^2 + 4x + 4} = \sqrt{(x+2)^2}$ Factor $x^2 + 4x + 4$.

$\qquad\qquad\quad\; = x + 2$

c. $\sqrt{x^2 + 2xy + y^2} = \sqrt{(x+y)^2}$ Factor $x^2 + 2xy + y^2$.

$\qquad\qquad\qquad\;\; = x + y$ ∎

EXERCISE 8.1

In Exercises 1–24, find the value of each expression.

1. $\sqrt{9}$	**2.** $\sqrt{16}$	**3.** $\sqrt{49}$	**4.** $\sqrt{100}$
5. $\sqrt{36}$	**6.** $\sqrt{4}$	**7.** $\sqrt{81}$	**8.** $\sqrt{121}$
9. $-\sqrt{25}$	**10.** $-\sqrt{49}$	**11.** $-\sqrt{81}$	**12.** $-\sqrt{36}$
13. $\sqrt{196}$	**14.** $\sqrt{169}$	**15.** $\sqrt{256}$	**16.** $\sqrt{225}$
17. $-\sqrt{289}$	**18.** $\sqrt{400}$	**19.** $\sqrt{10,000}$	**20.** $-\sqrt{2500}$
21. $\sqrt{324}$	**22.** $-\sqrt{625}$	**23.** $-\sqrt{3600}$	**24.** $\sqrt{1600}$

In Exercises 25–36, use a calculator or a table of square roots to find each square root to three decimal places.

25. $\sqrt{2}$	**26.** $\sqrt{3}$	**27.** $\sqrt{5}$	**28.** $\sqrt{10}$
29. $\sqrt{6}$	**30.** $\sqrt{8}$	**31.** $\sqrt{11}$	**32.** $\sqrt{17}$
33. $\sqrt{23}$	**34.** $\sqrt{53}$	**35.** $\sqrt{95}$	**36.** $\sqrt{99}$

In Exercises 37–48, use a calculator to find each square root to three decimal places.

37. $\sqrt{6428}$	**38.** $\sqrt{4444}$	**39.** $-\sqrt{9876}$	**40.** $-\sqrt{3619}$
41. $\sqrt{21.35}$	**42.** $\sqrt{13.78}$	**43.** $\sqrt{0.3588}$	**44.** $\sqrt{0.9999}$
45. $\sqrt{0.9925}$	**46.** $\sqrt{0.12345}$	**47.** $-\sqrt{0.8372}$	**48.** $-\sqrt{0.4279}$

In Exercises 49–56, indicate whether each number is rational, irrational, or imaginary.

49. $\sqrt{9}$	**50.** $\sqrt{17}$	**51.** $\sqrt{49}$	**52.** $\sqrt{-49}$
53. $-\sqrt{5}$	**54.** $\sqrt{0}$	**55.** $\sqrt{-100}$	**56.** $-\sqrt{225}$

In Exercises 57–72, find the value of each expression.

57. $\sqrt[3]{1}$	**58.** $\sqrt[3]{8}$	**59.** $\sqrt[3]{27}$	**60.** $\sqrt[3]{0}$
61. $\sqrt[3]{-8}$	**62.** $\sqrt[3]{-1}$	**63.** $\sqrt[3]{-64}$	**64.** $\sqrt[3]{-27}$
65. $\sqrt[3]{125}$	**66.** $\sqrt[3]{1000}$	**67.** $-\sqrt[3]{-1}$	**68.** $-\sqrt[3]{-27}$
69. $-\sqrt[3]{64}$	**70.** $-\sqrt[3]{343}$	**71.** $\sqrt[3]{729}$	**72.** $\sqrt[3]{512}$

In Exercises 73–80, find the value of each expression.

73. $\sqrt[4]{16}$	**74.** $\sqrt[4]{81}$	**75.** $-\sqrt[5]{32}$	**76.** $-\sqrt[5]{243}$
77. $\sqrt[6]{1}$	**78.** $\sqrt[5]{0}$	**79.** $\sqrt[5]{-32}$	**80.** $\sqrt[7]{-1}$

In Exercises 81–104, write each expression without using a radical sign. Assume that all variables represent positive numbers.

81. $\sqrt{x^2y^2}$ **82.** $\sqrt{x^2y^4}$ **83.** $\sqrt{x^4z^4}$ **84.** $\sqrt{y^6z^8}$

85. $-\sqrt{x^4y^2}$ **86.** $-\sqrt{x^6y^4}$ **87.** $\sqrt{4z^2}$ **88.** $\sqrt{9t^6}$

89. $-\sqrt{9x^4y^2}$ **90.** $-\sqrt{16x^2y^4}$ **91.** $\sqrt{x^2y^2z^2}$ **92.** $\sqrt{x^4y^6z^8}$

93. $-\sqrt{x^2y^2z^4}$ **94.** $-\sqrt{a^8b^6c^2}$ **95.** $-\sqrt{25x^4z^{12}}$ **96.** $-\sqrt{100a^6b^4}$

97. $\sqrt{36z^{36}}$ **98.** $\sqrt{64y^{64}}$ **99.** $-\sqrt{2^4z^2}$ **100.** $-\sqrt{3^6x^8y^2}$

101. $\sqrt[3]{27y^3z^6}$ **102.** $\sqrt[3]{64x^3y^6z^9}$ **103.** $\sqrt[3]{-8p^6q^3}$ **104.** $\sqrt[3]{-r^{12}s^3t^6}$

In Exercises 105–114, write each expression without using a radical sign. Assume that all variables represent positive numbers.

105. $\sqrt{(x+9)^2}$ **106.** $\sqrt{(x+6)^2}$ **107.** $\sqrt{(x+11)^2}$ **108.** $\sqrt{(x+13)^2}$

109. $\sqrt{x^2+6x+9}$ **110.** $\sqrt{x^2+10x+25}$ **111.** $\sqrt{x^2+20x+100}$ **112.** $\sqrt{x^2+16x+64}$

113. $\sqrt{x^2+14xy+49y^2}$ **114.** $\sqrt{x^2+50xy+625y^2}$

■■■■■■■■■■■■ REVIEW EXERCISES ■■■■■■■■■■■■

In Review Exercises 1–4, let $P(x) = -2x^2 + 3x - 5$. Evaluate each quantity.

1. $P(0)$ **2.** $P(4)$ **3.** $P(-1)$ **4.** $P(2t)$

In Review Exercises 5–8, let $y = f(x) = 2x^2 - x - 1$. Evaluate each quantity.

5. $f(0)$ **6.** $f(2)$ **7.** $f(-2)$ **8.** $f(-t)$

9. Express the words "y varies directly with x" as an equation.

10. Express the words "y varies inversely with x" as an equation.

11. The number of feet, s, which a body falls from rest in t seconds, varies directly with the square of t. If the body falls 64 feet in 2 seconds, how long will it take for it to fall 256 feet?

12. For a constant voltage, the resistance, R, in a circuit varies inversely with the amperage, I. If the resistance is 3 ohms when the current is 6 amperes, what will the amperage be when the resistance is 9 ohms?

8.2 SIMPLIFYING RADICAL EXPRESSIONS

In this section we shall discuss two properties involving radicals. We introduce the first of these with the following examples:

$$\sqrt{4 \cdot 25} = \sqrt{100} \qquad \sqrt{4}\sqrt{25} = 2 \cdot 5$$
$$= 10 \qquad \qquad \qquad = 10$$

In each case the answer is 10. Thus, $\sqrt{4 \cdot 25} = \sqrt{4} \cdot \sqrt{25}$. Likewise,

$$\sqrt{9 \cdot 16} = \sqrt{144} \qquad \sqrt{9}\sqrt{16} = 3 \cdot 4$$
$$= 12 \qquad \qquad \qquad = 12$$

In each case the answer is 12. Thus, $\sqrt{9 \cdot 16} = \sqrt{9} \cdot \sqrt{16}$. These results suggest the **multiplication property of radicals.**

> **The Multiplication Property of Radicals.**   If a and b are nonnegative numbers, then
> $$\sqrt{ab} = \sqrt{a}\sqrt{b}$$

The multiplication property of radicals points out that *the square root of the product of two nonnegative numbers is equal to the product of their square roots.*

The multiplication property of radicals can be used to simplify radical expressions. For example, we can simplify $\sqrt{12}$ by proceeding as follows:

$$\sqrt{12} = \sqrt{4 \cdot 3}$$ Factor 12 as $4 \cdot 3$.
$$= \sqrt{4} \cdot \sqrt{3}$$ Use the multiplication property of radicals.
$$= 2\sqrt{3}$$ Write $\sqrt{4}$ as 2.

To simplify radicals, it is useful to know the integers that are perfect squares. The number 400, for example, is a perfect square integer because $20^2 = 400$. The perfect square integers less than 400 are

1, 4, 9, 16, 25, 36, 49, 64, 81, 100, 121, 144, 169, 196, 225, 256, 289, 324, 361

Expressions containing variables such as x^4y^2 are also perfect squares because they can be written as the square of a quantity:

$$x^4y^2 = (x^2y)^2$$

A radical that represents a square root is in **simplified form** if the radicand has no perfect square factors. The radical $\sqrt{10}$, for example, is in simplified form because 10 factors as $2 \cdot 5$ and neither 2 nor 5 is a perfect square. However, the radical $\sqrt{12}$ is not in simplified form because 12 factors as $4 \cdot 3$ and 4 is a perfect square.

EXAMPLE 1   Simplify $\sqrt{72x^3}$. Assume that $x \geq 0$.

Solution   Factor $72x^3$ into two factors, one of which is the greatest perfect square that divides $72x^3$. Because the greatest perfect square that divides $72x^3$ is $36x^2$, such a factorization is $72x^3 = 36x^2 \cdot 2x$. Then use the multiplication property of radicals to write $\sqrt{72x^3}$ as $\sqrt{36x^2}\sqrt{2x}$ and simplify.

$$\sqrt{72x^3} = \sqrt{36x^2}\sqrt{2x}$$
$$= 6x\sqrt{2x} \qquad \sqrt{36x^2} = 6x.$$ ■

EXAMPLE 2   Simplify $\sqrt{45x^2y^3}$. Assume that $x \geq 0$ and $y \geq 0$.

Solution   Look for the greatest perfect square that divides $45x^2y^3$. Because 9 is the greatest perfect square that divides 45, x^2 is the greatest perfect square that divides x^2, and y^2 is the greatest perfect square that divides y^3, the quantity $9x^2y^2$ is the greatest perfect square that divides $45x^2y^3$. Use the multiplication property of radicals to write $\sqrt{45x^2y^3}$ as $\sqrt{9x^2y^2}\sqrt{5y}$ and simplify.

$$\sqrt{45x^2y^3} = \sqrt{9x^2y^2}\sqrt{5y}$$
$$= 3xy\sqrt{5y} \qquad \sqrt{9x^2y^2} = 3xy.$$

■

EXAMPLE 3 Simplify $3a\sqrt{288a^5b^7}$. Assume that $a \geq 0$ and $b \geq 0$.

Solution Look for the greatest perfect square that divides $288a^5b^7$. Because 144 is the greatest perfect square that divides 288, a^4 is the greatest perfect square that divides a^5, and b^6 is the greatest perfect square that divides b^7, the quantity $144a^4b^6$ is the greatest perfect square that divides $288a^5b^7$. Use the multiplication property of radicals to write $\sqrt{288a^5b^7}$ as $\sqrt{144a^4b^6}\sqrt{2ab}$ and simplify.

$$3a\sqrt{288a^5b^7} = 3a\sqrt{144a^4b^6}\sqrt{2ab}$$
$$= 3a(12a^2b^3\sqrt{2ab}) \qquad \sqrt{144a^4b^6} = 12a^2b^3.$$
$$= 36a^3b^3\sqrt{2ab}$$

■

The Division Property of Radicals

To find the second property involving radicals, we consider these examples.

$$\sqrt{\frac{100}{25}} = \sqrt{4} \qquad\qquad \frac{\sqrt{100}}{\sqrt{25}} = \frac{10}{5}$$
$$= 2 \qquad\qquad\qquad\qquad = 2$$

In each case the answer is 2. Thus, $\sqrt{\dfrac{100}{25}} = \dfrac{\sqrt{100}}{\sqrt{25}}$. Likewise,

$$\sqrt{\frac{36}{4}} = \sqrt{9} \qquad\qquad \frac{\sqrt{36}}{\sqrt{4}} = \frac{6}{2}$$
$$= 3 \qquad\qquad\qquad\qquad = 3$$

In each case the answer is 3. Thus, $\sqrt{\dfrac{36}{4}} = \dfrac{\sqrt{36}}{\sqrt{4}}$. These results suggest the **division property of radicals.**

> **The Division Property of Radicals.** If a and b are nonnegative numbers and $b \neq 0$, then
> $$\sqrt{\frac{a}{b}} = \frac{\sqrt{a}}{\sqrt{b}}$$

The division property of radicals points out that the *square root of the quotient of two numbers is the quotient of their square roots.*

The division property of radicals can also be used to simplify radical expressions. For example, we can simplify $\sqrt{\dfrac{59}{49}}$ by proceeding as follows:

$$\sqrt{\frac{59}{49}} = \frac{\sqrt{59}}{\sqrt{49}}$$

$$= \frac{\sqrt{59}}{7} \qquad \sqrt{49} = 7. \ \sqrt{59} \text{ does not simplify.}$$

EXAMPLE 4   Simplify $\sqrt{\dfrac{108}{25}}$.

Solution $\sqrt{\dfrac{108}{25}} = \dfrac{\sqrt{108}}{\sqrt{25}}$ Use the division property of radicals.

$\qquad\qquad = \dfrac{\sqrt{36 \cdot 3}}{5}$ Factor 108 using the factorization involving 36, the largest perfect square factor of 108, and write $\sqrt{25}$ as 5.

$\qquad\qquad = \dfrac{\sqrt{36}\sqrt{3}}{5}$ Use the multiplication property of radicals.

$\qquad\qquad = \dfrac{6\sqrt{3}}{5}$ $\sqrt{36} = 6.$

EXAMPLE 5   Simplify $\sqrt{\dfrac{44x^3}{9xy^2}}$. Assume that $x > 0$ and $y > 0$.

Solution $\sqrt{\dfrac{44x^3}{9xy^2}} = \sqrt{\dfrac{44x^2}{9y^2}}$ Simplify the fraction by dividing out the common factor of x.

$\qquad\qquad = \dfrac{\sqrt{44x^2}}{\sqrt{9y^2}}$ Use the division property of radicals.

$\qquad\qquad = \dfrac{\sqrt{4x^2}\sqrt{11}}{\sqrt{9y^2}}$

$\qquad\qquad = \dfrac{2x\sqrt{11}}{3y}$

Simplifying Cube Roots

The multiplication and division properties of radicals have been stated for square roots. They do extend to cube roots and higher. To simplify a cube root, as in the next example, it is helpful to know the cubes of the numbers from 1 to 10. These are the first ten perfect cubes.

1, 8, 27, 64, 125, 216, 343, 512, 729, 1000

Expressions containing variables such as x^6y^3 are also perfect cubes because they can be written as the cube of a quantity:

$$x^6y^3 = (x^2y)^3$$

EXAMPLE 6   Simplify **a.** $\sqrt[3]{16x^3y^4}$ and **b.** $\sqrt[3]{\dfrac{64n^4}{27m^3}}$.

Solution   **a.** Try to factor $16x^3y^4$ into two factors, one of which is the greatest perfect cube that divides $16x^3y^4$. Such a factorization is $16x^3y^4 = 8x^3y^3 \cdot 2y$. Then use the multiplication property of radicals and write $\sqrt[3]{16x^3y^4}$ as $\sqrt[3]{8x^3y^3}\sqrt[3]{2y}$ and simplify.

$$\sqrt[3]{16x^3y^4} = \sqrt[3]{8x^3y^3}\,\sqrt[3]{2y}$$
$$= 2xy\sqrt[3]{2y} \qquad \sqrt[3]{8x^3y^3} = 2xy.$$

b. $\sqrt[3]{\dfrac{64n^4}{27m^3}} = \dfrac{\sqrt[3]{64n^4}}{\sqrt[3]{27m^3}}$ Use the division property of radicals.

$\qquad = \dfrac{\sqrt[3]{64n^3}\,\sqrt[3]{n}}{3m}$ Use the multiplication property of radicals and write $\sqrt[3]{27m^3}$ as $3m$.

$\qquad = \dfrac{4n\sqrt[3]{n}}{3m}$ ∎

Warning: It is important to note that
$$\sqrt{a+b} \ne \sqrt{a} + \sqrt{b} \qquad \text{and} \qquad \sqrt{a-b} \ne \sqrt{a} - \sqrt{b}$$

To illustrate that this is true, we consider this correct simplification:
$$\sqrt{9 + 16} = \sqrt{25} = 5$$

However, it is incorrect to write
$$\sqrt{9 + 16} = \sqrt{9} + \sqrt{16} = 3 + 4 = 7$$

Likewise, $\sqrt{25 - 16} = \sqrt{9} = 3$. It is incorrect to write
$$\sqrt{25 - 16} = \sqrt{25} - \sqrt{16} = 5 - 4 = 1$$

EXERCISE 8.2

In Exercises 1–48, simplify each radical. Assume that all variables represent positive numbers.

1. $\sqrt{20}$	**2.** $\sqrt{18}$	**3.** $\sqrt{50}$	**4.** $\sqrt{75}$
5. $\sqrt{45}$	**6.** $\sqrt{54}$	**7.** $\sqrt{98}$	**8.** $\sqrt{27}$
9. $\sqrt{48}$	**10.** $\sqrt{128}$	**11.** $\sqrt{200}$	**12.** $\sqrt{300}$
13. $\sqrt{192}$	**14.** $\sqrt{250}$	**15.** $\sqrt{88}$	**16.** $\sqrt{275}$
17. $\sqrt{324}$	**18.** $\sqrt{405}$	**19.** $\sqrt{147}$	**20.** $\sqrt{722}$
21. $\sqrt{180}$	**22.** $\sqrt{320}$	**23.** $\sqrt{432}$	**24.** $\sqrt{720}$
25. $4\sqrt{288}$	**26.** $2\sqrt{800}$	**27.** $-7\sqrt{1000}$	**28.** $-3\sqrt{252}$

29. $2\sqrt{245}$ **30.** $3\sqrt{196}$ **31.** $-5\sqrt{162}$ **32.** $-4\sqrt{243}$

33. $\sqrt{25x}$ **34.** $\sqrt{36y}$ **35.** $\sqrt{a^2b}$ **36.** $\sqrt{rs^2}$

37. $\sqrt{9x^2y}$ **38.** $\sqrt{16xy^2}$ **39.** $8x^2y\sqrt{50x^2y^2}$ **40.** $3x^5y\sqrt{75x^3y^2}$

41. $12x\sqrt{16x^2y^3}$ **42.** $-4x^5y^3\sqrt{36x^3y^3}$ **43.** $-3xyz\sqrt{18x^3y^5}$ **44.** $15xy^2\sqrt{72x^2y^3}$

45. $\dfrac{3}{4}\sqrt{192a^3b^5}$ **46.** $-\dfrac{2}{9}\sqrt{162r^3s^3t}$ **47.** $-\dfrac{2}{5}\sqrt{80mn^2}$ **48.** $\dfrac{5}{6}\sqrt{180ab^2c}$

In Exercises 49–64, write each quotient as the quotient of two radicals and simplify.

49. $\sqrt{\dfrac{25}{9}}$ **50.** $\sqrt{\dfrac{36}{49}}$ **51.** $\sqrt{\dfrac{81}{64}}$ **52.** $\sqrt{\dfrac{121}{144}}$

53. $\sqrt{\dfrac{26}{25}}$ **54.** $\sqrt{\dfrac{17}{169}}$ **55.** $\sqrt{\dfrac{20}{49}}$ **56.** $\sqrt{\dfrac{50}{9}}$

57. $\sqrt{\dfrac{48}{81}}$ **58.** $\sqrt{\dfrac{27}{64}}$ **59.** $\sqrt{\dfrac{32}{25}}$ **60.** $\sqrt{\dfrac{75}{16}}$

61. $\sqrt{\dfrac{125}{121}}$ **62.** $\sqrt{\dfrac{250}{49}}$ **63.** $\sqrt{\dfrac{245}{36}}$ **64.** $\sqrt{\dfrac{500}{81}}$

In Exercises 65–72, simplify each expression. Assume that all variables represent positive numbers.

65. $\sqrt{\dfrac{72x^3}{y^2}}$ **66.** $\sqrt{\dfrac{108a^3b^2}{c^2d^4}}$ **67.** $\sqrt{\dfrac{125m^2n^5}{64n}}$ **68.** $\sqrt{\dfrac{72p^5q^7}{16pq^3}}$

69. $\sqrt{\dfrac{128m^3n^5}{36mn^7}}$ **70.** $\sqrt{\dfrac{75p^3q^2}{9p^5q^4}}$ **71.** $\sqrt{\dfrac{12r^7s^6t}{81r^5s^2t}}$ **72.** $\sqrt{\dfrac{36m^2n^9}{100mn^3}}$

In Exercises 73–86, simplify each cube root.

73. $\sqrt[3]{8x^3}$ **74.** $\sqrt[3]{27x^3y^3}$ **75.** $\sqrt[3]{-64x^5}$ **76.** $\sqrt[3]{-16x^4y^3}$

77. $\sqrt[3]{54x^3y^4z^6}$ **78.** $\sqrt[3]{-24x^5y^5z^4}$ **79.** $\sqrt[3]{-81x^2y^3z^4}$ **80.** $\sqrt[3]{1600xy^2z^3}$

81. $\sqrt[3]{\dfrac{27m^3}{8n^6}}$ **82.** $\sqrt[3]{\dfrac{125t^9}{27s^6}}$ **83.** $\sqrt[3]{\dfrac{16r^4s^5}{1000t^3}}$ **84.** $\sqrt[3]{\dfrac{54m^4n^3}{r^3s^6}}$

85. $\sqrt[3]{\dfrac{250a^3b^4}{16b}}$ **86.** $\sqrt[3]{\dfrac{81p^5q^3}{1000p^2q^6}}$

REVIEW EXERCISES

In Review Exercises 1–4, simplify each fraction.

1. $\dfrac{5xy^2z^3}{10x^2y^2z^4}$ **2.** $\dfrac{35a^3b^2c}{63a^2b^3c^2}$ **3.** $\dfrac{a^2-a-2}{a^2+a-6}$ **4.** $\dfrac{y^2+3y-18}{y^2-9}$

In Review Exercises 5–10, perform the indicated operations and simplify the result, if possible.

5. $\dfrac{t^2 - t - 6}{t^2 - 3t} \cdot \dfrac{t^2 - t}{t^2 + t - 2}$

6. $\dfrac{x + 3}{x^2 + x - 6} \div \dfrac{x - 2}{x^2 - 5x + 6}$

7. $\dfrac{2r}{r + 3} + \dfrac{6}{r + 3}$

8. $\dfrac{2(u - 4)}{u + 1} - \dfrac{2(u + 4)}{u + 1}$

9. $\dfrac{3a}{2b} + \dfrac{3b}{5a} - \dfrac{ab}{2c}$

10. $\dfrac{2}{y^2 - 1} - \dfrac{y}{y - 1}$

11. The number, n, of straight lines that can join P points is given by the formula

$$n = \frac{P(P - 1)}{2}$$

Find P when n is 21.

12. The height, h, that an object will reach in t seconds when it is thrown upward at 128 feet per second is given by the formula

$$h = 128t - 16t^2$$

At what times will the height of the object be 112 feet?

8.3 ADDING AND SUBTRACTING RADICAL EXPRESSIONS

When adding monomials, it is possible to combine *like terms*. For example,

$$3x + 5x = (3 + 5)x \qquad \text{Use the distributive property.}$$
$$= 8x$$

In like fashion, it is often possible to combine terms that contain **like radicals.**

DEFINITION. Radicals are called **like radicals** if they have the same index and the same radicand.

Because terms such as $3\sqrt{2}$ and $5\sqrt{2}$ contain like radicals, they are considered to be like terms. Such terms can be combined.

$$3\sqrt{2} + 5\sqrt{2} = (3 + 5)\sqrt{2} \qquad \text{Use the distributive property.}$$
$$= 8\sqrt{2}$$

Likewise,

$$2x\sqrt{3y} - x\sqrt{3y} = (2x - x)\sqrt{3y} \qquad \text{Use the distributive property.}$$
$$= x\sqrt{3y}$$

However, the terms in the expression

$$3\sqrt{2} + 5\sqrt{7}$$

cannot be combined because the terms $3\sqrt{2}$ and $5\sqrt{7}$ do not contain like radicals. The terms in the expression

$$2x\sqrt{5z} + 3y\sqrt{5z}$$

cannot be combined because the coefficients $2x$ and $3y$ have different variables.

Some radical expressions such as $3\sqrt{2}$ and $5\sqrt{8}$ can be simplified so that they contain like radicals. They can then be combined.

EXAMPLE 1   Simplify $3\sqrt{2} + 5\sqrt{8}$.

Solution   The radical $\sqrt{8}$ is not in simplified form because the radicand 8 has a perfect square factor. To simplify the radical and combine like terms, proceed as follows.

$$3\sqrt{2} + 5\sqrt{8} = 3\sqrt{2} + 5\sqrt{\mathbf{4 \cdot 2}} \qquad \text{Factor 8. Look for perfect square factors.}$$
$$= 3\sqrt{2} + 5\sqrt{4}\sqrt{2} \qquad \text{Use the multiplication property of radicals.}$$
$$= 3\sqrt{2} + 5(2)\sqrt{2} \qquad \sqrt{4} = 2.$$
$$= 3\boldsymbol{\sqrt{2}} + 10\boldsymbol{\sqrt{2}} \qquad \text{Simplify.}$$
$$= 13\boldsymbol{\sqrt{2}} \qquad \text{Combine like terms.} \qquad \blacksquare$$

EXAMPLE 2   Simplify $\sqrt{20} + \sqrt{45} + 3\sqrt{5}$.

Solution   Simplify the first two radicals. Then combine like terms.

$$\sqrt{20} + \sqrt{45} + 3\sqrt{5} = \sqrt{4 \cdot 5} + \sqrt{9 \cdot 5} + 3\sqrt{5} \qquad \text{Factor.}$$
$$= \sqrt{4}\sqrt{5} + \sqrt{9}\sqrt{5} + 3\sqrt{5} \qquad \text{Use the multiplication property of radicals.}$$
$$= 2\boldsymbol{\sqrt{5}} + 3\boldsymbol{\sqrt{5}} + 3\boldsymbol{\sqrt{5}} \qquad \text{Simplify.}$$
$$= 8\boldsymbol{\sqrt{5}} \qquad \text{Combine like terms.} \qquad \blacksquare$$

EXAMPLE 3   Simplify $\sqrt{8x^2y} + \sqrt{18x^2y}$. Assume that $x > 0$ and $y > 0$.

Solution   Simplify each of the radicals and combine like terms.

$$\sqrt{8x^2y} + \sqrt{18x^2y} = \sqrt{4 \cdot 2x^2y} + \sqrt{9 \cdot 2x^2y} \qquad \text{Factor.}$$
$$= \sqrt{4x^2}\sqrt{2y} + \sqrt{9x^2}\sqrt{2y} \qquad \text{Use the multiplication property of radicals.}$$
$$= 2x\boldsymbol{\sqrt{2y}} + 3x\boldsymbol{\sqrt{2y}} \qquad \text{Simplify.}$$
$$= 5x\boldsymbol{\sqrt{2y}} \qquad \text{Combine like terms.} \qquad \blacksquare$$

EXAMPLE 4   Simplify $\sqrt{28x^2y} - 2\sqrt{63y^3}$. Assume that $x > 0$ and $y > 0$.

Solution   Simplify each of the radicals and combine like terms.

$$\sqrt{28x^2y} - 2\sqrt{63y^3} = \sqrt{4 \cdot 7x^2y} - 2\sqrt{9 \cdot 7y^2y} \qquad \text{Factor.}$$
$$= \sqrt{4x^2}\sqrt{7y} - 2\sqrt{9y^2}\sqrt{7y} \qquad \text{Use the multiplication property of radicals.}$$
$$= 2x\sqrt{7y} - 2 \cdot 3y\sqrt{7y} \qquad \text{Simplify.}$$
$$= 2x\sqrt{7y} - 6y\sqrt{7y} \qquad \text{Simplify.}$$

Because the variables of the coefficients $2x$ and $6y$ are not the same, the expression does not simplify any further. $\qquad \blacksquare$

EXAMPLE 5   Simplify $\sqrt{27xy} + \sqrt{20xy}$. Assume that $x > 0$ and $y > 0$.

Solution   Simplify each radical.

$$\sqrt{27xy} + \sqrt{20xy} = \sqrt{9 \cdot 3xy} + \sqrt{4 \cdot 5xy} \qquad \text{Factor.}$$
$$= \sqrt{9}\sqrt{3xy} + \sqrt{4}\sqrt{5xy} \qquad \begin{array}{l}\text{Use the multiplication}\\ \text{property of radicals.}\end{array}$$
$$= 3\sqrt{3xy} + 2\sqrt{5xy} \qquad \text{Simplify.}$$

Because these terms do not contain like radicals, the expression does not simplify any further.

EXAMPLE 6   Simplify $\sqrt{8x} + \sqrt{3y} - \sqrt{50x} + \sqrt{27y}$.

Solution   Simplify each radical and combine like terms, if possible.

$$\sqrt{8x} + \sqrt{3y} - \sqrt{50x} + \sqrt{27y} = \sqrt{4 \cdot 2x} + \sqrt{3y} - \sqrt{25 \cdot 2x} + \sqrt{9 \cdot 3y}$$
$$= \sqrt{4}\sqrt{2x} + \sqrt{3y} - \sqrt{25}\sqrt{2x} + \sqrt{9}\sqrt{3y}$$
$$= 2\sqrt{2x} + \sqrt{3y} - 5\sqrt{2x} + 3\sqrt{3y}$$
$$= -3\sqrt{2x} + 4\sqrt{3y}$$

The next example illustrates that it is also possible to combine like terms containing like radicals other than square roots.

EXAMPLE 7   Simplify $\sqrt[3]{81x^4} - x\sqrt[3]{24x}$.

Solution   Simplify each radical and combine like terms, if possible.

$$\sqrt[3]{81x^4} - x\sqrt[3]{24x} = \sqrt[3]{27x^3 \cdot 3x} - x\sqrt[3]{8 \cdot 3x}$$
$$= \sqrt[3]{27x^3}\sqrt[3]{3x} - x\sqrt[3]{8}\sqrt[3]{3x}$$
$$= 3x\sqrt[3]{3x} - 2x\sqrt[3]{3x}$$
$$= x\sqrt[3]{3x}$$

═══════════════   **EXERCISE 8.3**   ═══════════════

In Exercises 1–24, find each indicated sum.

1. $\sqrt{12} + \sqrt{27}$ **2.** $\sqrt{20} + \sqrt{45}$ **3.** $\sqrt{48} + \sqrt{75}$ **4.** $\sqrt{48} + \sqrt{108}$

5. $\sqrt{45} + \sqrt{80}$ **6.** $\sqrt{80} + \sqrt{125}$ **7.** $\sqrt{125} + \sqrt{245}$ **8.** $\sqrt{36} + \sqrt{196}$

9. $\sqrt{20} + \sqrt{180}$ **10.** $\sqrt{80} + \sqrt{245}$ **11.** $\sqrt{160} + \sqrt{360}$ **12.** $\sqrt{12} + \sqrt{147}$

13. $3\sqrt{45} + 4\sqrt{245}$ **14.** $2\sqrt{28} + 7\sqrt{63}$

15. $2\sqrt{28} + 2\sqrt{112}$ **16.** $4\sqrt{63} + 6\sqrt{112}$

17. $5\sqrt{32} + 3\sqrt{72}$ **18.** $3\sqrt{72} + 2\sqrt{128}$

19. $3\sqrt{98} + 8\sqrt{128}$ **20.** $5\sqrt{90} + 7\sqrt{250}$

21. $\sqrt{20} + \sqrt{45} + \sqrt{80}$ **22.** $\sqrt{48} + \sqrt{27} + \sqrt{75}$

23. $\sqrt{24} + \sqrt{150} + \sqrt{240}$ **24.** $\sqrt{28} + \sqrt{63} + \sqrt{112}$

In Exercises 25–44, find each indicated difference.

25. $\sqrt{18} - \sqrt{8}$ **26.** $\sqrt{32} - \sqrt{18}$ **27.** $\sqrt{9} - \sqrt{50}$ **28.** $\sqrt{50} - \sqrt{32}$

29. $\sqrt{72} - \sqrt{32}$ **30.** $\sqrt{98} - \sqrt{72}$ **31.** $\sqrt{12} - \sqrt{48}$ **32.** $\sqrt{48} - \sqrt{75}$

33. $\sqrt{108} - \sqrt{75}$ **34.** $\sqrt{147} - \sqrt{48}$ **35.** $\sqrt{1000} - \sqrt{360}$ **36.** $\sqrt{180} - \sqrt{125}$

37. $2\sqrt{80} - 3\sqrt{125}$ **38.** $3\sqrt{245} - 2\sqrt{180}$ **39.** $8\sqrt{96} - 5\sqrt{24}$ **40.** $3\sqrt{216} - 3\sqrt{150}$

41. $\sqrt{288} - 3\sqrt{200}$ **42.** $\sqrt{392} - 2\sqrt{128}$ **43.** $5\sqrt{250} - 3\sqrt{160}$ **44.** $4\sqrt{490} - 3\sqrt{360}$

In Exercises 45–56, simplify each expression.

45. $\sqrt{12} + \sqrt{3} + \sqrt{27}$ **46.** $\sqrt{8} - \sqrt{50} + \sqrt{72}$ **47.** $\sqrt{200} - \sqrt{75} + \sqrt{48}$ **48.** $\sqrt{20} + \sqrt{80} - \sqrt{125}$

49. $\sqrt{24} - \sqrt{150} - \sqrt{54}$ **50.** $\sqrt{98} - \sqrt{300} + \sqrt{800}$ **51.** $\sqrt{200} + \sqrt{300} - \sqrt{75}$ **52.** $\sqrt{175} + \sqrt{125} - \sqrt{28}$

53. $\sqrt{48} - \sqrt{8} + \sqrt{27} - \sqrt{32}$ **54.** $\sqrt{162} + \sqrt{50} - \sqrt{75} - \sqrt{108}$

55. $\sqrt{147} + \sqrt{216} - \sqrt{108} - \sqrt{27}$ **56.** $\sqrt{180} - \sqrt{112} + \sqrt{45} - \sqrt{700}$

In Exercises 57–70, simplify each expression. Assume that all variables represent positive numbers.

57. $\sqrt{2x^2} + \sqrt{8x^2}$ **58.** $\sqrt{3y^2} - \sqrt{12y^2}$ **59.** $\sqrt{2x^3} + \sqrt{8x^3}$ **60.** $\sqrt{3y^3} - \sqrt{12y^3}$

61. $\sqrt{18x^2y} - \sqrt{27x^2y}$ **62.** $\sqrt{49xy} + \sqrt{xy}$ **63.** $\sqrt{32x^5} - \sqrt{18x^5}$ **64.** $\sqrt{27xy^3} - \sqrt{48xy^3}$

65. $3\sqrt{54x^2} + 5\sqrt{24x^2}$ **66.** $3\sqrt{24x^4y^3} + 2\sqrt{54x^4y^3}$

67. $y\sqrt{490y} - 2\sqrt{360y^3}$ **68.** $3\sqrt{20x} + 2\sqrt{63y}$

69. $\sqrt{20x^3y} + \sqrt{45x^5y^3} - \sqrt{80x^7y^5}$ **70.** $x\sqrt{48xy^2} - y\sqrt{27x^3} + \sqrt{75x^3y^2}$

In Exercises 71–82, simplify each expression.

71. $\sqrt[3]{16} + \sqrt[3]{54}$ **72.** $\sqrt[3]{24} - \sqrt[3]{81}$ **73.** $\sqrt[3]{81} - \sqrt[3]{24}$ **74.** $\sqrt[3]{32} + \sqrt[3]{108}$

75. $\sqrt[3]{40} + \sqrt[3]{125}$ **76.** $\sqrt[3]{3000} - \sqrt[3]{192}$ **77.** $\sqrt[3]{x^4} - \sqrt[3]{x^7}$ **78.** $\sqrt[3]{8x^5} + \sqrt[3]{27x^8}$

79. $\sqrt[3]{192x^4y^5} - \sqrt[3]{24x^4y^5}$ **80.** $\sqrt[3]{24a^5b^4} + \sqrt[3]{81a^5b^4}$

81. $\sqrt[3]{135x^7y^4} - \sqrt[3]{40x^7y^4}$ **82.** $\sqrt[3]{56a^4b^5} + \sqrt[3]{7a^4b^5}$

REVIEW EXERCISES

In Review Exercises 1–4, express each phrase as a ratio in lowest terms.

1. 3 to 8   **2.** 6 ounces to 18 ounces

3. 5 inches to 3 feet   **4.** 3 months to 10 years

In Review Exercises 5–10, solve each proportion.

5. $\dfrac{a-2}{8} = \dfrac{a+10}{24}$

6. $\dfrac{6}{t+12} = \dfrac{18}{4t}$

7. $\dfrac{-2}{x+14} = \dfrac{6}{x-6}$

8. $\dfrac{y-4}{4} = \dfrac{y+2}{12}$

9. $\dfrac{z+3}{4} = \dfrac{9}{2z}$

10. $\dfrac{s}{s+3} = \dfrac{2s}{s+2}$

11. The total surface area, A, of the cylindrical solid in Illustration 1 is given by the formula

$$A = \frac{44}{7}(r^2 + rh)$$

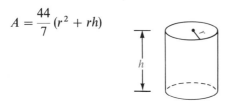

ILLUSTRATION 1

Solve the formula for h.

12. The total surface area, A, of the closed box in Illustration 2 is given by the formula

$$A = 2lw + 2wd + 2ld$$

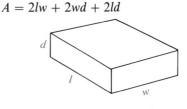

ILLUSTRATION 2

Solve the formula for d.

8.4 MULTIPLYING AND DIVIDING RADICAL EXPRESSIONS

The definition of square root implies that $\sqrt{5}$ is the number whose square is 5:

$$(\sqrt{5})^2 = 5 \quad \text{and} \quad \sqrt{5}\sqrt{5} = 5$$

In general, we have

$$(\sqrt{x})^2 = x \quad \text{and} \quad \sqrt{x}\sqrt{x} = x$$

Because of the multiplication property of radicals, the *product of the square roots of two nonnegative numbers is equal to the square root of the product of those numbers.* For example,

$$\sqrt{2}\sqrt{8} = \sqrt{2 \cdot 8} = \sqrt{16} = 4$$
$$\sqrt{3}\sqrt{27} = \sqrt{3 \cdot 27} = \sqrt{81} = 9$$

and

$$\sqrt{x}\sqrt{x^3} = \sqrt{x \cdot x^3} = \sqrt{x^4} = x^2 \quad (x \geq 0)$$

Likewise, the *product of the cube roots of two numbers is equal to the cube root of the product of those numbers.* For example,

$$\sqrt[3]{2}\sqrt[3]{4} = \sqrt[3]{2 \cdot 4} = \sqrt[3]{8} = 2$$
$$\sqrt[3]{4}\sqrt[3]{16} = \sqrt[3]{4 \cdot 16} = \sqrt[3]{64} = 4$$

and

$$\sqrt[3]{3x^2}\sqrt[3]{9x} = \sqrt[3]{3x^2 \cdot 9x} = \sqrt[3]{27x^3} = 3x$$

Multiplying Radical Expressions

To multiply monomials that contain radicals, we multiply the coefficients and multiply the radicals separately. We then simplify the result, if possible.

EXAMPLE 1   Multiply **a.** $3\sqrt{6}$ by $4\sqrt{3}$ and   **b.** $-2\sqrt[3]{7}$ by $6\sqrt[3]{49}$.

Solution   We use the commutative and associative properties to write the product so that we can multiply the integers and the radicals separately. Always remember to simplify any radicals that are in the product, if possible.

a. $3\sqrt{6} \cdot 4\sqrt{3} = 3(4)\sqrt{6}\sqrt{3}$ $\qquad$ **b.** $-2\sqrt[3]{7} \cdot 6\sqrt[3]{49} = -2(6)\sqrt[3]{7}\sqrt[3]{49}$

$$= 12\sqrt{18} \qquad\qquad\qquad\qquad\quad = -12\sqrt[3]{7 \cdot 49}$$

$$= 12\sqrt{9}\sqrt{2} \qquad\qquad\qquad\qquad = -12\sqrt[3]{343}$$

$$= 12(3)\sqrt{2} \qquad\qquad\qquad\qquad = -12(7)$$

$$= 36\sqrt{2} \qquad\qquad\qquad\qquad\quad = -84$$

To multiply a polynomial by a monomial, we use the distributive property to remove parentheses and then combine like terms, if possible.

EXAMPLE 2   Multiply **a.** $\sqrt{2}(\sqrt{6} + \sqrt{8})$ and   **b.** $\sqrt[3]{3}(\sqrt[3]{9} - 2)$.

Solution   **a.** $\sqrt{2}(\sqrt{6} + \sqrt{8}) = \sqrt{2}\sqrt{6} + \sqrt{2}\sqrt{8}$ $\qquad$ Use the distributive property to remove parentheses.

$$= \sqrt{12} + \sqrt{16} \qquad\qquad \text{Use the multiplication property of radicals.}$$

$$= \sqrt{4 \cdot 3} + \sqrt{16} \qquad\qquad \text{Factor 12.}$$

$$= \sqrt{4}\sqrt{3} + \sqrt{16} \qquad\qquad \text{Use the multiplication property of radicals.}$$

$$= 2\sqrt{3} + 4 \qquad\qquad\qquad \text{Simplify.}$$

b. $\sqrt[3]{3}(\sqrt[3]{9} - 2) = \sqrt[3]{3}\sqrt[3]{9} - 2\sqrt[3]{3}$ $\qquad$ Use the distributive property to remove parentheses.

$$= \sqrt[3]{27} - 2\sqrt[3]{3} \qquad\qquad \text{Use the multiplication property of radicals.}$$

$$= 3 - 2\sqrt[3]{3} \qquad\qquad\qquad \text{Simplify.}$$

The **FOIL** method is used to multiply one binomial by another.

EXAMPLE 3   Multiply and simplify $(\sqrt{3} + \sqrt{2})(\sqrt{3} - \sqrt{2})$.

Solution   This is the product of two binomial factors. One factor is the sum of two quantities, and the other factor is the difference of the same two quantities. Find the product by the **FOIL** method and note that the sum of the outer and inner products is zero.

$$(\sqrt{3} + \sqrt{2})(\sqrt{3} - \sqrt{2}) = \sqrt{3}\sqrt{3} - \sqrt{3}\sqrt{2} + \sqrt{2}\sqrt{3} - \sqrt{2}\sqrt{2}$$ Use the **FOIL** method.

$$= 3 - 2$$ Combine like terms and simplify.

$$= 1$$ Simplify.   ■

EXAMPLE 4   Multiply and simplify $(\sqrt{3x} + 1)(\sqrt{3x} + 2)$.

Solution $(\sqrt{3x} + 1)(\sqrt{3x} + 2) = \sqrt{3x}\sqrt{3x} + 2\sqrt{3x} + \sqrt{3x} + 2$ Use the **FOIL** method.

$$= 3x + 3\sqrt{3x} + 2$$ Combine like terms and simplify.   ■

EXAMPLE 5   Multiply and simplify $(\sqrt[3]{4x} - 3)(\sqrt[3]{2x^2} + 1)$.

Solution $(\sqrt[3]{4x} - 3)(\sqrt[3]{2x^2} + 1) = \sqrt[3]{4x}\sqrt[3]{2x^2} + 1\sqrt[3]{4x} - 3\sqrt[3]{2x^2} - 3$ Use the **FOIL** method.

$$= \sqrt[3]{8x^3} + \sqrt[3]{4x} - 3\sqrt[3]{2x^2} - 3$$ Use the multiplication property of radicals.

$$= 2x + \sqrt[3]{4x} - 3\sqrt[3]{2x^2} - 3$$ Simplify.   ■

Dividing Radical Expressions

To divide radical expressions, we can use the division property of radicals. For example, to divide $\sqrt{108}$ by $\sqrt{36}$, we proceed as follows:

$$\frac{\sqrt{108}}{\sqrt{36}} = \sqrt{\frac{108}{36}}$$

$$= \sqrt{3} \qquad \frac{108}{36} = \frac{36 \cdot 3}{36} = 3.$$

EXAMPLE 6   Simplify $\dfrac{\sqrt{22a^2b^7}}{\sqrt{99a^4b^3}}$. Assume that $a > 0$ and $b > 0$.

Solution $\dfrac{\sqrt{22a^2b^7}}{\sqrt{99a^4b^3}} = \sqrt{\dfrac{22a^2b^7}{99a^4b^3}}$

$$= \sqrt{\frac{2b^4}{9a^2}}$$ Simplify the radicand.

$$= \frac{\sqrt{2b^4}}{\sqrt{9a^2}}$$ Use the division property of radicals.

$$= \frac{\sqrt{b^4}\sqrt{2}}{\sqrt{9a^2}}$$ Use the multiplication property of radicals.

$$= \frac{b^2\sqrt{2}}{3a}$$ Simplify.   ■

Rationalizing the Denominator

It is often easier to work with fractions if radicals do not appear in their denominators. For example, it is impossible to perform the long division indicated by the fraction

$$\frac{1}{\sqrt{2}} \quad \text{or} \quad \frac{1}{1.4142135\ldots}$$

because of the unending decimal in the denominator. However, if the radical were not present in the denominator, the division would be possible. We can eliminate the radical in the denominator by multiplying both the numerator and the denominator by $\sqrt{2}$ because the product $\sqrt{2} \cdot \sqrt{2}$ is the rational number 2.

$$\frac{1}{\sqrt{2}} = \frac{1\sqrt{2}}{\sqrt{2}\sqrt{2}} \qquad \text{Multiply both numerator and denominator by } \sqrt{2}.$$

$$= \frac{\sqrt{2}}{2} \qquad \text{Simplify.}$$

The division indicated by the fraction

$$\frac{1.4142135\ldots}{2}$$

can easily be carried out to any desired number of decimal places.

The process of removing radicals that appear in the denominator of a fraction is called **rationalizing the denominator.**

EXAMPLE 7   Rationalize the denominator of the fractions **a.** $\dfrac{3}{\sqrt{3}}$ and **b.** $\dfrac{2}{\sqrt[3]{3}}$.

Solution   **a.** Multiply both the numerator and the denominator of the fraction by $\sqrt{3}$ and simplify.

$$\frac{3}{\sqrt{3}} = \frac{3\sqrt{3}}{\sqrt{3}\sqrt{3}} \qquad \text{Multiply both the numerator and the denominator by } \sqrt{3}.$$

$$= \frac{3\sqrt{3}}{3} \qquad \sqrt{3}\sqrt{3} = 3.$$

$$= \sqrt{3} \qquad \text{Divide out the common factor of 3.}$$

b. Because $\sqrt[3]{3} \cdot \sqrt[3]{9} = \sqrt[3]{27}$ and 27 is a perfect integer cube, multiply both the numerator and the denominator of the fraction by $\sqrt[3]{9}$ and simplify.

$$\frac{2}{\sqrt[3]{3}} = \frac{2\sqrt[3]{9}}{\sqrt[3]{3}\sqrt[3]{9}} = \frac{2\sqrt[3]{9}}{\sqrt[3]{27}} = \frac{2\sqrt[3]{9}}{3} \qquad ■$$

If a radical expression contains any radicals in a denominator, it is not considered to be in simplified form. The following examples show how to simplify fractions that contain radical expressions in their denominators.

EXAMPLE 8   Divide 5 by $\sqrt{20}$ by simplifying the fraction $\dfrac{5}{\sqrt{20}}$.

Solution   Begin by rationalizing the denominator. To do so, it is not necessary to multiply numerator and denominator by $\sqrt{20}$. It is sufficient to multiply by $\sqrt{5}$ because $5 \cdot 20$ is 100, which is a perfect integer square.

$$\frac{5}{\sqrt{20}} = \frac{5\sqrt{5}}{\sqrt{20}\sqrt{5}}$$ Multiply both the numerator and the denominator by $\sqrt{5}$.

$$= \frac{5\sqrt{5}}{\sqrt{100}}$$ Use the multiplication property of radicals.

$$= \frac{5\sqrt{5}}{10}$$ $\sqrt{100} = 10.$

$$= \frac{\sqrt{5}}{2}$$ Simplify the fraction.   ∎

EXAMPLE 9   Simplify the fraction $\dfrac{\sqrt{72x^5}}{\sqrt{45}}$. Assume that $x > 0$.

Solution $$\frac{\sqrt{72x^5}}{\sqrt{45}} = \frac{\sqrt{36 \cdot x^4 \cdot 2x}}{\sqrt{9 \cdot 5}}$$ Factor the radicands, looking for perfect square factors.

$$= \frac{\sqrt{36} \cdot \sqrt{x^4} \cdot \sqrt{2x}}{\sqrt{9} \cdot \sqrt{5}}$$ Use the multiplication property of radicals.

$$= \frac{6x^2\sqrt{2x}}{3\sqrt{5}}$$ Simplify.

$$= \frac{6x^2\sqrt{2x}\sqrt{5}}{3\sqrt{5}\sqrt{5}}$$ Multiply both the numerator and the denominator by $\sqrt{5}$ to rationalize the denominator.

$$= \frac{2x^2\sqrt{10x}}{5}$$ Simplify.   ∎

EXAMPLE 10   Simplify the fraction $\sqrt{\dfrac{3x^3y^2}{27xy^3}}$.

Solution $$\sqrt{\frac{3x^3y^2}{27xy^3}} = \sqrt{\frac{x^2}{9y}}$$ Simplify the fraction within the radical.

$$= \sqrt{\frac{x^2 \cdot y}{9y \cdot y}}$$ Multiply both the numerator and the denominator by y.

$$= \frac{\sqrt{x^2y}}{\sqrt{9y^2}}$$ Use the division property of radicals.

$$= \frac{x\sqrt{y}}{3y}$$ Simplify.   ∎

EXAMPLE 11 Simplify the fraction $\dfrac{2}{\sqrt{3}-1}$.

Solution Because the denominator of this fraction contains a radical, the fraction is not in simplified form. The denominator is a *binomial,* so multiplying the numerator and the denominator by $\sqrt{3}$ will not work. The key is to multiply the numerator and the denominator by the binomial $\sqrt{3}+1$. As you study the following solution, note that $(\sqrt{3}+1)(\sqrt{3}-1)$ is the product of the sum and difference of the same two numbers, and that the resulting product is free of radicals. Radical expressions such as $\sqrt{3}+1$ and $\sqrt{3}-1$ are called **conjugates** of each other.

$$\frac{2}{\sqrt{3}-1}=\frac{2(\sqrt{3}+1)}{(\sqrt{3}-1)(\sqrt{3}+1)}$$ Multiply both the numerator and the denominator by $\sqrt{3}+1$.

$$=\frac{2(\sqrt{3}+1)}{3-1}$$ Multiply the binomials in the denominator.

$$=\frac{2(\sqrt{3}+1)}{2}$$ Simplify.

$$=\sqrt{3}+1$$ Divide out the common factor of 2. ∎

EXAMPLE 12 Simplify the fraction $\dfrac{10\sqrt{7}}{\sqrt{7x}+\sqrt{2x}}$.

Solution Multiply both numerator and denominator of the fraction by $\sqrt{7x}-\sqrt{2x}$, which is the conjugate of the denominator. The product

$$(\sqrt{7x}+\sqrt{2x})(\sqrt{7x}-\sqrt{2x})$$

will be free of radicals.

$$\frac{10\sqrt{7}}{\sqrt{7x}+\sqrt{2x}}=\frac{10\sqrt{7}(\sqrt{7x}-\sqrt{2x})}{(\sqrt{7x}+\sqrt{2x})(\sqrt{7x}-\sqrt{2x})}$$ Multiply both the numerator and the denominator by $\sqrt{7x}-\sqrt{2x}$.

$$=\frac{10\sqrt{7}(\sqrt{7x}-\sqrt{2x})}{7x-2x}$$ Multiply the binomials.

$$=\frac{10\sqrt{7}(\sqrt{7x}-\sqrt{2x})}{5x}$$ Simplify.

$$=\frac{2\sqrt{7}(\sqrt{7x}-\sqrt{2x})}{x}$$ Simplify the fraction.

$$=\frac{14\sqrt{x}-2\sqrt{14x}}{x}$$ Remove parentheses. ∎

EXERCISE 8.4

In Exercises 1–62, perform each indicated multiplication. Assume that all variables represent positive numbers.

1. $\sqrt{3}\sqrt{3}$　　　　**2.** $\sqrt{7}\sqrt{7}$　　　　**3.** $\sqrt{2}\sqrt{8}$　　　　**4.** $\sqrt{27}\sqrt{3}$

5. $\sqrt{16}\sqrt{4}$　　　　**6.** $\sqrt{32}\sqrt{2}$　　　　**7.** $\sqrt[3]{8}\sqrt[3]{8}$　　　　**8.** $\sqrt[3]{4}\sqrt[3]{250}$

9. $\sqrt{x^3}\sqrt{x^3}$ **10.** $\sqrt{a^7}\sqrt{a^3}$ **11.** $\sqrt{b^8}\sqrt{b^6}$ **12.** $\sqrt{y^4}\sqrt{y^8}$

13. $(2\sqrt{5})(2\sqrt{3})$ **14.** $(4\sqrt{3})(2\sqrt{2})$ **15.** $(-5\sqrt{6})(4\sqrt{3})$ **16.** $(6\sqrt{3})(-7\sqrt{3})$

17. $(2\sqrt[3]{4})(3\sqrt[3]{16})$ **18.** $(-3\sqrt[3]{100})(\sqrt[3]{10})$ **19.** $(4\sqrt{x})(-2\sqrt{x})$ **20.** $(3\sqrt{y})(15\sqrt{y})$

21. $(-14\sqrt{50x})(-5\sqrt{20x})$ **22.** $(12\sqrt{24y})(-16\sqrt{2y})$

23. $\sqrt{8x^2}\sqrt{2x^2y}$ **24.** $\sqrt{27y}\sqrt{3y^3}$ **25.** $\sqrt{2}(\sqrt{2}+1)$ **26.** $\sqrt{3}(\sqrt{3}-2)$

27. $\sqrt{3}(\sqrt{27}-1)$ **28.** $\sqrt{2}(\sqrt{8}-1)$ **29.** $\sqrt{7}(\sqrt{7}-3)$ **30.** $\sqrt{5}(\sqrt{5}+2)$

31. $\sqrt{5}(3-\sqrt{5})$ **32.** $\sqrt{7}(2+\sqrt{7})$ **33.** $\sqrt{3}(\sqrt{6}+1)$ **34.** $\sqrt{2}(\sqrt{6}-2)$

35. $\sqrt[3]{7}(\sqrt[3]{49}-2)$ **36.** $\sqrt[3]{5}(\sqrt[3]{25}+3)$ **37.** $\sqrt{x}(\sqrt{3x}-2)$ **38.** $\sqrt{y}(\sqrt{y}+5)$

39. $2\sqrt{x}(\sqrt{9x}+3)$ **40.** $3\sqrt{z}(\sqrt{4z}-\sqrt{z})$ **41.** $3\sqrt{x}(2+\sqrt{x})$ **42.** $5\sqrt{y}(5-\sqrt{5y})$

43. $\sqrt{21x}(\sqrt{3x}+\sqrt{2x})$ **44.** $\sqrt{35y}(\sqrt{7y}-\sqrt{5y})$ **45.** $(\sqrt{2}+1)(\sqrt{2}-1)$ **46.** $(\sqrt{3}-1)(\sqrt{3}+1)$

47. $(\sqrt{5}+2)(\sqrt{5}-2)$ **48.** $(\sqrt{7}+5)(\sqrt{7}-5)$ **49.** $(\sqrt[3]{2}+1)(\sqrt[3]{2}+1)$ **50.** $(\sqrt[3]{5}-2)(\sqrt[3]{5}-2)$

51. $(\sqrt{7}-x)(\sqrt{7}+x)$ **52.** $(\sqrt{2}-\sqrt{x})(\sqrt{x}+\sqrt{2})$

53. $(\sqrt{2}-\sqrt{x})^2$ **54.** $(\sqrt{a}+\sqrt{3})^2$

55. $(\sqrt{6x}+\sqrt{7})(\sqrt{6x}-\sqrt{7})$ **56.** $(\sqrt{8y}+\sqrt{2z})(\sqrt{8y}-\sqrt{2z})$

57. $(\sqrt{2x}+3)(\sqrt{8x}-6)$ **58.** $(\sqrt{5y}-3)(\sqrt{20y}+6)$

59. $(\sqrt{8xy}+1)(\sqrt{8xy}+1)$ **60.** $(\sqrt{5x}+3\sqrt{y})(\sqrt{5x}-3\sqrt{y})$

61. $(\sqrt{16x}-\sqrt{x})(\sqrt{16x}-\sqrt{x})$ **62.** $(\sqrt{9xz}+2\sqrt{xz})(\sqrt{25xz}-\sqrt{xz})$

In Exercises 63–74, simplify each expression. Assume that all variables represent positive numbers.

63. $\dfrac{\sqrt{12x^3}}{\sqrt{27x}}$ **64.** $\dfrac{\sqrt{32}}{\sqrt{98x^2}}$ **65.** $\dfrac{\sqrt{18xy^2}}{\sqrt{25x}}$ **66.** $\dfrac{\sqrt{27y^3}}{\sqrt{75x^2y}}$

67. $\dfrac{\sqrt{196xy^3}}{\sqrt{49x^3y}}$ **68.** $\dfrac{\sqrt{50xyz^4}}{\sqrt{98xyz^2}}$ **69.** $\dfrac{\sqrt[3]{16x^6}}{\sqrt[3]{54x^3}}$ **70.** $\dfrac{\sqrt[3]{128a^6b^3}}{\sqrt[3]{16a^3b^6}}$

71. $\dfrac{\sqrt{3x^2y^3}}{\sqrt{27x}}$ **72.** $\dfrac{\sqrt{44x^2y^5}}{\sqrt{99x^4y}}$ **73.** $\dfrac{\sqrt{5x}\sqrt{10y^2}}{\sqrt{x^3y}}$ **74.** $\dfrac{\sqrt{7y}\sqrt{14x}}{\sqrt{8xy}}$

In Exercises 75–106, perform each indicated division by rationalizing a denominator and simplifying. Assume that all variables represent positive numbers.

75. $\dfrac{1}{\sqrt{3}}$ **76.** $\dfrac{1}{\sqrt{5}}$ **77.** $\dfrac{2}{\sqrt{7}}$ **78.** $\dfrac{3}{\sqrt{11}}$

79. $\dfrac{5}{\sqrt[3]{5}}$ **80.** $\dfrac{7}{\sqrt[3]{7}}$ **81.** $\dfrac{9}{\sqrt{27}}$ **82.** $\dfrac{4}{\sqrt{20}}$

83. $\dfrac{3}{\sqrt{32}}$ **84.** $\dfrac{5}{\sqrt{18}}$ **85.** $\dfrac{4}{\sqrt[3]{4}}$ **86.** $\dfrac{7}{\sqrt[3]{10}}$

87. $\dfrac{\sqrt{5}}{\sqrt{3}}$ **88.** $\dfrac{\sqrt{3}}{\sqrt{5}}$ **89.** $\dfrac{10}{\sqrt{x}}$ **90.** $\dfrac{12}{\sqrt{y}}$

91. $\dfrac{\sqrt{9}}{\sqrt{2x}}$ **92.** $\dfrac{\sqrt{4}}{\sqrt{3z}}$ **93.** $\dfrac{\sqrt{2x}}{\sqrt{9y}}$ **94.** $\dfrac{\sqrt{3xy}}{\sqrt{4x}}$

95. $\dfrac{2\sqrt{3}}{\sqrt{8x^2}}$ **96.** $\dfrac{3\sqrt{5}}{\sqrt{27y^2}}$ **97.** $\dfrac{5\sqrt{6x}}{\sqrt{50}}$ **98.** $\dfrac{8\sqrt{10y}}{\sqrt{40}}$

99. $\dfrac{\sqrt[3]{5}}{\sqrt[3]{2}}$ **100.** $\dfrac{\sqrt[3]{2}}{\sqrt[3]{5}}$ **101.** $\dfrac{\sqrt[3]{2x^2}}{\sqrt[3]{2x}}$ **102.** $\dfrac{\sqrt[3]{3y^4}}{\sqrt[3]{3y}}$

103. $\dfrac{2}{\sqrt[3]{4x^2y}}$ **104.** $\dfrac{3}{\sqrt[3]{9xy^2}}$ **105.** $\dfrac{-5}{\sqrt[3]{25a^2b^2}}$ **106.** $\dfrac{-4}{\sqrt[3]{4ab^2c^2}}$

In Exercises 107–130, perform each indicated division by rationalizing a denominator and simplifying. Assume that all variables represent positive numbers.

107. $\dfrac{3}{\sqrt{3}-1}$ **108.** $\dfrac{3}{\sqrt{5}-2}$ **109.** $\dfrac{3}{\sqrt{7}+2}$ **110.** $\dfrac{5}{\sqrt{8}+3}$

111. $\dfrac{12}{3-\sqrt{3}}$ **112.** $\dfrac{10}{5-\sqrt{5}}$ **113.** $\dfrac{\sqrt{2}}{\sqrt{2}+1}$ **114.** $\dfrac{\sqrt{3}}{\sqrt{3}-1}$

115. $\dfrac{-\sqrt{3}}{\sqrt{3}+1}$ **116.** $\dfrac{-\sqrt{2}}{\sqrt{2}-1}$ **117.** $\dfrac{5}{\sqrt{3}+\sqrt{2}}$ **118.** $\dfrac{3}{\sqrt{3}-\sqrt{2}}$

119. $\dfrac{\sqrt{8}}{\sqrt{5}-\sqrt{3}}$ **120.** $\dfrac{\sqrt{32}}{\sqrt{7}-\sqrt{3}}$ **121.** $\dfrac{\sqrt{3}-\sqrt{2}}{\sqrt{3}+\sqrt{2}}$ **122.** $\dfrac{\sqrt{5}+\sqrt{3}}{\sqrt{5}-\sqrt{3}}$

123. $\dfrac{\sqrt{3x}-1}{\sqrt{3x}+1}$ **124.** $\dfrac{\sqrt{5x}+3}{\sqrt{5x}-3}$ **125.** $\dfrac{\sqrt{2x}+5}{\sqrt{2x}+3}$ **126.** $\dfrac{\sqrt{3y}+3}{\sqrt{3y}-2}$

127. $\dfrac{1-\sqrt{2z}}{\sqrt{2z}+1}$ **128.** $\dfrac{1+\sqrt{3z}}{\sqrt{3z}-1}$ **129.** $\dfrac{y-\sqrt{15}}{y+\sqrt{15}}$ **130.** $\dfrac{x+\sqrt{17}}{x-\sqrt{17}}$

REVIEW EXERCISES

In Review Exercises 1–6, factor each polynomial.

1. $x^2-4x-21$ **2.** $y^2+6y-27$ **3.** $6x^2y-15xy$ **4.** $2x^3y^2-6x^2y$

5. $x^3 + 8$

6. $y^2 - 9$

In Review Exercises 7–10, solve each equation.

7. $x^2 - 13x + 30 = 0$ **8.** $2x^2 + x = 1$ **9.** $2x^2 - 8 = 0$ **10.** $3x^2 = 15x$

11. A 16-foot rope is to be cut into two pieces, one three times as long as the other. Find the length of each piece.

12. George gave twice as much money to charity as Bob, and Sally gave $6 more than Bob. The total amount given was $506. How much did George give?

8.5 SOLVING EQUATIONS CONTAINING RADICALS

In this section we shall discuss how to solve certain equations that contain radical expressions. To do so, we note that if two numbers are equal, then their squares are also equal. This fact is called the **squaring property of equality.**

> **The Squaring Property of Equality.**
>> If $a = b$, then $a^2 = b^2$.

The equation $x = 2$ has only one solution, the number 2. If we apply the squaring property by squaring both sides of the equation $x = 2$, we get the equation $x^2 = 4$. This new equation has *two* solutions. They are the numbers 2 and -2. Thus, squaring both sides of an equation can lead to another equation that has more solutions than the first. However, no solutions are lost when both sides are squared. The original solution 2 is still a solution of the squared equation.

To solve equations such as $\sqrt{x + 2} = 3$, we must square both sides to eliminate the radical. This might produce an equation that has more solutions than the original equation. Therefore, we *must* check every solution of the squared equation in the *original* equation.

EXAMPLE 1   Solve the equation $\sqrt{x + 2} = 3$.

Solution   Because this equation contains a radical, square both sides to eliminate it. Then proceed as follows:

$$\sqrt{x + 2} = 3$$
$$(\sqrt{x + 2})^2 = 3^2 \qquad \text{Square both sides.}$$
$$x + 2 = 9 \qquad \text{Simplify.}$$
$$x = 7 \qquad \text{Add } -2 \text{ to both sides.}$$

Check the solution by substituting 7 for x in the original equation.

$$\sqrt{x + 2} = 3$$
$$\sqrt{7 + 2} \stackrel{?}{=} 3 \qquad \text{Replace } x \text{ with 7.}$$
$$\sqrt{9} \stackrel{?}{=} 3$$
$$3 = 3$$

The solution checks. Because no solutions are lost in this process, 7 is the only solution of the original equation.   ∎

EXAMPLE 2   Solve the equation $\sqrt{x + 1} + 5 = 3$.

Solution   Rearrange the terms to isolate the radical on one side of the equation. Then proceed as follows:

$$\sqrt{x + 1} + 5 = 3$$
$$\sqrt{x + 1} = -2 \qquad \text{Add } -5 \text{ to both sides.}$$
$$(\sqrt{x + 1})^2 = (-2)^2 \qquad \text{Square both sides.}$$
$$x + 1 = 4 \qquad \text{Simplify.}$$
$$x = 3 \qquad \text{Add } -1 \text{ to both sides.}$$

Check this solution by substituting 3 for x in the original equation.

$$\sqrt{x + 1} + 5 = 3$$
$$\sqrt{3 + 1} + 5 \stackrel{?}{=} 3 \qquad \text{Replace } x \text{ with 3.}$$
$$\sqrt{4} + 5 \stackrel{?}{=} 3$$
$$2 + 5 \stackrel{?}{=} 3$$
$$7 = 3$$

Because $7 = 3$ is a *false* result, the number 3 is *not* a solution of the given equation. Thus, the original equation has *no* solution.   ∎

Example 2 illustrates that squaring both sides of an equation can lead to **extraneous solutions.** These solutions do not satisfy the original equation and must be discarded.

Follow these steps to solve an equation containing radical expressions.

---

1. Whenever possible, rearrange the terms to isolate a single radical on one side of the equation.
2. Square both sides of the equation and solve the resulting equation.
3. Check the solution in the original equation. This step is required.

EXAMPLE 3   Solve the equation $\sqrt{x + 12} = 3\sqrt{x + 4}$.

Solution   Square both sides to eliminate the radical expressions. Then proceed as follows:

$$\sqrt{x + 12} = 3\sqrt{x + 4}$$

$(\sqrt{x + 12})^2 = (3\sqrt{x + 4})^2$	Square both sides.
$x + 12 = 9(x + 4)$	Simplify.
$x + 12 = 9x + 36$	Remove parentheses.
$-24 = 8x$	Add $-x$ and -36 to both sides.
$-3 = x$	Divide both sides by 8.

To check the solution, replace x in the original equation with -3.

$$\sqrt{x + 12} = 3\sqrt{x + 4}$$

$\sqrt{-3 + 12} \overset{?}{=} 3\sqrt{-3 + 4}$	Replace x with -3.
$\sqrt{9} \overset{?}{=} 3\sqrt{1}$	
$3 = 3$	

The solution checks.

EXAMPLE 4   Solve the equation $x = \sqrt{2x + 10} - 1$.

Solution   Add 1 to both sides to isolate the radical on the right side. Then square both sides to eliminate the radical. Be sure to square the *entire* right side.

$x = \sqrt{2x + 10} - 1$	
$x + 1 = \sqrt{2x + 10}$	Add 1 to both sides to isolate the radical.
$(x + 1)^2 = (\sqrt{2x + 10})^2$	Square both sides.
$x^2 + 2x + 1 = 2x + 10$	Remove parentheses.
$x^2 - 9 = 0$	Add $-2x$ and -10 to both sides.
$(x - 3)(x + 3) = 0$	Factor.
$x - 3 = 0$ or $x + 3 = 0$	Set each factor equal to 0.
$x = 3$ $\mid$ $x = -3$	Solve each linear equation.

Check each possible solution.

For $x = 3$	**For $x = -3$**
$x = \sqrt{2x + 10} - 1$	$x = \sqrt{2x + 10} - 1$
$3 \overset{?}{=} \sqrt{2(3) + 10} - 1$	$-3 \overset{?}{=} \sqrt{2(-3) + 10} - 1$
$3 \overset{?}{=} \sqrt{16} - 1$	$-3 \overset{?}{=} \sqrt{4} - 1$
$3 \overset{?}{=} 4 - 1$	$-3 \overset{?}{=} 2 - 1$
$3 = 3$	$-3 = 1$

Thus, the given equation has only one solution, $x = 3$. The other value, $x = -3$, is not a solution because it does not check. The false solution -3 is an extraneous solution.

EXAMPLE 5   The square root of the sum of 3 and a certain number is 3 greater than the number. Find the number.

Solution   Let x represent the required number. Then $3 + x$ represents the sum of 3 and that required number. Translate the words of the problem into an equation.

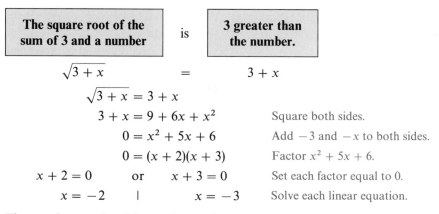

The square root of the sum of 3 and a number	is	3 greater than the number.
$\sqrt{3 + x}$	$=$	$3 + x$

$$\sqrt{3 + x} = 3 + x$$

$3 + x = 9 + 6x + x^2$ Square both sides.

$0 = x^2 + 5x + 6$ Add -3 and $-x$ to both sides.

$0 = (x + 2)(x + 3)$ Factor $x^2 + 5x + 6$.

$x + 2 = 0$ or $x + 3 = 0$ Set each factor equal to 0.

$x = -2$ | $x = -3$ Solve each linear equation.

The number can be either -2 or -3.

Check each solution in the words of the original problem. The square root of the sum of 3 and -2 is equal to 1, and 1 is 3 greater than -2. Thus, -2 checks. The square root of the sum of 3 and -3 is 0, and 0 is 3 greater than -3. Thus, -3 checks.   ■

EXERCISE 8.5

In Exercises 1–44, solve each equation. Check all solutions. If an equation has no solutions, so indicate.

1. $\sqrt{x} = 3$ **2.** $\sqrt{x} = 5$ **3.** $\sqrt{x} = 7$ **4.** $\sqrt{x} = 2$

5. $\sqrt{x} = -4$ **6.** $\sqrt{x} = -1$ **7.** $\sqrt{x + 3} = 2$ **8.** $\sqrt{x - 2} = 3$

9. $\sqrt{x - 5} = 5$ **10.** $\sqrt{x + 8} = 12$ **11.** $\sqrt{3 - x} = -2$ **12.** $\sqrt{5 - x} = 10$

13. $\sqrt{6 + 2x} = 4$ **14.** $\sqrt{7 + 3x} = -4$ **15.** $\sqrt{5x - 5} = 5$ **16.** $\sqrt{6x + 19} = 7$

17. $\sqrt{4x - 3} = 3$ **18.** $\sqrt{11x - 2} = 3$ **19.** $\sqrt{13x + 14} = 1$ **20.** $\sqrt{8x + 9} = 1$

21. $\sqrt{x + 3} + 5 = 12$ **22.** $\sqrt{x - 5} - 3 = 4$ **23.** $\sqrt{2x + 10} + 3 = 5$ **24.** $\sqrt{3x + 4} + 7 = 12$

25. $\sqrt{5x + 9} + 4 = 7$ **26.** $\sqrt{9x + 25} - 2 = 3$ **27.** $\sqrt{7 - 5x} + 4 = 3$ **28.** $\sqrt{7 + 6x} - 4 = -3$

29. $\sqrt{3x + 3} = 3\sqrt{x - 1}$ **30.** $2\sqrt{4x + 5} = 5\sqrt{x + 4}$

31. $2\sqrt{3x + 4} = \sqrt{5x + 9}$ **32.** $\sqrt{10 - 3x} = \sqrt{2x + 20}$

33. $\sqrt{3x + 6} = 2\sqrt{2x - 11}$ **34.** $2\sqrt{9x + 16} = \sqrt{3x + 64}$

35. $\sqrt{x + 1} = x - 1$ **36.** $\sqrt{x + 4} = x - 2$

37. $\sqrt{x + 1} = x + 1$ **38.** $\sqrt{x + 9} = x + 7$ **39.** $\sqrt{7x + 2} - 2x = 0$ **40.** $\sqrt{3x + 3} + 5 = x$

41. $x - 1 = \sqrt{x - 1}$ **42.** $x - 2 = \sqrt{x + 10}$ **43.** $x = \sqrt{3 - x} + 3$ **44.** $x = \sqrt{x - 4} + 4$

45. The square root of the sum of a certain number and 8 is 13. Find the number.

46. The square root of the sum of a certain number and 12 is 15. Find the number.

47. The square root of 3 less than a certain number is 5. Find the number.

48. The square root of 5 less than a certain number is 0. Find the number.

49. The square root of 7 more than a number is the square root of 14. Find the number.

50. The square root of twice a number is equal to the square root of 8 more than the number. Find the number.

51. The square root of twice a number is equal to the square root of 18 more than the number. Find the number.

52. Three times the square root of 2 greater than a number is equal to the square root of 4 more than 11 times that number. Find the number.

53. The square root of 4 more than a number is equal to 2 less than the number. Find the number.

54. Five times the square root of 1 less than a number is equal to 3 more than the number. Find the number. (*Hint:* There are two answers.)

55. What numbers are equal to their own square roots?

56. The sum of a number and its square root is equal to 0. Find the number.

57. The square root of the sum of two consecutive integers is 1 less than the smaller integer. Find the smallest integer.

58. The square root of the sum of two consecutive odd integers is 1 less than 3 times the smaller integer. Find the integers.

59. Can the sum of 4 and the principal square root of a number ever equal 0? Explain.

60. Can the difference of 4 and the principal square root of a number ever equal 0? Explain.

61. Einstein's Theory of Relativity predicts that an object moving at speed v will be shortened in the direction of its motion by a factor f given by

$$f = \sqrt{1 - \frac{v^2}{c^2}}$$

where c is the speed of light. Solve this equation for v^2.

62. Einstein's Theory of Relativity predicts that a clock moving at speed v will run slower by a factor f given by

$$f = \frac{1}{\sqrt{1 - \frac{v^2}{c^2}}}$$

where c is the speed of light. Solve this equation for v^2.

REVIEW EXERCISES

In Review Exercises 1–6, perform the indicated operations.

1. $3x^2y^3(-2xy^4)$

2. $-2a^4b^2c(5a^3bc^4)$

3. $(x + 4)(x - 3)$

4. $(2x - 3)(3x + 4)$

5. $(3x + 4)^2$

6. $(x + 1)(x + 2)(x + 3)$

In Review Exercises 7–10, solve each system of equations.

7. $\begin{cases} x + y = 5 \\ x - y = -1 \end{cases}$

8. $\begin{cases} 2x + y = 0 \\ x + 3y = 5 \end{cases}$

9. $\begin{cases} 2x + 3y = 0 \\ 3x - 2y = 13 \end{cases}$

10. $\begin{cases} 3x - 4y = 11 \\ 4x + y = -17 \end{cases}$

11. A woman has $2.75 in nickels, dimes, and quarters. She has twice as many dimes as nickels and 4 fewer quarters than dimes. How many nickels does she have?

12. John invested equal amounts in each of three accounts paying 6%, 7%, and 8% annual interest. His annual income from these investments is $105. How much did he invest at each rate?

8.6 RATIONAL EXPONENTS

A positive integer exponent indicates the number of times that a base is to be used as a factor in a product. For example, x^4 means that x is to be used as a factor four times.

$$x^4 = x \cdot x \cdot x \cdot x$$

A nonzero base with a negative integral exponent can be written as a quotient. For example,

$$x^{-4} = \frac{1}{x^4}$$

A nonzero base with an exponent of 0 is 1: $x^0 = 1$.

Exponents do not need to be integers. It is possible to raise certain bases to fractional powers. To give meaning to rational (fractional) exponents, we consider the number $\sqrt{7}$. Because $\sqrt{7}$ is the positive number whose square is 7, we have

$$(\sqrt{7})^2 = 7$$

We now consider the symbol $7^{1/2}$. If we demand that fractional exponents obey the same rules as integral exponents, then the square of $7^{1/2}$ must be 7 because

$$(7^{1/2})^2 = 7^{(1/2)2} = 7^1 = 7$$

Because the square of $7^{1/2}$ is 7, and the square of $\sqrt{7}$ is 7, it is reasonable to define $7^{1/2}$ to be $\sqrt{7}$. Similarly, we define

$$7^{1/3} \qquad \text{to be} \qquad \sqrt[3]{7}$$
$$7^{1/7} \qquad \text{to be} \qquad \sqrt[7]{7}$$

and so on. In general, we define the **rational exponent** $\frac{1}{n}$ as follows:

DEFINITION. If n is a positive integer greater than 1 and $\sqrt[n]{x}$ is a real number, then

$$x^{1/n} = \sqrt[n]{x}$$

EXAMPLE 1   Simplify **a.** $64^{1/2}$, **b.** $64^{1/3}$, **c.** $(-64)^{1/3}$, and **d.** $64^{1/6}$.

Solution   **a.** $64^{1/2} = \sqrt{64} = 8$ **b.** $64^{1/3} = \sqrt[3]{64} = 4$

c. $(-64)^{1/3} = \sqrt[3]{-64} = -4$ **d.** $64^{1/6} = \sqrt[6]{64} = 2$

The definition of $x^{1/n}$ can be extended to fractional exponents for which the numerator is not 1. For example, because $4^{3/2}$ can be written as $(4^{1/2})^3$, we have

$$4^{3/2} = (4^{1/2})^3 = (\sqrt{4})^3 = 2^3 = 8$$

Similarly, because $4^{3/2}$ can be written as $(4^3)^{1/2}$, we have

$$4^{3/2} = (4^3)^{1/2} = 64^{1/2} = \sqrt{64} = 8$$

In general, $x^{m/n}$ can be written as $(x^{1/n})^m$ or as $(x^m)^{1/n}$. Since $(x^{1/n})^m = (\sqrt[n]{x})^m$ and $(x^m)^{1/n} = \sqrt[n]{x^m}$, we make the following definition.

DEFINITION. If m and n are positive integers, x is nonnegative, and the fraction $\frac{m}{n}$ cannot be simplified, then

$$x^{m/n} = \sqrt[n]{x^m} = (\sqrt[n]{x})^m$$

EXAMPLE 2 Simplify **a.** $8^{2/3}$ and **b.** $(-27)^{4/3}$.

Solution **a.** $8^{2/3} = (\sqrt[3]{8})^2 = 2^2 = 4$ or $8^{2/3} = \sqrt[3]{8^2} = \sqrt[3]{64} = 4$

b. $(-27)^{4/3} = (\sqrt[3]{-27})^4 = (-3)^4 = 81$
or
$(-27)^{4/3} = \sqrt[3]{(-27)^4} = \sqrt[3]{531,441} = 81$ ∎

The work in Example 2 suggests that in order to avoid large numbers, it is usually easier to take the root of the base first.

EXAMPLE 3 Simplify **a.** $125^{4/3}$, **b.** $9^{5/2}$, **c.** $-25^{3/2}$, and **d.** $(-27)^{2/3}$.

Solution **a.** $125^{4/3} = (\sqrt[3]{125})^4 = (5)^4 = 625$ **b.** $9^{5/2} = (\sqrt{9})^5 = (3)^5 = 243$

c. $-25^{3/2} = -(\sqrt{25})^3 = -(5)^3 = -125$ **d.** $(-27)^{2/3} = (\sqrt[3]{-27})^2 = (-3)^2 = 9$ ∎

Because of the definition of $x^{1/n}$, the familiar rules of exponents are valid for rational exponents. The following example illustrates the use of each rule.

EXAMPLE 4 **a.** $4^{2/5}4^{1/5} = 4^{2/5 + 1/5} = 4^{3/5}$ $x^m x^n = x^{m+n}$

b. $(5^{2/3})^{1/2} = 5^{(2/3)(1/2)} = 5^{1/3}$ $(x^m)^n = x^{mn}$

c. $(3x)^{2/3} = 3^{2/3}x^{2/3}$ $(xy)^m = x^m y^m$

d. $\dfrac{4^{3/5}}{4^{2/5}} = 4^{3/5 - 2/5} = 4^{1/5}$ $\dfrac{x^m}{x^n} = x^{m-n}$

e. $\left(\dfrac{3}{2}\right)^{2/5} = \dfrac{3^{2/5}}{2^{2/5}}$ $\left(\dfrac{x}{y}\right)^n = \dfrac{x^n}{y^n}$

f. $4^{-2/3} = \dfrac{1}{4^{2/3}}$ $x^{-n} = \dfrac{1}{x^n}$

g. $5^0 = 1$ $x^0 = 1$ ∎

We can often use the rules of exponents to simplify expressions containing rational exponents.

EXAMPLE 5   Simplify **a.** $64^{-2/3}$, **b.** $(x^2)^{1/2}$, **c.** $(x^6y^4)^{1/2}$, and **d.** $(27x^{12})^{-1/3}$. Assume that $x > 0$ and $y > 0$.

Solution   **a.** $64^{-2/3} = \dfrac{1}{64^{2/3}}$

$= \dfrac{1}{(64^{1/3})^2}$

$= \dfrac{1}{4^2}$

$= \dfrac{1}{16}$

b. $(x^2)^{1/2} = x^{2(1/2)}$

$= x^1$

$= x$

c. $(x^6y^4)^{1/2} = x^{6(1/2)}y^{4(1/2)}$

$= x^3y^2$

d. $(27x^{12})^{-1/3} = \dfrac{1}{(27x^{12})^{1/3}}$

$= \dfrac{1}{27^{1/3}x^{12(1/3)}}$

$= \dfrac{1}{3x^4}$ ∎

EXAMPLE 6   Simplify **a.** $x^{1/3}x^{1/2}$, **b.** $\dfrac{3x^{2/3}}{6x^{1/5}}$, and **c.** $\dfrac{2x^{-1/2}}{x^{3/4}}$. Assume that $x > 0$.

Solution   **a.** $x^{1/3}x^{1/2} = x^{2/6}x^{3/6}$ Get a common denominator in the fractional exponents.

$= x^{5/6}$ Keep the base and add the exponents.

b. $\dfrac{3x^{2/3}}{6x^{1/5}} = \dfrac{3x^{10/15}}{6x^{3/15}}$ Get a common denominator in the fractional exponents.

$= \dfrac{1}{2}x^{10/15 - 3/15}$ Simplify $\frac{3}{6}$ and keep the base and subtract the exponents.

$= \dfrac{1}{2}x^{7/15}$

c. $\dfrac{2x^{-1/2}}{x^{3/4}} = \dfrac{2x^{-2/4}}{x^{3/4}}$ Get a common denominator in the fractional exponents.

$= 2x^{-2/4 - 3/4}$ Keep the base and subtract the exponents.

$= 2x^{-5/4}$ Simplify.

$= \dfrac{2}{x^{5/4}}$

$x^{-5/4} = \dfrac{1}{x^{5/4}}$ ∎

In Exercises 1–24, simplify each expression.

1. $81^{1/2}$ **2.** $100^{1/2}$ **3.** $-144^{1/2}$ **4.** $-400^{1/2}$

5. $\left(\dfrac{1}{4}\right)^{1/2}$ **6.** $\left(\dfrac{1}{25}\right)^{1/2}$ **7.** $\left(\dfrac{4}{49}\right)^{1/2}$ **8.** $\left(\dfrac{9}{64}\right)^{1/2}$

9. $27^{1/3}$ **10.** $8^{1/3}$ **11.** $-125^{1/3}$ **12.** $-1000^{1/3}$

13. $(-8)^{1/3}$ **14.** $(-125)^{1/3}$ **15.** $\left(\dfrac{1}{64}\right)^{1/3}$ **16.** $\left(\dfrac{1}{1000}\right)^{1/3}$

17. $\left(\dfrac{27}{64}\right)^{1/3}$ **18.** $\left(\dfrac{64}{125}\right)^{1/3}$ **19.** $16^{1/4}$ **20.** $81^{1/4}$

21. $32^{1/5}$ **22.** $-32^{1/5}$ **23.** $-243^{1/5}$ **24.** $\left(-\dfrac{1}{32}\right)^{1/5}$

In Exercises 25–44, simplify each expression.

25. $81^{3/2}$ **26.** $16^{3/2}$ **27.** $25^{3/2}$ **28.** $4^{5/2}$

29. $125^{2/3}$ **30.** $8^{4/3}$ **31.** $1000^{2/3}$ **32.** $27^{2/3}$

33. $(-8)^{2/3}$ **34.** $(-125)^{2/3}$ **35.** $(32)^{3/5}$ **36.** $-243^{3/5}$

37. $81^{3/4}$ **38.** $256^{3/4}$ **39.** $(-32)^{3/5}$ **40.** $243^{2/5}$

41. $\left(\dfrac{8}{27}\right)^{2/3}$ **42.** $\left(\dfrac{27}{64}\right)^{2/3}$ **43.** $\left(\dfrac{16}{625}\right)^{-3/4}$ **44.** $\left(\dfrac{49}{64}\right)^{-3/2}$

In Exercises 45–68, simplify each expression. Write each answer without using negative exponents.

45. $6^{3/5}6^{2/5}$ **46.** $3^{4/7}3^{3/7}$ **47.** $5^{2/3}5^{4/3}$ **48.** $2^{7/8}2^{9/8}$

49. $(7^{2/5})^{5/2}$ **50.** $(8^{1/3})^3$ **51.** $(5^{2/7})^7$ **52.** $(3^{3/8})^8$

53. $\dfrac{8^{3/2}}{8^{1/2}}$ **54.** $\dfrac{11^{9/7}}{11^{2/7}}$ **55.** $\dfrac{8^{5/3}}{8^{7/3}}$ **56.** $\dfrac{27^{13/15}}{27^{8/15}}$

57. $(2^{1/2}3^{1/2})^2$ **58.** $(3^{2/3}5^{1/3})^3$ **59.** $(4^{3/4}3^{1/4})^4$ **60.** $(2^{1/5}3^{2/5})^5$

61. $4^{-1/2}$ **62.** $8^{-1/3}$ **63.** $27^{-2/3}$ **64.** $36^{-3/2}$

65. $16^{-3/2}$ **66.** $100^{-5/2}$ **67.** $(-27)^{-4/3}$ **68.** $(-8)^{-4/3}$

In Exercises 69–88, simplify each expression. Assume that all variables represent positive numbers.

69. $(x^{1/2})^2$ **70.** $(x^9)^{1/3}$ **71.** $(x^{12})^{1/6}$ **72.** $(x^{18})^{1/9}$

73. $(x^{18})^{2/9}$ **74.** $(x^{12})^{3/4}$ **75.** $x^{5/6}x^{7/6}$ **76.** $x^{2/3}x^{7/3}$

77. $y^{4/7}y^{10/7}$ **78.** $y^{5/11}y^{6/11}$ **79.** $\dfrac{x^{3/5}}{x^{1/5}}$ **80.** $\dfrac{x^{4/3}}{x^{2/3}}$

81. $\dfrac{x^{1/7}x^{3/7}}{x^{2/7}}$ **82.** $\dfrac{x^{5/6}x^{5/6}}{x^{7/6}}$ **83.** $\left(\dfrac{x^{3/5}}{x^{2/5}}\right)^5$ **84.** $\left(\dfrac{x^{2/9}}{x^{1/9}}\right)^9$

85. $\left(\dfrac{y^{2/7}y^{3/7}}{y^{4/7}}\right)^{49}$ **86.** $\left(\dfrac{z^{3/5}z^{6/5}}{z^{2/5}}\right)^5$ **87.** $\left(\dfrac{y^{5/6}y^{7/6}}{y^{1/3}y}\right)^3$ **88.** $\left(\dfrac{t^{7/9}t^{5/9}}{tt^{2/9}}\right)^9$

In Exercises 89–100, simplify each expression. Assume that all variables represent positive numbers.

89. $x^{2/3}x^{3/4}$

90. $a^{3/5}a^{1/2}$

91. $(b^{1/2})^{3/5}$

92. $(x^{2/5})^{4/7}$

93. $\dfrac{t^{2/3}}{t^{2/5}}$

94. $\dfrac{p^{3/4}}{p^{1/3}}$

95. $\dfrac{x^{4/5}x^{1/3}}{x^{2/15}}$

96. $\dfrac{y^{2/3}y^{3/5}}{y^{1/5}}$

97. $\dfrac{a^{2/5}a^{1/5}}{a^{-1/3}}$

98. $\dfrac{q^{3/4}q^{4/5}}{q^{-2/3}}$

99. $\dfrac{12b^{-1/3}b^{-3/4}}{4b^{3/5}}$

100. $\dfrac{4c^{-3/4}c^{1/6}}{8c^{1/4}}$

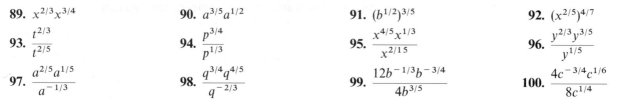

REVIEW EXERCISES

In Review Exercises 1–4, factor each expression.

1. $3x^3y^2z^4 - 6xyz^5 + 15x^2yz^2$

2. $3z^2 - 15tz + 12t^2$

3. $a^4 - b^4$

4. $30r^4 - 200r^2 + 5r^3$

In Review Exercises 5–6, solve each equation.

5. $\dfrac{x-5}{7} + \dfrac{2}{5} = \dfrac{7-x}{5}$

6. $\dfrac{t}{t+2} - 1 = \dfrac{1}{1-t}$

7. Find the slope m of the line represented by the equation $2x - 3y = 18$.

8. Write the equation of the line with slope -4 and y-intercept 7.

In Review Exercises 9–10, solve each system of equations.

9. $\begin{cases} 2x + y = 4 \\ 4x - 3y = 13 \end{cases}$

10. $\begin{cases} 3x + 4y = 2 \\ 12y = 3(2 - 3x) \end{cases}$

11. The **harmonic mean** of two numbers is equal to the quotient obtained when twice their product is divided by their sum. Find two consecutive positive integers with harmonic mean of $\frac{4}{3}$.

12. The **geometric mean** of two numbers is the square root of their product. Find two consecutive positive integers with geometric mean of $3\sqrt{10}$.

8.7   THE DISTANCE FORMULA

A triangle that contains a 90° angle is called a **right triangle.** The longest side of a right triangle is called the **hypotenuse,** which is the side opposite the right angle. The remaining two sides are called the **legs** of the triangle. In the right triangle shown in Figure 8-1, side c is the hypotenuse and sides a and b are the legs.

An important theorem from geometry, called the **Pythagorean theorem,** provides a formula relating the lengths of the three sides of any right triangle.

> **The Pythagorean Theorem.** If the length of the hypotenuse of a right triangle is c and the lengths of the two legs are a and b, then
> $$c^2 = a^2 + b^2$$

What makes the Pythagorean theorem useful is that equal positive numbers have equal positive square roots. This property is called the **square root property of equality.**

> **The Square Root Property of Equality.** If a and b are positive numbers, then
> $$\text{If } a = b, \text{ then } \sqrt{a} = \sqrt{b}.$$

Because the lengths of the sides of a triangle are positive numbers, we can use the square root property of equality and the Pythagorean theorem to calculate the lengths of the sides of a right triangle.

EXAMPLE 1   In the right triangle of Figure 8-1, let $a = 9$ feet and $b = 12$ feet. Find the length of the hypotenuse c.

Solution   Use the Pythagorean theorem, with $a = 9$ and $b = 12$.

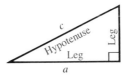

FIGURE 8-1

$$c^2 = a^2 + b^2$$
$$c^2 = 9^2 + 12^2 \qquad \text{Let } a = 9 \text{ and } b = 12.$$
$$c^2 = 81 + 144 \qquad \text{Simplify.}$$
$$c^2 = 225 \qquad \text{Simplify.}$$
$$\sqrt{c^2} = \sqrt{225} \qquad \text{Use the square root property of equality.}$$
$$c = 15 \qquad \text{Simplify.}$$

The hypotenuse of the triangle is 15 feet long.

EXAMPLE 2   In the right triangle of Figure 8-1, let $a = 5$ feet and the hypotenuse $c = 13$ feet. What is the length of leg b?

Solution   Use the Pythagorean theorem, with $a = 5$ and $c = 13$.

$$c^2 = a^2 + b^2$$
$$13^2 = 5^2 + b^2 \qquad \text{Let } a = 5 \text{ and } c = 13.$$
$$169 = 25 + b^2 \qquad \text{Simplify.}$$
$$169 - 25 = b^2 \qquad \text{Add } -25 \text{ to both sides.}$$
$$144 = b^2 \qquad \text{Simplify.}$$
$$\sqrt{144} = \sqrt{b^2} \qquad \text{Use the square root property of equality.}$$
$$12 = b \qquad \text{Simplify.}$$

The length of leg b is 12 feet.

EXAMPLE 3 A 26-foot ladder rests against the side of a building. The base of the ladder is 10 feet from the wall. How far up the side of the building does the ladder reach?

Solution The wall, the ground, and the ladder form a right triangle, as in Figure 8-2. By the Pythagorean theorem, the square of the hypotenuse of that triangle is equal to the sum of the squares of the two legs. The hypotenuse in this example is 26 feet, and one of the legs is the base-to-wall distance of 10 feet. Let d represent the other leg, which is the distance that the ladder reaches up the wall.

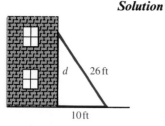

FIGURE 8-2

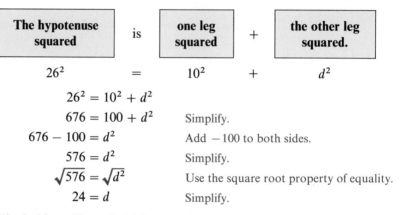

The hypotenuse squared	is	one leg squared	+	the other leg squared.
26^2	=	10^2	+	d^2

$$26^2 = 10^2 + d^2$$

$676 = 100 + d^2$	Simplify.
$676 - 100 = d^2$	Add -100 to both sides.
$576 = d^2$	Simplify.
$\sqrt{576} = \sqrt{d^2}$	Use the square root property of equality.
$24 = d$	Simplify.

The ladder will reach 24 feet up the side of the building. ∎

EXAMPLE 4 The gable end of a roof is an isosceles right triangle with a span of 48 feet (see Figure 8-3.) Find the distance from the eave to the peak.

Solution Recall that an isosceles triangle has two equal sides. In Figure 8-3, the two equal sides are the two legs of the right triangle, and the span of 48 is the length of the hypotenuse. Let x represent the length of each of the legs, the distance from eave to peak.

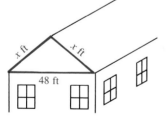

FIGURE 8-3

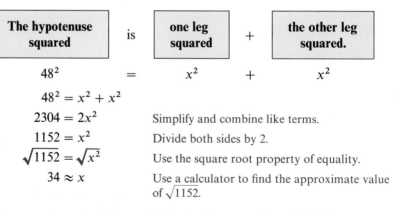

The hypotenuse squared	is	one leg squared	+	the other leg squared.
48^2	=	x^2	+	x^2

$$48^2 = x^2 + x^2$$

$2304 = 2x^2$	Simplify and combine like terms.
$1152 = x^2$	Divide both sides by 2.
$\sqrt{1152} = \sqrt{x^2}$	Use the square root property of equality.
$34 \approx x$	Use a calculator to find the approximate value of $\sqrt{1152}$.

The eave-to-peak distance of the roof is approximately 34 feet. ∎

We can use the Pythagorean theorem to derive a formula for finding the distance between two points $P(x_1, y_1)$ and $Q(x_2, y_2)$ plotted on a rectangular coordinate system. The distance d between points P and Q is the length of the hypotenuse of the triangle in Figure 8-4. The two legs have lengths $x_2 - x_1$ and $y_2 - y_1$.

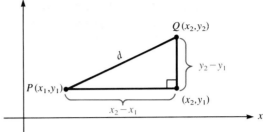

FIGURE 8-4

By the Pythagorean theorem, the square of the hypotenuse of this triangle is equal to the sum of the squares of the two legs. Thus, we have

$$d^2 = (x_2 - x_1)^2 + (y_2 - y_1)^2$$

To solve for d, we can use the square root property of equality to get

$$d = \sqrt{(x_2 - x_1)^2 + (y_2 - y_1)^2}$$

The result is called the **distance formula.**

> **The Distance Formula.**   The distance d between points $P(x_1, y_1)$ and $Q(x_2, y_2)$ is given by the formula
> $$d = \sqrt{(x_2 - x_1)^2 + (y_2 - y_1)^2}$$

EXAMPLE 5 Find the distance between the two points $P(1, 5)$ and $Q(4, 9)$.

Solution Let $(x_1, y_1) = (1, 5)$ and let $(x_2, y_2) = (4, 9)$. In other words, let

$$x_1 = 1, \qquad y_1 = 5, \qquad x_2 = 4, \qquad \text{and} \qquad y_2 = 9$$

Substitute these values into the distance formula and simplify.

$$\begin{aligned} d &= \sqrt{(x_2 - x_1)^2 + (y_2 - y_1)^2} \\ &= \sqrt{(4 - 1)^2 + (9 - 5)^2} \\ &= \sqrt{3^2 + 4^2} \\ &= \sqrt{9 + 16} \\ &= \sqrt{25} \\ &= 5 \end{aligned}$$

The distance between points P and Q is 5 units (see Figure 8-5).

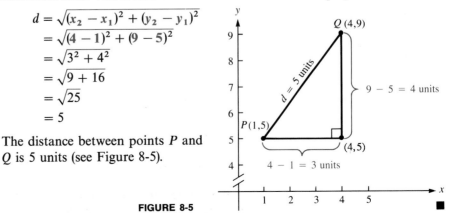

FIGURE 8-5

EXAMPLE 6   Find the distance between the points $P(-3, 6)$ and $Q(2, -6)$.

Solution   Let $(x_1, y_1) = (-3, 6)$ and let $(x_2, y_2) = (2, -6)$. In other words, let

$$x_1 = -3, \quad y_1 = 6, \quad x_2 = 2, \quad \text{and} \quad y_2 = -6$$

Substitute these values into the distance formula and simplify.

$$\begin{aligned}
d &= \sqrt{(x_2 - x_1)^2 + (y_2 - y_1)^2} \\
&= \sqrt{[2 - (-3)]^2 + (-6 - 6)^2} \\
&= \sqrt{5^2 + (-12)^2} \\
&= \sqrt{25 + 144} \\
&= \sqrt{169} \\
&= 13
\end{aligned}$$

The distance between points P and Q is 13 units.   ■

EXERCISE 8.7

In Exercises 1–10, refer to the right triangle of Illustration 1. Find the length of the unknown side.

1. $a = 4$ and $b = 3$. Find c.

2. $a = 6$ and $b = 8$. Find c.

3. $a = 5$ and $b = 12$. Find c.

4. $a = 15$ and $c = 17$. Find b.

5. $a = 21$ and $c = 29$. Find b.

6. $b = 16$ and $c = 34$. Find a.

7. $b = 45$ and $c = 53$. Find a.

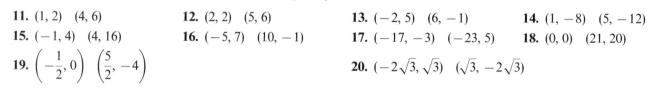

ILLUSTRATION 1

8. $a = 7$ and $b = 1$. Find c.

9. $a = 5$ and $c = 9$. Find b.

10. $a = 1$ and $c = \sqrt{2}$. Find b.

In Exercises 11–20, find the distance between the given points.

11. $(1, 2)$ $(4, 6)$

12. $(2, 2)$ $(5, 6)$

13. $(-2, 5)$ $(6, -1)$

14. $(1, -8)$ $(5, -12)$

15. $(-1, 4)$ $(4, 16)$

16. $(-5, 7)$ $(10, -1)$

17. $(-17, -3)$ $(-23, 5)$

18. $(0, 0)$ $(21, 20)$

19. $\left(-\dfrac{1}{2}, 0\right)$ $\left(\dfrac{5}{2}, -4\right)$

20. $(-2\sqrt{3}, \sqrt{3})$ $(\sqrt{3}, -2\sqrt{3})$

21. A 20-foot ladder reaches a window 16 feet above the ground. How far from the wall is the base of the ladder?

22. A 150-foot-tall tower is secured by three guy wires fastened at the top and to anchors 15 feet from the base of the tower. How long is each guy wire?

23. A 34-foot-long wire reaches from the top of a telephone pole to a point on the ground 16 feet from the base of the pole. How tall is the telephone pole?

24. A rectangular garden has sides of 28 and 45 feet. How long is a path that extends from one corner to the opposite corner?

25. The legs of a certain right triangle are equal, and the hypotenuse is $2\sqrt{2}$ units long. Find the length of each leg.

26. The sides of a square are 3 feet long. Find the length of each diagonal of the square.

27. The diagonal of a square is 3 feet long. Find its perimeter.

28. A woman drives 4.2 miles east and then 4.0 miles north. How far is she from her starting point?

29. A man drives 7 miles north, 8 miles west, and 5 miles north. How far is he from his starting point?

30. The entrance to a one-way tunnel is a rectangle with a semicircular roof. Its dimensions are given in Illustration 2. How high can a 10-foot-wide truck be, without getting stuck in the tunnel?

31. The square in Illustration 3 is inscribed in a circle. The sides of the square are 6 inches long. Find the area of the circle.

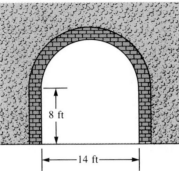

8 ft

──14 ft──

ILLUSTRATION 2

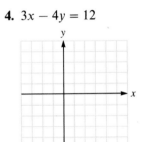

6 in.
6 in. 6 in.
6 in.

ILLUSTRATION 3

32. Will a square with area 40 square inches fit inside a circle with an area of 120 square inches?

REVIEW EXERCISES

In Review Exercises 1–4, graph each equation.

1. $x = 3$

2. $y = -3$

3. $-2x + y = 4$

4. $3x - 4y = 12$

In Review Exercises 5–8, graph each inequality.

5. $3y \le 9$

6. $-x > 3$

7. $4x - y > 4$

8. $2x - 5y \le 10$

9. Write 7.2×10^6 in standard notation.   **10.** Write 0.435×10^{-6} in standard notation.

11. Two cars leave Malta at the same time, one heading east at 55 miles per hour and the other heading west at 50 miles per hour. How long will it take for the cars to be 315 miles apart?

12. Lemon drops worth $1.10 per pound are mixed with jelly beans worth $1.25 per pound to make 30 pounds of a mixture worth $1.15 per pound. How many pounds of jelly beans should be used?

CHAPTER SUMMARY

Key Words

conjugate (8.4)
cube root (8.1)
distance formula (8.7)
division property of
    radicals (8.2)
extraneous solution (8.5)
hypotenuse (8.7)
imaginary number (8.1)
index (8.1)
irrational number (8.1)
leg of a right triangle (8.7)

like radicals (8.3)
multiplication property of
    radicals (8.2)
nth root of a number (8.1)
perfect integer square (8.1)
principal square root (8.1)
Pythagorean theorem (8.7)
radical sign (8.1)
radicand (8.1)
rational exponent (8.6)
rational number (8.1)

rationalizing the
    denominator (8.4)
real number (8.1)
right triangle (8.7)
simplified form of a radical (8.2)
square root (8.1)
square root property of
    equality (8.7)
squaring property of
    equality (8.5)

Key Ideas

(8.1) The number b is a square root of a if $b^2 = a$.

If a is a positive number, then the principal square root of a, denoted by $\sqrt{a}$, is the positive square root of a. The principal square root of 0 is 0.

If a is a positive integer and not a perfect square, then $\sqrt{a}$ is an irrational number.

The square root of a negative number is not a real number.

The cube root of a is denoted by $\sqrt[3]{a}$, and $\sqrt[3]{a} = b$ if $b^3 = a$.

The nth root of a is denoted by $\sqrt[n]{a}$, and $\sqrt[n]{a} = b$ if $b^n = a$.

(8.2) If a and b are nonnegative numbers, then

$$\sqrt{ab} = \sqrt{a}\sqrt{b} \quad \text{and} \quad \text{if } b \neq 0, \text{ then } \sqrt{\frac{a}{b}} = \frac{\sqrt{a}}{\sqrt{b}}$$

To simplify an expression involving square roots, use the multiplication and division properties of radicals to remove perfect square factors from the radicands.

(8.3) Radical expressions can be added or subtracted if they contain like radicals. Often radicals can be converted to like radicals and then added.

(8.4) If a square root appears as a monomial in the denominator of a fraction, rationalize the denominator by multiplying both the numerator and the denominator of the fraction by some appropriate square root.

If the denominator of a fraction contains radicals within a binomial, multiply the numerator and the denominator by the conjugate of the denominator.

(8.5) If $a = b$, then $a^2 = b^2$.

To solve an equation that involves square roots, rearrange the terms of the equation so that no more than one radical appears on one side of the equation. Then square both sides of the equation and solve the resulting equation. Finally, *check the solution.*

(8.6) $x^{1/n} = \sqrt[n]{x}$ $x^{m/n} = \sqrt[n]{x^m} = (\sqrt[n]{x})^m$

(8.7) **The Pythagorean theorem.** In any right triangle, the sum of the squares of the two legs is equal to the square of the hypotenuse.

Let a and b be positive numbers. If $a = b$, then $\sqrt{a} = \sqrt{b}$.

The distance formula. The distance between the points (x_1, y_1) and (x_2, y_2) is given by the formula

$$d = \sqrt{(x_2 - x_1)^2 + (y_2 - y_1)^2}$$

CHAPTER 8 REVIEW EXERCISES

(8.1) *In Review Exercises 1–12, find the value of each expression.*

1. $\sqrt{25}$

2. $\sqrt{64}$

3. $-\sqrt{144}$

4. $-\sqrt{289}$

5. $\sqrt{256}$

6. $-\sqrt{64}$

7. $\sqrt{169}$

8. $-\sqrt{225}$

9. $-\sqrt[3]{-27}$

10. $-\sqrt[3]{125}$

11. $\sqrt[4]{81}$

12. $\sqrt[5]{32}$

In Review Exercises 13–16, use a calculator to find each value to three decimal places.

13. $\sqrt{21}$

14. $-\sqrt{15}$

15. $-\sqrt{57.3}$

16. $\sqrt{751.9}$

(8.2) *In Review Exercises 17–36, simplify each expression. Assume that all variables represent positive numbers.*

17. $\sqrt{32}$

18. $\sqrt{50}$

19. $\sqrt{500}$

20. $\sqrt{112}$

21. $\sqrt{80x^2}$

22. $\sqrt{63y^2}$

23. $-\sqrt{250t^3}$

24. $-\sqrt{700z^5}$

25. $\sqrt{200x^2y}$

26. $\sqrt{75y^2z}$

27. $\sqrt[3]{8x^2y^3}$

28. $\sqrt[3]{250x^4y^3}$

29. $\sqrt{\dfrac{16}{25}}$

30. $\sqrt{\dfrac{100}{49}}$

31. $\sqrt[3]{\dfrac{1000}{27}}$

32. $\sqrt[3]{\dfrac{16}{64}}$

33. $\sqrt{\dfrac{60}{49}}$

34. $\sqrt{\dfrac{80}{225}}$

35. $\sqrt{\dfrac{242x^4}{169x^2}}$

36. $\sqrt{\dfrac{450a^6}{196a^2}}$

(8.3–8.4) *In Review Exercises 37–60, perform the indicated operations. Assume that all variables represent positive numbers.*

37. $\sqrt{2} + \sqrt{8} - \sqrt{18}$

38. $\sqrt{3} + \sqrt{27} - \sqrt{12}$

39. $3\sqrt{5} + 5\sqrt{45}$

40. $5\sqrt{28} - 3\sqrt{63}$

41. $3\sqrt{2x^2y} + 2x\sqrt{2y}$

42. $3y\sqrt{5xy^3} - y^2\sqrt{20xy}$

43. $\sqrt[3]{16} + \sqrt[3]{54}$

44. $\sqrt[3]{2000x^3} - \sqrt[3]{128x^3}$

45. $(3\sqrt{2})(-2\sqrt{3})$

46. $(-5\sqrt{x})(-2\sqrt{x})$

47. $(3\sqrt{3x})(4\sqrt{6x})$

48. $(-2\sqrt{27y^3})(y\sqrt{2y})$

49. $(\sqrt[3]{4})(2\sqrt[3]{4})$

50. $(-2\sqrt[3]{32x^2})(3\sqrt[3]{2x^2})$

51. $\sqrt{2}(\sqrt{8} - \sqrt{18})$

52. $\sqrt{6y}(\sqrt{2y} + \sqrt{75})$

53. $(\sqrt{3} + \sqrt{5})(\sqrt{3} - \sqrt{5})$

54. $(\sqrt{15} + 3x)(\sqrt{15} + 3x)$

55. $(\sqrt[3]{3} + 2)(\sqrt[3]{3} - 1)$

56. $(\sqrt[3]{5} - 1)(\sqrt[3]{5} + 1)$

57. $(3\sqrt{5} + 2)^2$

58. $(2\sqrt{3} - 1)^2$

59. $(\sqrt{x} - \sqrt{2})^2$

60. $(\sqrt{7} + \sqrt{a})^2$

In Review Exercises 61–68, rationalize each denominator. Assume that all variables represent positive numbers.

61. $\dfrac{1}{\sqrt{7}}$

62. $\dfrac{3}{\sqrt{18}}$

63. $\dfrac{8}{\sqrt[3]{16}}$

64. $\dfrac{10}{\sqrt[3]{32}}$

65. $\dfrac{\sqrt{7}}{\sqrt{35}}$

66. $\dfrac{3}{\sqrt{3} - 1}$

67. $\dfrac{2\sqrt{5}}{\sqrt{5} + \sqrt{3}}$

68. $\dfrac{\sqrt{7x} + \sqrt{x}}{\sqrt{7x} - \sqrt{x}}$

(8.5) *In Review Exercises 69–76, solve each equation. Check all solutions. If an equation has no solution, so indicate.*

69. $\sqrt{x + 3} = 3$

70. $\sqrt{2x + 10} = 2$

71. $\sqrt{3x + 4} = -2\sqrt{x}$

72. $\sqrt{2(x + 4)} - \sqrt{4x} = 0$

73. $\sqrt{x + 5} = x - 1$

74. $\sqrt{2x + 9} = x - 3$

75. $\sqrt{2x + 5} - 1 = x$

76. $\sqrt{4a + 13} + 2 = a$

(8.6) *In Review Exercises 77–92, simplify each expression. Write each answer without using negative exponents.*

77. $49^{1/2}$

78. $(-1000)^{1/3}$

79. $36^{3/2}$

80. $\left(\dfrac{4}{9}\right)^{5/2}$

81. $8^{2/3}8^{4/3}$

82. $\dfrac{5^{17/7}}{5^{3/7}}$

83. $\dfrac{x^{4/5}x^{3/5}}{(x^{1/5})^2}$

84. $\left(\dfrac{r^{1/3}r^{2/3}}{r^{4/3}}\right)^3$

85. $6^{5/3}6^{-2/3}$

86. $\dfrac{5^{2/3}}{5^{-1/3}}$

87. $\dfrac{x^{2/5}x^{1/5}}{x^{-2/5}}$

88. $(a^4b^8)^{-1/2}$

89. $x^{1/3}x^{2/5}$

90. $\dfrac{t^{3/4}}{t^{2/3}}$

91. $\dfrac{x^{-4/5}x^{1/3}}{x^{1/3}}$

92. $\dfrac{r^{1/4}r^{1/3}}{r^{5/6}}$

(8.7) *In Review Exercises 93–96, refer to the right triangle of Illustration 1. Find the length of the unknown side.*

93. $a = 21$ and $b = 28$. Find c.

94. $a = 25$ and $c = 65$. Find b.

95. $a = 1$ and $c = \sqrt{2}$. Find b.

96. $b = 5$ and $c = 7$. Find a.

ILLUSTRATION 1

In Review Exercises 97–100, find the distance between the given points.

97. $(-7, 12)$ $(-4, 8)$

98. $(-15, -3)$ $(-10, -15)$

99. $(1, 1)$ $(-1, 1)$

100. $(-10, 11)$ $(10, -10)$

101. A window frame is 32 inches by 60 inches. How long is the frame's diagonal?

102. A 53-foot wire runs from the top of a building to a point 28 feet from the base of the building. How tall is the building?

CHAPTER 8 TEST

In Problems 1–4, write each expression without a radical sign. Assume that $x > 0$.

_____   **1.** $\sqrt{100}$

_____   **2.** $-\sqrt{400}$

_____   **3.** $\sqrt[3]{-27}$

_____   **4.** $\sqrt{3x}\sqrt{27x}$

In Problems 5–10, simplify each expression. Assume that $x > 0$ and $y > 0$.

_____   **5.** $\sqrt{8x^2}$

_____   **6.** $\sqrt{54x^3y}$

_____   7. $\sqrt{\dfrac{320}{10}}$

_____   8. $\sqrt{\dfrac{18x^2y^3}{2xy}}$

_____   9. $\sqrt[3]{x^6y^6}$

_____   10. $\sqrt[4]{\dfrac{16x^8}{y^4}}$

In Problems 11–15, perform each operation. Give all answers in simplified form. Assume that all variables are positive numbers.

_____   11. $\sqrt{12} + \sqrt{27}$
_____   12. $\sqrt{8x^3} - x\sqrt{18x}$
_____   13. $(-2\sqrt{8x})(3\sqrt{12x})$
_____   14. $\sqrt{3}(\sqrt{8} + \sqrt{6})$
_____   15. $(\sqrt{2} + \sqrt{3})(\sqrt{2} - \sqrt{3})$

In Problems 16–19, rationalize each denominator.

_____   16. $\dfrac{2}{\sqrt{2}}$

_____   17. $\sqrt{\dfrac{3xy^3}{48x^2}}$

_____   18. $\dfrac{2}{\sqrt{5} - 2}$

_____   19. $\dfrac{\sqrt{3x}}{\sqrt{x} + 2}$

In Problems 20–24, solve each equation.

_____   20. $\sqrt{x} + 3 = 9$
_____   21. $\sqrt{x - 2} - 2 = 6$
_____   22. $\sqrt{3x + 9} = 2\sqrt{x + 1}$
_____   23. $3\sqrt{x - 3} = \sqrt{2x + 8}$
_____   24. $\sqrt{3x + 1} = x - 1$

In Problems 25–30, simplify each expression. Write all answers without using negative exponents. Assume that all variables represent positive numbers.

_____   25. $121^{1/2}$
_____   26. $27^{-4/3}$
_____   27. $(y^{15})^{2/5}$

a^{10} **28.** $\left(\dfrac{a^{5/3}a^{4/3}}{(a^{1/3})^2 a^{2/3}}\right)^6$

$p^{17/12}$ **29.** $p^{2/3}p^{3/4}$

$\dfrac{1}{x^{4/15}}$ **30.** $\dfrac{x^{2/3}x^{-4/5}}{x^{2/15}}$

13 in.   **31.** Find the length of the hypotenuse of a right triangle with legs of 5 inches and 12 inches.

10   **32.** Find the distance between the points (1, 4) and (7, 12).

5 units   **33.** Find the distance between the points $(-2, -3)$ and $(-5, 1)$.

10 ft   **34.** A 26-foot ladder reaches a point on a wall 24 feet above the ground. How far from the wall is the ladder's base?

9

Quadratic Equations

In this chapter we shall develop general techniques for solving quadratic equations—equations of the form $ax^2 + bx + c = 0$, where a, b, and c are real numbers and $a \neq 0$.

9.1   SOLVING EQUATIONS OF THE FORM $x^2 = c$

Previously we have solved quadratic equations by the **factoring method.**

The Factoring Method.   To solve a quadratic equation by factoring, we

1. write the equation in $ax^2 + bx + c = 0$ form (called **quadratic form**),
2. factor the trinomial on the left-hand side of the equation,
3. use the zero-factor theorem to set each factor equal to 0, and
4. solve each resulting linear equation.

We review the factoring method in Example 1.

EXAMPLE 1   Use the factoring method to solve the equations **a.** $6x^2 - 3x = 0$ and **b.** $6x^2 - x - 2 = 0$.

Solution   **a.** $6x^2 - 3x = 0$

$3x(2x - 1) = 0$ Factor out $3x$.

$3x = 0$ or $2x - 1 = 0$ Set each factor equal to 0.

$x = 0$ $2x = 1$ Solve each linear equation.

$$x = \frac{1}{2}$$

Check: **For $x = 0$**   **For $x = \dfrac{1}{2}$**

$6x^2 - 3x = 0$ $6x^2 - 3x = 0$

$6(0)^2 - 3(0) \overset{?}{=} 0$ $6\left(\dfrac{1}{2}\right)^2 - 3\left(\dfrac{1}{2}\right) \overset{?}{=} 0$

$6(0) - 0 \overset{?}{=} 0$ $6\left(\dfrac{1}{4}\right) - \dfrac{3}{2} \overset{?}{=} 0$

$0 - 0 \overset{?}{=} 0$ $\dfrac{3}{2} - \dfrac{3}{2} \overset{?}{=} 0$

$0 = 0$ $0 = 0$

b. $6x^2 - x - 2 = 0$

$(3x - 2)(2x + 1) = 0$ Factor the trinomial.

$3x - 2 = 0$ or $2x + 1 = 0$ Set each factor equal to 0.

$3x = 2$ $2x = -1$ Solve each linear equation.

$$x = \frac{2}{3} \qquad\qquad x = -\frac{1}{2}$$

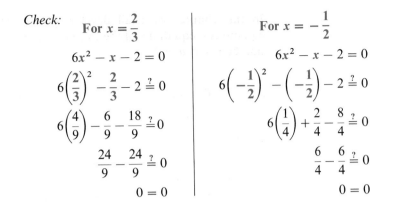

Check:

For $x = \dfrac{2}{3}$	For $x = -\dfrac{1}{2}$
$6x^2 - x - 2 = 0$	$6x^2 - x - 2 = 0$
$6\left(\dfrac{2}{3}\right)^2 - \dfrac{2}{3} - 2 \overset{?}{=} 0$	$6\left(-\dfrac{1}{2}\right)^2 - \left(-\dfrac{1}{2}\right) - 2 \overset{?}{=} 0$
$6\left(\dfrac{4}{9}\right) - \dfrac{6}{9} - \dfrac{18}{9} \overset{?}{=} 0$	$6\left(\dfrac{1}{4}\right) + \dfrac{2}{4} - \dfrac{8}{4} \overset{?}{=} 0$
$\dfrac{24}{9} - \dfrac{24}{9} \overset{?}{=} 0$	$\dfrac{6}{4} - \dfrac{6}{4} \overset{?}{=} 0$
$0 = 0$	$0 = 0$

Unfortunately, the factoring method does not work on all quadratic equations. For example, the quadratic equation $x^2 + 5x + 1 = 0$ cannot be factored by using only integers for coefficients. Thus, we must develop other methods for solving such quadratic equations. We begin by discussing a method for solving **incomplete quadratic equations** of the form $x^2 = c$, where c is a nonnegative number.

If $x^2 = c$, then x must be a number whose square is c. There are two such numbers: $\sqrt{c}$ and $-\sqrt{c}$. Hence, there are *two* solutions for the quadratic equation $x^2 = c$. They are

$$x = \sqrt{c} \qquad \text{and} \qquad x = -\sqrt{c}$$

Both of these solutions check because $(\sqrt{c})^2 = c$ and $(-\sqrt{c})^2 = c$. Thus, we have the following theorem.

Theorem.   If $c > 0$, the quadratic equation $x^2 = c$ has two real solutions. They are

$$x = \sqrt{c} \qquad \text{or} \qquad x = -\sqrt{c}$$

It is common to write the previous result using double sign notation. The equation $x = \pm\sqrt{c}$ (read as x equals plus or minus $\sqrt{c}$) means that either $x = \sqrt{c}$ or $x = -\sqrt{c}$.

EXAMPLE 2   Solve the equation $x^2 = 16$.

Solution   The equation $x^2 = 16$ has two solutions. They are

$$x = \sqrt{16} \qquad \text{or} \qquad x = -\sqrt{16}$$
$$= 4 \qquad \qquad \qquad = -4$$

The solution can also be written as $x = \pm 4$.

Check: For $x = 4$ | For $x = -4$

$$x^2 = 16$$

$$4^2 \stackrel{?}{=} 16$$

$$16 = 16$$

$$x^2 = 16$$

$$(-4)^2 \stackrel{?}{=} 16$$

$$16 = 16$$

The solutions check.   ∎

The method used in Example 2 is called the **square root method.** Example 2 can also be solved with factoring.

$$x^2 = 16$$

$$x^2 - 16 = 0$$

$$(x + 4)(x - 4) = 0$$

$$x + 4 = 0 \qquad \text{or} \qquad x - 4 = 0$$

$$x = -4 \qquad | \qquad x = 4$$

EXAMPLE 3   Solve the equation $3x^2 - 9 = 0$.

Solution   Solve this equation by the square root method.

$$3x^2 - 9 = 0$$

$$3x^2 = 9 \qquad \text{Add 9 to both sides.}$$

$$x^2 = 3 \qquad \text{Divide both sides by 3.}$$

This incomplete quadratic equation has two solutions. They are

$$x = \sqrt{3} \qquad \text{or} \qquad x = -\sqrt{3}$$

These solutions can be written as $x = \pm\sqrt{3}$.

Check: For $x = \sqrt{3}$ | For $x = -\sqrt{3}$

$$3x^2 - 9 = 0$$

$$3(\sqrt{3})^2 - 9 \stackrel{?}{=} 0$$

$$3(3) - 9 \stackrel{?}{=} 0$$

$$9 - 9 \stackrel{?}{=} 0$$

$$0 = 0$$

$$3x^2 - 9 = 0$$

$$3(-\sqrt{3})^2 - 9 \stackrel{?}{=} 0$$

$$3(3) - 9 \stackrel{?}{=} 0$$

$$9 - 9 \stackrel{?}{=} 0$$

$$0 = 0$$

The solutions check.   ∎

EXAMPLE 4   Solve the equation $(x + 1)^2 - 9 = 0$.

Solution
$$(x + 1)^2 - 9 = 0$$

$$(x + 1)^2 = 9 \qquad \text{Add 9 to both sides.}$$

There are two possibilities. They are

$$x + 1 = \sqrt{9} \qquad \text{or} \qquad x + 1 = -\sqrt{9}$$

$$x + 1 = +3 \qquad | \qquad x + 1 = -3$$

$$x = 2 \qquad | \qquad x = -4$$

Check: For $x = 2$ or   For $x = -4$

$$(x + 1)^2 - 9 = 0$$

$$(2 + 1)^2 - 9 \overset{?}{=} 0$$

$$3^2 - 9 \overset{?}{=} 0$$

$$9 - 9 \overset{?}{=} 0$$

$$0 = 0$$

$$(x + 1)^2 - 9 = 0$$

$$(-4 + 1)^2 - 9 \overset{?}{=} 0$$

$$(-3)^2 - 9 \overset{?}{=} 0$$

$$9 - 9 \overset{?}{=} 0$$

$$0 = 0$$

The solutions check.   ∎

EXAMPLE 5   Solve the equation $(x - 2)^2 - 18 = 0$.

Solution $(x - 2)^2 - 18 = 0$

$(x - 2)^2 = 18$ Add 18 to both sides.

The two solutions are

$$x - 2 = \sqrt{18} \quad \text{or} \quad x - 2 = -\sqrt{18}$$

$$x = 2 + \sqrt{18} \qquad\qquad x = 2 - \sqrt{18}$$

$$x = 2 + 3\sqrt{2} \qquad\qquad x = 2 - 3\sqrt{2} \qquad \sqrt{18} = \sqrt{9}\sqrt{2} = 3\sqrt{2}.$$

Both solutions check.   ∎

EXAMPLE 6   Solve the equation $3x^2 - 4 = 2(x^2 + 2)$.

Solution $3x^2 - 4 = 2(x^2 + 2)$

$3x^2 - 4 = 2x^2 + 4$ Remove parentheses.

$3x^2 = 2x^2 + 8$ Add 4 to both sides.

$x^2 = 8$ Add $-2x^2$ to both sides.

The two solutions of this equation are

$$x = \sqrt{8} \quad \text{or} \quad x = -\sqrt{8}$$

$$= 2\sqrt{2} \qquad\qquad = -2\sqrt{2} \qquad \text{Simplify the radical.}$$

Both solutions check.   ∎

EXERCISE 9.1

In Exercises 1–12, use the factoring method to solve each equation.

1. $x^2 - 9 = 0$ **2.** $x^2 + x = 0$ **3.** $3x^2 + 9x = 0$ **4.** $2x^2 - 8 = 0$

5. $x^2 - 5x + 6 = 0$ **6.** $x^2 + 7x + 12 = 0$ **7.** $3x^2 + x - 2 = 0$ **8.** $2x^2 - x - 6 = 0$

9. $6x^2 + 11x + 3 = 0$ **10.** $5x^2 + 13x - 6 = 0$ **11.** $10x^2 + x - 2 = 0$ **12.** $6x^2 + 37x + 6 = 0$

In Exercises 13–24, use the square root method to solve each equation.

13. $x^2 = 1$ **14.** $x^2 = 4$ **15.** $x^2 = 9$ **16.** $x^2 = 32$

17. $x^2 = 20$ **18.** $x^2 = 0$ **19.** $3x^2 = 27$ **20.** $4x^2 = 64$

21. $4x^2 = 16$ **22.** $5x^2 = 125$ **23.** $x^2 = a$ **24.** $x^2 = 4b$

In Exercises 25–34, use the square root method to solve each equation for x.

25. $(x + 1)^2 = 25$ **26.** $(x - 1)^2 = 49$ **27.** $(x + 2)^2 = 81$ **28.** $(x + 3)^2 = 16$

29. $(x - 2)^2 = 8$ **30.** $(x + 2)^2 = 50$ **31.** $(x - a)^2 = 4a^2$ **32.** $(x + y)^2 = 9y^2$

33. $(x + b)^2 = 16c^2$ **34.** $(x - c)^2 = 25b^2$

In Exercises 35–40, use the square root method to solve each equation. (Hint: Factor the perfect trinomial square first.)

35. $x^2 + 4x + 4 = 4$ **36.** $x^2 - 6x + 9 = 9$

37. $9x^2 - 12x + 4 = 16$ **38.** $4x^2 - 20x + 25 = 36$

39. $4x^2 + 4x + 1 = 20$ **40.** $9x^2 + 12x + 4 = 12$

In Exercises 41–46, solve each equation.

41. $6(x^2 - 1) = 4(x^2 + 3)$ **42.** $5(x^2 - 2) = 2(x^2 + 1)$

43. $8(x^2 - 6) = 4(x^2 + 13)$ **44.** $8(x^2 - 1) = 5(x^2 + 10) + 50$

45. $5(x + 1)^2 = (x + 1)^2 + 32$ **46.** $6(x - 4)^2 = 4(x - 4)^2 + 36$

REVIEW EXERCISES

In Review Exercises 1–10, write each expression without using parentheses.

1. $(y - 1)^2$ **2.** $(z + 2)^2$ **3.** $(x + y)^2$ **4.** $(a - b)^2$

5. $(2r - s)^2$ **6.** $(m + 3n)^2$ **7.** $(3a + 2b)^2$ **8.** $(2p - 5q)^2$

9. $(5r - 8s)^2$ **10.** $(7y + 9z)^2$

11. Some robbers leave the scene of a crime by car and proceed out of town at 60 miles per hour. One-half hour later the police follow in a helicopter traveling at 120 miles per hour. How long will it take the police to overtake the robbers?

12. A grocer mixes 40 pounds of a 60 cents-per-pound candy with 60 pounds of a 45 cents-per-pound candy. How much should he charge per pound for the mixture?

9.2 COMPLETING THE SQUARE

If the polynomial in a quadratic equation factors easily, the factoring method is usually the best way to solve the equation. If the polynomial does not factor easily, a different method, called **completing the square,** can be used. In this method, the left-hand side of a quadratic equation such as $x^2 - 4x - 12 = 0$ is rewritten so that it becomes a perfect trinomial square that can be easily factored. The resulting quadratic equation is then easy to solve.

The method of completing the square is based on two special products. They are

$$x^2 + 2bx + b^2 = (x + b)^2$$

and

$$x^2 - 2bx + b^2 = (x - b)^2$$

The trinomials $x^2 + 2bx + b^2$ and $x^2 - 2bx + b^2$ are both perfect trinomial squares because each factors as the square of a binomial. In each of these trinomials, if we take one-half of the coefficient of the x in the middle term and square it, we get the third term.

$$x^2 + 2b \cdot x + b^2 \qquad\qquad x^2 - 2b \cdot x + b^2$$

$$\left[\frac{1}{2}(2b)\right]^2 = (b)^2 = b^2 \qquad \left[\frac{1}{2}(-2b)\right]^2 = (-b)^2 = b^2$$

Thus, to make a binomial such as $x^2 + 12x$ into a perfect trinomial square, we must take one-half of 12, square it, and add it to $x^2 + 12x$.

$$x^2 + 12x + \left[\frac{1}{2}(12)\right]^2 = x^2 + 12x + (6)^2$$

$$= x^2 + 12x + 36$$

Note that $x^2 + 12x + 36$ is a perfect trinomial square because it is equal to $(x + 6)^2$.

EXAMPLE 1 Add the square of one-half of the coefficient of x to make **a.** $x^2 + 4x$, **b.** $x^2 - 6x$, and **c.** $x^2 - 5x$ into perfect trinomial squares.

Solution **a.** $x^2 + 4x + \left[\frac{1}{2}(4)\right]^2 = x^2 + 4x + (2)^2$

$$= x^2 + 4x + 4$$

Note that $x^2 + 4x + 4 = (x + 2)^2$.

b. $x^2 - 6x + \left[\frac{1}{2}(-6)\right]^2 = x^2 - 6x + (-3)^2$

$$= x^2 - 6x + 9$$

Note that $x^2 - 6x + 9 = (x - 3)^2$.

c. $x^2 - 5x + \left[\frac{1}{2}(-5)\right]^2 = x^2 - 5x + \left(-\frac{5}{2}\right)^2$

$$= x^2 - 5x + \frac{25}{4}$$

Note that $x^2 - 5x + \frac{25}{4} = \left(x - \frac{5}{2}\right)^2$. ∎

EXAMPLE 2   Use completing the square to solve the equation $x^2 - 4x - 12 = 0$.

Solution   Note that the coefficient of x^2 is 1. This is a necessary condition that must be true before we attempt to complete the square. Proceed as follows:

$$x^2 - 4x - 12 = 0$$
$$x^2 - 4x = 12 \qquad \text{Add 12 to both sides.}$$

Add the square of one-half of the coefficient of x to both sides of the equation. Then the left-hand side becomes a perfect trinomial square.

$$x^2 - 4x + \left[\frac{1}{2}(-4)\right]^2 = 12 + \left[\frac{1}{2}(-4)\right]^2$$

$$x^2 - 4x + 4 = 12 + 4 \qquad \text{Simplify.}$$
$$(x - 2)^2 = 16 \qquad \text{Factor } x^2 - 4x + 4 \text{ and simplify.}$$
$$x - 2 = \pm\sqrt{16} \qquad \text{Solve the quadratic equation for } x - 2.$$
$$x = 2 \pm 4 \qquad \text{Add 2 to both sides and simplify.}$$

Thus there are two solutions: either

$$\begin{array}{ccc} x = 2 + 4 & \text{or} & x = 2 - 4 \\ = 6 & | & = -2 \end{array}$$

Check both solutions and verify that they are correct. Note that this equation can be solved by factoring. ∎

EXAMPLE 3   Use completing the square to solve the equation $4x^2 + 4x - 3 = 0$.

Solution   First divide both sides of the equation by 4 to make the coefficient of x^2 equal to 1. Then proceed as follows.

$$4x^2 + 4x - 3 = 0$$

$$x^2 + x - \frac{3}{4} = 0 \qquad \text{Divide both sides by 4.}$$

$$x^2 + x = \frac{3}{4} \qquad \text{Add } \tfrac{3}{4} \text{ to both sides.}$$

$$x^2 + x + \left(\frac{1}{2}\right)^2 = \frac{3}{4} + \left(\frac{1}{2}\right)^2 \qquad \text{Add } (\tfrac{1}{2})^2 \text{ to both sides to complete the square.}$$

$$\left(x + \frac{1}{2}\right)^2 = 1 \qquad \text{Factor and simplify.}$$

$$x + \frac{1}{2} = \pm 1 \qquad \text{Solve the quadratic equation for } x + \tfrac{1}{2}.$$

$$x = -\frac{1}{2} \pm 1 \qquad \text{Add } -\tfrac{1}{2} \text{ to both sides.}$$

$$x = -\frac{1}{2} + 1 \quad \text{or} \quad x = -\frac{1}{2} - 1$$

$$= \frac{1}{2} \qquad\qquad\qquad = -\frac{3}{2}$$

Check both solutions. Note that this equation can be solved by factoring. ■

EXAMPLE 4 Use completing the square to solve the equation $2x^2 - 5x - 3 = 0$.

Solution First divide both sides of the equation by 2 to make the coefficient of x^2 equal to 1. Then proceed as follows.

$$2x^2 - 5x - 3 = 0$$

$$x^2 - \frac{5}{2}x - \frac{3}{2} = 0 \qquad\qquad \text{Divide both sides by 2.}$$

$$x^2 - \frac{5}{2}x = \frac{3}{2} \qquad\qquad \text{Add } \tfrac{3}{2} \text{ to both sides.}$$

$$x^2 - \frac{5}{2}x + \left[\frac{1}{2}\left(-\frac{5}{2}\right)\right]^2 = \frac{3}{2} + \left[\frac{1}{2}\left(-\frac{5}{2}\right)\right]^2 \qquad \begin{array}{l}\text{Add } [\tfrac{1}{2}(-\tfrac{5}{2})]^2 \text{ to both} \\ \text{sides to complete} \\ \text{the square.}\end{array}$$

$$x^2 - \frac{5}{2}x + \frac{25}{16} = \frac{3}{2} + \frac{25}{16} \qquad\qquad \text{Simplify.}$$

$$\left(x - \frac{5}{4}\right)^2 = \frac{24}{16} + \frac{25}{16} \qquad\qquad \begin{array}{l}\text{Factor on the left-hand} \\ \text{side and get a common} \\ \text{denominator on the} \\ \text{right-hand side.}\end{array}$$

$$\left(x - \frac{5}{4}\right)^2 = \frac{49}{16} \qquad\qquad \text{Add the fractions.}$$

$$x - \frac{5}{4} = \pm\frac{7}{4} \qquad\qquad \begin{array}{l}\text{Solve the quadratic} \\ \text{equation for } x - \tfrac{5}{4}.\end{array}$$

$$x = \frac{5}{4} \pm \frac{7}{4} \qquad\qquad \text{Add } \tfrac{5}{4} \text{ to both sides.}$$

$$x = \frac{5}{4} + \frac{7}{4} \quad \text{or} \quad x = \frac{5}{4} - \frac{7}{4}$$

$$= \frac{12}{4} \qquad\qquad\qquad = -\frac{2}{4}$$

$$= 3 \qquad\qquad\qquad\quad = -\frac{1}{2}$$

Check both solutions. Note that this equation can be solved by factoring. ■

EXAMPLE 5   Use completing the square to solve the equation $2x^2 + 4x - 2 = 0$.

Solution   This equation cannot be solved by factoring. Thus, we must complete the square.

$$2x^2 + 4x - 2 = 0$$
$$x^2 + 2x - 1 = 0 \qquad \text{Divide both sides by 2.}$$
$$x^2 + 2x = 1 \qquad \text{Add 1 to both sides.}$$
$$x^2 + 2x + (1)^2 = 1 + (1)^2 \qquad \text{Add } (1)^2 \text{ to both sides to complete the square.}$$
$$(x + 1)^2 = 2 \qquad \text{Factor and simplify.}$$
$$x + 1 = \pm\sqrt{2} \qquad \text{Solve the quadratic equation for } x + 1.$$
$$x = -1 \pm \sqrt{2} \qquad \text{Add } -1 \text{ to both sides.}$$
$$x = -1 + \sqrt{2} \qquad \text{or} \qquad x = -1 - \sqrt{2}$$

Both solutions check.   ■

To solve an equation by completing the square, follow these steps.

---

1. Make sure that the coefficient of x^2 is 1. If it is not, make it 1 by dividing both sides of the equation by the coefficient of x^2.
2. If necessary, add a number to both sides of the equation to get the constant term on the right-hand side of the equation.
3. Complete the square.
   a. Identify the coefficient of x.
   b. Take half the coefficient of x.
   c. Square half the coefficient of x.
   d. Add that square to both sides of the equation.
4. Factor the trinomial square and combine terms.
5. Solve the resulting quadratic equation.
6. Check each solution.

EXERCISE 9.2

In Exercises 1–12, complete the square to make each binomial into a perfect trinomial square. Factor each trinomial answer to show that it is the square of a binomial.

1. $x^2 + 4x$

2. $x^2 + 6x$

3. $x^2 - 10x$

4. $x^2 - 8x$

5. $x^2 + 11x$

6. $x^2 + 21x$

7. $a^2 - 3a$

8. $b^2 - 13b$

9. $b^2 + \dfrac{2}{3}b$

10. $a^2 + \dfrac{8}{5}a$ **11.** $c^2 - \dfrac{5}{2}c$ **12.** $c^2 - \dfrac{11}{3}c$

In Exercises 13–30, solve each quadratic equation by completing the square. In some equations, you may have to rearrange some terms.

13. $x^2 + 6x + 8 = 0$ **14.** $x^2 + 8x + 12 = 0$ **15.** $x^2 - 8x + 12 = 0$ **16.** $x^2 - 4x + 3 = 0$

17. $x^2 - 2x - 15 = 0$ **18.** $x^2 - 2x - 8 = 0$ **19.** $x^2 - 7x + 12 = 0$ **20.** $x^2 - 7x + 10 = 0$

21. $x^2 + 5x - 6 = 0$ **22.** $x^2 = 14 - 5x$ **23.** $2x^2 = 4 - 2x$ **24.** $3x^2 + 9x + 6 = 0$

25. $3x^2 + 48 = -24x$ **26.** $3x^2 = 3x + 6$ **27.** $2x^2 = 3x + 2$ **28.** $3x^2 = 2 - 5x$

29. $4x^2 = 2 - 7x$ **30.** $2x^2 = 5x + 3$

In Exercises 31–38, use completing the square to solve each equation.

31. $x^2 + 4x + 1 = 0$ **32.** $x^2 + 6x + 2 = 0$ **33.** $x^2 - 2x - 4 = 0$ **34.** $x^2 - 4x - 2 = 0$

35. $x^2 = 4x + 3$ **36.** $x^2 = 6x - 3$ **37.** $2x^2 = 2 - 4x$ **38.** $3x^2 = 12 - 6x$

In Exercises 39–44, write each equation in quadratic form and solve it by completing the square.

39. $2x(x + 3) = 8$ **40.** $3x(x - 2) = 9$ **41.** $6(x^2 - 1) = 5x$ **42.** $2(3x^2 - 2) = 5x$

43. $x(x + 3) - \dfrac{1}{2} = -2$ **44.** $x[(x - 2) + 3] = 3\left(x - \dfrac{2}{9}\right)$

REVIEW EXERCISES

In Review Exercises 1–6, solve each equation.

1. $\dfrac{3t(2t + 1)}{2} + 6 = 3t^2$ **2.** $\dfrac{2(x + 2)}{4} - 4x = 8$ **3.** $20r^2 - 11r - 3 = 0$ **4.** $\dfrac{2}{3x} - \dfrac{5}{9} = -\dfrac{1}{x}$

5. $\sqrt{x + 12} = \sqrt{3x}$ **6.** $\dfrac{1}{2}\sqrt{3(t + 2)} = \sqrt{2t - 11}$

In Review Exercises 7–10, simplify each radical expression. Assume all variables represent positive numbers.

7. $\sqrt{80}$ **8.** $12\sqrt{x^3 y^2}$ **9.** $\dfrac{x}{\sqrt{7x}}$ **10.** $\dfrac{\sqrt{x} + 2}{\sqrt{x} - 2}$

11. The bus for a biology field trip costs $195. If 2 more students had signed up for the trip, it would have cost each student $2 less. How many students took the trip?

12. The team members will share equally in the $120 cost of new equipment. When two more join the team, each member's share decreases by $2. How much is each paying now?

9.3   THE QUADRATIC FORMULA

The method of completing the square can be used to solve the **general quadratic equation** $ax^2 + bx + c = 0$, with $a \neq 0$. To do so, we begin by dividing both sides of the equation by a and adding $-\frac{c}{a}$ to both sides.

$$ax^2 + bx + c = 0$$

$$\frac{ax^2}{a} + \frac{bx}{a} + \frac{c}{a} = \frac{0}{a}$$

$$x^2 + \frac{b}{a}x + \frac{c}{a} = 0$$

$$x^2 + \frac{b}{a}x \qquad = -\frac{c}{a}$$

We then complete the square on x by adding $\left(\frac{1}{2} \cdot \frac{b}{a}\right)^2$, or $\frac{b^2}{4a^2}$, to both sides:

$$x^2 + \frac{b}{a}x + \frac{b^2}{4a^2} = \frac{b^2}{4a^2} - \frac{c}{a}$$

We now factor the trinominal on the left-hand side of the equation and add the fractions on the right-hand side to obtain

$$\left(x + \frac{b}{2a}\right)\left(x + \frac{b}{2a}\right) = \frac{b^2}{4a^2} - \frac{4ac}{4aa}$$

$$\left(x + \frac{b}{2a}\right)^2 = \frac{b^2 - 4ac}{4a^2}$$

We solve this equation by the square root method. Its solutions are

$$x + \frac{b}{2a} = \sqrt{\frac{b^2 - 4ac}{4a^2}} \qquad \text{and} \qquad x + \frac{b}{2a} = -\sqrt{\frac{b^2 - 4ac}{4a^2}}$$

$$x + \frac{b}{2a} = \frac{\sqrt{b^2 - 4ac}}{\sqrt{4a^2}} \qquad\qquad\qquad x + \frac{b}{2a} = -\frac{\sqrt{b^2 - 4ac}}{\sqrt{4a^2}}$$

$$x = -\frac{b}{2a} + \frac{\sqrt{b^2 - 4ac}}{2a} \qquad\qquad x = -\frac{b}{2a} - \frac{\sqrt{b^2 - 4ac}}{2a}$$

$$x = \frac{-b + \sqrt{b^2 - 4ac}}{2a} \qquad\qquad\qquad x = \frac{-b - \sqrt{b^2 - 4ac}}{2a}$$

These two solutions are usually written as a single expression called the **quadratic formula.**

> **The Quadratic Formula.**   The solutions of the general quadratic equation $ax^2 + bx + c = 0$, where $a \neq 0$, are
> $$x = \frac{-b \pm \sqrt{b^2 - 4ac}}{2a}$$

The quadratic formula can be used to solve specific quadratic equations.

EXAMPLE 1   Use the quadratic formula to solve the equation $x^2 + 5x + 6 = 0$.

Solution   In this example, $a = 1$, $b = 5$, and $c = 6$. Substitute these values into the quadratic formula and simplify.

$$x = \frac{-b \pm \sqrt{b^2 - 4ac}}{2a}$$

$$= \frac{-5 \pm \sqrt{5^2 - 4(1)(6)}}{2(1)} \qquad \text{Substitute 1 for } a, 5 \text{ for } b, \text{ and } 6 \text{ for } c.$$

$$= \frac{-5 \pm \sqrt{25 - 24}}{2}$$

$$= \frac{-5 \pm \sqrt{1}}{2}$$

$$= \frac{-5 \pm 1}{2}$$

Thus,

$$x = \frac{-5 + 1}{2} \qquad \text{and} \qquad x = \frac{-5 - 1}{2}$$

$$= \frac{-4}{2} \qquad\qquad\qquad = \frac{-6}{2}$$

$$= -2 \qquad\qquad\qquad = -3$$

The solutions of the equation are -2 and -3. Check both solutions.   ■

EXAMPLE 2   Use the quadratic formula to solve the equation $2x^2 = 5x + 3$.

Solution   Begin by writing the given equation in quadratic form.

$$2x^2 = 5x + 3$$

$$2x^2 - 5x - 3 = 0 \qquad \text{Add } -5x \text{ and } -3 \text{ to both sides.}$$

In this example, $a = 2$, $b = -5$, and $c = -3$. Substitute these values into the quadratic formula and simplify.

$$x = \frac{-b \pm \sqrt{b^2 - 4ac}}{2a}$$

$$= \frac{-(-5) \pm \sqrt{(-5)^2 - 4(2)(-3)}}{2(2)} \qquad \begin{array}{l}\text{Substitute 2 for } a, -5 \text{ for } b,\\ \text{and } -3 \text{ for } c.\end{array}$$

$$= \frac{5 \pm \sqrt{25 + 24}}{4}$$

$$= \frac{5 \pm \sqrt{49}}{4}$$

$$= \frac{5 \pm 7}{4}$$

Thus,

$$x = \frac{5 + 7}{4} \quad \text{or} \quad x = \frac{5 - 7}{4}$$

$$= \frac{12}{4} \qquad\qquad = \frac{-2}{4}$$

$$= 3 \qquad\qquad = -\frac{1}{2}$$

The solutions of the equation are 3 and $-\frac{1}{2}$. Check both solutions.   ■

In the next example the solutions are both irrational numbers.

EXAMPLE 3   Use the quadratic formula to solve the equation $3x^2 = 2x + 4$.

Solution   Begin by writing the given equation in quadratic form.

$$3x^2 = 2x + 4$$

$$3x^2 - 2x - 4 = 0 \qquad \text{Add } -2x \text{ and } -4 \text{ to both sides.}$$

In this example, $a = 3$, $b = -2$, and $c = -4$. Substitute these values into the quadratic formula and simplify.

$$x = \frac{-b \pm \sqrt{b^2 - 4ac}}{2a}$$

$$= \frac{-(-2) \pm \sqrt{(-2)^2 - 4(3)(-4)}}{2(3)} \qquad \text{Substitute 3 for } a, -2 \text{ for } b,\ \text{and } -4 \text{ for } c.$$

$$= \frac{2 \pm \sqrt{4 + 48}}{6}$$

$$= \frac{2 \pm \sqrt{52}}{6}$$

$$= \frac{2 \pm 2\sqrt{13}}{6} \qquad\qquad \sqrt{52} = \sqrt{4 \cdot 13} = \sqrt{4}\sqrt{13} = 2\sqrt{13}.$$

$$= \frac{2(1 \pm \sqrt{13})}{6} \qquad\qquad \text{Factor out 2.}$$

$$= \frac{1 \pm \sqrt{13}}{3} \qquad\qquad \text{Divide out the common factor of 2.}$$

Thus,

$$x = \frac{1}{3} + \frac{\sqrt{13}}{3} \quad \text{or} \quad x = \frac{1}{3} - \frac{\sqrt{13}}{3}$$

Both solutions check.   ■

EXERCISE 9.3

In Exercises 1–12, write each equation in quadratic form, if necessary. Then determine the value of a, b, and c for each quadratic equation. **Do not solve the equation.**

1. $x^2 + 4x + 3 = 0$

2. $x^2 - x - 4 = 0$

3. $3x^2 - 2x + 7 = 0$

4. $4x^2 + 7x - 3 = 0$

5. $4y^2 = 2y - 1$

6. $2x = 3x^2 + 4$

7. $x(3x - 5) = 2$

8. $y(5y + 10) = 8$

9. $7(x^2 + 3) = -14x$

10. $5(a^2 + 5) = -4a$

11. $(2a + 3)(a - 2) = (a + 1)(a - 1)$

12. $(3a + 2)(a - 1) = (2a + 7)(a - 1)$

In Exercises 13–36, use the quadratic formula to solve each equation.

13. $x^2 - 5x + 6 = 0$

14. $x^2 + 5x + 4 = 0$

15. $x^2 + 7x + 12 = 0$

16. $x^2 - x - 12 = 0$

17. $2x^2 - x - 1 = 0$

18. $2x^2 + 3x - 2 = 0$

19. $3x^2 + 5x + 2 = 0$

20. $3x^2 - 4x + 1 = 0$

21. $4x^2 + 4x - 3 = 0$

22. $4x^2 + 3x - 1 = 0$

23. $5x^2 - 8x - 4 = 0$

24. $6x^2 - 8x + 2 = 0$

25. $x^2 + 3x + 1 = 0$

26. $x^2 + 3x - 2 = 0$

27. $x^2 + 5x - 3 = 0$

28. $x^2 + 5x + 3 = 0$

29. $2x^2 + x = 5$

30. $3x^2 - x = 1$

31. $x^2 + 1 = -4x$

32. $x^2 + 1 = -8x$

33. $x^2 = 1 - 2x$

34. $x^2 = 2 - 2x$

35. $3x^2 = 6x + 2$

36. $3x^2 = -8x - 2$

In Exercises 37–38, use these facts: The two solutions of the quadratic equation $ax^2 + bx + c = 0$ (with $a \neq 0$) are

$$x_1 = \frac{-b + \sqrt{b^2 - 4ac}}{2a} \quad \text{and} \quad x_2 = \frac{-b - \sqrt{b^2 - 4ac}}{2a}$$

37. Show that $x_1 + x_2 = -\dfrac{b}{a}$.

38. Show that $x_1 x_2 = \dfrac{c}{a}$.

REVIEW EXERCISES

In Review Exercises 1–2, solve each equation for the indicated variable.

1. $A = p + prt$; for r

2. $F = \dfrac{GMm}{d^2}$; for M

In Review Exercises 3–4, graph each equation.

3. $y = -2x + 1$

4. $x = 3y - 2$

In Review Exercises 5–6, write the equation of the line with the given properties in general form.

5. Slope of $\frac{3}{5}$ and passing through (0, 12)

6. Passing through (6, 8) and the origin

In Review Exercises 7–8, graph each inequality.

7. $y < 2x$

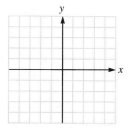

8. $2x + y \geq 4$

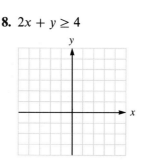

In Review Exercises 9–10, let $f(x) = -x^2 + 7x - 3$. Calculate each value.

9. $f(0)$

10. $f(-3)$

11. The distance between two locations, measured in nautical miles, is directly proportional to that distance, measured in statute miles. If 4.6 statute miles equal 4 nautical miles, how many nautical miles are in 230 statute miles?

12. The illumination, I, measured at a distance d from a light source varies inversely as the square of the distance. If $I = 300$ when $d = 4$, what is I when $d = 6$?

9.4   COMPLEX NUMBERS

So far, all of the work with quadratic expressions has involved real numbers only. The solutions of some quadratic equations are not real numbers. Consider the following example.

EXAMPLE 1   Solve the quadratic equation $x^2 + x + 1 = 0$.

Solution   Because the factoring method does not work conveniently, use the quadratic formula, with $a = 1$, $b = 1$, and $c = 1$

$$x = \frac{-b \pm \sqrt{b^2 - 4ac}}{2a}$$

$$= \frac{-1 \pm \sqrt{1^2 - 4(1)(1)}}{2(1)}$$

$$= \frac{-1 \pm \sqrt{1 - 4}}{2}$$

$$= \frac{-1 \pm \sqrt{-3}}{2}$$

$$x = \frac{-1 + \sqrt{-3}}{2} \quad \text{or} \quad x = \frac{-1 - \sqrt{-3}}{2} \qquad \blacksquare$$

Each solution in Example 1 involves the number $\sqrt{-3}$. This number is not a real number, because the square of no real number is -3; the square of a real number is never negative. For years, mathematicians believed that numbers like $\sqrt{-3}$, $\sqrt{-1}$, and $\sqrt{-9}$ were nonsense, as illegal as division by zero. Even the great English mathematician Sir Isaac Newton (1642–1727) called them impossible. In the seventeenth century, these numbers were named **imaginary numbers** by René Descartes (1596–1650).

Mathematicians no longer think of imaginary numbers as being fictitious or ridiculous. In fact, imaginary numbers have important uses, such as describing the behavior of alternating current in electronics.

The imaginary number $\sqrt{-1}$ occurs often enough to warrant a special symbol; the letter i is used to denote $\sqrt{-1}$. Because i represents the square root of -1, it follows that

$$i^2 = -1$$

The powers of the imaginary number i produce an interesting pattern:

$$i = \sqrt{-1} = i \qquad\qquad i^5 = i^4 \cdot i = 1 \cdot i = i$$
$$i^2 = \sqrt{-1}\sqrt{-1} = -1 \qquad i^6 = i^4 \cdot i^2 = 1(-1) = -1$$
$$i^3 = i^2 \cdot i = -1 \cdot i = -i \qquad i^7 = i^4 \cdot i^3 = 1(-i) = -i$$
$$i^4 = i^2 \cdot i^2 = (-1)(-1) = 1 \qquad i^8 = i^4 \cdot i^4 = (1)(1) = 1$$

The pattern continues: $i, -1, -i, 1, \ldots$.

If we assume that multiplication of imaginary numbers is commutative and associative, then

$$(2i)^2 = 2^2 i^2 = 4(-1) = -4$$

Because $(2i)^2 = -4$, it follows that $2i$ is a square root of -4, and we write

$$\sqrt{-4} = 2i$$

Note that this result could have been obtained by the following process:

$$\sqrt{-4} = \sqrt{4(-1)} = \sqrt{4}\sqrt{-1} = 2i$$

Leonhard Euler (1707–1783)
Euler first used the letter i to represent $\sqrt{-1}$, the letter e for the base of natural logarithms, and the symbol Σ for summation. Euler was one of the most prolific mathematicians of all time, contributing to almost all areas of mathematics. Much of his work was accomplished after he became blind.

Similarly, we have

$$\sqrt{-25} = \sqrt{25(-1)} = \sqrt{25}\sqrt{-1} = 5i$$

$$\sqrt{-\frac{1}{9}} = \sqrt{\frac{1}{9}(-1)} = \sqrt{\frac{1}{9}}\sqrt{-1} = \frac{1}{3}i$$

and

$$\sqrt{\frac{-100}{49}} = \sqrt{\frac{100}{49}(-1)} = \frac{\sqrt{100}}{\sqrt{49}}\sqrt{-1} = \frac{10}{7}i$$

The previous examples illustrate the following rule.

> If at least one of a and b is a nonnegative real number and if there are no divisions by 0, then
>
> $$\sqrt{ab} = \sqrt{a}\sqrt{b} \qquad \text{and} \qquad \sqrt{\frac{a}{b}} = \frac{\sqrt{a}}{\sqrt{b}}$$

Complex Numbers

Imaginary numbers such as $\sqrt{-3}$, $\sqrt{-1}$, and $\sqrt{-9}$ form a subset of a broader set of numbers called **complex numbers.**

> **DEFINITION.**   A **complex number** is any number that can be written in the form $a + bi$, where a and b are real numbers, and $i = \sqrt{-1}$. The number a is called the **real part** and the number b is called the **imaginary part** of the complex number $a + bi$.

If $b = 0$, the complex number $a + bi$ is a real number. If $b \neq 0$ and $a = 0$, the complex number $0 + bi$ (or just bi) is an imaginary number.

We now discuss some properties of complex numbers.

> **DEFINITION.**   The complex numbers $a + bi$ and $c + di$ are equal if and only if $a = c$ and $b = d$.

EXAMPLE 2   **a.** $2 + 3i = \sqrt{4} + \frac{6}{2}i$, because $2 = \sqrt{4}$ and $3 = \frac{6}{2}$.

b. $4 - 5i = \frac{12}{3} - \sqrt{25}i$, because $4 = \frac{12}{3}$ and $-5 = -\sqrt{25}$.

c. $x + yi = 4 + 7i$ if and only if $x = 4$ and $y = 7$.   ∎

> **DEFINITION.** Complex numbers are added and subtracted as if they were binomials:
>
> $$(a + bi) + (c + di) = (a + c) + (b + d)i$$

EXAMPLE 3　**a.** $(8 + 4i) + (12 + 8i) = 8 + 4i + 12 + 8i = 20 + 12i$

b. $(7 - 4i) + (9 + 2i) = 7 - 4i + 9 + 2i = 16 - 2i$

c. $(-6 + i) - (3 - 4i) = -6 + i - 3 + 4i = -9 + 5i$

d. $(2 - 4i) - (-4 + 3i) = 2 - 4i + 4 - 3i = 6 - 7i$ ∎

To multiply a complex number by an imaginary number, we use the distributive property to remove parentheses and then simplify. For example,

$$
\begin{aligned}
-5i(4 - 8i) &= -5i(4) - (-5i)(8i) \\
&= -20i + 40i^2 \\
&= -40 - 20i \qquad \text{Remember that } i^2 = -1.
\end{aligned}
$$

To multiply two complex numbers, we use the following definition:

> **DEFINITION.** Complex numbers are multiplied as if they were binomials, with $i^2 = -1$:
>
> $$
> \begin{aligned}
> (a + bi)(c + di) &= ac + adi + bci + bdi^2 \\
> &= (ac - bd) + (ad + bc)i
> \end{aligned}
> $$

EXAMPLE 4　**a.** $(2 + 3i)(3 - 2i) = 6 - 4i + 9i - 6i^2$

$$
\begin{aligned}
&= 6 + 5i + 6 \\
&= 12 + 5i
\end{aligned}
$$

b. $(3 + i)(1 + 2i) = 3 + 6i + i + 2i^2$

$$
\begin{aligned}
&= 3 + 7i - 2 \\
&= 1 + 7i
\end{aligned}
$$

c. $(-4 + 2i)(2 + i) = -8 - 4i + 4i + 2i^2$

$$
\begin{aligned}
&= -8 - 2 \\
&= -10
\end{aligned}
$$

d. $(-1 - i)(4 - i) = -4 + i - 4i + i^2$

$$
\begin{aligned}
&= -4 - 3i - 1 \\
&= -5 - 3i
\end{aligned}
$$ ∎

The next example shows how to write several complex numbers in $a + bi$ form. When writing answers, it is common practice to accept the form $a - bi$ as a substitute for the form $a + (-b)i$.

EXAMPLE 5   **a.** $7 = 7 + 0i$

b. $3i = 0 + 3i$

c. $4 - \sqrt{-16} = 4 - \sqrt{-1(16)} = 4 - \sqrt{16}\sqrt{-1} = 4 - 4i$

d. $5 + \sqrt{-11} = 5 + \sqrt{-1(11)} = 5 + \sqrt{11}\sqrt{-1} = 5 + \sqrt{11}i$

e. $2i^2 + 4i^3 = 2(-1) + 4(-i) = -2 - 4i$

f. $\dfrac{3}{2i} = \dfrac{3}{2i} \cdot \dfrac{i}{i} = \dfrac{3i}{2i^2} = \dfrac{3i}{2(-1)} = \dfrac{3i}{-2} = 0 - \dfrac{3}{2}i$

g. $-\dfrac{5}{i} = -\dfrac{5}{i} \cdot \dfrac{i^3}{i^3} = -\dfrac{5(-i)}{1} = 5i = 0 + 5i$ ■

We must rationalize denominators to write complex numbers such as

$$\frac{1}{3+i}, \qquad \frac{3-i}{2+i}, \qquad \text{and} \qquad \frac{5+i}{5-i}$$

in $a + bi$ form. To this end, we make the following definition.

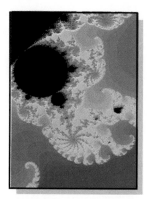

Dr. Bernoit B. Mandelbrot has used mathematics to study unpredictable and irregular events in the physical, behavioral, and biological sciences. His study uses properties of the complex numbers to produce intricate images, such as this portion of the Mandelbrot set.

From CHAOS AND FRACTALS; THE MATHEMATICS BEHIND COMPUTER GRAPHICS, Vol. 39, Devaney and Keen. Reprinted by permission of the American Mathematical Society.

> **DEFINITION.**   The complex numbers $a + bi$ and $a - bi$ are called **complex conjugates** of each other.

For example,

$3 + 4i$ and $3 - 4i$ are complex conjugates.
$5 - 7i$ and $5 + 7i$ are complex conjugates.
$8 + 17i$ and $8 - 17i$ are complex conjugates.

EXAMPLE 6   Find the product of the complex number $3 + i$ and its complex conjugate.

Solution   The complex conjugate of $3 + i$ is $3 - i$. Find the product of these two binomials as follows:

$$
\begin{aligned}
(3 + i)(3 - i) &= 9 - 3i + 3i - i^2 \\
&= 9 - i^2 \\
&= 9 - (-1) \qquad \text{Because } i^2 = -1. \\
&= 10
\end{aligned}
$$ ■

In general, the product of the complex number $a + bi$ and its complex conjugate $a - bi$ is the real number $a^2 + b^2$, as the following work shows:

$$(a + bi)(a - bi) = a^2 - abi + abi - b^2i^2 = a^2 - b^2(-1) = a^2 + b^2$$

Thus, we have

$$(a + bi)(a - bi) = a^2 + b^2$$

EXAMPLE 7   Write $\dfrac{1}{3 + i}$ in $a + bi$ form.

Solution   Because the product of $3 + i$ and its conjugate is a real number, rationalize the denominator by multiplying both the numerator and the denominator of the fraction by the complex conjugate of the denominator, and simplify.

$$\frac{1}{3 + i} = \frac{1}{3 + i} \cdot \frac{3 - i}{3 - i}$$

$$= \frac{3 - i}{9 - 3i + 3i - i^2}$$

$$= \frac{3 - i}{9 - (-1)}$$

$$= \frac{3 - i}{10}$$

$$= \frac{3}{10} - \frac{1}{10}i \qquad\qquad ■$$

EXAMPLE 8   Write $\dfrac{3 - i}{2 + i}$ in $a + bi$ form.

Solution   Rationalize the denominator by multiplying the numerator and the denominator of the fraction by the complex conjugate of the denominator, and simplify.

$$\frac{3 - i}{2 + i} = \frac{3 - i}{2 + i} \cdot \frac{2 - i}{2 - i}$$

$$= \frac{6 - 3i - 2i + i^2}{4 - 2i + 2i - i^2}$$

$$= \frac{5 - 5i}{4 - (-1)}$$

$$= \frac{5(1 - i)}{5} \qquad\qquad \text{Factor out 5 in the numerator.}$$

$$= 1 - i \qquad\qquad \text{Simplify.} \qquad\qquad ■$$

EXAMPLE 9   Divide $5 + i$ by $5 - i$ and express the quotient in $a + bi$ form.

Solution   The quotient obtained when dividing $5 + i$ by $5 - i$ can be expressed as the fraction $\frac{5+i}{5-i}$. To express this quotient in $a + bi$ form, rationalize the denominator by multiplying both the numerator and the denominator by the complex conjugate of the denominator. Then simplify.

$$\frac{5+i}{5-i} = \frac{5+i}{5-i} \cdot \frac{5+i}{5+i}$$

$$= \frac{25 + 5i + 5i + i^2}{25 + 5i - 5i - i^2}$$

$$= \frac{25 + 10i - 1}{25 - (-1)}$$

$$= \frac{24 + 10i}{26}$$

$$= \frac{2(12 + 5i)}{26} \qquad \text{Factor out 2 in the numerator.}$$

$$= \frac{12 + 5i}{13} \qquad \text{Simplify.}$$

$$= \frac{12}{13} + \frac{5}{13} i \qquad\qquad\qquad\qquad\qquad\qquad ■$$

In most cases, the complex numbers we encounter will not be in $a + bi$ form. To avoid mistakes, always put complex numbers in $a + bi$ form before doing any arithmetic involving the numbers.

EXAMPLE 10   Write $\dfrac{4 + \sqrt{-16}}{2 + \sqrt{-4}}$ in $a + bi$ form.

Solution
$$\frac{4 + \sqrt{-16}}{2 + \sqrt{-4}} = \frac{4 + 4i}{2 + 2i}$$

$$= \frac{2(2 + 2i)}{2 + 2i} \qquad \text{Factor out 2 in the numerator and simplify.}$$

$$= 2 + 0i \qquad\qquad\qquad\qquad\qquad\qquad\qquad ■$$

Just as with real numbers, it is possible to define the absolute value of a complex number.

> **DEFINITION.** The **absolute value** of the complex number $a + bi$ is $\sqrt{a^2 + b^2}$. In symbols,
> $$|a + bi| = \sqrt{a^2 + b^2}$$

EXAMPLE 11   **a.** $|3 + 4i| = \sqrt{3^2 + 4^2} = \sqrt{9 + 16} = \sqrt{25} = 5$

b. $|5 - 12i| = \sqrt{5^2 + (-12)^2} = \sqrt{25 + 144} = \sqrt{169} = 13$

c. $|1 + i| = \sqrt{1^2 + 1^2} = \sqrt{1 + 1} = \sqrt{2}$

d. $|a + 0i| = \sqrt{a^2 + 0^2} = \sqrt{a^2} = |a|$

Note that the absolute value of any complex number is a nonnegative real number. Note also that the result of part **d** is consistent with the definition of the absolute value of a real number.   ■

EXAMPLE 12   If a and b are both negative numbers, is the formula $\sqrt{a}\sqrt{b} = \sqrt{ab}$ still true?

Solution   Let $a = -4$ and $b = -1$. Then compute $\sqrt{a}\sqrt{b}$ and $\sqrt{ab}$ to see if their values are equal:

$$\sqrt{a}\sqrt{b} = \sqrt{-4}\sqrt{-1} = 2i \cdot i = 2i^2 = -2$$

On the other hand, we have

$$\sqrt{ab} = \sqrt{(-4)(-1)} = \sqrt{4} = 2$$

Because their values are different, the formula $\sqrt{ab} = \sqrt{a}\sqrt{b}$ is *not* true if both a and b are negative.   ■

EXERCISE 9.4

In Exercises 1–10, solve each quadratic equation. Write all roots in bi or a + bi form.

1. $x^2 + 9 = 0$ **2.** $x^2 + 16 = 0$ **3.** $3x^2 = -16$ **4.** $2x^2 = -25$

5. $x^2 + 2x + 2 = 0$ **6.** $x^2 + 3x + 3 = 0$

7. $2x^2 + x + 1 = 0$ **8.** $3x^2 + 2x + 1 = 0$

9. $3x^2 - 4x + 2 = 0$ **10.** $2x^2 - 3x + 2 = 0$

In Exercises 11–18, simplify each expression.

11. i^{21} **12.** i^{19} **13.** i^{27} **14.** i^{22}

15. i^{100} **16.** i^{42} **17.** i^{97} **18.** i^{200}

In Exercises 19–60, express each number in a + bi form, if necessary, and perform the indicated operations. Give all answers in a + bi form.

19. $(3 + 4i) + (5 - 6i)$ **20.** $(5 + 3i) - (6 - 9i)$

21. $(7 - 3i) - (4 + 2i)$ **22.** $(8 + 3i) + (-7 - 2i)$

23. $(8 + \sqrt{-25}) + (7 + \sqrt{-4})$ **24.** $(-7 + \sqrt{-81}) - (-2 - \sqrt{-64})$

25. $(-8 - \sqrt{-3}) - (7 - \sqrt{-27})$ **26.** $(2 + \sqrt{-8}) + (-3 - \sqrt{-2})$

27. $3i(2 - i)$ **28.** $-4i(3 + 4i)$ **29.** $(2 + 3i)(3 - i)$ **30.** $(4 - i)(2 + i)$

31. $(2 - 4i)(3 + 2i)$ **32.** $(3 - 2i)(4 - 3i)$

33. $(2 + \sqrt{-2})(3 - \sqrt{-2})$ **34.** $(5 + \sqrt{-3})(2 - \sqrt{-3})$

35. $(-2 - \sqrt{-16})(1 + \sqrt{-4})$ **36.** $(-3 - \sqrt{-81})(-2 + \sqrt{-9})$

37. $(2 + \sqrt{-3})(3 - \sqrt{-2})$ **38.** $(1 + \sqrt{-5})(2 - \sqrt{-3})$

39. $(8 - \sqrt{-5})(-2 - \sqrt{-7})$

40. $(-1 + \sqrt{-6})(2 - \sqrt{-3})$

41. $\dfrac{1}{i}$

42. $\dfrac{1}{i^3}$

43. $\dfrac{4}{5i^3}$

44. $\dfrac{3}{2i}$

45. $\dfrac{3i}{8\sqrt{-9}}$

46. $\dfrac{5i^3}{2\sqrt{-4}}$

47. $\dfrac{-3}{5i^5}$

48. $\dfrac{-4}{6i^7}$

49. $\dfrac{-6}{\sqrt{-32}}$

50. $\dfrac{5}{\sqrt{-125}}$

51. $\dfrac{3}{5 + i}$

52. $\dfrac{-2}{2 - i}$

53. $\dfrac{-12}{7 - \sqrt{-1}}$

54. $\dfrac{4}{3 + \sqrt{-1}}$

55. $\dfrac{5i}{6 + 2i}$

56. $\dfrac{-4i}{2 - 6i}$

57. $\dfrac{3 - 2i}{3 + 2i}$

58. $\dfrac{2 + 3i}{2 - 3i}$

59. $\dfrac{3 + \sqrt{-2}}{2 + \sqrt{-5}}$

60. $\dfrac{2 - \sqrt{-5}}{3 + \sqrt{-7}}$

In Exercises 61–70, find each indicated value.

61. $|6 + 8i|$

62. $|12 + 5i|$

63. $|12 - 5i|$

64. $|3 - 4i|$

65. $|5 + 7i|$

66. $|6 - 5i|$

67. $|4 + \sqrt{-2}|$

68. $|3 + \sqrt{-3}|$

69. $|8 + \sqrt{-5}|$

70. $|7 - \sqrt{-6}|$

REVIEW EXERCISES

In Review Exercises 1–6, factor each polynomial.

1. $3x^2 - 27$

2. $2x^2 - 8$

3. $2x^2 + x - 1$

4. $3x^2 + 7x + 4$

5. $-x^2 - 4x + 21$

6. $-x^2 - x + 6$

In Review Exercises 7–12, find the volume of each figure.

7.

7 ft   13 ft

8.

8 ft
5 ft
12 ft

9.

14 m

10. 2.1 ft   3.2 ft

11.3 ft

11.

3 cm   7 cm

12.

5 in.
3 in.
2 in.
10 in.

9.5   GRAPHING QUADRATIC FUNCTIONS

In Chapter 6 we constructed graphs of linear equations. These graphs were straight lines. In this section we shall graph equations of the form $y = ax^2 + bx + c$, where $a \neq 0$. These graphs will not be straight lines but curves that are called **parabolas.**

EXAMPLE 1   Graph $y = x^2$.

Solution   First find several ordered pairs (x, y) that satisfy the equation. To do so, pick several numbers x and find the corresponding values of y. Begin by letting x be some number such as 3.

$$y = x^2$$
$$y = 3^2 \qquad \text{Substitute 3 for } x.$$
$$y = 9$$

The ordered pair $(3, 9)$ and several others that satisfy the equation appear in the table of values shown in Figure 9-1. To graph the equation, plot each ordered pair given in the table and draw a smooth curve that passes through each of the plotted points. The resulting curve, called a **parabola,** is the graph of the equation $y = x^2$. The lowest point on this graph is called the **vertex of the parabola.** Thus, the vertex of this parabola is the point $(0, 0)$.

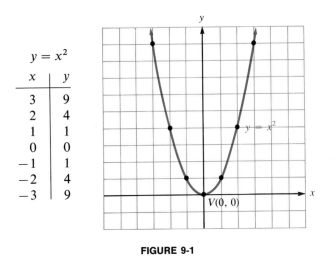

$y = x^2$

x	y
3	9
2	4
1	1
0	0
−1	1
−2	4
−3	9

FIGURE 9-1

EXAMPLE 2   Graph $y = x^2 - 4x + 4$ and find its vertex.

Solution   To construct a table of values such as the one in Figure 9-2, pick several numbers x and find the corresponding values of y. Then plot the points and join each one with a smooth curve.

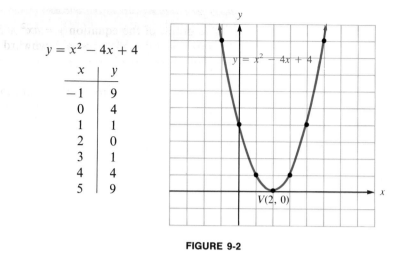

$$y = x^2 - 4x + 4$$

x	y
-1	9
0	4
1	1
2	0
3	1
4	4
5	9

FIGURE 9-2

Again, the graph is a parabola that opens upward. Thus, the vertex is the lowest point on the graph. The vertex of this parabola is the point (2, 0).   ■

EXAMPLE 3   Graph $y = -x^2 + 2x - 1$ and find its vertex.

Solution   Construct a table of values such as in Figure 9-3, plot the points, and draw the graph.

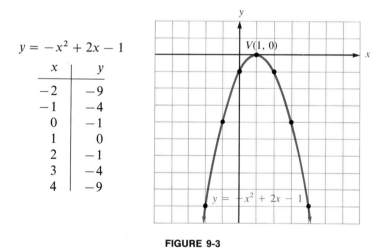

$$y = -x^2 + 2x - 1$$

x	y
-2	-9
-1	-4
0	-1
1	0
2	-1
3	-4
4	-9

FIGURE 9-3

The graph is a parabola that opens downward. Because it opens downward, its vertex is its highest point. The vertex is the point (1, 0).   ■

The results of these first three examples suggest the following fact.

The graph of the equation $y = ax^2 + bx + c$ is a parabola. It opens upward if $a > 0$, and it opens downward if $a < 0$.

It is usually easier to graph a parabola if we know the coordinates of its vertex. We can determine the coordinates of the vertex of the graph of an equation such as

$$y = x^2 - 6x + 8$$

if we complete the square in the following way.

$$y = x^2 - 6x + 8$$
$$y = x^2 - 6x + 9 - 9 + 8 \qquad \text{Add 9 to complete the square on } x^2 - 6x \text{ and subtract 9.}$$
$$y = (x - 3)^2 - 1 \qquad \text{Factor } x^2 - 6x + 9 \text{ and combine like terms.}$$

Since $a = +1$ in the equation $y = x^2 - 6x + 8$, the graph of this equation is a parabola that opens upward. Thus, its vertex will be its lowest point, and this occurs when y is its smallest possible value. Because $(x - 3)^2$ is always a nonnegative number, the smallest value of y will occur when $(x - 3)^2 = 0$ or when $x = 3$. To find the corresponding value of y, we substitute 3 for x in the equation $y = (x - 3)^2 - 1$ and simplify.

$$y = (x - 3)^2 - 1$$
$$y = (3 - 3)^2 - 1 \qquad \text{Substitute 3 for } x.$$
$$y = 0^2 - 1$$
$$y = -1$$

Thus, the vertex of the graph of $y = x^2 - 6x + 8$, or $y = (x - 3)^2 - 1$, is the point $(3, -1)$.

A generalization of this discussion leads to the following fact.

The graph of an equation of the form
$$y = a(x - h)^2 + k$$
is a parabola with its vertex at the point with coordinates (h, k). The parabola opens upward if $a > 0$, and it opens downward if $a < 0$.

EXAMPLE 4   Find the vertex of the parabola determined by $y = -4(x - 3)^2 + 2$. Will the parabola open upward or downward?

Solution   In the equation $y = a(x - h)^2 + k$, the coordinates of the vertex are given by the ordered pair (h, k). In the equation $y = -4(x - 3)^2 + 2$, 3 takes the place of h, 2 takes the place of k, and -4 takes the place of a. Thus, the vertex is the point $(h, k) = (3, 2)$. Because $a = -4$ and $-4 < 0$, the parabola opens downward. ∎

EXAMPLE 5   Find the vertex of the parabola determined by $y = 5(x + 1)^2 + 4$. Will the parabola open upward or downward?

Solution   The equation

$$y = 5(x + 1)^2 + 4$$

is equivalent to the equation

$$y = 5[x - (-1)]^2 + 4$$

In this equation, $h = -1$, $k = 4$, and $a = 5$. Thus, the vertex is the point $(h, k) = (-1, 4)$. Because $a = 5$ and $5 > 0$, the parabola opens upward. ■

EXAMPLE 6   Find the vertex of the parabola determined by $y = 2x^2 + 8x + 2$ and graph the parabola.

Solution   Recall that to complete the square, the coefficient of the term involving x^2 must be 1. To make this so, factor 2 out of the binomial $2x^2 + 8x$. Then proceed as follows:

$$
\begin{aligned}
y &= 2x^2 + 8x + 2 \\
&= 2(x^2 + 4x) + 2 & &\text{Factor 2 out of } 2x^2 + 8x. \\
&= 2(x^2 + 4x + 4 - 4) + 2 & &\text{Complete the square on } x^2 + 4x. \\
&= 2[(x + 2)^2 - 4] + 2 & &\text{Factor } x^2 + 4x + 4. \\
&= 2(x + 2)^2 - 2 \cdot 4 + 2 & &\text{Distribute the multiplication by 2.} \\
&= 2(x + 2)^2 - 6 & &\text{Simplify and combine terms.}
\end{aligned}
$$

or

$$y = 2[x - (-2)]^2 + (-6)$$

Because $h = -2$ and $k = -6$, the vertex of the parabola is the point with coordinates $(h, k) = (-2, -6)$. Because $a = 2$, the parabola opens upward. Pick some numbers x on either side of $x = -2$ and construct a table of values such as the one in Figure 9-4. An easy point to find is the y-intercept of the graph. To do so, substitute 0 for x in the given equation, and determine y: when $x = 0$, $y = 2$. Thus, the y-intercept of the graph is 2. Find and plot some other ordered pairs and draw the parabola.

$y = 2x^2 + 8x + 2$

x	y
0	2
−1	−4
−2	−6
−3	−4
−4	2

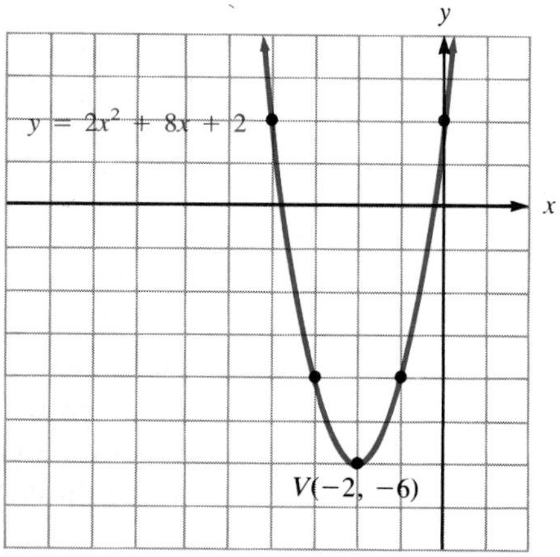

FIGURE 9-4

■

Much can be determined about the graph of the equation $y = ax^2 + bx + c$ from the coefficients a, b, and c. The y-intercept of the graph is the value of y attained when $x = 0$: The y-intercept is $y = c$. The x-intercepts (if any) of the graph are those values of x that cause y to be 0. To find them, we can solve the quadratic equation $ax^2 + bx + c = 0$. Finally, by using the methods of Example 6, we could complete the square on the equation $y = ax^2 + bx + c$ and determine the coordinates of the vertex of the parabola. We summarize these results as follows:

Graphing the Parabola $y = ax^2 + bx + c$.

The y-intercept is $y = c$.

The x-intercepts (if any) are the roots of the quadratic equation
$$ax^2 + bx + c = 0$$

The x-coordinate of the vertex of the parabola $y = ax^2 + bx + c$ is
$$x = -\frac{b}{2a}$$

To find the y-coordinate of the vertex, substitute $-\dfrac{b}{2a}$ for x in the equation $y = ax^2 + bx + c$ and solve for y.

EXAMPLE 7   Graph the equation $y = x^2 - 2x - 3$.

Solution   The equation is written in the form $y = ax^2 + bx + c$, with $a = 1$, $b = -2$, and $c = -3$. Because $a > 0$, the parabola opens upward. To find the x-coordinate of the vertex, substitute the values for a and b into the formula $x = -\frac{b}{2a}$.

$$x = -\frac{b}{2a}$$
$$x = -\frac{-2}{2(1)}$$
$$= 1$$

Thus, the x-coordinate of the vertex is $x = 1$. To find the y-coordinate, substitute $x = 1$ into the equation of the parabola and solve for y.

$$y = x^2 - 2x - 3$$
$$y = 1^2 - 2 \cdot 1 - 3$$
$$= 1 - 2 - 3$$
$$= -4$$

Thus, the vertex of the parabola is the point $(1, -4)$.

To graph the parabola, find several other points with coordinates that satisfy the equation. One easy point to find is the y-intercept of the graph. It is the value $y = c$, attained when $x = 0$. Thus, the parabola passes through the point $(0, -3)$.

To find the x-intercepts of the graph, set y equal to 0 and solve the resulting quadratic equation:

$$y = x^2 - 2x - 3$$
$$0 = x^2 - 2x - 3$$
$$0 = (x - 3)(x + 1) \qquad \text{Factor.}$$
$$x - 3 = 0 \quad \text{or} \quad x + 1 = 0 \qquad \text{Set each factor equal to 0.}$$
$$x = 3 \quad | \quad x = -1$$

Thus, the x-intercepts of the graph are 3 and -1. The graph passes through the points $(3, 0)$ and $(-1, 0)$. The graph appears in Figure 9-5.   ■

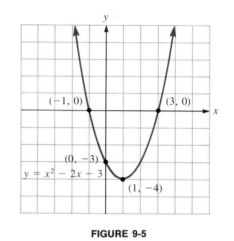

FIGURE 9-5

EXAMPLE 8   An electronics firm manufactures a type of radio with limited demand. Over the past ten years, the firm has learned that it can sell x radios at a price of $(200 - \frac{1}{5}x)$ dollars. How many radios should the firm manufacture and sell to maximize its revenue? Find the maximum revenue.

Solution   The revenue obtained by the firm is the product of the number of radios that it sells (x) and the price of each radio $(200 - \frac{1}{5}x)$. Thus, the revenue, R, is given by the formula

$$R = x\left(200 - \frac{1}{5}x\right) \qquad \text{or} \qquad R = -\frac{1}{5}x^2 + 200x$$

Since the graph of this equation is a parabola that opens downward, the maximum value of R will be the value of R determined by the vertex of the parabola. Because the x-coordinate of the vertex is $x = \frac{-b}{a}$, we have

$$x = \frac{-b}{2a} = \frac{-200}{2(-\frac{1}{5})} = \frac{-200}{-\frac{2}{5}} = (-200)\left(-\frac{5}{2}\right) = 500$$

Thus, the firm should manufacture 500 radios. The maximum revenue is

$$R = -\frac{1}{5}x^2 + 200x = -\frac{1}{5}(500)^2 + 200(500) = 50,000$$

The electronics firm should manufacture 500 radios to gain a maximum revenue of $50,000. ∎

EXERCISE 9.5

In Exercises 1–12, graph each equation.

1. $y = x^2 + 1$

2. $y = x^2 - 4$

3. $y = -x^2$

4. $y = -(x - 1)^2$

5. $y = x^2 + x$

6. $y = x^2 - 2x$

7. $y = -x^2 - 4x$

8. $y = -x^2 + 2x$

9. $y = x^2 + 4x + 4$

10. $y = x^2 - 6x + 9$

11. $y = x^2 - 4x + 6$

12. $y = x^2 + 2x - 3$

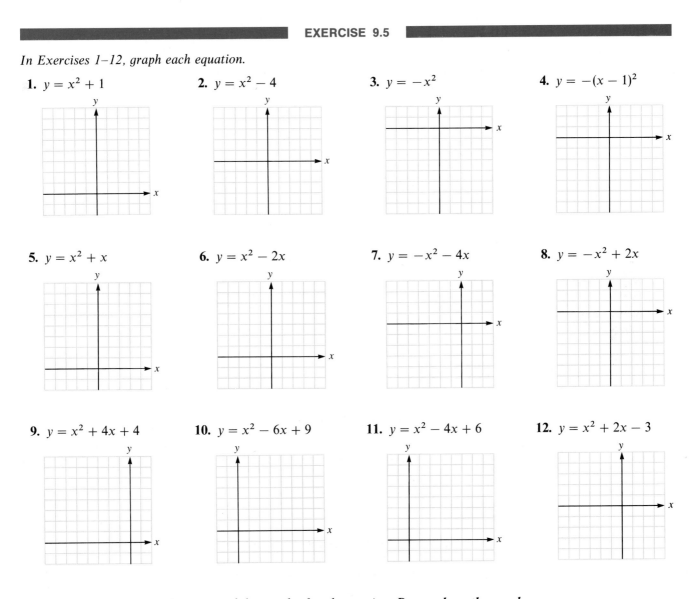

In Exercises 13–24, find the vertex of the graph of each equation. ***Do not draw the graph.***

13. $y = -3(x - 2)^2 + 4$

14. $y = 4(x - 3)^2 + 2$

15. $y = 5(x + 1)^2 - 5$

16. $y = 4(x + 3)^2 + 1$

17. $y = (x - 1)^2$ **18.** $y = -(x - 2)^2$ **19.** $y = -7x^2 + 4$ **20.** $y = 5x^2 - 2$

21. $y = x^2 + 2x + 5$ **22.** $y = x^2 + 4x + 1$ **23.** $y = x^2 - 6x - 12$ **24.** $y = x^2 - 8x - 20$

In Exercises 25–36, complete the square, if necessary, to determine the vertex of the graph of each equation. Then graph the equation.

25. $y = x^2 - 4x + 4$ **26.** $y = x^2 + 6x + 9$ **27.** $y = -x^2 - 2x - 1$ **28.** $y = -x^2 + 2x - 1$

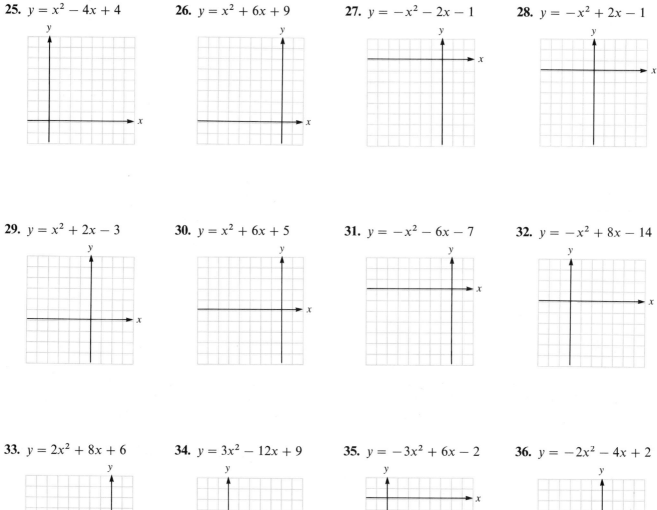

29. $y = x^2 + 2x - 3$ **30.** $y = x^2 + 6x + 5$ **31.** $y = -x^2 - 6x - 7$ **32.** $y = -x^2 + 8x - 14$

33. $y = 2x^2 + 8x + 6$ **34.** $y = 3x^2 - 12x + 9$ **35.** $y = -3x^2 + 6x - 2$ **36.** $y = -2x^2 - 4x + 2$

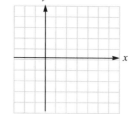

In Exercises 37–42, determine the vertex and the x- and y-intercepts of the graph of each equation. Then graph each equation.

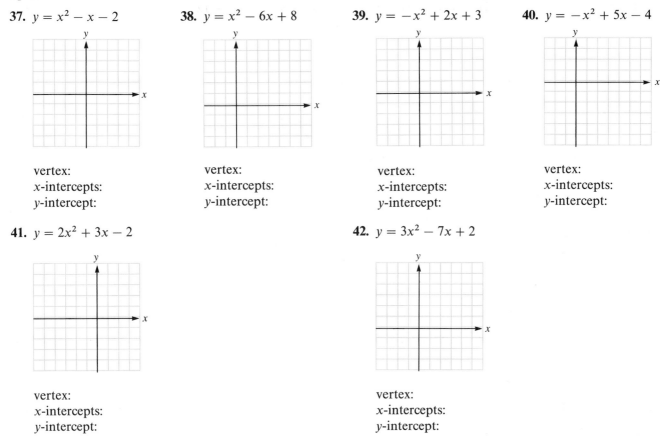

37. $y = x^2 - x - 2$

vertex:
x-intercepts:
y-intercept:

38. $y = x^2 - 6x + 8$

vertex:
x-intercepts:
y-intercept:

39. $y = -x^2 + 2x + 3$

vertex:
x-intercepts:
y-intercept:

40. $y = -x^2 + 5x - 4$

vertex:
x-intercepts:
y-intercept:

41. $y = 2x^2 + 3x - 2$

vertex:
x-intercepts:
y-intercept:

42. $y = 3x^2 - 7x + 2$

vertex:
x-intercepts:
y-intercept:

43. A company has found that it can sell x TVs at a price of $(450 - \frac{1}{6}x)$. How many TVs must the company sell to maximize its revenue? What will be the revenue?

44. A wholesaler sells CD players for \$150 each. However, the wholesaler gives volume discounts on purchases of between 500 and 1000 units according to the formula $(150 - \frac{1}{10}n)$, where n represents the number of units purchased. How many units would a retailer have to buy for the wholesaler to obtain maximum revenue?

REVIEW EXERCISES

In Review Exercises 1–10, simplify each expression.

1. $\sqrt{8}$

2. $\sqrt[3]{125}$

3. $\sqrt{24}$

4. $\sqrt[3]{128}$

5. $\sqrt{12} + \sqrt{27}$

6. $3\sqrt{6y}(-4\sqrt{3y})$

7. $(\sqrt{3} + 1)(\sqrt{3} - 1)$

8. $\dfrac{1}{\sqrt{5}}$

9. $\dfrac{2}{\sqrt{5} - 1}$

10. $\dfrac{x + \sqrt{3}}{\sqrt{3} - \sqrt{2}}$

11. The base of a 41-foot ladder is 9 feet from a vertical wall. How far up the wall does the ladder reach? (See Illustration 1.)

12. A rectangular garden is 20 feet wide by 21 feet long. How long is a diagonal path joining opposite corners? (See Illustration 2.)

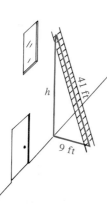

ILLUSTRATION 1

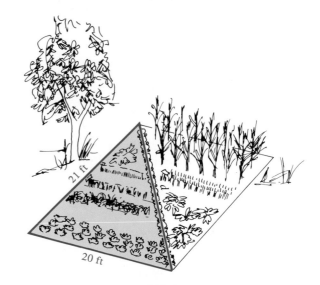

ILLUSTRATION 2

CHAPTER SUMMARY

Key Words

absolute value of a complex number (9.4)
completing the square (9.2)
complex conjugates (9.4)
complex number (9.4)
factoring method (9.1)

general quadratic equation (9.3)
imaginary number (9.4)
incomplete quadratic equation (9.1)

parabola (9.5)
quadratic form (9.1)
quadratic formula (9.3)
square root method (9.1)
vertex of a parabola (9.5)

Key Ideas

(9.1) The two solutions of the incomplete quadratic equation $x^2 = c$ are $+\sqrt{c}$ and $-\sqrt{c}$.

To make a binomial $x^2 + 2bx$ into a perfect trinomial square, add the square of one-half of the coefficient of x:

$$x^2 + 2bx + b^2 = (x + b)^2$$

(9.2) To solve a quadratic equation by completing the square, follow these steps:

1. If necessary, divide both sides of the equation by the coefficient of x^2 to make its coefficient 1.

2. If necessary, get the constant on the right-hand side of the equation.

3. Complete the square.

4. Solve the resulting incomplete quadratic equation.

5. Check each solution.

(9.3) The solutions of $ax^2 + bx + c = 0$, $a \neq 0$, can be found by using the quadratic formula

$$x = \frac{-b \pm \sqrt{b^2 - 4ac}}{2a}$$

(9.4) **Properties of complex numbers:** If a, b, c, and d are real numbers and $i^2 = -1$, then

$a + bi = c + di$ if and only if $a = c$ and $b = d$

$(a + bi) + (c + di) = (a + c) + (b + d)i$

$(a + bi)(c + di) = (ac - bd) + (ad + bc)i$

$|a + bi| = \sqrt{a^2 + b^2}$

(9.5) The graph of the equation $y = ax^2 + bc + c$ is a parabola. It opens upward if $a > 0$, and it opens downward if $a < 0$.

The graph of an equation of the form $y = a(x - h)^2 + k$ is a parabola with its vertex at the point (h, k). The parabola opens upward if $a > 0$, and it opens downward if $a < 0$.

The x-coordinate of the vertex of the parabola $y = ax^2 + bx + c$ is $x = -\dfrac{b}{2a}$.

To find the y-coordinate of the vertex, substitute $-\dfrac{b}{2a}$ for x in the equation of the parabola, and determine y.

CHAPTER 9 REVIEW EXERCISES

(9.1) *In Review Exercises 1–6, use the square root method to solve each incomplete quadratic equation.*

1. $x^2 = 25$ **2.** $x^2 = 36$ **3.** $2x^2 = 18$ **4.** $4x^2 = 9$

5. $x^2 = 8$ **6.** $x^2 = 75$

In Review Exercises 7–12, use the square root method to solve each equation.

7. $(x - 1)^2 = 25$ **8.** $(x + 3)^2 = 36$ **9.** $2(x + 1)^2 = 18$ **10.** $4(x - 2)^2 = 9$

11. $(x - 8)^2 = 8$ **12.** $(x + 5)^2 = 75$

In Review Exercises 13–16, solve each equation. Note that each trinomial is a perfect square.

13. $x^2 + 2x + 1 = 9$ **14.** $x^2 - 6x + 9 = 4$ **15.** $x^2 - 8x + 16 = 20$ **16.** $x^2 - 2x + 1 = 18$

(9.2) *In Review Exercises 17–24, solve each quadratic equation by completing the square.*

17. $x^2 + 5x - 14 = 0$ **18.** $x^2 - 8x + 15 = 0$ **19.** $x^2 + 4x - 77 = 0$ **20.** $x^2 - 2x - 1 = 0$

21. $x^2 + 4x - 3 = 0$ **22.** $x^2 - 6x + 4 = 0$ **23.** $2x^2 + 5x - 3 = 0$ **24.** $2x^2 - 2x - 1 = 0$

(9.3) *In Review Exercises 25–32, use the quadratic formula to solve each quadratic equation.*

25. $x^2 - 2x - 15 = 0$ **26.** $x^2 - 6x - 7 = 0$ **27.** $x^2 - 15x + 26 = 0$ **28.** $2x^2 - 7x + 3 = 0$

29. $6x^2 - 7x - 3 = 0$ **30.** $x^2 + 4x + 1 = 0$ **31.** $x^2 - 6x + 7 = 0$ **32.** $x^2 + 3x = 0$

(9.4) *In Review Exercises 33–48, perform the indicated operations. Give all answers in a + bi form.*

33. $(5 + 4i) + (7 - 12i)$

34. $(-6 - 40i) - (-8 + 28i)$

35. $7i(-3 + 4i)$

36. $6i(2 + i)$

37. $(5 - 3i)(-6 + 2i)$

38. $(2 + \sqrt{128}i)(3 - \sqrt{98}i)$

39. $\dfrac{3}{4i}$

40. $\dfrac{-2}{5i^3}$

41. $\dfrac{6}{2 + i}$

42. $\dfrac{7}{3 - i}$

43. $\dfrac{4 + i}{4 - i}$

44. $\dfrac{3 - i}{3 + i}$

45. $\dfrac{3}{5 + \sqrt{-4}}$

46. $\dfrac{2}{3 - \sqrt{-9}}$

47. $|9 + 12i|$

48. $|24 - 10i|$

(9.5) *In Review Exercises 49–52, find the vertex of the graph of each equation.* **Do not draw the graph.**

49. $y = 5(x - 6)^2 + 7$

50. $y = 3(x + 3)^2 - 5$

51. $y = 2x^2 - 4x + 7$

52. $y = -3x^2 + 18x - 11$

In Review Exercises 53–54, graph each equation.

53. $y = x^2 + 8x + 10$

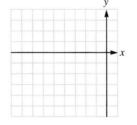

54. $y = -2x^2 - 4x - 6$

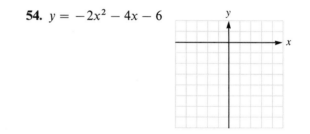

CHAPTER 9 TEST

_____   **1.** Solve by factoring: $6x^2 + x - 1 = 0$.

_____   **2.** Use the square root method to solve $x^2 = 16$.

_____   **3.** Solve $(x - 2)^2 = 3$.

_____   **4.** Solve $2(x^2 - 6) = x^2 + x - 6$.

In Problems 5–6, find the number required to complete the square.

_____   **5.** $x^2 + 14x$

_____   **6.** $x^2 - 7x$

_____   **7.** Use the method of completing the square to solve $3a^2 + 6a - 12 = 0$.

_____   **8.** Write the quadratic formula.

_____   **9.** Use the quadratic formula to solve $x^2 + 3x - 10 = 0$.

_____   **10.** Use the quadratic formula to solve $2x^2 - 5x = 12$.

_____   **11.** Solve: $2x^2 + 5x + 1 = 2$.

_____   **12.** Write the number $\sqrt{-49}$ in bi form.

_____   **13.** Add: $(3 + 4i) + (-2 + 5i)$.

_____   **14.** Subtract: $(4 - 3i) - (-5 + 2i)$.

_____   **15.** Multiply: $(-2 - 5i)(-3 + 2i)$.

_____   **16.** Rationalize the denominator: $\dfrac{-2}{3 + i}$.

_____   **17.** Rationalize the denominator: $\dfrac{2 + i}{2 - i}$.

_____   **18.** Evaluate: $|3 - 4i|$.

_____   **19.** Find the vertex of the parabola determined by the equation $y = -4(x + 5)^2 - 4$.

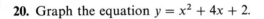

_____   **20.** Graph the equation $y = x^2 + 4x + 2$.

Sample Final Examination

1. How many prime numbers are there between 20 and 30?
   a. 1 **b.** 2 **c.** 3 **d.** 4 **e.** none of the above

2. If $x = 3$, $y = -2$, and $z = -1$, find the value of $\dfrac{x + z}{y}$.
   a. 1 **b.** -1 **c.** -2 **d.** 2 **e.** none of the above

3. If $x = -2$, $y = -3$, and $z = -4$, find the value of $\dfrac{|x - z|}{|y|}$.
   a. -2 **b.** 2 **c.** $\frac{2}{3}$ **d.** $-\frac{2}{3}$ **e.** none of the above

4. The distributive property is written symbolically as
   a. $a + b = b + a$ **b.** $ab = ba$ **c.** $a(b + c) = ab + ac$ **d.** $(a + b) + c = a + (b + c)$
   e. none of the above

5. Solve for x: $7x - 4 = 24$
   a. $x = 4$ **b.** $x = 5$ **c.** $x = -4$ **d.** $x = -5$ **e.** none of the above

6. Solve for z: $6z - (9 - 3z) = -3(z + 2)$
   a. $\frac{4}{3}$ **b.** $\frac{3}{4}$ **c.** $-\frac{4}{3}$ **d.** $-\frac{1}{4}$ **e.** none of the above

7. Solve for r: $\dfrac{r}{5} - \dfrac{r - 3}{10} = 0$
   a. -3 **b.** -2 **c.** 3 **d.** 1 **e.** none of the above

8. Solve for x: $\dfrac{ax}{b} + c = 4$
   a. $\frac{b(4 - x)}{c}$ **b.** $\frac{4b - 4c}{b}$ **c.** $\frac{b(4 - c)}{a}$ **d.** $\frac{b}{a}$ **e.** none of the above

9. A man bought 25 pencils, some at 10 cents and some at 15 cents. The 25 pencils cost $3. How many 10-cent pencils did he buy?
   a. 5 **b.** 10 **c.** 15 **d.** 20 **e.** none of the above

10. Solve the inequality: $-3(x - 2) + 3 \geq 6$
    a. $x \geq -1$ **b.** $x \geq 1$ **c.** $x \leq -1$ **d.** $x \leq 1$ **e.** none of the above

11. Simplify: $x^2x^3x^7$
 a. $12x$ b. x^{12} c. x^{42} d. x^{35} e. none of the above

12. Simplify: $\dfrac{(x^2)^7}{x^3x^4}$
 a. 0   b. 1   c. x^7 d. x^2 e. none of the above

13. Simplify: $\dfrac{x^{-2}y^3}{xy^{-1}}$
 a. x^3y^4 b. $\dfrac{x^3}{y^4}$ c. xy d. $\dfrac{y^4}{x^3}$ e. none of the above

14. Write 73,000,000 in scientific notation.
 a. 7.3×10^7 b. 7.3×10^{-7} c. 73×10^6 d. 0.73×10^9 e. none of the above

15. If $P(x) = 2x^2 + 3x - 4$, find $P(-2)$.
 a. 2   b. -2 c. -18 d. 10   e. none of the above

16. Simplify: $2(y + 3) - 3(y - 2)$
 a. $-y$ b. $5y$ c. $5y + 12$ d. $-y + 12$ e. none of the above

17. Multiply: $-3x^2y^2(2xy^3)$
 a. $6x^3y^5$ b. $5x^2y^5$ c. $-6x^3y^5$ d. $-6xy^5$ e. none of the above

18. Multiply: $2a^3b^2(3a^2b - 2ab^2)$
 a. $6a^6b^3 - 4a^4b^4$ b. $6a^5b^3 - 4a^4b^4$ c. $6a^5b^3 - 4a^4b^2$ d. $6a^5b^4 - 4a^4b^4$ e. none of the above

19. Multiply: $(x + 7)(2x - 3)$
 a. $2x^2 - 11x - 21$ b. $2x^2 + 11x + 21$ c. $2x^2 + 11x - 21$ d. $-2x^2 + 11x - 21$
 e. none of the above

20. Divide: $(x^2 + 5x - 14) \div (x + 7)$
 a. $x + 2$ b. $x - 2$ c. $x + 1$ d. $x - 1$ e. none of the above

21. Factor completely: $r^2h + r^2a$
 a. $r(rh + ra)$ b. $r^2h(1 + a)$ c. $r^2a(h + 1)$ d. $r^2(h + a)$ e. none of the above

22. Factor completely: $m^2n^4 - 49$
 a. $(mn^2 + 7)(mn^2 + 7)$ b. $(mn^2 - 7)(mn^2 - 7)$ c. $(mn + 7)(mn - 7)$ d. $(mn^2 + 7)(mn^2 - 7)$
 e. none of the above

23. One of the factors of $x^2 - 5x + 6$ is
 a. $x + 3$ b. $x + 2$ c. $x - 6$ d. $x - 2$ e. none of the above

24. One of the factors of $36x^2 + 12x + 1$ is
 a. $6x - 1$ b. $6x + 1$ c. $x + 6$ d. $x - 6$ e. none of the above

25. One of the factors of $2x^2 + 7xy + 6y^2$ is
 a. $2x + y$ b. $x - 3y$ c. $2x - y$ d. $x + 6y$ e. none of the above

26. One of the factors of $8x^3 - 27$ is
 a. $2x + 3$ b. $4x^2 - 6x + 9$ c. $4x^2 + 12x + 9$ d. $4x^2 + 6x + 9$ e. none of the above

27. One of the factors of $2x^2 + 2xy - 3x - 3y$ is
 a. $x - y$ b. $2x + 3$ c. $x - 3$ d. $2x - 3$ e. none of the above

28. Solve for x: $x^2 + x - 6 = 0$
 a. $x = -2, x = 3$ b. $x = 2, x = -3$ c. $x = 2, x = 3$ d. $x = -2, x = -3$ e. none of the above

29. Solve for x: $6x^2 - 7x - 3 = 0$

    a. $x = \frac{3}{2}, x = \frac{1}{3}$ **b.** $x = -\frac{3}{2}, x = \frac{1}{3}$ **c.** $x = \frac{3}{2}, x = -\frac{1}{3}$ **d.** $x = -\frac{3}{2}, x = -\frac{1}{3}$ **e.** none of the above

30. Simplify the fraction: $\dfrac{x^2 - 16}{x^2 - 8x + 16}$

    a. $\frac{x+4}{x-4}$ **b.** 1   **c.** $\frac{x-4}{x+4}$ **d.** $-\frac{1}{8x}$ **e.** none of the above

31. Multiply: $\dfrac{x^2 + 11x - 12}{x - 5} \cdot \dfrac{x^2 - 5x}{x - 1}$

    a. $-x(x + 12)$ **b.** $x(x + 12)$ **c.** $\frac{x^2+11x-12}{(x-5)(x-1)}$ **d.** 1   **e.** none of the above

32. Divide: $\dfrac{t^2 + 7t}{t^2 + 5t} \div \dfrac{t^2 + 4t - 21}{t - 3}$

    a. $\frac{1}{t+5}$ **b.** $t + 5$ **c.** 1   **d.** $\frac{(t+7)(t+7)}{t+5}$ **e.** none of the above

33. Simplify: $\dfrac{3x}{2} - \dfrac{x}{4}$

    a. $-x$ **b.** $\frac{5}{4}$ **c.** $2x$ **d.** $\frac{5x}{4}$ **e.** none of the above

34. Simplify: $\dfrac{a + 3}{2a - 6} - \dfrac{2a + 3}{a^2 - 3a} + \dfrac{3}{4}$

    a. $\frac{5a+4}{4a}$ **b.** $\frac{4a+4}{5}$ **c.** 0   **d.** $\frac{9-a}{5a-a^2-2}$ **e.** none of the above

35. Simplify: $\dfrac{x + \dfrac{1}{y}}{\dfrac{1}{x} + y}$

    a. $\frac{x^2y+x}{y+xy^2}$ **b.** 1   **c.** $\frac{xy+x}{y+xy}$ **d.** $\frac{x}{y}$ **e.** none of the above

36. Solve for x: $\dfrac{1}{x} + \dfrac{1}{2x} = \dfrac{1}{4}$

    a. 4   **b.** 5   **c.** 6   **d.** 7   **e.** none of the above

37. Solve for s: $\dfrac{2}{s + 1} + \dfrac{1 - s}{s} = \dfrac{1}{s^2 + s}$

    a. 1   **b.** 2   **c.** 3   **d.** 4   **e.** none of the above

38. Find the x-intercept of the graph of $3x - 4y = 12$.

    a. 3   **b.** -3 **c.** 4   **d.** -4 **e.** none of the above

39. The graph of $y = -3x + 12$ does not pass through

    a. quadrant I   **b.** quadrant II   **c.** quadrant III   **d.** quadrant IV   **e.** none of the above

40. Find the slope of the line passing through $P(-2, 4)$ and $Q(8, -6)$.

    a. -1 **b.** 1   **c.** $-\frac{1}{3}$ **d.** 3   **e.** none of the above

41. The equation of the line passing through $P(-2, 4)$ and $Q(8, -6)$ is

    a. $y = -x + 2$ **b.** $y = -x + 6$ **c.** $y = -x - 6$ **d.** $y = -x - 2$ **e.** none of the above

42. If $f(x) = x^2 - 2x$, find $f(a + 2)$.

    a. $a^2 - 2a$ **b.** $a^2 + 2a$ **c.** $a + 2$ **d.** $a + 2a^2$ **e.** none of the above

43. Solve the system

$$\begin{cases} 2x - 5y = 5 \\ 3x - 2y = -16 \end{cases}$$

for x.

a. $x = 8$ **b.** $x = -8$ **c.** $x = 4$ **d.** $x = -4$ **e.** none of the above

44. Solve the system

$$\begin{cases} 8x - y = 29 \\ 2x + y = 11 \end{cases}$$

for y.

a. $y = -4$ **b.** $y = 4$ **c.** $y = 3$ **d.** $y = -3$ **e.** none of the above

45. Simplify: $\sqrt{12}$

a. $2\sqrt{3}$ **b.** $4\sqrt{3}$ **c.** $6\sqrt{2}$ **d.** $4\sqrt{2}$ **e.** none of the above

46. Simplify: $\sqrt{\dfrac{3}{4}}$

a. $\frac{\sqrt{3}}{4}$ **b.** $\frac{3}{2}$ **c.** $\frac{\sqrt{3}}{2}$ **d.** $\frac{9}{16}$ **e.** none of the above

47. Simplify: $\sqrt{75x^3}$

a. $5\sqrt{x^3}$ **b.** $5x\sqrt{x}$ **c.** $x\sqrt{75x}$ **d.** $25x\sqrt{3x}$ **e.** none of the above

48. Simplify: $3\sqrt{5} - \sqrt{20}$

a. $3\sqrt{-15}$ **b.** $2\sqrt{5}$ **c.** $-\sqrt{5}$ **d.** $\sqrt{5}$ **e.** none of the above

49. Rationalize the denominator: $\dfrac{11}{\sqrt{11}}$

a. $\frac{1}{11}$ **b.** $\sqrt{11}$ **c.** $\frac{\sqrt{11}}{11}$ **d.** 1 **e.** none of the above

50. Rationalize the denominator: $\dfrac{7}{3 - \sqrt{2}}$

a. $3 - \sqrt{2}$ **b.** $7(3 - \sqrt{2})$ **c.** $3 + \sqrt{2}$ **d.** $7(3 + \sqrt{2})$ **e.** none of the above

51. Solve for x: $\sqrt{\dfrac{3x - 1}{5}} = 2$

a. 7 **b.** 4 **c.** -7 **d.** -4 **e.** none of the above

52. Solve for n: $3\sqrt{n} - 1 = 1$

a. $\frac{2}{3}$ **b.** $\sqrt{\frac{2}{3}}$ **c.** $\frac{4}{9}$ **d.** $n\sqrt{n+1}$ **e.** none of the above

53. Simplify: $(a^6b^4)^{1/2}$

a. $(ab)^5$ **b.** a^3b^2 **c.** $\frac{1}{a^3b^2}$ **d.** $\frac{1}{a^6b^4}$ **e.** none of the above

54. Simplify: $\left(\dfrac{8}{125}\right)^{2/3}$

a. $\frac{4}{25}$ **b.** $\frac{25}{4}$ **c.** $\frac{2}{5}$ **d.** $\frac{5}{2}$ **e.** none of the above

55. What number must be added to $x^2 + 12x$ to make it a perfect trinomial square?

a. 6 **b.** 12 **c.** 144 **d.** 36 **e.** none of the above

56. Write the quadratic formula.

a. $x = \dfrac{b \pm \sqrt{b^2 - 4ac}}{2a}$ **b.** $x = \dfrac{-b \pm \sqrt{b^2 - 4ac}}{2a}$ **c.** $x = \dfrac{-b \pm \sqrt{b^2 + 4ac}}{2a}$ **d.** $x = \dfrac{-b \pm \sqrt{b^2 - 4ac}}{2b}$

e. none of the above

57. One solution of the equation $x^2 - 2x - 2 = 0$ is

a. $2\sqrt{3}$ **b.** $2 + 2\sqrt{3}$ **c.** $1 - \sqrt{3}$ **d.** $2 - 2\sqrt{3}$ **e.** none of the above

58. The vertex of the graph of the equation $y = x^2 - 2x + 1$ is

a. $(1, 0)$ **b.** $(0, 1)$ **c.** $(-1, 0)$ **d.** $(0, -1)$ **e.** none of the above

59. The graph of $y = -x^2 + 2x - 1$

a. does not intersect the y-axis   **b.** does not intersect the x-axis   **c.** passes through the origin

d. passes through $(2, 2)$ **e.** none of the above

60. The number $\sqrt{-36}$ written in bi form is

a. $36i$ **b.** -6 **c.** $6i$ **d.** $-6i$ **e.** none of the above

Set Notation

In mathematics, any collection of objects is called a **set.** For example, the collection of natural numbers is referred to as the *set* of natural numbers. The objects contained within a set are called the **members** or the **elements** of the set. The symbol

$$\in$$

read as "is an element of," is used to indicate membership in a set. Elements of a set are often listed between **braces.**

EXAMPLE 1 The set of even numbers between 2 and 10 is designated by the notation

$$\{4, 6, 8\}$$

To indicate that 6 is an element of this set, write

$$6 \in \{4, 6, 8\}$$

To indicate that 7 is *not* an element of the set, write

$$7 \notin \{4, 6, 8\} \qquad \text{Read } \notin \text{ as "is not an element of."}$$

To economize on symbols, we can designate the set $\{4, 6, 8\}$ by a capital letter. Thus, if $A = \{4, 6, 8\}$, we can write

$$6 \in A \qquad \text{and} \qquad 7 \notin A \qquad\qquad ■$$

It is often difficult or even impossible to describe a set by listing its elements. For example, it is impossible to describe the set of composite numbers by listing them all, because there is no end to the list of composite numbers. To designate such sets, a notation called **set-builder notation** is used, which provides a rule that determines membership in the set. For example,

$$C = \{x \,|\, x \text{ is a composite number}\}$$

is read as "C is the set of all numbers x such that x is a composite number." The vertical bar in this notation means "such that." Set C is the set of composite numbers.

EXAMPLE 2   **a.** The set

$$P = \{2, 3, 5, 7\}$$

can also be described as

$$P = \{x \mid x \text{ is a prime number less than } 10\}$$

b. The set

$$V = \{a, e, i, o, u\}$$

can also be described as

$$V = \{x \mid x \text{ is a vowel}\}$$ ■

A set can have several elements, or it can have no elements at all. A set that has no elements is called the **empty set,** and it is designated by either of the symbols

$$\{ \ \} \qquad \text{or} \qquad \varnothing$$

The set of all odd numbers that are even and the set of all unicorns are examples of the empty set.

DEFINITION.   If every element of set A is also an element of set B, then we say that A **is a subset of B.** The notation

$$A \subset B$$

indicates that A is a subset of B.

EXAMPLE 3   **a.** $\{a, b, c\}$ is a subset of $\{a, b, c, d\}$ because every element of the first set is also an element of the second set.

b. $\{3, 5, 7\}$ is *not* a subset of $\{1, 2, 3, 4, 5\}$ because $7 \notin \{1, 2, 3, 4, 5\}$.

c. If J is the set of integers and Q is the set of rational numbers, then

$$J \subset Q \qquad \text{Read } \subset \text{ as "is a subset of."}$$

because every integer is also a rational number.

d. Suppose that

$$O = \{x \mid x \text{ is an odd natural number}\}$$
$$P = \{x \mid x \text{ is a prime number}\}$$

Then

$$O \not\subset P \qquad \text{Read } \not\subset \text{ as "is not a subset of."}$$

because some odd numbers (such as the number 9) are not prime numbers.

It is also true that

$$P \not\subset O$$

because one prime number (the number 2) is not an odd number.

e. Let A be any set. Because the empty set contains no elements that are not found in A, the empty set is a subset of A. If A is any set, then

$$\emptyset \subset A$$

In other words, the empty set is a subset of *any* set.

f. Note the difference between "is an element of" and "is a subset of":

$$a \in \{a, b, c\}, \quad \text{but} \quad \{a\} \notin \{a, b, c\}$$
$$\{a\} \subset \{a, b, c\} \quad \text{but} \quad a \not\subset \{a, b, c\} \qquad \blacksquare$$

There are two operations that are used to combine sets. They are the operations of **union** and **intersection.**

> **DEFINITION.**   The **union of sets A and B** is the set that contains all of the elements that belong to either set A or set B or both. The notation $A \cup B$ designates the union of A and B.

If A and B are two sets, the notation $A \cup B$ is read as "the union of A and B," or just "A union B."

EXAMPLE 4   **a.** If $A = \{1, 2, 3\}$, and $B = \{3, 4\}$, then

$$A \cup B = \{1, 2, 3, 4\}$$

b. If $E = \{x \,|\, x \text{ is an even integer}\}$,
$O = \{x \,|\, x \text{ is an odd integer}\}$, and
$J = \{x \,|\, x \text{ is an integer}\}$, then

$$E \cup O = J$$

c. If A is *any* set, then

$$\emptyset \cup A = A$$

In other words, the only elements in $\emptyset \cup A$ are those elements in set A. $\blacksquare$

> **DEFINITION.**   The **intersection of sets A and B** is the set that contains only those elements that are in *both* sets A and B. The notation $A \cap B$ designates the intersection of sets A and B.

If A and B are two sets, the expression $A \cap B$ is read as "the intersecti[on] A and B," or just "A intersect B."

EXAMPLE 5   **a.** If $A = \{1, 2, 3\}$ and $B = \{3, 4\}$, then

$$A \cap B = \{3\}$$

b. If $C = \{1, 2, 3, 4, 5\}$, and $D = \{2, 4, 6, 8\}$, then

$$C \cap D = \{2, 4\}$$

c. If $E = \{x \mid x$ is an even integer$\}$, and $O = \{x \mid x$ is an odd integer$\}$, then

$$E \cap O = \varnothing$$

d. If A is any set, then

$$\varnothing \cap A = \varnothing$$

In other words, there are no elements common to the empty set and set A. ∎

EXERCISE II.1

In Exercises 1–8, designate each set by listing its elements.

1. The set of even numbers between 1 and 9.

2. The set of odd numbers between 1 and 9.

3. The set of the names of days of the week that begin with S.

4. The set of states bordering the Mississippi river.

5. The set of integers between 5 and 6.

6. The set of all eight-legged giraffes.

7. The set of prime numbers less than 23.

8. The set of all even prime numbers.

In Exercises 9–14, describe each set by listing its members.

9. $\{x \mid x$ is the largest state of the United States$\}$

10. $\{x \mid x$ is a letter in the word *algebra*$\}$

11. $\{x \mid x$ is a numeral on the face of a clock$\}$

12. $\{x \mid x$ is a natural number less than 10$\}$

13. $\{x \mid x$ is an even prime number$\}$

14. $\{x \mid x$ is a perfect square less than 20$\}$

In Exercises 15–20, describe each set by using set-builder notation.

~ $\{1, 2, 3, 4, 5, 6, 7, 8, 9\}$

16. $\{0\}$

~ $36, 49, 64, 81\}$

18. $\{$I, II, III, IV, V, VI, VII, VIII, IX, X$\}$

n, May, July, August, October, December$\}$

In Exercises 21–32, let A = {1, 2, 3} and let B = {2, 4, 6}. Make each a true statement by inserting one of the symbols $\in$, $\notin$, $\subset$, *or* $\not\subset$ *between the expressions.*

21. 1   *A*

22. 4   *B*

23. 3   *B*

24. 4   *A*

25. {2, 3}   *B*

26. {1, 4}   *A*

27. {1}   *A*

28. {3}   *B*

29. $\varnothing$ *A*

30. { }   *B*

31. *A*   *A*

32. *A*   *B*

In Exercises 33–40, let A = {a, b, c}, let B = {a, d, e, f}, and let C = {e, f, g}. List the elements of each set.

33. $A \cup B$

34. $A \cup C$

35. $A \cap C$

36. $A \cap B$

37. $B \cup C$

38. $B \cap C$

39. $A \cup (B \cup C)$

40. $A \cap (B \cap C)$

Powers and Roots

n	n^2	$\sqrt{n}$	n^3	$\sqrt[3]{n}$	n	n^2	$\sqrt{n}$	n^3	$\sqrt[3]{n}$
1	1	1.000	1	1.000	51	2,601	7.141	132,651	3.708
2	4	1.414	8	1.260	52	2,704	7.211	140,608	3.733
3	9	1.732	27	1.442	53	2,809	7.280	148,877	3.756
4	16	2.000	64	1.587	54	2,916	7.348	157,464	3.780
5	25	2.236	125	1.710	55	3,025	7.416	166,375	3.803
6	36	2.449	216	1.817	56	3,136	7.483	175,616	3.826
7	49	2.646	343	1.913	57	3,249	7.550	185,193	3.849
8	64	2.828	512	2.000	58	3,364	7.616	195,112	3.871
9	81	3.000	729	2.080	59	3,481	7.681	205,379	3.893
10	100	3.162	1,000	2.154	60	3,600	7.746	216,000	3.915
11	121	3.317	1,331	2.224	61	3,721	7.810	226,981	3.936
12	144	3.464	1,728	2.289	62	3,844	7.874	238,328	3.958
13	169	3.606	2,197	2.351	63	3,969	7.937	250,047	3.979
14	196	3.742	2,744	2.410	64	4,096	8.000	262,144	4.000
15	225	3.873	3,375	2.466	65	4,225	8.062	274,625	4.021
16	256	4.000	4,096	2.520	66	4,356	8.124	287,496	4.041
17	289	4.123	4,913	2.571	67	4,489	8.185	300,763	4.062
18	324	4.243	5,832	2.621	68	4,624	8.246	314,432	4.082
19	361	4.359	6,859	2.668	69	4,761	8.307	328,509	4.102
20	400	4.472	8,000	2.714	70	4,900	8.367	343,000	4.121
21	441	4.583	9,261	2.759	71	5,041	8.426	357,911	4.141
22	484	4.690	10,648	2.802	72	5,184	8.485	373,248	4.160
23	529	4.796	12,167	2.844	73	5,329	8.544	389,017	4.179
24	576	4.899	13,824	2.884	74	5,476	8.602	405,224	4.198
25	625	5.000	15,625	2.924	75	5,625	8.660	421,875	4.217
26	676	5.099	17,576	2.962	76	5,776	8.718	438,976	4.236
27	729	5.196	19,683	3.000	77	5,929	8.775	456,533	4.254
28	784	5.292	21,952	3.037	78	6,084	8.832	474,552	4.273
29	841	5.385	24,389	3.072	79	6,241	8.888	493,039	4.291
30	900	5.477	27,000	3.107	80	6,400	8.944	512,000	4.309
31	961	5.568	29,791	3.141	81	6,561	9.000	531,441	4.327
32	1,024	5.657	32,768	3.175	82	6,724	9.055	551,368	4.344
33	1,089	5.745	35,937	3.208	83	6,889	9.110	571,787	4.362
34	1,156	5.831	39,304	3.240	84	7,056	9.165	592,704	4.380
35	1,225	5.916	42,875	3.271	85	7,225	9.220	614,125	4.397
36	1,296	6.000	46,656	3.302	86	7,396	9.274	636,056	4.414
37	1,369	6.083	50,653	3.332	87	7,569	9.327	658,503	4.431
38	1,444	6.164	54,872	3.362	88	7,744	9.381	681,472	4.448
39	1,521	6.245	59,319	3.391	89	7,921	9.434	704,969	4.465
40	1,600	6.325	64,000	3.420	90	8,100	9.487	729,000	4.481
41	1,681	6.403	68,921	3.448	91	8,281	9.539	753,571	4.498
42	1,764	6.481	74,088	3.476	92	8,464	9.592	778,688	4.514
43	1,849	6.557	79,507	3.503	93	8,649	9.644	804,357	4.531
44	1,936	6.633	85,184	3.530	94	8,836	9.695	830,584	4.547
45	2,025	6.708	91,125	3.557	95	9,025	9.747	857,375	4.563
46	2,116	6.782	97,336	3.583	96	9,216	9.798	884,736	4.579
47	2,209	6.856	103,823	3.609	97	9,409	9.849	912,673	4.595
48	2,304	6.928	110,592	3.634	98	9,604	9.899	941,192	4.610
49	2,401	7.000	117,649	3.659	99	9,801	9.950	970,299	4.626
50	2,500	7.071	125,000	3.684	100	10,000	10.000	1,000,000	4.642

Answers to Selected Exercises

Exercise 1.1 (page 5)

1. 1, 2, 4, 13, 15 **3.** 1, 13, 15 **5.** 4, 15 **7.** 2 **9.** 0, 1, 2, 6, 11, 12 **11.** 2, 6, 12 **13.** 11 **15.** 0, 1

17.

19.

21.

23.

25.

27. = **29.** < **31.** > **33.** = **35.** =

37. = **39.** < **41.** $7 > 3$ **43.** $17 \leq 17$ **45.** $3 + 4 = 7$ **47.** $7 \geq 3$ **49.** $0 < 6$ **51.** $8 < 3 + 8$

53. $10 - 4 > 6 - 2$ **55.** $3 \cdot 4 > 2 \cdot 3$ **57.** $\frac{24}{6} > \frac{12}{4}$ **59.** 6 is the greater; 6 lies to the right of 3.

61. 11 is the greater; 11 lies to the right of 6.

63. 2 is the greater; 2 lies to the right of 0.

65. 8 is the greater; 8 lies to the right of 0.

67. By adding 1, we can always obtain a larger natural number.

69. A natural number can either be divided by 2 or it can't.

REVIEW EXERCISES (page 7)

1. 250 **3.** 148 **5.** 16,606 **7.** 105 **9.** 726 **11.** $2480

Exercise 1.2 (page 17)

1. $\frac{1}{2}$ **3.** $\frac{1}{4}$ **5.** $\frac{3}{4}$ **7.** $\frac{4}{3}$ **9.** $\frac{9}{8}$ **11.** in lowest terms **13.** $\frac{3}{10}$ **15.** $\frac{20}{99}$ **17.** $\frac{8}{5}$ **19.** $\frac{3}{2}$ **21.** $\frac{1}{4}$ **23.** 1

25. 10 **27.** $\frac{20}{3}$ **29.** $\frac{9}{10}$ **31.** $\frac{44}{21}$ **33.** $\frac{5}{8}$ **35.** $\frac{1}{4}$ **37.** $\frac{14}{5}$ **39.** 1 **41.** 28 **43.** $\frac{1}{5}$ **45.** $\frac{6}{5}$ **47.** $\frac{1}{13}$

49. $\frac{2}{17}$ **51.** $\frac{5}{24}$ **53.** $\frac{19}{15}$ **55.** $\frac{17}{12}$ **57.** $\frac{22}{35}$ **59.** $\frac{1}{6}$ **61.** $\frac{9}{4}$ **63.** $\frac{29}{3}$ **65.** $\frac{4}{5}$ **67.** $\frac{3}{20}$ **69.** $\frac{67}{120}$ **71.** $\frac{3}{10}$

73. $5\frac{1}{5}$ **75.** $1\frac{2}{3}$ **77.** $1\frac{1}{4}$ **79.** $\frac{5}{9}$ **81.** $7\frac{2}{7}$ cm **83.** $53\frac{1}{6}$ ft

REVIEW EXERCISES (page 19)

1. yes   **3.** 13, prime number   **5.** 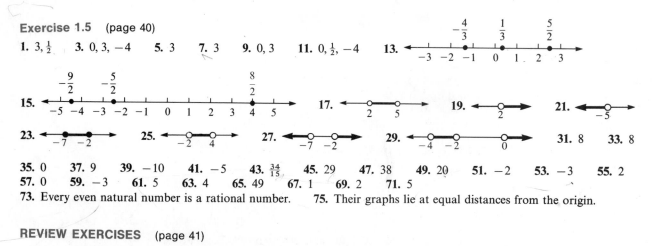   **7.** false   **9.** false   **11.** 20

Exercise 1.3 (page 22)

1. $x + y$ **3.** $x(2y)$ or $2xy$ **5.** $y - x$ **7.** $\frac{y}{x}$ **9.** $\frac{x}{y} + z$ **11.** $z - xy$ **13.** $3xy$ **15.** $\frac{x+y}{y+z}$ **17.** $xy + \frac{y}{z}$
19. $(c + 4)$ hr   **21.** $22t$¢ or $0.22t$ **23.** $\frac{x}{5}$ ft   **25.** $(3d + 5)$ dollars   **27.** the sum of a number x and 3
29. the quotient obtained when x is divided by y **31.** twice the product of x and y
33. the quotient obtained when 5 is divided by the sum of x and y
35. the quotient obtained when the sum of 3 and x is divided by y
37. the quotient obtained when the product of x and y is divided by the sum of x and y **39.** 10   **41.** 2   **43.** 24
45. 48   **47.** 1   **49.** 16   **51.** 1 term; 6   **53.** 3 terms; 1   **55.** 4 terms; 3   **57.** 3 terms; 4   **59.** 4 terms; 3
61. 19 and x **63.** 29, x, y, and z **65.** 3, x, y, and z **67.** 2, 3, 3, x, and z **69.** 5, 1, and 8   **71.** x and y
73. 3, 1, and 25; $3(1)(25) = 75$ **75.** x and y

REVIEW EXERCISES (page 25)

1. 1; odd natural number   **3.** 7; odd natural and prime number   **5.** $\frac{1}{5}$ **7.** $2 \cdot 3 \cdot 5 \cdot 5$ **9.** 8   **11.** $\frac{29}{30}$

Exercise 1.4 (page 32)

1. 64   **3.** 36   **5.** $\frac{1}{10,000}$ **7.** xx **9.** $3zzzz$ **11.** $(5t)(5t)$ **13.** $5(2x)(2x)(2x)$ **15.** 36   **17.** 1000   **19.** 18
21. 216   **23.** 11   **25.** 3   **27.** 28   **29.** 64   **31.** 13   **33.** 16   **35.** 90   **37.** 28   **39.** 21   **41.** 56   **43.** 1
45. 1   **47.** 28   **49.** 60   **51.** 17   **53.** 9   **55.** 8   **57.** 8   **59.** $\frac{1}{144}$ **61.** 11   **63.** 1   **65.** $\frac{8}{9}$ **67.** 4
69. 4   **71.** 12   **73.** 4   **75.** 11   **77.** 24   **79.** 12   **81.** 25   **83.** 1   **85.** 28   **87.** 35   **89.** 1   **91.** 2
93. $(3 \cdot 8) + (5 \cdot 3)$ **95.** $(3 \cdot 8 + 5)3$ **97.** $(4 + 3)(5 - 3)$ **99.** $(4 + 3)5 - 3$ **101.** 16 in.   **103.** 15 m
105. 25 sq m   **107.** 60 sq ft   **109.** 88 m   **111.** 1386 sq ft   **113.** 6 cu cm   **115.** 288π cu m or $\frac{6336}{7}$ cu m
117. 256π cu m   **119.** $12,348\pi$ cu ft   **121.** 360 cu ft per student

REVIEW EXERCISES (page 34)

1. $>$ **3.** $\geq$ **5.** 14   **7.** 4   **9.** $2 \cdot 67$ **11.** 8

Exercise 1.5 (page 40)

1. $3, \frac{1}{2}$ **3.** $0, 3, -4$ **5.** 3   **7.** 3   **9.** $0, 3$ **11.** $0, \frac{1}{2}, -4$ **13.**

15.   **17.**   **19.**   **21.**

23.   **25.**   **27.**   **29.**   **31.** 8   **33.** 8

35. 0   **37.** 9   **39.** -10 **41.** -5 **43.** $\frac{34}{15}$ **45.** 29   **47.** 38   **49.** 20   **51.** -2 **53.** -3 **55.** 2
57. 0   **59.** -3 **61.** 5   **63.** 4   **65.** 49   **67.** 1   **69.** 2   **71.** 5
73. Every even natural number is a rational number.   **75.** Their graphs lie at equal distances from the origin.

REVIEW EXERCISES (page 41)

1. 52   **3.** 36   **5.** $=$ **7.** $>$ **9.** 1125 lb   **11.** 51.3127 cubic units

Exercise 1.6   (page 47)

1. 12   **3.** -10 **5.** 2   **7.** -2 **9.** -12 **11.** -1.3 **13.** $\frac{12}{35}$ **15.** $-\frac{4}{12}$ **17.** 1   **19.** 2.2   **21.** 7
23. -1 **25.** -7 **27.** -8 **29.** -18 **31.** $\frac{1}{5}$ **33.** 1.3   **35.** -1 **37.** 3   **39.** 10   **41.** -3 **43.** -1
45. 9   **47.** 1   **49.** 7   **51.** 4   **53.** 12   **55.** -17 **57.** 5   **59.** $\frac{1}{2}$ **61.** $-\frac{22}{5}$ **63.** $-8\frac{3}{4}$ **65.** -4.2
67. 4   **69.** -7 **71.** 10   **73.** 0   **75.** 8   **77.** 64   **79.** 3   **81.** 2.45   **83.** 1   **85.** 1   **87.** -3 **89.** -15
91. 4   **93.** 15   **95.** -45 **97.** 1   **99.** $-\frac{3}{5}$ **101.** 3   **103.** -1 **105.** -1 **107.** $\frac{7}{6}$ **109.** $8   **111.** $+4°$
113. $+6°$ **115.** $351.20   **117.** 0 yards gained   **119.** The Dow was 2162.   **121.** 700 shares   **123.** $5.50
125. 18°   **127.** 2000 years   **129.** $-$55

REVIEW EXERCISES   (page 50)

1. $\frac{8}{9}$ **3.** 15   **5.** 13   **7.** $x + 2y$ **9.** $|x + y|$ **11.** $(36 + 10x + 25y)$ cents

Exercise 1.7   (page 55)

1. 3   **3.** 18   **5.** 32   **7.** -32 **9.** -54 **11.** -16 **13.** 2   **15.** 1   **17.** 72   **19.** -24 **21.** -420
23. -96 **25.** 4   **27.** -9 **29.** -2 **31.** 5   **33.** -3 **35.** -8 **37.** -8 **39.** 6   **41.** 36   **43.** 18
45. 5   **47.** 7   **49.** -4 **51.** 2   **53.** -4 **55.** -2 **57.** 2   **59.** 1   **61.** -6 **63.** -30 **65.** 7
67. -10 **69.** -66 **71.** 6   **73.** -10 **75.** 14   **77.** -81 **79.** 88   **81.** -30 **83.** -21 **85.** $-\frac{1}{6}$
87. $-\frac{11}{12}$ **89.** $-\frac{7}{36}$ **91.** $-\frac{11}{48}$ **93.** $+6°$ **95.** $-$450   **97.** $+720$ gal   **99.** $-$7   **101.** -5 hr

REVIEW EXERCISES   (page 57)

1. $=$ **3.** $<$ **5.** $>$ **7.** $>$ **9.** $=$ **11.**

Exercise 1.8   (page 61)

1. 10   **3.** -24 **5.** 144   **7.** 3   **9.** Both equal 12.   **11.** Both equal 29.   **13.** Both equal 60.   **15.** Both equal 0.
17. Both equal -6.   **19.** Both equal -12.   **21.** $3x + 3y$ **23.** $x^2 + 3x$ **25.** $-ax - bx$ **27.** $4x^2 + 4x$
29. $-5t - 10$ **31.** $-2ax - 2a^2$ **33.** $-2, \frac{1}{2}$ **35.** $-\frac{1}{3}, 3$ **37.** 0, no multiplicative inverse.   **39.** $\frac{5}{2}, -\frac{2}{5}$
41. $0.2, -5$ **43.** $-\frac{4}{3}, \frac{3}{4}$ **45.** commutative property of addition   **47.** commutative property of multiplication
49. distributive property   **51.** commutative property of addition   **53.** multiplicative identity property
55. additive inverse property   **57.** $3x + 3 \cdot 2$ **59.** xy^2 **61.** $(y + x)z$ **63.** $x(yz)$ **65.** x

REVIEW EXERCISES   (page 63)

1. $x + y^2$ **3.** $(x + 3)^2$ **5.** x **7.** $-y$ **9.** 5   **11.** $-\frac{5}{2}$

CHAPTER 1 REVIEW EXERCISES   (page 64)

1. 1, 2, 3, 4, 5   **3.** 1, 3, 5   **5.**

7. $>$ **9.** $\frac{5}{3}$ **11.** $\frac{1}{3}$
13. $\frac{10}{21}$ **15.** $2x + 3$ **17.** 7   **19.** 3   **21.** 1   **23.** 81   **25.** 22   **27.** 85 sq cm   **29.** $\frac{1372}{3}\pi$ cu m   **31.** $<$
33. $>$ **35.** 0   **37.**

39. -5 **41.** -4 **43.** -2 **45.** 4   **47.** -6 **49.** 2   **51.** -7
53. 6   **55.** -8 **57.** closure property of addition   **59.** associative property of addition
61. commutative property of addition   **63.** commutative property of addition   **65.** additive inverse property

CHAPTER 1 TEST   (page 66)

1. 31, 37, 41, 43, 47   **3.**

5. $5y - (x + y)$ **7.** $<$ **9.** 1   **11.** $\frac{9}{2}$ **13.** 3   **15.** -23
17. 64 sq cm   **19.** 5   **21.** 12   **23.** 1   **25.** 0   **27.** commutative property of multiplication
29. commutative property of addition   **31.** $(24x + 14y)$ mi

Exercise 2.1   (page 72)

1. an equation   **3.** not an equation   **5.** an equation   **7.** an equation   **9.** a solution   **11.** a solution
13. not a solution   **15.** not a solution   **17.** not a solution   **19.** a solution   **21.** a solution   **23.** a solution
25. a solution   **27.** $x = 10$ **29.** $a = -3$ **31.** $b = 7$ **33.** $x = 4$ **35.** $y = 13$ **37.** $x = 32$ **39.** $x = 740$
41. $x = 0$ **43.** $z = \frac{7}{6}$ **45.** $h = \frac{7}{3}$ **47.** $x = -6$ **49.** $y = 5$ **51.** $t = 9$ **53.** $x = -36$ **55.** $c = -1$
57. $r = -62$ **59.** $x = 0$ **61.** $d = \frac{5}{6}$ **63.** $w = \frac{14}{3}$ **65.** $r = -\frac{5}{2}$ **67.** $x = -10$ **69.** $y = 12$ **71.** $a = 16$
73. $x = 21$ **75.** $z = 74$ **77.** $b = -28$ **79.** $x = 806$ **81.** $x = 460$ **83.** $x = 0$ **85.** $x = \frac{46}{35}$ **87.** $x = \frac{47}{4}$
89. $y = -\frac{46}{17}$ **91.** 5   **93.** 18   **95.** 62   **97.** \$45,200   **99.** 3 hot dogs   **101.** \$28   **103.** \$39.50   **105.** \$190

REVIEW EXERCISES   (page 74)

1. 15   **3.** 12   **5.** -1 **7.** 1   **9.** -18 **11.** -124

Exercise 2.2   (page 80)

1. $x = 1$ **3.** $x = \frac{5}{2}$ **5.** $y = -3$ **7.** $y = 5$ **9.** $z = -2$ **11.** $z = 1$ **13.** $x = 3$ **15.** $x = -2$ **17.** $t = 2$
19. $s = -5$ **21.** $x = 3$ **23.** $y = -2$ **25.** $x = -2$ **27.** $a = 0$ **29.** $b = -9$ **31.** $s = -33$ **33.** $b = 28$
35. $r = 5$ **37.** $u = 7$ **39.** $x = -8$ **41.** $x = 10$ **43.** $a = -6$ **45.** $x = 12$ **47.** $k = 3$ **49.** $n = \frac{9}{2}$
51. $x = -2$ **53.** $q = 6$ **55.** $r = 35$ **57.** $z = -\frac{2}{3}$ **59.** $k = 0$ **61.** $x = 6$ **63.** $w = 3$ **65.** 5   **67.** 5
69. 7   **71.** 8   **73.** 7   **75.** 3   **77.** -4 **79.** 117   **81.** \$2000   **83.** 18 years   **85.** 61 years   **87.** 3 hr
89. 1200 gal

REVIEW EXERCISES   (page 81)

1. 50 cm   **3.** 27 sq cm   **5.** 635.664 cu cm   **7.** closure property of addition   **9.** commutative property of addition
11. additive inverse property

Exercise 2.3   (page 86)

1. $20x$ **3.** $3x^2$ **5.** cannot be simplified   **7.** $7x + 6$ **9.** $7z - 15$ **11.** $12x + 121$ **13.** $-22x^2 + 46$
15. $x^3 - 10$ **17.** $5x^2 + 24x$ **19.** $6y + 62$ **21.** $-2x + 7y$ **23.** $18y$ **25.** $2x - y + 2$ **27.** $5x + 7$
29. $5x + 35$ **31.** $x = -2$ **33.** $x = -3$ **35.** $y = 1$ **37.** $y = 1$ **39.** $a = 6$ **41.** $b = 35$ **43.** $x = -1$
45. $a = \frac{3}{2}$ **47.** $x = -2$ **49.** $x = -9$ **51.** $x = 0$ **53.** $a = -20$ **55.** $x = -41$ **57.** $x = -6$ **59.** $x = -9$
61. $b = \frac{37}{2}$ **63.** $t = 9$ **65.** $s = -1$ **67.** $x = 8$ **69.** $x = 2$ **71.** $x = 5$ **73.** $y = 4$ **75.** $x = -3$
77. $x = 1$ **79.** $x = -5$ **81.** $a = \frac{8}{3}$ **83.** $x = 0$ **85.** $y = -3$ **87.** $x = 2$ **89.** an identity
91. an impossible equation   **93.** $t = 16$ **95.** an impossible equation   **97.** an identity   **99.** an identity

REVIEW EXERCISES   (page 87)

1. 0   **3.** -122 **5.** 2   **7.** $x = 2$ **9.** $w = 9$ **11.** -17

Exercise 2.4   (page 91)

1. $I = \dfrac{E}{R}$ **3.** $w = \dfrac{V}{lh}$ **5.** $b = P - a - c$ **7.** $w = \dfrac{P - 2l}{2}$ **9.** $t = \dfrac{A - P}{Pr}$ **11.** $r = \dfrac{C}{2\pi}$ **13.** $w = \dfrac{2Kg}{v^2}$

15. $R = \dfrac{P}{I^2}$ **17.** $g = \dfrac{wv^2}{2K}$ **19.** $M = \dfrac{Fd^2}{Gm}$ **21.** $d^2 = \dfrac{GMm}{F}$ **23.** $r = \dfrac{G}{2b} + 1$ or $r = \dfrac{G + 2b}{2b}$ **25.** $t = \dfrac{d}{r}$; $t = 3$

27. $t = \dfrac{i}{pr}$; $t = 2$ **29.** $c = P - a - b$; $c = 13$ **31.** $h = \dfrac{2K}{a + b}$; $h = 8$ **33.** $I = \dfrac{E}{R}$, 4 amp   **35.** $r = \dfrac{C}{2\pi}$; 2.71 ft

37. $R = \dfrac{P}{I^2}$; 13.78 ohms   **39.** $m = \dfrac{Fd^2}{GM}$ **41.** $D = \dfrac{L - 3.25r - 3.25R}{2}$; 6 ft

REVIEW EXERCISES   (page 93)

1. $-2x$ **3.** unlike terms   **5.** $7y^3 - 2x^2 + 5$ **7.** $-x - 13$ **9.** $x = 21$ **11.** $x = 6$

Exercise 2.5   (page 97)

1. 3   **3.** 2   **5.** 11   **7.** 35   **9.** 3   **11.** 0   **13.** 4 ft and 8 ft   **15.** 5 m, 15 m, 25 m   **17.** 7 ft, 14 ft, 14 ft
19. $316   **21.** 3 first-place, 11 second-place, 22 third-place   **23.** 250 calories   **25.** $10   **27.** 125 lb   **29.** 300 g

REVIEW EXERCISES   (page 98)

1. $3 + 2 = 2 + 3$ **3.** $4(5 \cdot 6) = (4 \cdot 5)6$ **5.** $-(-7) = 7$ **7.** $|-12| = 12$ **9.** $-(-12) = 12$ **11.** $12 + (-12) = 0$

Exercise 2.6   (page 103)

1. 26 and 28   **3.** 39, 40, and 41   **5.** 7   **7.** 13   **9.** 19 ft   **11.** 29 m by 18 m   **13.** 17 in. by 39 in.   **15.** 60°
17. 2   **19.** 6 nickels, 6 dimes, 9 quarters   **21.** 9   **23.** 7   **25.** 960 pairs   **27.** 300   **29.** 7500 gal   **31.** A

REVIEW EXERCISES   (page 105)

1. 200 cu cm   **3.** $\frac{1372}{3}\pi$ cu m   **5.** 27   **7.** $7x - 6$ **9.** $-\frac{3}{2}$ **11.** x

Exercise 2.7   (page 109)

1. $4500 at 9%, $19,500 at 6%   **3.** $3750 in each account   **5.** $2350 at 5%, $4700 at 10%
7. $2500 at 10%, $5000 at 12%   **9.** 3   **11.** 6.5 hr   **13.** 7.5   **15.** 500 mph   **17.** 20   **19.** 50   **21.** 7.5 oz
23. 40 lb lemon drops and 60 lb jelly beans   **25.** $1.20   **27.** 80

REVIEW EXERCISES   (page 110)

1.   **3.**   **5.**   **7.**   **9.** $1488

11. $3

Exercise 2.8   (page 116)

1.   **3.**   **5.**   **7.**   **9.**

11.   **13.**   **15.**   **17.**   **19.**

21.   **23.**   **25.**   **27.**   **29.**

31.   **33.**   **35.**   **37.**   **39.**

41.   **43.**   **45.**   **47.**   **49.**

51.   **53.**   **55.**   **57.**   **59.**

61. $0 \text{ ft} < s \leq 19 \text{ ft}$ **63.** $0.1 \text{ mi} \leq x \leq 2.5 \text{ mi}$ **65.** $3.3 \text{ mi} < x < 4.1 \text{ mi}$ **67.** $66.2° < F < 71.6°$
69. $37.052 \text{ in.} < C < 38.308 \text{ in.}$ **71.** $68.18 \text{ kg} < w < 86.36 \text{ kg}$ **73.** $140 \text{ lb} \leq c \leq 150 \text{ lb}$ **75.** $5 \text{ ft} < w < 9 \text{ ft}$

REVIEW EXERCISES   (page 118)

1. $5x^2 - 2y^2$ **3.** $-x^2 + 2xy$ **5.** $-x + 14$ **7.** $\frac{1}{3}$ **9.** 3   **11.** $12,500

CHAPTER 2 REVIEW EXERCISES   (page 119)

1. solution   **3.** not a solution   **5.** solution   **7.** solution   **9.** $x = -4$ **11.** $x = 9$ **13.** $y = 3$ **15.** $z = -1$
17. $t = 1$ **19.** $a = 2$ **21.** $x = -2$ **23.** $b = 5$ **25.** $x = 13$ **27.** $y = 5$ **29.** $x = 15$ **31.** $a = 15$ **33.** $14x$
35. $5b$ **37.** $5x + 7y$ **39.** $4y^2 - 6$ **41.** $9x$ **43.** 0 **45.** $x = -7$ **47.** $a = 1$ **49.** $x = 7$ **51.** $x = -3$
53. $t = 9$ **55.** $x = 1$ **57.** $R = \dfrac{E}{I}$ **59.** $R = \dfrac{P}{I^2}$ **61.** $h = \dfrac{V}{lw}$ **63.** $h = \dfrac{V}{\pi r^2}$ **65.** $G = \dfrac{Fd^2}{Mm}$
67. $V = \frac{T}{n} + 3$ or $V = \frac{T + 3n}{n}$ **69.** -8 **71.** \$4.85   **73.** 85 ft   **75.** $15°$ **77.** 13 in.
79. 40 units on each machine, 80 units total   **81.** 20   **83.** \$17,500   **85.** ![number line]   **87.** ![number line]
89. ![number line]   **91.** ![number line]   **93.** ![number line]

CHAPTER 2 TEST   (page 121)

1. solution   **3.** not a solution   **5.** $x = -2$ **7.** $x = -3$ **9.** $x = -2$ **11.** $6x - 15$ **13.** $3x^2 - 6x$
15. $-18x$ **17.** $t = \frac{d}{r}$ **19.** $h = \frac{A}{2\pi r}$ **21.** $v = \frac{RT}{P}$ **23.** 165   **25.** \$5250   **27.** $7\frac{1}{2}$ **29.** ![number line]

Exercise 3.1   (page 128)

1. base of 4, exponent of 3   **3.** base of x, exponent of 5   **5.** base of $2y$, exponent of 3   **7.** base of x, exponent of 4
9. base of x, exponent of 1   **11.** base of x, exponent of 3   **13.** $5 \cdot 5 \cdot 5$ **15.** $xxxxxx$ **17.** $-4xxxxx$
19. $(3t)(3t)(3t)(3t)(3t)$ **21.** 2^3 **23.** x^4 **25.** $(2x)^3$ **27.** $-4t^4$ **29.** 625   **31.** 13   **33.** 561   **35.** -725
37. x^7 **39.** x^7 **41.** t^3 **43.** a^{12} **45.** y^9 **47.** $12x^7$ **49.** $-4y^5$ **51.** $12x^9$ **53.** $6a^6$ **55.** 3^8 **57.** y^{15}
59. a^{21} **61.** x^{25} **63.** $243z^{30}$ **65.** x^{31} **67.** r^{36} **69.** s^{33} **71.** x^3y^3 **73.** r^6s^4 **75.** $16a^2b^4$
77. $-8r^6s^9t^3$ **79.** $\dfrac{a^3}{b^3}$ **81.** $\dfrac{x^{10}}{y^{15}}$ **83.** $\dfrac{-32a^5}{b^5}$ **85.** a^{16} **87.** $\dfrac{y^2}{4}$ **89.** $-\dfrac{8r^2}{2}$ **91.** x^2 **93.** y^4 **95.** $3a$
97. ab^4 **99.** $\dfrac{10r^{13}s^3}{3}$ **101.** $\dfrac{x^{12}y^{16}}{2}$

REVIEW EXERCISES   (page 129)

1. ![number line]   **3.** $\frac{19}{18}$ **5.** 1   **7.** three times the sum of x and y

9. the absolute value of the difference obtained when y is subtracted from x **11.** $|2x| + 3$

Exercise 3.2   (page 132)

1. 8   **3.** 1   **5.** 1   **7.** 512   **9.** 2   **11.** 1   **13.** 1   **15.** -2 **17.** $\dfrac{1}{x^2}$ **19.** $\dfrac{1}{b^5}$ **21.** $\dfrac{1}{16y^4}$ **23.** $\dfrac{1}{a^3b^6}$
25. $\dfrac{1}{y}$ **27.** $\dfrac{1}{r^6}$ **29.** y^5 **31.** 1   **33.** $\dfrac{1}{a^2b^4}$ **35.** $\dfrac{1}{x^6y^3}$ **37.** $\dfrac{1}{x^3}$ **39.** $\dfrac{1}{y^2}$ **41.** a^8b^{12} **43.** $-\dfrac{y^{10}}{32x^{15}}$
45. a^{14} **47.** $\dfrac{1}{b^{14}}$ **49.** $\dfrac{256x^{28}}{81}$ **51.** $\dfrac{16y^{14}}{z^{10}}$ **53.** $\dfrac{x^{14}}{128y^{28}}$ **55.** $\dfrac{16u^4v^8}{81}$ **57.** $\dfrac{1}{9a^2b^2}$ **59.** $\dfrac{c^{15}}{216a^9b^3}$ **61.** $\dfrac{1}{512}$
63. $\dfrac{17y^{27}z^5}{x^{35}}$ **65.** x^{3m} **67.** u^{5m} **69.** y^{2m+2} **71.** y^m **73.** $\dfrac{1}{x^{3n}}$ or x^{-3n} **75.** x^{2m+2} **77.** x^{-12+8n} or x^{8n-12}
79. y^{4n-8}

REVIEW EXERCISES   (page 133)

1. 2   **3.** $x = 8$ **5.** 9   **7.** $s = \dfrac{f(P - L)}{i}$ or $s = \dfrac{fP - fL}{i}$ **9.** ![number line]   **11.** \$13,000 and \$28,400

Exercise 3.3 (page 136)

1. 2.3×10^4 **3.** 1.7×10^6 **5.** 6.2×10^{-2} **7.** 5.1×10^{-6} **9.** 4.25×10^3 **11.** 2.5×10^{-2} **13.** 230
15. 812,000   **17.** 0.00115   **19.** 0.000976   **21.** 25,000,000   **23.** 0.00051   **25.** 2.52×10^{13} mi   **27.** 114,000,000 mi
29. 6.22×10^{-3} mi   **31.** 714,000   **33.** 30,000   **35.** 200,000   **37.** 1.9008×10^{11} ft   **39.** 3.3×10^{-1} km/sec

REVIEW EXERCISES (page 137)

1. 11, 13, 17, 19, 23, 29   **3.** 3   **5.** commutative property of addition   **7.** $x = 6$ **9.**

11. 11 ft $\leq s \leq$ 15 ft

Exercise 3.4 (page 141)

1. binomial   **3.** trinomial   **5.** monomial   **7.** binomial   **9.** trinomial   **11.** not a monomial, binomial, or trinomial
13. 4th degree   **15.** 3rd degree   **17.** 8th degree   **19.** 6th degree   **21.** 12th degree   **23.** 0th degree
25. $P(2) = 7$ **27.** $P(-1) = -8$ **29.** $P(w) = 5w - 3$ **31.** $P(-y) = -5y - 3$ **33.** $Q(0) = -4$ **35.** $Q(-1) = -5$
37. $Q(r) = -r^2 - 4$ **39.** $Q(3s) = -9s^2 - 4$ **41.** $R(0) = 3$ **43.** $R(-2) = 11$ **45.** $R(-b) = b^2 + 2b + 3$
47. $R(-\frac{1}{4}w) = \frac{1}{16}w^2 + \frac{1}{2}w + 3$ **49.** $P(\frac{1}{5}) = -1$ **51.** $P(u^2) = 5u^2 - 2$ **53.** $P(-4z^6) = -20z^6 - 2$
55. $P(x^2y^2) = 5x^2y^2 - 2$ **57.** $P(x + h) = 5x + 5h - 2$ **59.** $P(x) + P(h) = 5x + 5h - 4$ **61.** $P(2y + z) = 10y + 5z - 2$
63. $P(2y) + P(z) = 10y + 5z - 4$

REVIEW EXERCISES (page 141)

1. $u = 8$ **3.**   **5.** 560 kwh   **7.** x^{18} **9.** y^9 **11.** 1,900,000,000,000,000,000,000,000,000 kg

Exercise 3.5 (page 145)

1. like terms; $7y$ **3.** unlike terms   **5.** like terms; $13x^3$ **7.** like terms; $8x^3y^2$ **9.** like terms; $65t^6$ **11.** unlike terms
13. $9y$ **15.** $-12t^2$ **17.** $16u^3$ **19.** $7x^5y^2$ **21.** $14rst$ **23.** $-6a^2bc$ **25.** $15x^2$ **27.** $4x^2y^2$ **29.** $95x^8y^4$
31. $7x + 4$ **33.** $2a + 7$ **35.** $7x - 7y$ **37.** $-19x - 4y$ **39.** $6x^2 + x - 5$ **41.** $7b + 4$ **43.** $3x + 1$
45. $5(x + 3)$ or $5x + 15$ **47.** $3(x - y)$ or $3x - 3y$ **49.** $5(x^2 - 5x - 4)$ or $5x^2 - 25x - 20$ **51.** $5x^2 + x + 11$
53. $-7x^3 - 7x^2 - x - 1$ **55.** $2x^2y + xy + 13y^2$ **57.** $5x^2 + 6x - 8$ **59.** $-x^3 + 6x^2 + x + 14$
61. $-12x^2y^2 - 13xy + 36y^2$ **63.** $6x^2 - 2x - 1$ **65.** $t^3 + 3t^2 + 6t - 5$ **67.** $-3x^2 + 5x - 7$ **69.** $6x - 2$
71. $-5x^2 - 8x - 19$ **73.** $4y^3 - 12y^2 + 8y + 8$ **75.** $3a^2b^2 - 6ab + b^2 - 6ab^2$ **77.** $-6x^2y^2 + 4xy^2z - 20xy^3 + 2y$
79. $6x + 3h - 10$

REVIEW EXERCISES (page 147)

1.   **3.** -8 **5.** -9 **7.** $a = -3$ **9.** 315 sq ft   **11.** $m = \dfrac{2K}{v^2}$

Exercise 3.6 (page 152)

1. $12x^5$ **3.** $-24b^6$ **5.** $6x^5y^5$ **7.** $-3x^4y^7z^8$ **9.** $x^{10}y^{15}$ **11.** $a^5b^4c^7$ **13.** $3x + 12$ **15.** $-4t - 28$
17. $3x^2 - 6x$ **19.** $-6x^4 + 2x^3$ **21.** $3x^2y + 3xy^2$ **23.** $6x^4 + 8x^3 - 14x^2$ **25.** $2x^7 - x^2$ **27.** $-6r^3t^2 + 2r^2t^3$
29. $-3x^5y^6 + 3x^6y^5 - 3x^4y^5$ **31.** $-a^4bc^3 - ab^4c^3 + abc^6$ **33.** $-6x^4y^4 - 6x^3y^5$ **35.** $a^2 + 9a + 20$
37. $3x^2 + 10x - 8$ **39.** $6a^2 + 2a - 20$ **41.** $6x^2 - 7x - 5$ **43.** $2x^2 + 3x - 9$ **45.** $6t^2 + 7st - 3s^2$
47. $x^2 - xy - 2y^2$ **49.** $-4r^2 - 20rs - 21s^2$ **51.** $-6a^2 - 16ab - 8b^2$ **53.** $-12t^2 + 7tu - u^2$
55. $x^2 + xz + yx + yz$ **57.** $u^2 + 2tu + uv + 2tv$ **59.** $4x^2 + 11x + 6$ **61.** $12x^2 + 14xy - 10y^2$ **63.** $x^3 - 1$
65. $x^2 + 8x + 16$ **67.** $t^2 - 6t + 9$ **69.** $r^2 - 16$ **71.** $x^2 + 10x + 25$ **73.** $4s^2 + 4s + 1$ **75.** $16x^2 - 25$
77. $9r^2 + 24rs + 16s^2$ **79.** $x^2 - 4xy + 4y^2$ **81.** $4a^2 - 12ab + 9b^2$ **83.** $16x^2 - 25y^2$ **85.** $2x^2 - 6x - 8$
87. $3a^3 - 3ab^2$ **89.** $-6y^4z - 9y^3z^2 + 6y^2z^3$ **91.** $4t^3 + 11t^2 + 18t + 9$ **93.** $-3x^3 + 25x^2y - 56xy^2 + 16y^3$

95. $x^3 - 8y^3$ **97.** $5t^2 - 11t$ **99.** $x^2y + 3xy^2 + 2x^2$ **101.** $2x^2 + xy - y^2$ **103.** $8x$ **105.** $5s^2 - 7s - 9$
107. $s = -3$ **109.** $z = -8$ **111.** $x = -1$ **113.** $a = 0$ **115.** $y = 1$ **117.** 6   **119.** 9 and 7   **121.** 90 ft
123. $\frac{3}{2}$ in.

REVIEW EXERCISES   (page 155)

1. $-\frac{1}{2}$ **3.** 4   **5.** distributive property   **7.** commutative property of multiplication   **9.** $y = 0$ **11.** 4.8×10^{18} m

Exercise 3.7   (page 157)

1. $\frac{1}{3}$ **3.** $-\frac{5}{3}$ **5.** $\frac{3}{4}$ **7.** 1   **9.** $-\frac{1}{4}$ **11.** $\frac{42}{19}$ **13.** $\frac{x}{z}$ **15.** $\frac{r^2}{s}$ **17.** $\frac{2x^2}{y}$ **19.** $\frac{-3u^3}{v^2}$ **21.** $\frac{4r}{y^2}$ **23.** $-\frac{13}{3rs}$

25. $\frac{x^4}{y^6}$ **27.** a^8b^8 **29.** $-\frac{3r}{s^9}$ **31.** $-\frac{x^3}{4y^3}$ **33.** $\frac{125}{8b^3}$ **35.** $\frac{xy^2}{3}$ **37.** a^8 **39.** z^3 **41.** $\frac{2}{y} + \frac{3}{x}$ **43.** $\frac{1}{5y} - \frac{2}{5x}$

45. $\frac{1}{y^2} + \frac{2y}{x^2}$ **47.** $3a - 2b$ **49.** $\frac{1}{y} - \frac{1}{2x} + \frac{2z}{xy}$ **51.** $3x^2y - 2x - \frac{1}{y}$ **53.** $5x - 6y + 1$ **55.** $\frac{10x^2}{y} - 5x$

57. $-\frac{4x}{3} + \frac{3x^2}{2}$ **59.** $xy - 1$ **61.** $\frac{x}{y} - \frac{11}{6} + \frac{y}{2x}$ **63.** 2

REVIEW EXERCISES   (page 159)

1. $P(4) = 52$ **3.** 2.65×10^{-4} **5.** 1   **7.** $\frac{2y^2}{x}$ **9.** $x = -2$ **11.** 19 and 20

Exercise 3.8   (page 164)

1. $x + 2$ **3.** $y + 12$ **5.** $a + b$ **7.** $3a - 2$ **9.** $b + 3$ **11.** $x - 3y$ **13.** $2x + 1$ **15.** $x - 7$ **17.** $3x + 2y$
19. $2x - y$ **21.** $x + 5y$ **23.** $x - 5y$ **25.** $x^2 + 2x - 1$ **27.** $2x^2 + 2x + 1$ **29.** $x^2 + xy + y^2$
31. $x + 1 + \frac{-1}{2x + 3}$ **33.** $2x + 2 + \frac{-3}{2x + 1}$ **35.** $x^2 + 2x + 1$ **37.** $x^2 + 2x - 1 + \frac{6}{2x + 3}$

39. $2x^2 + 8x + 14 + \frac{31}{x - 2}$ **41.** $x + 1$ **43.** $2x - 3$ **45.** $x^2 - x + 1$ **47.** $x^2 - 3x + 10 + \frac{-30}{x + 3}$

49. $5x^2 - x + 4 + \frac{16}{3x - 4}$

REVIEW EXERCISES   (page 165)

1. 21, 22, 24, 25, 26, 27, 28   **3.** -2 **5.** 25   **7.** -2 **9.** $8x^2 - 6x + 1$ **11.** 880 sq in.

CHAPTER 3 REVIEW EXERCISES   (page 167)

1. 125   **3.** 64   **5.** 13   **7.** 162   **9.** x^5 **11.** y^{10} **13.** $2b^{12}$ **15.** $256s^3$ **17.** x^{15} **19.** 9   **21.** x^4
23. $\frac{y^2}{x^2}$ **25.** x **27.** x^{10} **29.** $\frac{1}{x^4}$ **31.** $\frac{1}{8s^3}$ **33.** 7.28×10^2 **35.** 1.36×10^{-2} **37.** 7.61×10^0
39. 1.2×10^{-4} **41.** 726,000   **43.** 2.68   **45.** 7.31   **47.** 7th degree   **49.** 5th degree   **51.** $P(3) = 11$
53. $P(-2) = -4$ **55.** $P(3) = 402$ **57.** $P(-2) = 82$ **59.** $7x$ **61.** $4x^2y^2$ **63.** $5x^2$ **65.** $8x^2 - 6x$
67. $9x^2 + 3x + 9$ **69.** $10x^3y^5$ **71.** $5x + 15$ **73.** $3x^4 - 5x^2$ **75.** $-x^2y^3 + x^3y^2$ **77.** $x^2 + 5x + 6$
79. $6a^2 - 6$ **81.** $2a^2 - ab - b^2$ **83.** $-9a^2 + b^2$ **85.** $y^2 - 4$ **87.** $y^2 - 6y + 9$ **89.** $3x^3 + 7x^2 + 5x + 1$

91. $x = 1$ **93.** $x = 7$ **95.** $x = 1$ **97.** $\frac{3}{2y} + \frac{3}{x}$ **99.** $-3a - 4b + 5c$ **101.** $x + 1 + \frac{3}{x + 2}$ **103.** $2x + 1$

105. $3x^2 + 2x + 1 + \frac{2}{2x - 1}$

CHAPTER 3 TEST   (page 169)

1. $2x^3y^4$ **3.** y^6 **5.** $32x^{21}$ **7.** 3   **9.** y^3 **11.** 2.8×10^4 **13.** 7400   **15.** binomial   **17.** 0   **19.** $-7x + 2y$
21. $5x^3 + 2x^2 + 2x - 5$ **23.** $-4x^5y$ **25.** $6x^2 - 7x - 20$ **27.** $2x^3 - 7x^2 + 14x - 12$ **29.** $\dfrac{y}{2x}$ **31.** $x - 2$

Exercise 4.1   (page 174)

1. $2^2 \cdot 3$ **3.** $3 \cdot 5$ **5.** $2^3 \cdot 5$ **7.** $2 \cdot 7^2$ **9.** $3^2 \cdot 5^2$ **11.** $2^5 \cdot 3^2$ **13.** $3(x + 2)$ **15.** $x(y - z)$ **17.** $t(t + 2)$
19. $2r^2(r^2 - 2)$ **21.** $a^2b^3z^2(az - 1)$ **23.** $8xy^2z^3(3xyz + 1)$ **25.** $6uvw^2(2w - 3v)$ **27.** $3(x + y - 2z)$
29. $a(b + c - d)$ **31.** $2y(2y + 4 - x)$ **33.** $3r(4r - s + 3rs^2)$ **35.** $abx(1 - b + x)$ **37.** $2xyz^2(2xy - 3y + 6)$
39. $7a^2b^2c^2(10a + 7bc - 3)$ **41.** $-(a + b)$ **43.** $-(2x - 5y)$ **45.** $-(2a - 3b)$ **47.** $-(3m + 4n - 1)$
49. $-(3xy - 2z - 5w)$ **51.** $-(3ab + 5ac - 9bc)$ **53.** $-3xy(x + 2y)$ **55.** $-4a^2b^2(b - 3a)$
57. $-2ab^2c(2ac - 7a + 5c)$ **59.** $-7ab(2a^5b^5 - 7ab^2 + 3)$ **61.** $-5a^2b^3c(1 - 3abc + 5a^2)$

REVIEW EXERCISES   (page 174)

1. $6xy + 3x + 4y + 2$ **3.** $2ab - 2a + b - 1$ **5.** $6pr - 3pq - 2qr + q^2$ **7.** $3x + 3y + ax + ay$
9. $x^2 + 4x + 3 - xy - y$ **11.** $3x^3 - 6x - x^2y + 2y + 3x^2 - 6$

Exercise 4.2   (page 177)

1. $(x + y)(2 + b)$ **3.** $(x + y)(3 - a)$ **5.** $(r - 2s)(3 - x)$ **7.** $(x - 3)(x - 2)$ **9.** $2(a^2 + b)(x + y)$
11. $3(r + 3s)(x^2 - 2y^2)$ **13.** $(a + b + c)(3x - 2y)$ **15.** $7xy(r + 2s - t)(2x - 3)$ **17.** $(x + 1)(x + 3 - y)$
19. $(x^2 - 2)(3x - y + 1)$ **21.** $(x + y)(2 + a)$ **23.** $(r + s)(7 - k)$ **25.** $(r + s)(x + y)$ **27.** $(2x + 3)(a + b)$
29. $(b + c)(2a + 3)$ **31.** $(x + y)(2x - 3)$ **33.** $(v - 3w)(3t + u)$ **35.** $(3p + q)(3m - n)$ **37.** $(m - n)(p - 1)$
39. $(a - b)(x - y)$ **41.** $x^2(a + b)(x + 2y)$ **43.** $4a(b + 3)(a - 2)$ **45.** $(x^2 + 1)(x + 2)$ **47.** $y(x^2 - y)(x - 1)$
49. $(x + 2)(x + y + 1)$ **51.** $(m - n)(a + b + c)$ **53.** $(d + 3)(a - b - c)$ **55.** $(a + b + c)(x^2 - y)$ **57.** $(r - s)(2 + b)$
59. $(x + y)(a + b)$ **61.** $(a - b)(c - d)$ **63.** $r(r + s)(a - b)$ **65.** $(b + 1)(a + 3)$ **67.** $(r - s)(p - q)$

REVIEW EXERCISES   (page 178)

1. u^9 **3.** $\dfrac{a}{b}$ **5.** 4.5×10^{-4} **7.** $a^2 - b^2$ **9.** $9x^2 - 4y^2$ **11.** $\dfrac{x + y}{xy}$

Exercise 4.3   (page 180)

1. $(x + 4)(x - 4)$ **3.** $(y + 7)(y - 7)$ **5.** $(2y + 7)(2y - 7)$ **7.** $(3x + y)(3x - y)$ **9.** $(5t + 6u)(5t - 6u)$
11. $(4a + 5b)(4a - 5b)$ **13.** prime polynomial   **15.** $(a^2 + 2b)(a^2 - 2b)$ **17.** $(7y + 15z^2)(7y - 15z^2)$
19. $(14x^2 + 13y)(14x^2 - 13y)$ **21.** $8(x + 2y)(x - 2y)$ **23.** $2(a + 2y)(a - 2y)$ **25.** $3(r + 2s)(r - 2s)$
27. $x(x + y)(x - y)$ **29.** $x(2a + 3b)(2a - 3b)$ **31.** $3m(m + n)(m - n)$ **33.** $x^2(2x + y)(2x - y)$
35. $2ab(a + 11b)(a - 11b)$ **37.** $(x^2 + 9)(x + 3)(x - 3)$ **39.** $(a^2 + 4)(a + 2)(a - 2)$ **41.** $(a^2 + b^2)(a + b)(a - b)$
43. $(9r^2 + 16s^2)(3r + 4s)(3r - 4s)$ **45.** $(a^2 + b^4)(a + b^2)(a - b^2)$ **47.** $(x^4 + y^4)(x^2 + y^2)(x + y)(x - y)$
49. $2(x^2 + y^2)(x + y)(x - y)$ **51.** $b(a^2 + b^2)(a + b)(a - b)$ **53.** $3n(4m^2 + 9n^2)(2m + 3n)(2m - 3n)$ **55.** $3ay(a^4 + 2y^4)$
57. $3a^2(a^4 + b^2)(a^2 + b)(a^2 - b)$ **59.** $2y^2(x^4 + 4y^2)(x^2 + 2y)(x^2 - 2y)$ **61.** $a^2b^2(a^2 + b^2c^2)(a + bc)(a - bc)$
63. $a^2b^3(b^2 + 25)(b + 5)(b - 5)$ **65.** $3rs(9r^2 + 4s^2)(3r + 2s)(3r - 2s)$ **67.** $(4x - 4y + 3)(4x - 4y - 3)$
69. $(a + 3)(a + 3)(a - 3)$ **71.** $(y + 4)(y - 4)(y - 3)$ **73.** $3(x + 2)(x - 2)(x + 1)$ **75.** $3(m + n)(m - n)(m + a)$
77. $2(m + 4)(m - 4)(mn^2 + 4)$

REVIEW EXERCISES   (page 182)

1. $x^2 + 12x + 36$ **3.** $a^2 - 6a + 9$ **5.** $x^2 + 9xy + 20y^2$ **7.** $m^2 + mn - 6n^2$ **9.** $u^2 - 8uv + 15v^2$
11. $p = w\left(k - h - \dfrac{v^2}{2g}\right)$

Exercise 4.4   (page 187)

1. $(x + 3)(x + 3)$ **3.** $(y - 4)(y - 4)$ **5.** $(t + 10)(t + 10)$ **7.** $(u - 9)(u - 9)$ **9.** $(x + 2y)(x + 2y)$
11. $(r - 5s)(r - 5s)$ **13.** $(x + 2)(x + 1)$ **15.** $(a - 5)(a + 1)$ **17.** $(z + 11)(z + 1)$ **19.** $(t - 7)(t - 2)$
21. prime polynomial   **23.** $(y - 6)(y + 5)$ **25.** $(a + 8)(a - 2)$ **27.** $(t - 10)(t + 5)$ **29.** prime polynomial
31. $(y + z)(y + z)$ **33.** $(x + 2y)(x + 2y)$ **35.** $(m + 5n)(m - 2n)$ **37.** $(a - 6b)(a + 2b)$ **39.** $(u + 5v)(u - 3v)$
41. $-(x + 5)(x + 2)$ **43.** $-(y + 5)(y - 3)$ **45.** $-(t + 17)(t - 2)$ **47.** $-(r - 10)(r - 4)$ **49.** $-(a + 3b)(a + b)$
51. $-(x - 7y)(x + y)$ **53.** $(x - 4)(x - 1)$ **55.** $(y + 9)(y + 1)$ **57.** $(c + 5)(c - 1)$ **59.** $-(r - 2s)(r + s)$
61. $(r + 3x)(r + x)$ **63.** $(a - 2b)(a - b)$ **65.** $2(x + 3)(x + 2)$ **67.** $3y(y + 1)(y + 1)$ **69.** $-5(a - 3)(a - 2)$
71. $3(z - 4t)(z - t)$ **73.** $4y(x + 6)(x - 3)$ **75.** $-4x(x + 3y)(x - 2y)$ **77.** $(x + 2)(ax + 2a + b)$
79. $(a + 5)(a + 3 + b)$ **81.** $(a + b + 2)(a + b - 2)$ **83.** $(b + y + 2)(b - y - 2)$

REVIEW EXERCISES   (page 188)

1. $6x^2 + 7x + 2$ **3.** $8t^2 + 6t - 9$ **5.** $6m^2 - 13mn + 6n^2$ **7.** $20u^2 - 7uv - 6v^2$ **9.** $15x^4 + 11x^2y + 2y^2$
11. $3x^4 - 4x^2y - 15y^2$

Exercise 4.5   (page 194)

1. $(2x - 1)(x - 1)$ **3.** $(3a + 1)(a + 4)$ **5.** $(z + 3)(4z + 1)$ **7.** $(3y + 2)(2y + 1)$ **9.** $(3x - 2)(2x - 1)$
11. $(3a + 2)(a - 2)$ **13.** $(2x + 1)(x - 2)$ **15.** $(2m - 3)(m + 4)$ **17.** $(5y + 1)(2y - 1)$ **19.** $(3y - 2)(4y + 1)$
21. $(5t + 3)(t + 2)$ **23.** $(8m - 3)(2m - 1)$ **25.** $(3x - y)(x - y)$ **27.** $(2u + 3v)(u - v)$ **29.** $(2a - b)(2a - b)$
31. $(3r + 2s)(2r - s)$ **33.** $(2x + 3y)(2x + y)$ **35.** $(4a - 3b)(a - 3b)$ **37.** $(3x + 2)(x - 5)$ **39.** $(2a - 5)(4a - 3)$
41. $(4y - 3)(3y - 4)$ **43.** prime polynomial   **45.** $(2a + 3b)(a + b)$ **47.** $(3p - q)(2p + q)$ **49.** prime polynomial
51. $(4x - 5y)(3x - 2y)$ **53.** $(3x - 5y)(2x - 3y)$ **55.** $(5a + 8b)(5a - 2b)$ **57.** $2(2x - 1)(x + 3)$ **59.** $y(y + 12)(y + 1)$
61. $3x(2x + 1)(x - 3)$ **63.** $m(2m - 3)(m + 1)$ **65.** $2a^2(3a - 5)(a + 4)$ **67.** $3r^3(5r - 2)(2r + 5)$ **69.** $4(a - 2b)(a + b)$
71. $4(2x + y)(x - 2y)$ **73.** $x^2y^2(2x - y)(x + y)$ **75.** $-2mn(4m + 3n)(2m + n)$ **77.** $-2uv^3(7u - 3v)(2u - v)$
79. $3x(7x + 4)(5x - 3)$ **81.** $(2x + y + 4)(2x + y - 4)$ **83.** $(3 + a + 2b)(3 - a - 2b)$
85. $(2x + y + a + b)(2x + y - a - b)$ **87.** $2z(x - y + 3z)(x - y - 3z)$ **89.** $(2x + y)(2x + y + 3)$ **91.** $(x + 5)(x + 4)$
93. $(2r + 5)(r + 2)$ **95.** $(3x - 5)(2x + 1)$ **97.** $(3t + 4)(4t - 1)$ **99.** $(2x + 3y)(x - 2y)$

REVIEW EXERCISES   (page 195)

1. $x^3 - 27$ **3.** $y^3 + 64$ **5.** $a^3 - b^3$ **7.** $x^3 + 8y^3$ **9.** $r^3 + s^3$ **11.** $n = \dfrac{l - f - d}{d}$

Exercise 4.6   (page 198)

1. $(y + 1)(y^2 - y + 1)$ **3.** $(a - 3)(a^2 + 3a + 9)$ **5.** $(2 + x)(4 - 2x + x^2)$ **7.** $(s - t)(s^2 + st + t^2)$
9. $(3x + y)(9x^2 - 3xy + y^2)$ **11.** $(a + 2b)(a^2 - 2ab + 4b^2)$ **13.** $(4x - y)(16x^2 + 4xy + y^2)$
15. $(3x - 5y)(9x^2 + 15xy + 25y^2)$ **17.** $(a^2 - b)(a^4 + a^2b + b^2)$ **19.** $(x^2 + y^2)(x^4 - x^2y^2 + y^4)$
21. $2(x + 3)(x^2 - 3x + 9)$ **23.** $-(x - 6)(x^2 + 6x + 36)$ **25.** $8x(2m - n)(4m^2 + 2mn + n^2)$
27. $xy(x + 6y)(x^2 - 6xy + 36y^2)$ **29.** $3rs^2(3r - 2s)(9r^2 + 6rs + 4s^2)$ **31.** $a^3b^2(5a + 4b)(25a^2 - 20ab + 16b^2)$
33. $yz(y^2 - z)(y^4 + y^2z + z^2)$ **35.** $2mp(p + 2q)(p^2 - 2pq + 4q^2)$ **37.** $(x + 1)(x^2 - x + 1)(x - 1)(x^2 + x + 1)$
39. $(x^2 + y)(x^4 - x^2y + y^2)(x^2 - y)(x^4 + x^2y + y^2)$ **41.** $(x + y)(x^2 - xy + y^2)(3 - z)$
43. $(m + 2n)(m^2 - 2mn + 4n^2)(1 + x)$ **45.** $(a + 3)(a^2 - 3a + 9)(a - b)$ **47.** $(x + 2)(x - 2)(y + z)$
49. $(r + s)(r - s)(x - a)$ **51.** $(x - 1)(x + 1)$ **53.** $(y + 1)(y - 1)(y - 3)(y^2 + 3y + 9)$

REVIEW EXERCISES   (page 199)

1. $x = 4$ **3.** $x = -3$ **5.** $a = \frac{2}{3}$ **7.** 1   **9.** 1   **11.** 0.0000000000001 cm

Exercise 4.7   (page 201)

1. $3(2x + 1)$ **3.** $(x - 7)(x + 1)$ **5.** $(3t - 1)(2t + 3)$ **7.** $(2x + 5)(2x - 5)$ **9.** $(t - 1)(t - 1)$
11. $(a - 2)(a^2 + 2a + 4)$ **13.** $(y^2 - 2)(x + 1)(x - 1)$ **15.** $7p^4q^2(10q - 5 + 7p)$ **17.** $2a(b + 6)(b - 2)$
19. $-4p^2q^3(2pq^4 + 1)$ **21.** $(2a - b + 3)(2a - b - 3)$ **23.** prime polynomial   **25.** $-2x^2(x + 4)(x^2 - 4x + 16)$
27. $2t^2(3t - 5)(t + 4)$ **29.** $(x - a)(a + b)(a - b)$ **31.** $(2p^2 - 3q^2)(4p^4 + 6p^2q^2 + 9q^4)$
33. $(5p - 4y)(25p^2 + 20py + 16y^2)$ **35.** $-x^2y^2z(16x^2 - 24x^3yz^3 + 15yz^6)$ **37.** $(9p^2 + 4q^2)(3p + 2q)(3p - 2q)$
39. prime polynomial   **41.** $2(3x + 5y^2)(9x^2 - 15xy^2 + 25y^4)$ **43.** prime polynomial   **45.** $t(7t - 1)(3t - 1)$
47. $(x + y)(x - y)(x + y)(x^2 - xy + y^2)$ **49.** $2(a + b)(a - b)(c + 2d)$

REVIEW EXERCISES   (page 202)

1. $2x^2 + 4x$ **3.** $8a^3$ **5.** $4x^2 - 12x + 9$ **7.** $\dfrac{a^{12}}{4b^8}$ **9.** $t = 4$

Exercise 4.8   (page 206)

1. $2, -3$ **3.** $4, -1$ **5.** $\frac{5}{2}, -2$ **7.** $1, -2, 3$ **9.** $-2, 4, -7$ **11.** $4, -6, \frac{3}{2}, \frac{2}{3}$ **13.** $12, 1$ **15.** $5, -3$
17. $\frac{2}{3}, -\frac{1}{2}$ **19.** $\frac{1}{4}, -\frac{2}{3}$ **21.** $0, -3$ **23.** $0, 8$ **25.** $\frac{5}{4}, -1$ **27.** $4, -4$ **29.** $7, -7$ **31.** $0, 6$ **33.** $5, -\frac{2}{5}$
35. $-3, \frac{2}{7}$ **37.** $-3, -5$ **39.** $2, -4$ **41.** $-3, -\frac{1}{3}$ **43.** $-3, -3$ **45.** $1, -\frac{3}{2}$ **47.** $5, 3$ **49.** $1, -2, -3$
51. $0, 3, -1$ **53.** $0, 4, -4$ **55.** $0, 3, -\frac{1}{2}$ **57.** $0, \frac{1}{3}, \frac{1}{7}$ **59.** $3, -3, \frac{2}{3}, -\frac{2}{3}$

REVIEW EXERCISES   (page 207)

1. $x(x + 3)$ **3.** $(y + 3)(y - 3)$ **5.** $(x + 12)(x + 1)$ **7.** $4s$ in.   **9.** 4 cm by 8 cm   **11.** 9%

Exercise 4.9   (page 210)

1. 5 and 7   **3.** 9   **5.** 9 sec   **7.** $\frac{15}{4}$ sec and 10 sec   **9.** 4 m by 9 m   **11.** 48 ft   **13.** $h = 5$ m, $b = 12$ m
15. 9 square units   **17.** 20 cm   **19.** 3 cm   **21.** 4 cm by 7 cm

REVIEW EXERCISES   (page 212)

1.   **3.**   **5.**   **7.**   **9.**

11. $T_2 = T_1(1 - E)$

CHAPTER 4 REVIEW EXERCISES   (page 213)

1. $5 \cdot 7$ **3.** $2^5 \cdot 3$ **5.** $3 \cdot 29$ **7.** $2 \cdot 5^2 \cdot 41$ **9.** $3(x + 3y)$ **11.** $7x(x + 2)$ **13.** $2x(x^2 + 2x - 4)$
15. $a(x + y - 1)$ **17.** $5a(a + b^2 + 2cd - 3)$ **19.** $(x + y)(a + b)$ **21.** $2x(x + 2)(x + 3)$ **23.** $(p + 3q)(3 + a)$
25. $(x + a)(x + b)$ **27.** $y(3x - y)(x - 2)$ **29.** $(x + 3)(x - 3)$ **31.** $(x + 2 + y)(x + 2 - y)$ **33.** $6y(x + 2y)(x - 2y)$
35. $(x + 3)(x + 7)$ **37.** $(x + 6)(x - 4)$ **39.** $(2x + 1)(x - 3)$ **41.** $(2x + 3)(3x - 1)$ **43.** $x(x + 3)(6x - 1)$
45. $(x + a + y)(x + a - y)$ **47.** $(x + y)(a + b)$ **49.** $(c - 3)(c^2 + 3c + 9)$ **51.** $2(x + 3)(x^2 - 3x + 9)$ **53.** $0, -2$
55. $3, -3$ **57.** $3, 4$ **59.** $-4, 6$ **61.** $3, -\frac{1}{2}$ **63.** $\frac{1}{2}, -\frac{1}{2}$ **65.** $0, 3, 4$ **67.** $0, \frac{1}{2}, -3$ **69.** 5 and 7
71. 6 ft by 8 ft   **73.** 3 ft by 9 ft   **75.** 4 units

CHAPTER 4 TEST   (page 214)

1. $2^2 \cdot 7^2$ **3.** $5a(12b^2c^3 + 6a^2b^2c - 5)$ **5.** $(x + y)(a + b)$ **7.** $3(a + 3b)(a - 3b)$ **9.** $(x + 3)(x + 1)$
11. $(x + 9y)(x + y)$ **13.** $(3x + 1)(x + 4)$ **15.** $(2x - y)(x + 2y)$ **17.** $6(2a - 3b)(a + 2b)$ **19.** $8(3 + a)(9 - 3a + a^2)$
21. $16(r + 2s)(r^2 - 2rs + 4s^2)$ **23.** $-1, -\frac{3}{2}$ **25.** $3, -6$ **27.** 12 sec

Exercise 5.1   (page 222)

1. $\dfrac{2}{3}$ **3.** $\dfrac{4}{5}$ **5.** $\dfrac{2}{13}$ **7.** $\dfrac{2}{9}$ **9.** $-\dfrac{1}{3}$ **11.** $2x$ **13.** $-\dfrac{x}{3}$ **15.** $\dfrac{5}{a}$ **17.** $\dfrac{2}{z}$ **19.** $\dfrac{a}{3}$ **21.** $\dfrac{2}{3}$ **23.** $\dfrac{3}{2}$

25. in lowest terms   **27.** $\dfrac{3x}{y}$ **29.** $\dfrac{7x}{8y}$ **31.** $\dfrac{1}{3}$ **33.** 5   **35.** $\dfrac{x}{2}$ **37.** $\dfrac{3x}{5y}$ **39.** $\dfrac{2}{3}$ **41.** -1 **43.** -1

45. -1 **47.** $\dfrac{x+1}{x-1}$ **49.** $\dfrac{x-5}{x+2}$ **51.** $\dfrac{2x}{x-2}$ **53.** $\dfrac{x}{y}$ **55.** $\dfrac{x+2}{x^2}$ **57.** $\dfrac{x-4}{x+4}$ **59.** $\dfrac{2(x+2)}{x-1}$

61. in lowest terms   **63.** $\dfrac{3-x}{3+x}$ or $-\dfrac{x-3}{x+3}$ **65.** $\dfrac{4}{3}$ **67.** $\dfrac{1}{x+3}$ **69.** $x+1$ **71.** $a-2$ **73.** $\dfrac{b+2}{b+1}$ **75.** $\dfrac{y+3}{x-3}$

REVIEW EXERCISES   (page 223)

1. If a and b are real numbers, then
$a+b$ is a real number
$a-b$ is a real number
ab is a real number
and if $b \neq 0$, then

$\dfrac{a}{b}$ is a real number

3. If a, b, and c are real numbers, then
$$(a+b)+c = a+(b+c)$$
$$(ab)c = a(bc)$$

5. If a is a real number, then 0 and 1 are unique numbers such that
$$a+0 = a$$
$$a \cdot 1 = a$$

7. 0   **9.** 10   **11.** 0

Exercise 5.2   (page 225)

1. $\dfrac{8}{15}$ **3.** $\dfrac{45}{91}$ **5.** $\dfrac{2}{5}$ **7.** $-\dfrac{2}{3}$ **9.** $\dfrac{3}{11}$ **11.** $\dfrac{15}{4}$ **13.** $\dfrac{5}{7}$ **15.** $\dfrac{3x}{2}$ **17.** $\dfrac{yx}{z}$ **19.** $\dfrac{3y}{10}$ **21.** $\dfrac{14}{9}$ **23.** 26

25. $x^2 y^2$ **27.** $2xy^2$ **29.** $-3y^2$ **31.** $\dfrac{b^3 c}{a^4}$ **33.** $\dfrac{r^3 t^4}{s}$ **35.** $\dfrac{(z+7)(z+2)}{7z}$ **37.** x **39.** $\dfrac{x}{5}$ **41.** $x+2$

43. $\dfrac{3}{2x}$ **45.** $x-2$ **47.** x **49.** $\dfrac{(x-2)^2}{x}$ **51.** $\dfrac{(m-2)(m-3)}{2(m+2)}$ **53.** 1   **55.** $\dfrac{1}{3}$ **57.** $\dfrac{c^2}{ab}$ **59.** $\dfrac{x+1}{2(x-2)}$

61. 1   **63.** $\dfrac{1}{x-4}$ **65.** $\dfrac{x^2 - 2x + 4}{x-2}$ **67.** $\dfrac{-1}{(x+1)(x+3)}$ **69.** $\dfrac{x+y}{x(x-y)}$ **71.** $\dfrac{-(x-y)(x^2 + xy + y^2)}{a+b}$

REVIEW EXERCISES   (page 227)

1. $-6x^5 y^6 z$ **3.** $\dfrac{1}{81y^4}$ **5.** $\dfrac{1}{x^m}$ **7.** 9.3×10^7 **9.** 23 women   **11.** 9.5% and 10.5%

Exercise 5.3   (page 230)

1. $\dfrac{2}{3}$ **3.** $\dfrac{3}{10}$ **5.** $\dfrac{6}{5}$ **7.** $\dfrac{16}{35}$ **9.** $\dfrac{3}{5}$ **11.** $\dfrac{7}{5}$ **13.** $\dfrac{7}{3}$ **15.** $\dfrac{3x}{2}$ **17.** $\dfrac{3}{2y}$ **19.** 3   **21.** $\dfrac{6}{y}$ **23.** 6   **25.** $\dfrac{2x}{3}$

27. $\dfrac{2y^2}{15z}$ **29.** $\dfrac{2}{y}$ **31.** $\dfrac{2}{3x}$ **33.** $\dfrac{2(z-2)}{z}$ **35.** $\dfrac{5z(z-7)}{z+2}$ **37.** $\dfrac{x+2}{3}$ **39.** 1   **41.** $\dfrac{x-2}{x-3}$ **43.** $x+5$ **45.** 1

47. $\dfrac{3}{7}$ **49.** $\dfrac{9}{2x}$ **51.** $\dfrac{x}{36}$ **53.** $\dfrac{(x+1)(x-1)}{5(x-3)}$ **55.** 2   **57.** $\dfrac{2x(1-x)}{5(x-2)}$ **59.** $\dfrac{y^2}{3}$ **61.** $\dfrac{x+2}{x-2}$ **63.** 1

65. $\dfrac{1}{(x+1)^2}$ **67.** $-(x+y)$ or $-x-y$ **69.** $a+2$ **71.** $\dfrac{-p}{m+n}$

REVIEW EXERCISES (page 232)

1. $4y^3 + 4y^2 - 8y + 32$ **3.** $2r^3 - 3r^2 + 10r - 15$ **5.** $-6m^3 + 5m^2 + 3m - 2$ **7.** $5y^2 + 22y + 114 + \dfrac{569}{y - 5}$ **9.** 84

11. 29,500 regular admission; 6250 student admission

Exercise 5.4 (page 235)

1. $\dfrac{2}{3}$ **3.** $\dfrac{3}{5}$ **5.** $\dfrac{1}{3}$ **7.** 2   **9.** 2   **11.** $\dfrac{12}{7}$ **13.** $\dfrac{25}{4}$ **15.** $4x$ **17.** $-\dfrac{2y}{3}$ **19.** 0   **21.** $\dfrac{4x}{y}$ **23.** $\dfrac{2y}{x}$

25. $\dfrac{x^2}{2y}$ **27.** $\dfrac{2y + 6}{5z}$ **29.** 9   **31.** $\dfrac{1}{7}$ **33.** $-\dfrac{4}{3}$ **35.** $\dfrac{2}{13}$ **37.** $-\dfrac{24}{23}$ **39.** 1   **41.** $-\dfrac{17}{41}$ **43.** $\dfrac{10}{7}$ **45.** $\dfrac{x}{y}$

47. $\dfrac{y}{x}$ **49.** $\dfrac{x}{2}$ **51.** $\dfrac{-2}{5z}$ **53.** $\dfrac{1}{y}$ **55.** $\dfrac{-1}{z}$ **57.** $\dfrac{x + 3}{xy}$ **59.** 1   **61.** 0   **63.** $\dfrac{8}{5}$ **65.** $\dfrac{4x}{3}$ **67.** $\dfrac{2x}{3y}$

69. $\dfrac{4x - 2y}{y + 2}$ **71.** $\dfrac{2x + 10}{x - 2}$ **73.** $\dfrac{xy}{x - y}$ **75.** $\dfrac{-1}{a - b}$

REVIEW EXERCISES (page 236)

1. 7^2 **3.** $2^3 \cdot 17$ **5.** $2 \cdot 3 \cdot 17$ **7.** $(x - 5)(x + 3)$ **9.** $2(x + 2)(x - 2)$ **11.** $(x + y)(a - 5)$

Exercise 5.5 (page 241)

1. $\dfrac{4}{6}$ **3.** $\dfrac{125}{20}$ **5.** $\dfrac{2x}{x^2}$ **7.** $\dfrac{5x}{xy}$ **9.** $\dfrac{8xy}{x^2y}$ **11.** $\dfrac{3x(x + 1)}{(x + 1)^2}$ **13.** $\dfrac{2y(x + 1)}{x^2 + x}$ **15.** $\dfrac{z(z + 1)}{z^2 - 1}$ **17.** $\dfrac{(x + 2)^2}{x^2 - 4}$

19. $\dfrac{2(x + 2)}{x^2 + 3x + 2}$ **21.** 60   **23.** 42   **25.** $6x$ **27.** x^2y^2 **29.** $18xy$ **31.** $x^2 - 1$ **33.** $x^2 + 6x$

35. $(x + 1)(x - 2)^2$ **37.** $(x + 1)(x + 5)(x - 5)$ **39.** $\dfrac{7}{6}$ **41.** $-\dfrac{1}{6}$ **43.** $\dfrac{5y}{9}$ **45.** $\dfrac{7a}{60}$ **47.** $\dfrac{53x}{42}$ **49.** $\dfrac{4xy + 6x}{3y}$

51. $\dfrac{2 - 3x^2}{x}$ **53.** $\dfrac{3x^2 + 2xy}{6y^2}$ **55.** $\dfrac{4y + 10}{15y}$ **57.** $\dfrac{x^2 + 4x + 1}{x^2y}$ **59.** $\dfrac{2x^2 - 1}{x(x + 1)}$ **61.** $\dfrac{-x^2 + 3x + 1}{x - 2}$

63. $\dfrac{2xy + x - y}{xy}$ **65.** $\dfrac{x + 2}{x - 2}$ **67.** $\dfrac{2x^2 + 2}{(x - 1)(x + 1)}$ **69.** $\dfrac{10x + 4}{(x - 2)(x + 2)}$ **71.** $\dfrac{2x^2 - 4x + 8}{(x - 2)(x - 2)(x + 2)}$ **73.** $\dfrac{x}{x - 2}$

75. $\dfrac{5x + 3}{x + 1}$ **77.** $\dfrac{-10y + 18}{y(y - 3)}$ **79.** $\dfrac{-1}{2(x - 2)}$ **81.** $\dfrac{3}{x - 3}$

REVIEW EXERCISES (page 243)

1. 10   **3.** -8 **5.** 20   **7.** A prime number is a natural number greater than 1 that is only divisible by itself and 1.

9. A composite number is natural number greater than 1 that is not prime.   **11.** $\dfrac{x}{xy + 3}$

Exercise 5.6 (page 247)

1. $\dfrac{8}{9}$ **3.** $\dfrac{3}{8}$ **5.** $\dfrac{5}{4}$ **7.** $\dfrac{5}{7}$ **9.** $\dfrac{x^2}{y}$ **11.** $\dfrac{5t^2}{27}$ **13.** $\dfrac{1 - 3x}{5 + 2x}$ **15.** $\dfrac{1 + x}{2 + x}$ **17.** $\dfrac{3 - x}{x - 1}$ **19.** $\dfrac{1}{x + 2}$ **21.** $\dfrac{1}{x + 3}$

23. $\dfrac{xy}{y + x}$ **25.** $\dfrac{y}{x - 2y}$ **27.** $\dfrac{x^2}{x^2 - 2x + 1}$ **29.** $\dfrac{7x + 3}{-x - 3}$ **31.** $\dfrac{-x + 2}{-x - 3}$ or $\dfrac{x - 2}{x + 3}$ **33.** -1 **35.** $\dfrac{y}{x^2}$ **37.** $\dfrac{x + 1}{1 - x}$

39. $\dfrac{a^2 - a + 1}{a^2}$ **41.** 2   **43.** $\dfrac{y - 5}{y + 5}$

REVIEW EXERCISES   (page 248)

1. base of 3; exponent of 4   **3.** $abbbb$ **5.** t^9 **7.** $-2r^7$ **9.** $\dfrac{81}{256r^8}$ **11.** $\dfrac{r^{10}}{9}$

Exercise 5.7   (page 252)

1. $x = 4$ **3.** $y = -20$ **5.** $x = 6$ **7.** $x = 60$ **9.** $a = -12$ **11.** $x = 0$ **13.** $z = -7$ **15.** $x = -1$
17. $c = 12$ **19.** $x = 0$ **21.** $x = -3$ **23.** $x = 3$ **25.** no solution; $a = 0$ is extraneous   **27.** $y = \frac{3}{7}$ **29.** $x = 5$
31. no solution; $a = -2$ is extraneous   **33.** no solution; $x = 5$ is extraneous   **35.** $r = -1$ **37.** $x = 6$ **39.** $u = 2$
41. $x = -3$ **43.** $q = 1$ **45.** no solution; $y = -2$ is extraneous   **47.** $y = 1; y = 2$ **49.** $y = 3; y = -3$ is extraneous
51. $x = 3, x = -4$ **53.** $z = 1$ **55.** $n = 0$ **57.** $b = -2, b = 1$

REVIEW EXERCISES   (page 253)

1. $(a + 5)(a^2 - 5a + 25)$ **3.** $(2x - 5y^2)(4x^2 + 10xy^2 + 25y^4)$ **5.** $(b + 3)(a + 2)$ **7.** $(r + s)(m + n)$ **9.** $(2 - a)(a + b)$
11. $6r^2s^4 + 11rs^2 - 10$

Exercise 5.8   (page 258)

1. $b = \dfrac{10}{9}$ **3.** $a = \dfrac{b}{b - 1}$ **5.** 24 m   **7.** $d_1 = \dfrac{fd_2}{d_2 - f}$ **9.** 2   **11.** 5   **13.** $\dfrac{2}{3}, \dfrac{3}{2}$ **15.** $2\frac{2}{9}$ hr   **17.** $2\frac{6}{11}$ days
19. 4 mph   **21.** 4 mph   **23.** 7% and 8%   **25.** 5 workers   **27.** 30 calculators

REVIEW EXERCISES   (page 259)

1. $x = 3, x = 2$ **3.** $t = -2, t = -3, t = -4$ **5.** $y = 0, y = 0, y = 1$ **7.** $x = 1, x = -1, x = 2, x = -2$ **9.** 52 ft
11. 150 mi

Exercise 5.9   (page 263)

1. $\frac{5}{7}$ **3.** $\frac{1}{2}$ **5.** $\frac{2}{3}$ **7.** $\frac{1}{3}$ **9.** $\frac{1}{5}$ **11.** $\frac{3}{7}$ **13.** $\frac{1}{18}$ **15.** $\frac{3}{4}$ **17.** $\frac{6}{5}$ **19.** $\frac{25}{1}$ **21.** not a proportion
23. a proportion   **25.** not a proportion   **27.** not a proportion   **29.** a proportion   **31.** a proportion
33. a proportion   **35.** $x = 4$ **37.** $c = 6$ **39.** $x = -3$ **41.** $x = 9$ **43.** $x = 0$ **45.** $x = -17$ **47.** $x = -\frac{3}{2}$
49. $x = -\frac{3}{5}$ **51.** $x = 65$ **53.** $x = 39$ **55.** $x = -\frac{5}{2}, x = -1$ **57.** $x = -5, x = -1$
59. $x = 4; x = 0$ is extraneous   **61.** \$17   **63.** \$6.50   **65.** $7\frac{1}{2}$ gal   **67.** \$309   **69.** $49\frac{7}{24}$ ft   **71.** 162 teachers
73. 48 to 1 is close enough   **75.** $21\frac{3}{7}$ ft   **77.** 528 ft   **79.** 880 ft high

REVIEW EXERCISES   (page 266)

1. 24 in.   **3.** 22 cm   **5.** 36 sq in.   **7.** $\frac{65}{2}$ sq cm   **9.** 125π cu ft   **11.** 9408π cu cm

CHAPTER 5 REVIEW EXERCISES   (page 267)

1. $\dfrac{2}{5}$ **3.** $-\dfrac{1}{3}$ **5.** $\dfrac{1}{2x}$ **7.** $\dfrac{x}{x + 1}$ **9.** 2   **11.** $\dfrac{x + 3}{x - 5}$ **13.** $\dfrac{x}{x - 1}$ **15.** $\dfrac{3x}{y}$ **17.** 1   **19.** $\dfrac{3y}{2}$ **21.** $x + 2$

23. 8   **25.** $3x^2y^2$ **27.** $(x + 2)(x - 3)$ **29.** 1   **31.** $\dfrac{x^2 + x - 1}{x(x - 1)}$ **33.** $\dfrac{x - 2}{x(x + 1)}$ **35.** $\dfrac{x + 1}{x}$ **37.** $\dfrac{9}{4}$ **39.** $\dfrac{1 + x}{1 - x}$

41. $x^2 + 3$ **43.** $x = 3$ **45.** $x = 3$ **47.** $x = -2$ **49.** $x = \dfrac{y}{y + 1}$ **51.** 5 mph   **53.** $\dfrac{1}{2}$ **55.** $\dfrac{2}{3}$ **57.** $x = \dfrac{9}{2}$

59. $x = 7$ **61.** 30 tons   **63.** 0.05 g   **65.** 132.5 ft

CHAPTER 5 TEST (page 269)

1. $\dfrac{8x}{9y}$ **3.** 3 **5.** $\dfrac{x+1}{3(x-2)}$ **7.** $\dfrac{x^2}{3}$ **9.** $\dfrac{10x-1}{x-1}$ **11.** $\dfrac{2x^2+x+1}{x(x+1)}$ **13.** $\dfrac{2x^3}{y^3}$ **15.** $x=-5$ **17.** $x=4$

19. yes **21.** $3\dfrac{15}{16}$ hr **23.** 8050 ft

Exercise 6.1 (page 280)

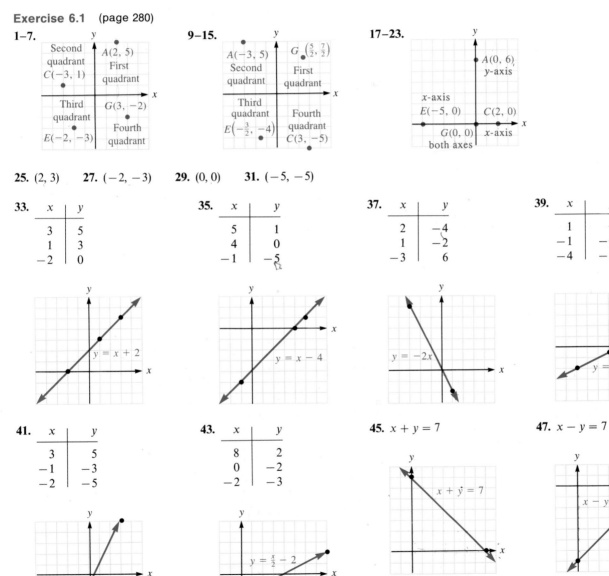

1–7.

9–15.

17–23.

25. $(2, 3)$ **27.** $(-2, -3)$ **29.** $(0, 0)$ **31.** $(-5, -5)$

33.

x	y
3	5
1	3
-2	0

$y = x + 2$

35.

x	y
5	1
4	0
-1	-5

$y = x - 4$

37.

x	y
2	-4
1	-2
-3	6

$y = -2x$

39.

x	y
1	$\frac{1}{2}$
-1	$-\frac{1}{2}$
-4	-2

$y = \frac{x}{2}$

41.

x	y
3	5
-1	-3
-2	-5

$y = 2x - 1$

43.

x	y
8	2
0	-2
-2	-3

$y = \frac{x}{2} - 2$

45. $x + y = 7$

$x + y = 7$

47. $x - y = 7$

$x - y = 7$

49. $2x + y = 5$ **51.** $2x + 3y = 12$ **53.** $3x - 4y = -12$ **55.** $5x + 2y = -10$

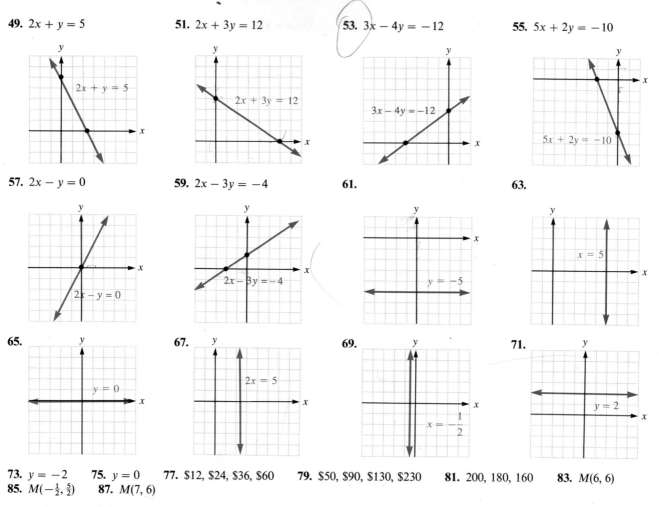

57. $2x - y = 0$ **59.** $2x - 3y = -4$ **61.**   **63.**

65.   **67.**   **69.**   **71.**

73. $y = -2$ **75.** $y = 0$ **77.** $12, $24, $36, $60   **79.** $50, $90, $130, $230   **81.** 200, 180, 160   **83.** $M(6, 6)$
85. $M(-\frac{1}{2}, \frac{5}{2})$ **87.** $M(7, 6)$

REVIEW EXERCISES   (page 284)

1. 5   **3.** -10 **5.** $y = -1$ **7.** $h = \dfrac{2A}{b}$ **9.** 22, 24, and 26   **11.** $\dfrac{4y}{3} + 2x$

Exercise 6.2   (page 292)

1. 1   **3.** -1 **5.** 1   **7.** 2   **9.** $\frac{1}{5}$ **11.** 0   **13.** -1 **15.** $\frac{1}{2}$

17.   **19.**   **21.**   **23.**

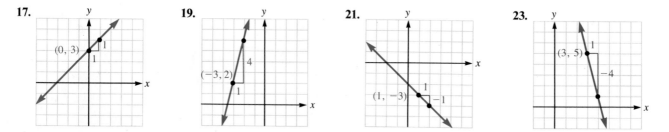

25.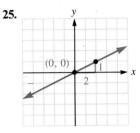

27.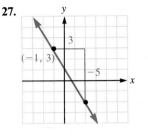

29. slope 3; y-intercept 3

31. slope 5; y-intercept 1

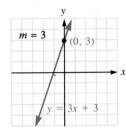

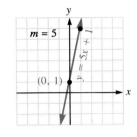

33. slope -3; y-intercept 0

35. slope 3; y-intercept -2

37. slope $\frac{1}{3}$; y-intercept 0

39. slope $\frac{5}{3}$; y-intercept $\frac{1}{2}$

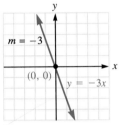

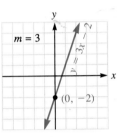

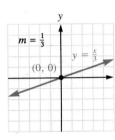

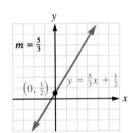

41. $y = -3x + 7$; slope -3; y-intercept 7   **43.** $y = \frac{5}{2}x - 13$; slope $\frac{5}{2}$; y-intercept -13 **45.** $y = 2$; slope 0; y-intercept 2
47. $y = x - \frac{7}{2}$; slope 1; y-intercept $-\frac{7}{2}$ **49.** $y = \frac{2}{3}x + 7$; slope $\frac{2}{3}$; y-intercept 7   **51.** $y = \frac{1}{3}x - \frac{5}{3}$; slope $\frac{1}{3}$; y-intercept $-\frac{5}{3}$
53. $\frac{1}{220}$ **55.** $\frac{5}{12}$ pitch   **57.** \$26,750   **59.** \$6000   **61.** \$150

REVIEW EXERCISES   (page 295)

1. $3x(x - 2)$ **3.** $(2z + 1)(z - 3)$ **5.** $(3u + 4)(3u + 4)$ **7.** $y = 0, y = -2$ **9.** $x = 1, x = 6$ **11.** $(a + b)(z + 3)$

Exercise 6.3   (page 302)

1. $4x - y = -5$ **3.** $7x + y = -2$ **5.** $x + 2y = -1$ **7.** $3x + 5y = -2$ **9.** $y = 3$ **11.** $y = 0$ **13.** $2x - y = 1$
15. $5x - y = 6$ **17.** $2x + y = 0$ **19.** $x + 2y = 6$ **21.** $7x - 5y = 19$ **23.** $y = 3$ **25.** neither   **27.** parallel
29. perpendicular   **31.** parallel   **33.** parallel   **35.** neither   **37.** perpendicular   **39.** perpendicular   **41.** parallel
43. $2x - 15y = -120$ **45.** $4x + 9y = -2$ **47.** $5x - y = 0$ **49.** $5x - y = 4$ **51.** $5x - 2y = 8$ **53.** $2x - y = -1$
55. $y = 2$ **57.** $5x + y = 15$ **59.** $x + 5y = 0$ **61.** $5x + y = 3$ **63.** \$800   **65.** \$8000

REVIEW EXERCISES   (page 304)

1. $\dfrac{x + 1}{x}$ **3.** $\dfrac{6}{x + 1}$ **5.** $\dfrac{4x}{(x + 1)(x - 1)}$ **7.** 7.3×10^{10} **9.** 3.72×10^{-1} **11.** $x^3 - 12x^2 + 48x - 64$

Exercise 6.4   (page 308)

1.   **3.**   **5.**   **7.**

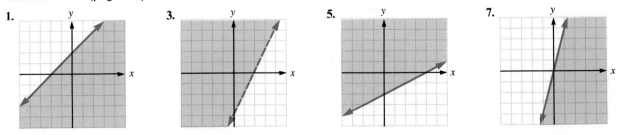

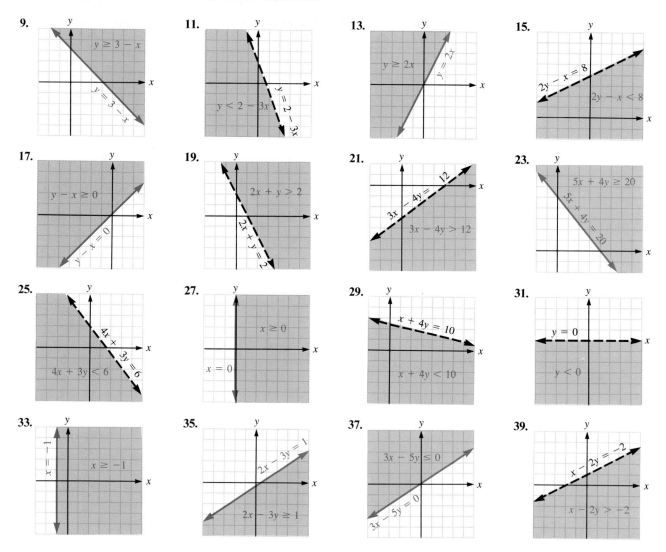

REVIEW EXERCISES (page 311)

1. 41, 43, 47 **3.** If a and b are real numbers, then $ab = ba$. **5.** 1 **7.** $P(2) = 7$ **9.** $P(-x) = 3x^2 + 4x + 3$
11. $P(x^2) = 3x^4 - 4x^2 + 3$

Exercise 6.5 (page 314)

1. a function **3.** a function **5.** a function **7.** a function
9. not a function; if $x = 2$, y could be 2, or any number less than 2 **11.** a function
13. not a function; if $x = 2$, y is either 2 or -2 **15.** $f(3) = 9$, $f(0) = 0$, $f(-1) = -3$
17. $f(3) = 3$, $f(0) = -3$, $f(-1) = -5$ **19.** $f(3) = 22$, $f(0) = 7$, $f(-1) = 2$ **21.** $f(3) = 3$, $f(0) = 9$, $f(-1) = 11$
23. $f(1) = 1$, $f(-2) = 4$, $f(3) = 9$ **25.** $f(1) = 0$, $f(-2) = -9$, $f(3) = 26$ **27.** $f(1) = 4$, $f(-2) = 1$, $f(3) = 16$
29. $f(1) = 1$, $f(-2) = 10$, $f(3) = 15$ **31.** $f(2) = 4$, $f(1) = 3$, $f(-2) = 4$ **33.** $f(2) = 2$, $f(1) = -1$, $f(-2) = 2$
35. $f(2) = \frac{1}{5}$, $f(1) = \frac{1}{4}$, $f(-2) = 1$ **37.** $f(2) = -2$, $f(1) = -\frac{1}{2}$, $f(-2) = \frac{2}{5}$ **39.** $g(w) = 2w$, $g(w + 1) = 2w + 2$

41. $g(w) = 3w - 5$, $g(w + 1) = 3w - 2$ **43.** $g(w) = w^2 + w$, $g(w + 1) = w^2 + 3w + 2$
45. $g(w) = w^2 - 1$, $g(w + 1) = w^2 + 2w$ **47.** 12   **49.** $2b - 2a$ **51.** $2b$ **53.** 1

REVIEW EXERCISES   (page 315)

1. $y = -2$ **3.** $x = -\frac{3}{5}$ **5.** $x = 6$ **7.** ⟵————○⟶ 5   **9.** ⟵——●——○⟶ 9 14   **11.** ⟵————●⟶ $-\frac{1}{6}$

Exercise 6.6   (page 320)

1. $d = kn$ **3.** $l = \dfrac{k}{w}$ **5.** $A = kr^2$ **7.** $D = kst$ **9.** $I = \dfrac{kV}{R}$ **11.** $y = 35$ **13.** $r = 42$ **15.** $s = 675$

17. $y = 1$ **19.** $r = \dfrac{80}{3}$ **21.** $y = 24$ **23.** $y = 4$ **25.** $D = 80$ **27.** $y = \dfrac{5}{2}$ **29.** 576 ft   **31.** $2450   **33.** $2\frac{1}{2}$ hr

35. $53\frac{1}{3}$ cu m   **37.** $270   **39.** $3\frac{9}{11}$ amp

REVIEW EXERCISES   (page 322)

1. $8x + 5$ **3.** $-3x^3 + 6x^2 - 6x$ **5.** $\dfrac{y + 2}{y^2}$ **7.** $\dfrac{1 + t}{t - 1}$ **9.** $x + 4$ **11.** 8 in. by 8 in.

CHAPTER 6 REVIEW EXERCISES   (page 323)

1, 3, 5.

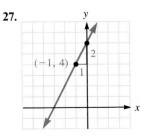

7. $A(3, 1)$ **9.** $C(-3, -4)$ **11.** $E(0, 0)$ **13.** $G(-5, 0)$ **15.**

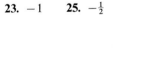

17.

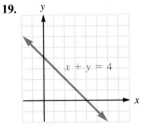

19.

21.

23. -1 **25.** $-\frac{1}{2}$

27.

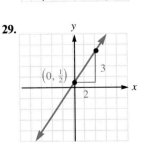

29.

31. slope 5, y-intercept 2   **33.** slope 0, y-intercept -3

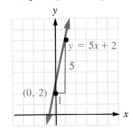

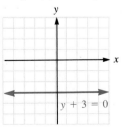

35. $3x + y = 2$ **37.** $7x - y = 0$ **39.** $3x - y = 0$ **41.** $x - 9y = -9$ **43.** neither   **45.** perpendicular
47. neither   **49.** parallel   **51.** $7x - y = 9$ **53.** $5x + 2y = 0$

55.   **57.**

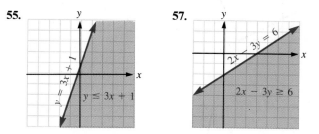

59. a function   **61.** not a function; if $x = 1$, then y is either 1 or -1 **63.** 1   **65.** 7   **67.** -2 **69.** $a^2 - a + 2$
71. $s = 400$ **73.** $R = 81$

CHAPTER 6 TEST   (page 326)

1.   **3.**   **5.** $\frac{4}{3}$ **7.** -2 **9.** $\frac{3}{5}$ **11.** $x - 2y = -6$ **13.** $x = -3$

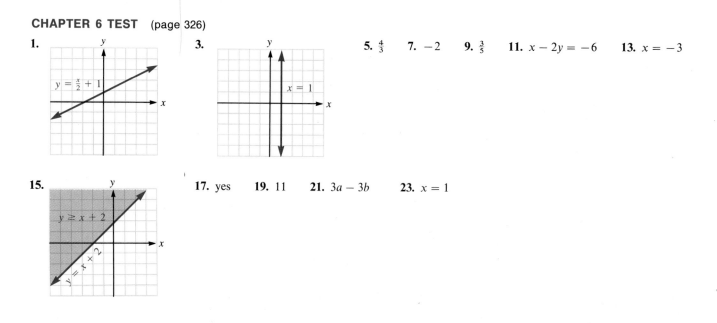

15.   **17.** yes   **19.** 11   **21.** $3a - 3b$ **23.** $x = 1$

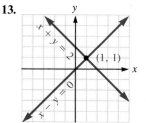

Exercise 7.1   (page 334)

1. a solution   **3.** a solution   **5.** not a solution   **7.** a solution   **9.** not a solution   **11.** not a solution

13.   **15.**   **17.**   **19.**

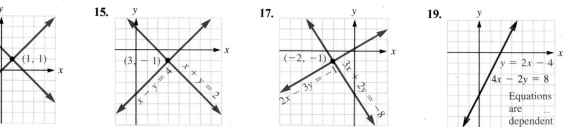

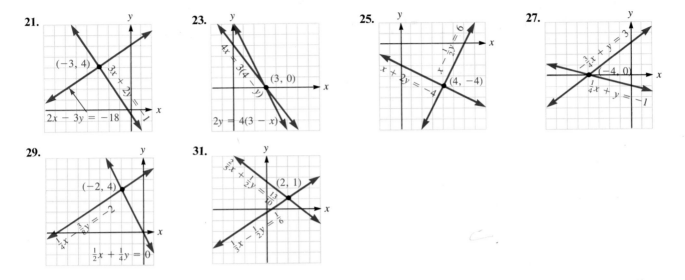

REVIEW EXERCISES   (page 336)

1. x^{12} **3.** 1   **5.** $8x^2 - 9x + 12$ **7.** $-12x^3y^3 - 9x^4y^3$ **9.** $3ab + 4b - 5a$ **11.** $V = \frac{9}{2}$

Exercise 7.2   (page 340)

1. $(2, 4)$ **3.** $(3, 0)$ **5.** $(-3, -1)$ **7.** inconsistent   **9.** $(-2, 3)$ **11.** $(3, 2)$ **13.** $(3, -2)$ **15.** $(-1, 2)$
17. $(-1, -1)$ **19.** dependent   **21.** $(1, 1)$ **23.** $(4, -2)$ **25.** $(-3, -1)$ **27.** $(-1, -3)$ **29.** $(\frac{1}{2}, \frac{1}{3})$ **31.** $(1, 4)$
33. $(4, 2)$ **35.** $(-5, -5)$ **37.** $(-6, 4)$ **39.** $(\frac{1}{5}, 4)$ **41.** $(5, 5)$

REVIEW EXERCISES   (page 341)

1. $8xy(xy - 4yz + 2z^2)$ **3.** $(a^3 + 5)(a^3 - 5)$ **5.** $(r + 5s)(r - 3s)$ **7.** $\dfrac{3b}{2c}$ **9.** $-\dfrac{1}{2}$ **11.** $I = 2$

Exercise 7.3   (page 345)

1. $(1, 4)$ **3.** $(-2, 3)$ **5.** $(-1, 1)$ **7.** $(2, 5)$ **9.** $(-3, 4)$ **11.** $(0, 8)$ **13.** $(2, 3)$ **15.** $(3, -2)$ **17.** $(2, 7)$
19. inconsistent   **21.** dependent   **23.** $(4, 0)$ **25.** $(\frac{10}{3}, \frac{10}{3})$ **27.** $(5, -6)$ **29.** $(-5, 0)$ **31.** $(-1, 2)$ **33.** $(1, -\frac{5}{2})$
35. $(-1, 2)$ **37.** $(0, 1)$ **39.** $(-2, 3)$ **41.** $(2, 2)$

REVIEW EXERCISES   (page 346)

1. $x = 4$ **3.** $z = \frac{2}{3}, z = 1$ **5.** $y = -\frac{5}{2}, y = \frac{2}{5}$ **7.** $w = \dfrac{P - 2l}{2}$ **9.** 105, 106, 107   **11.** \$4000

Exercise 7.4   (page 353)

1. 32 and 64   **3.** 8 and 5   **5.** Paint costs \$15; a brush costs \$5.   **7.** Cleaner costs \$5.40; soaking solution costs \$6.20.
9. 10 ft and 15 ft   **11.** 25 ft by 30 ft   **13.** 60 sq ft   **15.** 40 quarters and 40 dimes   **17.** 10 dimes and 10 quarters
19. Bill invested \$2000; Janette invested \$3000.   **21.** 250 student tickets, 100 nonstudent tickets   **23.** 10 mph
25. 50 mph   **27.** 5 liters of 40% solution, 10 liters of 55% solution   **29.** 32 lb peanuts, 16 lb cashews
31. Lisa bought 5 \$7 gifts and 2 \$9 gifts.   **33.** 15 \$87 radios and 10 \$119 radios   **35.** 9% and 10%

REVIEW EXERCISES (page 355)

1.
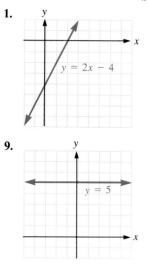
$y = 2x - 4$

3.
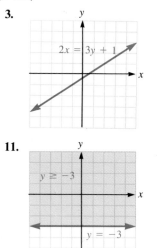
$2x = 3y + 1$

5.
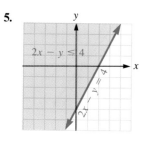
$2x - y \leq 4$
$2x - y = 4$

7.
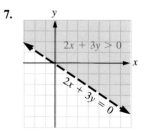
$2x + 3y > 0$
$2x + 3y = 0$

9.
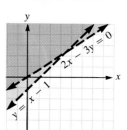
$y = 5$

11.
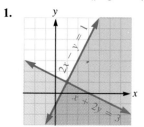
$y \geq -3$
$y = -3$

Exercise 7.5 (page 359)

1.
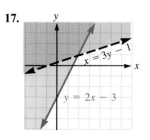
$2x - y = 1$
$x + 2y = 3$

3.
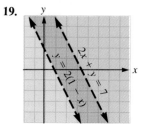
$x + y = -1$
$x - y = -1$

5.
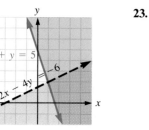
$2x - y = 4$
$x + y = -1$

7.
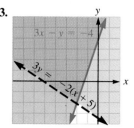
$y = 3$
$x = 2$

9.
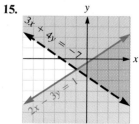
$2x - 3y = 0$
$y = x - 1$

11. no solutions
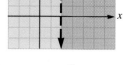
$x + y = 3$
$x + y = 1$

13.
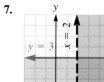
$y = 0$
$x = 0$

15.
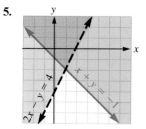
$3x + 4y = -7$
$2x - 3y = 1$

17.
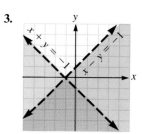
$x = 3y - 1$
$y = 2x - 3$

19.
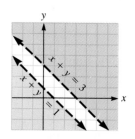
$2x + y = 7$
$y = 2(1 - x)$

21.
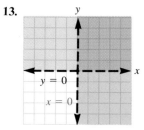
$3x + y = 5$
$y = -6$
$2x - 4y$

23.
$3x - y = -4$
$3y = -2(x + 5)$

25. 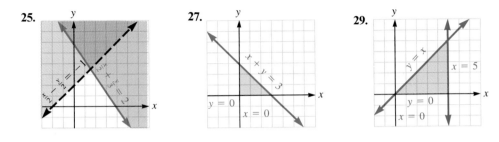   **27.**   **29.**

REVIEW EXERCISES   (page 361)

1. x^7 **3.** a^{10} **5.** $\dfrac{1}{z^3}$ **7.** $\dfrac{81m^8}{n^{12}}$ **9.** $\dfrac{xz}{y^2}$ **11.** $3x(5x-9)$

Exercise 7.6   (page 365)

1. $(1, 1, 2)$ **3.** $(0, 2, 2)$ **5.** $(3, 2, 1)$ **7.** inconsistent   **9.** dependent; one solution is $(\frac{1}{2}, \frac{5}{3}, 0)$ **11.** $(2, 6, 9)$
13. $-2, 4, 16$ **15.** $9, 10, 11$ **17.** 10 nickels, 5 dimes, 2 quarters
19. 50 expensive footballs, 75 middle-priced footballs, 1000 cheap footballs
21. 250 five-dollar tickets, 375 three-dollar tickets, 125 two-dollar tickets

REVIEW EXERCISES   (page 367)

1. -16 **3.** 6   **5.** $x = \frac{5}{3}y$ **7.** $a = 2$ **9.** $-\frac{5}{3}$

CHAPTER 7 REVIEW EXERCISES   (page 368)

1. a solution   **3.** a solution   **5.**   **7.** Equations are dependent.   **9.** $(-1, -2)$ **11.** $(1, -1)$

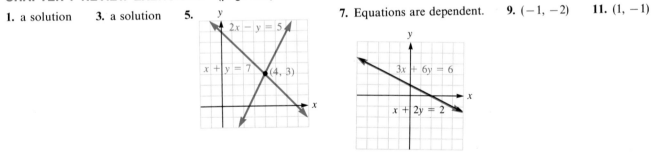

13. $(3, -5)$ **15.** $(-1, 7)$ **17.** $(0, 9)$ **19.** Equations are dependent.   **21.** 3 and 15
23. Orange costs 35 cents; grapefruit costs 50 cents.   **25.** Milk costs $1.69; eggs cost $1.14.

27.   **29.**   **31.** $x = 1, y = 0, z = -3$

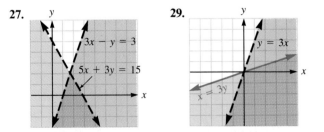

33. dependent equations; possible solution is $(3, 1, 3)$

CHAPTER 7 TEST (page 369)

1. a solution **3.** **5.** $(-2, -3)$ **7.** $(2, 4)$ **9.** The system is inconsistent. **11.** 65

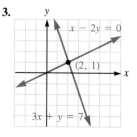

13. **15.** $x = 1, y = 3, z = 0$

Exercise 8.1 (page 376)

1. 3 **3.** 7 **5.** 6 **7.** 9 **9.** -5 **11.** -9 **13.** 14 **15.** 16 **17.** -17 **19.** 100 **21.** 18 **23.** -60
25. 1.414 **27.** 2.236 **29.** 2.449 **31.** 3.317 **33.** 4.796 **35.** 9.747 **37.** 80.175 **39.** -99.378 **41.** 4.621
43. 0.599 **45.** 0.996 **47.** -0.915 **49.** rational number **51.** rational number **53.** irrational number
55. imaginary number **57.** 1 **59.** 3 **61.** -2 **63.** -4 **65.** 5 **67.** 1 **69.** -4 **71.** 9 **73.** 2
75. -2 **77.** 1 **79.** -2 **81.** xy **83.** x^2z^2 **85.** $-x^2y$ **87.** $2z$ **89.** $-3x^2y$ **91.** xyz **93.** $-xyz^2$
95. $-5x^2z^6$ **97.** $6z^{18}$ **99.** $-4z$ **101.** $3yz^2$ **103.** $-2p^2q$ **105.** $x + 9$ **107.** $x + 11$ **109.** $x + 3$
111. $x + 10$ **113.** $x + 7y$

REVIEW EXERCISES (page 377)

1. $P(0) = -5$ **3.** $P(-1) = -10$ **5.** $f(0) = -1$ **7.** $f(-2) = 9$ **9.** $y = kx$ **11.** 4 sec

Exercise 8.2 (page 381)

1. $2\sqrt{5}$ **3.** $5\sqrt{2}$ **5.** $3\sqrt{5}$ **7.** $7\sqrt{2}$ **9.** $4\sqrt{3}$ **11.** $10\sqrt{2}$ **13.** $8\sqrt{3}$ **15.** $2\sqrt{22}$ **17.** 18 **19.** $7\sqrt{3}$
21. $6\sqrt{5}$ **23.** $12\sqrt{3}$ **25.** $48\sqrt{2}$ **27.** $-70\sqrt{10}$ **29.** $14\sqrt{5}$ **31.** $-45\sqrt{2}$ **33.** $5\sqrt{x}$ **35.** $a\sqrt{b}$ **37.** $3x\sqrt{y}$
39. $40x^3y^2\sqrt{2}$ **41.** $48x^2y\sqrt{y}$ **43.** $-9x^2y^3z\sqrt{2xy}$ **45.** $6ab^2\sqrt{3ab}$ **47.** $-\dfrac{8n\sqrt{5m}}{5}$ **49.** $\dfrac{5}{3}$ **51.** $\dfrac{9}{8}$ **53.** $\dfrac{\sqrt{26}}{5}$
55. $\dfrac{2\sqrt{5}}{7}$ **57.** $\dfrac{4\sqrt{3}}{9}$ **59.** $\dfrac{4\sqrt{2}}{5}$ **61.** $\dfrac{5\sqrt{5}}{11}$ **63.** $\dfrac{7\sqrt{5}}{6}$ **65.** $\dfrac{6x\sqrt{2x}}{y}$ **67.** $\dfrac{5mn^2\sqrt{5}}{8}$ **69.** $\dfrac{4m\sqrt{2}}{3n}$ **71.** $\dfrac{2rs^2\sqrt{3}}{9}$
73. $2x$ **75.** $-4x\sqrt[3]{x^2}$ **77.** $3xyz^2\sqrt[3]{2y}$ **79.** $-3yz\sqrt[3]{3x^2z}$ **81.** $\dfrac{3m}{2n^2}$ **83.** $\dfrac{rs\sqrt[3]{2rs^2}}{5t}$ **85.** $\dfrac{5ab}{2}$

REVIEW EXERCISES (page 382)

1. $\dfrac{1}{2xz}$ **3.** $\dfrac{a + 1}{a + 3}$ **5.** 1 **7.** 2 **9.** $\dfrac{15a^2c + 6b^2c - 5a^2b^2}{10abc}$ **11.** 7

Exercise 8.3   (page 385)

1. $5\sqrt{3}$ **3.** $9\sqrt{3}$ **5.** $7\sqrt{5}$ **7.** $12\sqrt{5}$ **9.** $8\sqrt{5}$ **11.** $10\sqrt{10}$ **13.** $37\sqrt{5}$ **15.** $12\sqrt{7}$ **17.** $38\sqrt{2}$
19. $85\sqrt{2}$ **21.** $9\sqrt{5}$ **23.** $7\sqrt{6}+4\sqrt{15}$ **25.** $\sqrt{2}$ **27.** $3-5\sqrt{2}$ **29.** $2\sqrt{2}$ **31.** $-2\sqrt{3}$ **33.** $\sqrt{3}$
35. $4\sqrt{10}$ **37.** $-7\sqrt{5}$ **39.** $22\sqrt{6}$ **41.** $-18\sqrt{2}$ **43.** $13\sqrt{10}$ **45.** $6\sqrt{3}$ **47.** $10\sqrt{2}-\sqrt{3}$ **49.** $-6\sqrt{6}$
51. $10\sqrt{2}+5\sqrt{3}$ **53.** $7\sqrt{3}-6\sqrt{2}$ **55.** $6\sqrt{6}-2\sqrt{3}$ **57.** $3x\sqrt{2}$ **59.** $3x\sqrt{2x}$ **61.** $3x\sqrt{2y}-3x\sqrt{3y}$
63. $x^2\sqrt{2x}$ **65.** $19x\sqrt{6}$ **67.** $-5y\sqrt{10y}$ **69.** $2x\sqrt{5xy}+3x^2y\sqrt{5xy}-4x^3y^2\sqrt{5xy}$ **71.** $5\sqrt[3]{2}$ **73.** $\sqrt[3]{3}$
75. $2\sqrt[3]{5}+5$ **77.** $x\sqrt[3]{x}-x^2\sqrt[3]{x}$ **79.** $2xy\sqrt[3]{3xy^2}$ **81.** $x^2y\sqrt[3]{5xy}$

REVIEW EXERCISES   (page 386)

1. $\dfrac{3}{8}$ **3.** $\dfrac{5}{36}$ **5.** $a=8$ **7.** $x=-9$ **9.** $z=3, z=-6$ **11.** $h=\dfrac{\dfrac{7A}{44}-r^2}{r}=\dfrac{7A}{44r}-r$

Exercise 8.4   (page 392)

1. 3   **3.** 4   **5.** 8   **7.** 4   **9.** x^3 **11.** b^7 **13.** $4\sqrt{15}$ **15.** $-60\sqrt{2}$ **17.** 24   **19.** $-8x$ **21.** $700x\sqrt{10}$
23. $4x^2\sqrt{y}$ **25.** $2+\sqrt{2}$ **27.** $9-\sqrt{3}$ **29.** $7-3\sqrt{7}$ **31.** $3\sqrt{5}-5$ **33.** $3\sqrt{2}+\sqrt{3}$ **35.** $7-2\sqrt[3]{7}$
37. $x\sqrt{3}-2\sqrt{x}$ **39.** $6x+6\sqrt{x}$ **41.** $6\sqrt{x}+3x$ **43.** $3x\sqrt{7}+x\sqrt{42}$ **45.** 1   **47.** 1   **49.** $\sqrt[3]{4}+2\sqrt[3]{2}+1$
51. $7-x^2$ **53.** $2-2\sqrt{2x}+x$ **55.** $6x-7$ **57.** $4x-18$ **59.** $8xy+4\sqrt{2xy}+1$ **61.** $9x$ **63.** $\dfrac{2x}{3}$
65. $\dfrac{3y\sqrt{2}}{5}$ **67.** $\dfrac{2y}{x}$ **69.** $\dfrac{2x}{3}$ **71.** $\dfrac{y\sqrt{xy}}{3}$ **73.** $\dfrac{5\sqrt{2y}}{x}$ **75.** $\dfrac{\sqrt{3}}{3}$ **77.** $\dfrac{2\sqrt{7}}{7}$ **79.** $\sqrt[3]{25}$ **81.** $\sqrt{3}$ **83.** $\dfrac{3\sqrt{2}}{8}$
85. $2\sqrt[3]{2}$ **87.** $\dfrac{\sqrt{15}}{3}$ **89.** $\dfrac{10\sqrt{x}}{x}$ **91.** $\dfrac{3\sqrt{2x}}{2x}$ **93.** $\dfrac{\sqrt{2xy}}{3y}$ **95.** $\dfrac{\sqrt{6}}{2x}$ **97.** $\sqrt{3x}$ **99.** $\dfrac{\sqrt[3]{20}}{2}$ **101.** $\sqrt[3]{x}$
103. $\dfrac{\sqrt[3]{2xy^2}}{xy}$ **105.** $\dfrac{-\sqrt[3]{5ab}}{ab}$ **107.** $\dfrac{3(\sqrt{3}+1)}{2}$ **109.** $\sqrt{7}-2$ **111.** $6+2\sqrt{3}$ **113.** $2-\sqrt{2}$ **115.** $\dfrac{\sqrt{3}-3}{2}$
117. $5\sqrt{3}-5\sqrt{2}$ **119.** $\sqrt{10}+\sqrt{6}$ **121.** $5-2\sqrt{6}$ **123.** $\dfrac{3x-2\sqrt{3x}+1}{3x-1}$ **125.** $\dfrac{2x+2\sqrt{2x}-15}{2x-9}$
127. $\dfrac{2\sqrt{2z}-1-2z}{2z-1}$ **129.** $\dfrac{y^2-2y\sqrt{15}+15}{y^2-15}$

REVIEW EXERCISES   (page 394)

1. $(x-7)(x+3)$ **3.** $3xy(2x-5)$ **5.** $(x+2)(x^2-2x+4)$ **7.** $x=3, x=10$ **9.** $x=2, x=-2$
11. 4 ft and 12 ft

Exercise 8.5   (page 398)

1. $x=9$ **3.** $x=49$ **5.** no solution   **7.** $x=1$ **9.** $x=30$ **11.** no solution   **13.** $x=5$ **15.** $x=6$
17. $x=3$ **19.** $x=-1$ **21.** $x=46$ **23.** $x=-3$ **25.** $x=0$ **27.** no solution   **29.** $x=2$ **31.** $x=-1$
33. $x=10$ **35.** $x=3$ **37.** $x=0$ or $x=-1$ **39.** $x=2$ **41.** $x=2, x=1$ **43.** $x=3$ **45.** 161   **47.** 28
49. 7   **51.** 18   **53.** 5   **55.** 0 and 1   **57.** 4
59. No. The number 4 is a positive number, and the principal square root of a number is a positive number or 0. The sum of a positive and a nonnegative cannot equal 0.
61. $v^2=c^2-f^2c^2$

REVIEW EXERCISES   (page 399)

1. $-6x^3y^7$ **3.** x^2+x-12 **5.** $9x^2+24x+16$ **7.** $x=2, y=3$ **9.** $x=3, y=-2$ **11.** 5

Exercise 8.6   (page 403)

1. 9   **3.** -12 **5.** $\frac{1}{2}$ **7.** $\frac{2}{7}$ **9.** 3   **11.** -5 **13.** -2 **15.** $\frac{1}{4}$ **17.** $\frac{3}{4}$ **19.** 2   **21.** 2   **23.** -3

25. 729   **27.** 125   **29.** 25   **31.** 100   **33.** 4   **35.** 8   **37.** 27   **39.** -8 **41.** $\frac{4}{9}$ **43.** $\frac{125}{8}$ **45.** 6

47. 25   **49.** 7   **51.** 25   **53.** 8   **55.** $\frac{1}{4}$ **57.** 6   **59.** 192   **61.** $\frac{1}{2}$ **63.** $\frac{1}{9}$ **65.** $\frac{1}{64}$ **67.** $\frac{1}{81}$ **69.** x

71. x^2 **73.** x^4 **75.** x^2 **77.** y^2 **79.** $x^{2/5}$ **81.** $x^{2/7}$ **83.** x **85.** y^7 **87.** y^2 **89.** $x^{17/12}$ **91.** $b^{3/10}$

93. $t^{4/15}$ **95.** x **97.** $a^{14/15}$ **99.** $\dfrac{3}{b^{101/60}}$

REVIEW EXERCISES   (page 404)

1. $3xyz^2(x^2yz^2 - 2z^3 + 5x)$ **3.** $(a^2 + b^2)(a + b)(a - b)$ **5.** $x = 5$ **7.** $m = \frac{2}{3}$ **9.** $x = \frac{5}{2}, y = -1$ **11.** 1 and 2

Exercise 8.7   (page 408)

1. $c = 5$ **3.** $c = 13$ **5.** $b = 20$ **7.** $a = 28$ **9.** $b = 2\sqrt{14}$ **11.** 5   **13.** 10   **15.** 13   **17.** 10   **19.** 5
21. 12 ft   **23.** 30 ft   **25.** 2 units   **27.** $6\sqrt{2}$ ft   **29.** $4\sqrt{13}$ mi   **31.** 18π sq in.

REVIEW EXERCISES   (page 409)

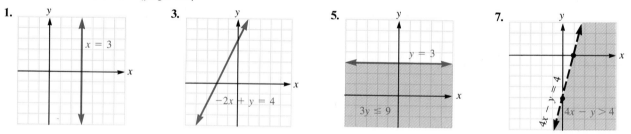

1.   **3.** $-2x + y = 4$ **5.** $y = 3$, $3y \leq 9$ **7.** $4x - y = 4$, $4x - y > 4$

9. 7,200,000   **11.** 3 hours

CHAPTER 8 REVIEW EXERCISES   (page 411)

1. 5   **3.** -12 **5.** 16   **7.** 13   **9.** 3   **11.** 3   **13.** 4.583   **15.** -7.570 **17.** $4\sqrt{2}$ **19.** $10\sqrt{5}$ **21.** $4x\sqrt{5}$

23. $-5t\sqrt{10t}$ **25.** $10x\sqrt{2y}$ **27.** $2y\sqrt[3]{x^2}$ **29.** $\frac{4}{5}$ **31.** $\frac{10}{3}$ **33.** $\frac{2\sqrt{15}}{7}$ **35.** $\frac{11x\sqrt{2}}{13}$ **37.** 0   **39.** $18\sqrt{5}$

41. $5x\sqrt{2y}$ **43.** $5\sqrt[3]{2}$ **45.** $-6\sqrt{6}$ **47.** $36x\sqrt{2}$ **49.** $4\sqrt[3]{2}$ **51.** -2 **53.** -2 **55.** $\sqrt[3]{9} + \sqrt[3]{3} - 2$

57. $49 + 12\sqrt{5}$ **59.** $x - 2\sqrt{2x} + 2$ **61.** $\frac{\sqrt{7}}{7}$ **63.** $2\sqrt[3]{4}$ **65.** $\frac{\sqrt{5}}{5}$ **67.** $5 - \sqrt{15}$ **69.** $x = 6$ **71.** no solution

73. $x = 4$; $x = -1$ is extraneous   **75.** $x = 2$; $x = -2$ is extraneous   **77.** 7   **79.** 216   **81.** 64   **83.** x **85.** 6

87. x **89.** $x^{11/15}$ **91.** $\dfrac{1}{x^{4/5}}$ **93.** $c = 35$ **95.** $b = 1$ **97.** 5 units   **99.** 2 units   **101.** 68 in.

CHAPTER 8 TEST   (page 412)

1. 10   **3.** -3 **5.** $2x\sqrt{2}$ **7.** $4\sqrt{2}$ **9.** x^2y^2 **11.** $5\sqrt{3}$ **13.** $-24x\sqrt{6}$ **15.** -1 **17.** $\dfrac{y\sqrt{xy}}{4x}$

19. $\dfrac{x\sqrt{3} - 2\sqrt{3x}}{x - 4}$ **21.** $x = 66$ **23.** $x = 5$ **25.** 11   **27.** y^6 **29.** $p^{17/12}$ **31.** 13 in.   **33.** 5 units

Exercise 9.1   (page 419)

1. $3, -3$ **3.** $0, -3$ **5.** $2, 3$ **7.** $-1, \dfrac{2}{3}$ **9.** $-\dfrac{1}{3}, -\dfrac{3}{2}$ **11.** $\dfrac{2}{5}, -\dfrac{1}{2}$ **13.** $1, -1$ **15.** $3, -3$ **17.** $2\sqrt{5}, -2\sqrt{5}$

19. $3, -3$ **21.** $2, -2$ **23.** $\sqrt{a}, -\sqrt{a}$ **25.** $-6, 4$ **27.** $7, -11$ **29.** $2 + 2\sqrt{2}, 2 - 2\sqrt{2}$ **31.** $3a, -a$

33. $-b + 4c, -b - 4x$ **35.** $0, -4$ **37.** $2, -\dfrac{2}{3}$ **39.** $\dfrac{-1 + 2\sqrt{5}}{2}, \dfrac{-1 - 2\sqrt{5}}{2}$ **41.** $3, -3$ **43.** $5, -5$

45. $-1 + 2\sqrt{2}, -1 - 2\sqrt{2}$

REVIEW EXERCISES   (page 420)

1. $y^2 - 2y + 1$ **3.** $x^2 + 2xy + y^2$ **5.** $4r^2 - 4rs + s^2$ **7.** $9a^2 + 12ab + 4b^2$ **9.** $25r^2 - 80rs + 64s^2$ **11.** $\frac{1}{2}$ hr

Exercise 9.2   (page 424)

1. $x^2 + 4x + 4, (x + 2)^2$ **3.** $x^2 - 10x + 25, (x - 5)^2$ **5.** $x^2 + 11x + \dfrac{121}{4}, \left(x + \dfrac{11}{2}\right)^2$ **7.** $a^2 - 3a + \dfrac{9}{4}, \left(a - \dfrac{3}{2}\right)^2$

9. $b^2 + \dfrac{2}{3}b + \dfrac{1}{9}, \left(b + \dfrac{1}{3}\right)^2$ **11.** $c^2 - \dfrac{5}{2}c + \dfrac{25}{16}, \left(c - \dfrac{5}{4}\right)^2$ **13.** $-2, -4$ **15.** $2, 6$ **17.** $5, -3$ **19.** $3, 4$

21. $1, -6$ **23.** $1, -2$ **25.** $-4, -4$ **27.** $2, -\dfrac{1}{2}$ **29.** $-2, \dfrac{1}{4}$ **31.** $-2 + \sqrt{3}, -2 - \sqrt{3}$ **33.** $1 + \sqrt{5}, 1 - \sqrt{5}$

35. $2 + \sqrt{7}, 2 - \sqrt{7}$ **37.** $-1 + \sqrt{2}, -1 - \sqrt{2}$ **39.** $1, -4$ **41.** $\dfrac{3}{2}, -\dfrac{2}{3}$ **43.** $\dfrac{-3 + \sqrt{3}}{2}, \dfrac{-3 - \sqrt{3}}{2}$

REVIEW EXERCISES   (page 425)

1. $t = -4$ **3.** $r = \dfrac{3}{4}, r = -\dfrac{1}{5}$ **5.** $x = 6$ **7.** $4\sqrt{5}$ **9.** $\dfrac{\sqrt{7x}}{7}$ **11.** 13

Exercise 9.3   (page 429)

1. $a = 1, b = 4, c = 3$ **3.** $a = 3, b = -2, c = 7$ **5.** $a = 4, b = -2, c = 1$ **7.** $a = 3, b = -5, c = -2$

9. $a = 7, b = 14, c = 21$ **11.** $a = 1, b = -1, c = -5$ **13.** $2, 3$ **15.** $-3, -4$ **17.** $1, -\dfrac{1}{2}$ **19.** $-1, -\dfrac{2}{3}$

21. $\dfrac{1}{2}, -\dfrac{3}{2}$ **23.** $2, -\dfrac{2}{5}$ **25.** $\dfrac{-3 + \sqrt{5}}{2}, \dfrac{-3 - \sqrt{5}}{2}$ **27.** $\dfrac{-5 + \sqrt{37}}{2}, \dfrac{-5 - \sqrt{37}}{2}$ **29.** $\dfrac{-1 + \sqrt{41}}{4}, \dfrac{-1 - \sqrt{41}}{4}$

31. $-2 + \sqrt{3}, -2 - \sqrt{3}$ **33.** $-1 + \sqrt{2}, -1 - \sqrt{2}$ **35.** $\dfrac{3 + \sqrt{15}}{3}, \dfrac{3 - \sqrt{15}}{3}$

REVIEW EXERCISES   (page 429)

1. $r = \dfrac{A - p}{pt}$ **3.**   **5.** $3x - 5y = -60$ **7.**   **9.** $f(0) = -3$ **11.** 200

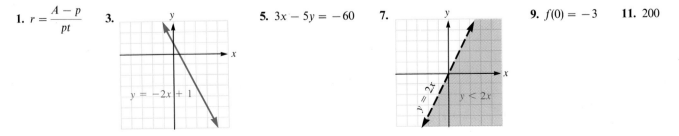

Exercise 9.4   (page 437)

1. $3i, -3i$ **3.** $\dfrac{4\sqrt{3}}{3}i, -\dfrac{4\sqrt{3}}{3}i$ **5.** $-1+i, -1-i$ **7.** $-\dfrac{1}{4}+\dfrac{\sqrt{7}}{4}i, -\dfrac{1}{4}-\dfrac{\sqrt{7}}{4}i$ **9.** $\dfrac{2}{3}+\dfrac{\sqrt{2}}{3}i, \dfrac{2}{3}-\dfrac{\sqrt{2}}{3}i$ **11.** i

13. $-i$ **15.** 1 **17.** i **19.** $8-2i$ **21.** $3-5i$ **23.** $15+7i$ **25.** $-15+2\sqrt{3}i$ **27.** $3+6i$ **29.** $9+7i$
31. $14-8i$ **33.** $8+\sqrt{2}i$ **35.** $6-8i$ **37.** $6+\sqrt{6}+(3\sqrt{3}-2\sqrt{2})i$ **39.** $-16-\sqrt{35}+(2\sqrt{5}-8\sqrt{7})i$ **41.** $0-i$

43. $0+\dfrac{4}{5}i$ **45.** $\dfrac{1}{8}+0i$ **47.** $0+\dfrac{3}{5}i$ **49.** $0+\dfrac{3\sqrt{2}}{4}i$ **51.** $\dfrac{15}{26}-\dfrac{3}{26}i$ **53.** $-\dfrac{42}{25}-\dfrac{6}{25}i$ **55.** $\dfrac{1}{4}+\dfrac{3}{4}i$

57. $\dfrac{5}{13}-\dfrac{12}{13}i$ **59.** $\dfrac{6+\sqrt{10}}{9}+\dfrac{2\sqrt{2}-3\sqrt{5}}{9}i$ **61.** 10 **63.** 13 **65.** $\sqrt{74}$ **67.** $3\sqrt{2}$ **69.** $\sqrt{69}$

REVIEW EXERCISES   (page 438)

1. $3(x+3)(x-3)$ **3.** $(2x-1)(x+1)$ **5.** $-(x-3)(x+7)$ **7.** 637π cu ft   **9.** $\dfrac{10,976}{3}\pi$ cu m   **11.** 21π cu cm

Exercise 9.5   (page 445)

1.
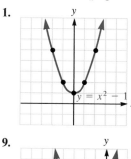

3.
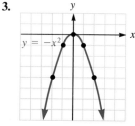

5.
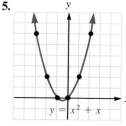

7.
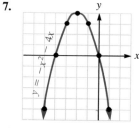

9.
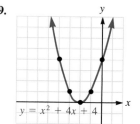

11.
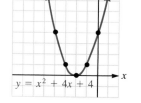

13. $(2, 4)$ **15.** $(-1, -5)$ **17.** $(1, 0)$ **19.** $(0, 4)$

21. $(-1, 4)$ **23.** $(3, -21)$ **25.**
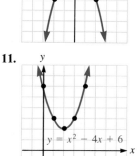

27.
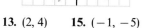

29.
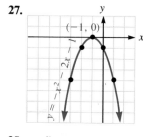

31.
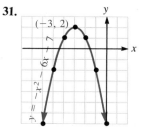

33.
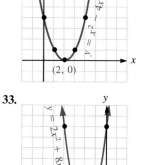

35.
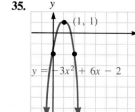
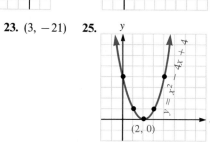
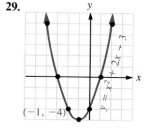

37. vertex at $(\frac{1}{2}, -\frac{49}{4})$;
x-intercepts at $(4, 0)$ and $(-3, 0)$;
y-intercept at $(0, -12)$

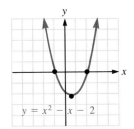

39. vertex at $(-\frac{1}{2}, \frac{49}{4})$;
x-intercepts at $(-4, 0)$ and $(3, 0)$;
y-intercept at $(0, 12)$

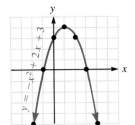

41. vertex at $(\frac{11}{4}, -\frac{81}{8})$;
x-intercepts at $(\frac{1}{2}, 0)$ and $(5, 0)$;
y-intercept at $(0, 5)$

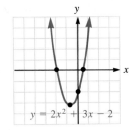

43. 1350; $303,750

REVIEW EXERCISES   (page 447)

1. $2\sqrt{2}$ **3.** $2\sqrt{6}$ **5.** $5\sqrt{3}$ **7.** 2   **9.** $\dfrac{\sqrt{5}+1}{2}$ **11.** 40 ft

CHAPTER 9 REVIEW EXERCISES   (page 449)

1. 5, −5   **3.** 3, −3   **5.** $2\sqrt{2}, -2\sqrt{2}$ **7.** −4, 6   **9.** 2, −4   **11.** $8 + 2\sqrt{2}, 8 - 2\sqrt{2}$ **13.** 2, −4
15. $4 + 2\sqrt{5}, 4 - 2\sqrt{5}$ **17.** 2, −7   **19.** 7, −11   **21.** $-2 + \sqrt{7}, -2 - \sqrt{7}$ **23.** $\frac{1}{2}$, −3   **25.** 5, −3   **27.** 13, 2
29. $\frac{3}{2}, -\frac{1}{3}$ **31.** $3 + \sqrt{2}, 3 - \sqrt{2}$ **33.** $12 - 8i$ **35.** $-28 - 21i$ **37.** $-24 + 28i$ **39.** $0 - \frac{3}{4}i$ **41.** $\frac{12}{5} - \frac{6}{5}i$
43. $\frac{15}{17} + \frac{8}{17}i$ **45.** $\frac{15}{29} - \frac{6}{29}i$ **47.** 15   **49.** (6, 7)   **51.** (1, 5)   **53.**

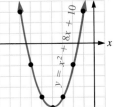

CHAPTER 9 TEST   (page 450)

1. $\dfrac{1}{3}, -\dfrac{1}{2}$ **3.** $2 + \sqrt{3}, 2 - \sqrt{3}$ **5.** 49   **7.** $-1 + \sqrt{5}, -1 - \sqrt{5}$ **9.** 2, −5   **11.** $\dfrac{-5 + \sqrt{33}}{4}, \dfrac{-5 - \sqrt{33}}{4}$

13. $1 + 9i$ **15.** $16 + 11i$ **17.** $\dfrac{3}{5} + \dfrac{4}{5}i$ **19.** $(-5, -4)$

ANSWERS TO SAMPLE FINAL EXAMINATION   (page 452)

1. b   **3.** c   **5.** a   **7.** a   **9.** c   **11.** b   **13.** d   **15.** b   **17.** c   **19.** c   **21.** d   **23.** d   **25.** e   **27.** d
29. c   **31.** b   **33.** d   **35.** d   **37.** b   **39.** c   **41.** a   **43.** e   **45.** a   **47.** e   **49.** b   **51.** a   **53.** b
55. d   **57.** c   **59.** e

SET NOTATION   (page 460)

1. 2, 4, 6, 8   **3.** Sunday, Saturday   **5.** no elements   **7.** 2, 3, 5, 7, 11, 13, 17, 19   **9.** Alaska
11. 1, 2, 3, 4, 5, 6, 7 8, 9, 10, 11, 12   **13.** 2   **15.** $\{x \mid x$ is a natural number less than 10$\}$
17. $\{x \mid x$ is the square of a natural number less than 10$\}$ **19.** $\{x \mid x$ is a month with 31 days$\}$ **21.** ∈   **23.** ∉   **25.** ⊄
27. ⊂   **29.** ⊂   **31.** ⊂   **33.** a, b, c, d, e, f **35.** no elements   **37.** a, d, e, f, g **39.** a, b, c, d, e, f, g

INDEX